Continental and Oceanic Rifts

Geodynamics Series

The Final Reports of the International Geodynamics Program sponsored by the Inter-Union Commission on Geodynamics.

Geodynamics Series

Geodynamic Series

The Final Reports of the International Geodynamics Program sponsored by the Inter-Union Commission on Geodynamics

1 Dynamics of Plate Interiors

A. W. Bally, P. L. Bender, T. R. McGetchin, and R. I. Walcott (Editors)

2 Paleoreconstruction of the Continents

M. W. McElhinny and D. A. Valencio (Editors)

3 Zagros · Hindu Kush · Himalaya Geodynamic Evolution

H. K. Gupta and F. M. Delany (Editors)

4 Anelasticity in the Earth

F. D. Stacey, M. S. Patterson, and A. Nicholas (Editors)

5 Evolution of the Earth

R. J. O'Connell and W. S. Fyfe (Editors)

6 Dynamics of Passive Margins

R. A. Scrutton (Editor)

7 Alpine-Mediterranean Geodynamics

H. Berckhemer and K. Hsü (Editors)

Continental and Oceanic Rifts

Edited by G. Pálmason

Geodynamics Series
Volume 8

American Geophysical Union
Washington, D.C.
Geological Society of America
Boulder, Colorado
1982

Final Report of Working Group 4, Continental and Oceanic Rifts, coordinated by G. Palmason on behalf of the Bureau of Inter-Union Commission on Geodynamics

American Geophysical Union, 2000 Florida Avenue, N.W.
Washington, D.C. 20009

Geological Society of America, 3300 Penrose Place; P.O. Box 9140
Boulder, Colorado 80301

Library of Congress Cataloging in Publication Data

Inter-Union Commission on Geodynamics. Working Group 4.
Continental and oceanic rifts.

(Geodynamics series ; v. 8)
Includes bibliographies.
1. Rifts (Geology)--Addresses, essays, lectures.
I. Pálmason, Gudmunder. II. Title. III. Series.
QE606.I57 1982 551.8'7 82-3991
ISBN 0-87590-504-8 AACR2

Printed in the United States of America

CONTENTS

FOREWORD

After a decade of intense and productive scientific cooperation between geologists, geophysicists and geochemists the International Geodynamics Program formally ended on July 31, 1980. The scientific accomplishments of the program are represented in more than seventy scientific reports and in this series of Final Report volumes.

The concept of the Geodynamics Program, as a natural successor to the Upper Mantle Project, developed during 1970 and 1971. The International Union of Geological Sciences (IUGS) and the International Union of Geodesy and Geophysics (IUGG) then sought support for the new program from the International Council of Scientific Unions (ICSU). As a result the Inter-Union Commission on Geodynamics was established by ICSU to manage the International Geodynamics Program.

The governing body of the Inter-Union Commission on Geodynamics was a Bureau of seven members, three appointed by IUGG, three by IUGS and one jointly by the two Unions. The President was appointed by ICSU and a Secretary-General by the Bureau from among its members. The scientific work of the Program was coordinated by the Commission, composed of the Chairmen of the Working Groups and the representatives of the national committees for the International Geodynamics Program. Both the Bureau and the Commission met annually, often in association with the Assembly of one of the Unions, or one of the constituent Associations of the Unions.

Initially the Secretariat of the Commission was in Paris with support from France through BRGM, and later in Vancouver with support from Canada through DEMR and NRC.

The scientific work of the Program was coordinated by ten Working Groups.

WG 1 Geodynamics of the Western Pacific-Indonesian Region

WG 2 Geodynamics of the Eastern Pacific Region, Caribbean and Scotia Arcs

WG 3 Geodynamics of the Alpine-Himalayan Region, West

WG 4 Geodynamics of Continental and Oceanic Rifts

WG 5 Properties and Processes of the Earth's Interior

WG 6 Geodynamics of the Alpine-Himalayan Region, East

WG 7 Geodynamics of Plate Interiors

WG 8 Geodynamics of Seismically Inactive Margins

WG 9 History and Interaction of Tectonic, Metamorphic and Magmatic Processes

WG 10 Global Syntheses and Paleoreconstruction

These Working Groups held discussion meetings and sponsored symposia. The papers given at the symposia were published in a series of Scientific Reports. The scientific studies were all organized and financed at the national level by national committees even when multinational programs were involved. It is to the national committees, and to those who participated in the studies organized by those committees, that the success of the Program must be attributed.

Financial support for the symposia and the meetings of the Commission was provided by subventions from IUGG, IUGS, UNESCO and ICSU.

Information on the activities of the Commission and its Working Groups is available in a series of 17 publications: Geodynamics Reports, 1-8, edited by F. Delany, published by BRGM; Geodynamics Highlights, 1-4, edited by F. Delany, published by BRGM; and Geodynamics International, 13-17, edited by R. D. Russell. Geodynamics International was published by World Data Center A for Solid Earth Geophysics, Boulder, Colorado 80308, USA. Copies of these publications, which contain lists of the Scientific Reports, may be obtained from WDC A. In some cases only microfiche copies are now available.

This volume is one of a series of Final Reports summarizing the work of the Commission. The Final Report volumes, organized by the Working Groups, represent in part a statement of what has been accomplished during the Program and in part an analysis of problems still to be solved. This volume from Working Group 4 (Chairman, J. Sutton) was edited by G. Pálmason.

At the end of the Geodynamics Program it is clear that the kinematics of the major plate movements during the past 200 million years is well understood, but there is much less understanding of the dynamics of the processes which cause these movements.

Perhaps the best measure of the success of the Program is the enthusiasm with which the Unions and national committees have joined in the establishment of a successor program to be known as:

Dynamics and evolution of the lithosphere: The framework for earth resources and the reduction of the hazards.

To all of those who have contributed their time so generously to the Geodynamics Program we tender our thanks.

C. L. Drake, President ICG, 1971-1975

A. L. Hales, President ICG, 1975-1980

Members of Working Group 4:

G. Pálmason
F. Aumento
R.J. Bridwell
R.W. Decker
A. Dimitriev
L.V. Dmitriev
N.A. Florensov
J. Francheteau
J.G. Francis
T. Francis
I.G. Gass
R.W. Girdler
P. Gouin
J.H. Illies
D.E. Karig
N. Logatchev
P.A. Mohr
D. Peck
I.A. Ramberg
L. Rykunov
R. Schlich
G.E. Sigvaldason
M. Talwani
L.A.J. Williams

PREFACE

During the past decade of the International Geodynamics Program many publications dealing with specific rift areas of the Earth's surface have appeared. Most of them followed symposia that were held in the respective rift zones. Among the rifts thus treated are the Rhinegraben, Afar, Iceland, the Baikal rift, the Oslo paleorift, the Rio Grande rift and the Dead Sea rift. Furthermore, publications dealing with studies of the ocean floor have discussed many oceanic rift areas.

In view of the recent growth of publications dealing with specific rift areas, the Working Group 4 (Continental and Oceanic Rifts) decided at its meeting in Santa Fe, New Mexico, in 1978, that this volume should emphasize comparative studies on topical subjects rather than focus on geographical areas. It is hoped that this will prove to be a useful approach to reviewing the present state of knowledge of the rift zones of the Earth.

The papers in this volume deal with many aspects of rift systems that have received attention in recent years. An introductory paper gives a much needed overview of the history of rift studies from the early work in East Africa at the end of the 19th century to the present time, when the main features of the World Rift System have been recognized. Subsequent papers deal with various physical aspects of rifts, including ground deformation, seismicity, gravity and heat flow. One paper deals with hydrothermal systems, which seem to be as integral a part of active rift systems as e.g. volcanism and seismicity. Magmatism is the subject of three papers, and metallogenesis in relation to rifting is discussed in one paper. Two papers deal with certain aspects of the Iceland rift zone and the Rhinegraben system. Finally, one paper gives some thoughts on unsolved problems of the continental rifts.

It is hoped that this collection of papers will prove a valuable source of information for rift studies during the decade of the International Lithosphere Program.

Gudmundur Pálmason
Chairman, ICG Working Group 4, Geodynamics of Continental and Oceanic Rifts

HISTORY OF RIFT STUDIES

C. L. Drake

Faculty of Arts and Sciences, Dartmouth College,

Hanover, N.H. 03755, U.S.A.

and

R. W. Girdler

School of Physics, The University, Newcastle upon

Tyne, NE1 7 RU, U.K.

Introduction

The great fault systems of the continents have been known for more than a century but it is only in recent years that the rifts in the ocean floor have been reasonably well mapped. The years following the second world war saw a spate of ocean exploration with many new instruments. Previously, there had been the suggestion of Wiseman and Sewell (1937) giving the results of the John Murray expedition that the rift along the Carlsberg Ridge in the Indian Ocean might be connected to the rifts of East Africa via the Gulf of Aden. This connexion turned out to be fundamental and today most people accept that the continental and oceanic rifts are closely interconnected and that continental rifts can evolve into ocean rifts and eventually into ocean basins.

The evolution of rifts into ocean basins implies large horizontal movements of the lithosphere. As is well known, the whole concept of continental drift was for many years unfashionable. It was not until the advent of palaeomagnetism in the 1950's that it became at all respectable. In some ways, this was surprising as the observational evidence for large scale relative movements between continental blocks has been documented for a long time, for example, the movements along the Great Glen fault of Scotland, the Dead-Sea - Jordan rift valley, the San Andreas fault of California and the Alpine fault of New Zealand. It is difficult to ascertain when horizontal displacements were first recognized but the large shear along the Dead Sea rift was discussed by Lartet (1869).

Strangely, the problem of accepting continental drift seems to have been associated with the mechanism rather than the evidence for it. The Polflucht forces of Wegener were hopelessly inadequate. The large scale mantle convection currents advocated by Arthur Holmes and Vening Meinesz were difficult to conceive in a so called solid mantle and the expanding Earth of Halm and Hilgenberg did not explain the large horizontal shear along adjacent pieces of continent.

The situation may be summarised in the words of Sir Edward Bullard (1965) : "There are phenomena, such as ice ages and thunderstorms, for whose occurrence there is incontrovertible evidence, but for which there is no theory that is not open to substantial objections. Difficulties of accounting for a phenomenon do not provide a proof of its non-existence, though they may give a strong indication that the evidence is being misinterpreted".

Early Studies (Pre-1940)

Early studies of rifting were necessarily confined to land areas. Field observations indicated normal faults, often with throws which seemed enormous. For example, Joseph LeConte (1889) in an article on the origin of normal faults and of the structure of the basin region of North America states "The vertical displacement on the north side of the Uinta Mountains according to Powell is 20,000 ft. (6.1 km)., that on the west side of the Wahsatch according to King is 40,000 ft. (12.2 km). In the plateau region according to Dutton there are faults extending for 200 miles (322 km) with a vertical displacement of 2000 ft. (0.6 km) to 12000 ft. (3.6 km)."

To the early workers, including LeConte, it was impossible to account for such faults unless there is a "subcrust liquid" and the idea became prevalent of a series of crustal blocks floating at different levels in accord with Archimedes' principle. It is always difficult to find the earliest references to such ideas but some of the earliest must be Mr W. Hopkins' "Researches in Physical Geology" published in the Philosophical Transactions of the Royal Society. In the third series (1842), Hopkins speaks of subterranean reservoirs and an elevatory force, produced by expansion of the fluid matter to raise the superincumbent solid mass and to form in it a system of fissures. The plane of these fissures will scarcely ever be parallel and diagrams are shown to demonstrate how crustal blocks float at different levels according to their shape.

With these observations and ideas on normal faults and the observations of Lartet (1869) of horizontal shear along the fault system in Palestine, it might be said that the embryo of the so called modern plate tectonics was already born in the middle of the last century.

The enormous horizontal extent of rift systems became appreciated with the works of John Walter Gregory. Gregory first visited East Africa in 1892-3 and his monumental book (405 pages) entitled "The Great Rift Valley" appeared in 1896. In this he describes the Great Rift Valley as extending from Palestine in the north to beyond the Zambesi in the south, a distance of some 5000 miles. He attributes the recognition of the extension of the Red Sea fractures into Africa to H. Douvillé (1886). Gregory visited Africa again in 1919 and as a result his more often referenced book appeared in 1921. It is perhaps worth quoting its full title viz. "The Rift Valleys and Geology of East Africa, an account of the origin and history of the rift valleys and their relations to the contemporary earth-movements which transformed the geography of the world with some account of the prehistoric stone implements, soils, water supply and mineral resources of the Kenya Colony." A quote from Gregory's preface is also appropriate viz: "Pioneer geology has to choose between the rashness of using imperfect evidence or the sterility of uncorrelated, unexplained facts." Gregory was the first to suggest the name "Rift Valley" using the term to denote the sinking of material between parallel fractures. He cites a long, comprehensive bibliography but he seems to have been unaware of the works of Hopkins, independently formulating his own theory for the origin of rifts which was very similar. He envisaged the formation of a long, low arch which ruptured as the lateral supports gave way. The top then sank "as the keystone of a bridge sinks if its buttresses slip or settle. The sinking of the keystone of the East African arch into the plastic layer below forced some of it up the adjacent cracks, through which the material was discharged in the volcanic eruptions. Each renewal of the subsidence was followed by fresh eruptions".

At the turn of the century, there was also some indication of the larger extent of the rifts in the ocean floor. The Danish "INGOLF" expedition of 1895-1896 discovered and named the Reykjanes Ridge running for 1100 Km south west of Iceland. A report of the expedition by Wandel (1898) suggested that the ridge must be young and of volcanic origin. The arguments were (a) earthquakes had been felt by ships crossing the ridge and (b) volcanic activity must have taken place after the Ice age as not a single ice strewn boulder was recovered in dredge hauls on the axis of the ridge whereas the regions to the east and west of the ridge were strewn with boulders.

The significance of this discovery was commented upon by Th. Thoroddsen (1901 , 1908) who pointed out that the Reykjanes Ridge seemed closely related to the active rift zone on land. In particular, many volcanic eruptions occurred in the vicinity of Eldeyjar and the ridge was narrow and volcanic with its surface covered by volcanic debris and scoriae. He also noticed that the volcanic line across Iceland has the same trend (S.W. - N.E.) and the volcanic island of Jan Mayen is probably located on its continuation although at that time no volcanic ridge was known to extend north of Iceland as far as Jan Mayen. Some twenty years later, Sieberg (1923) on his seismicity map, shows the Reykjanes Ridge and the mid-Atlantic Ridge as seismically active but not in continuity.

Some of the most remarkable work in Iceland was that of Niels Nielsen from Denmark (Nielsen 1929, 1930, 1933) who was a supporter of the ideas of Wegener at a time when they were hotly debated. Nielsen envisaged the fissure eruptions and fault structures as being "the result of a pull from west to east which has simply split the land into innumerable fissures". He recognised two profoundly different types of tectonics, the Alpine pressure tectonics and the Icelandic tensional tectonics. His concept of tensional processes causing rifting and magma intrusions is very similar to that envisaged in recent models.

The tension hypothesis seems to have been almost generally accepted (e.g. Suess, 1904 Cloos, 1939) but in the 1920's an alternative hypothesis relating the rifts to compression was proposed. the chief proponents were E.J. Wayland (1929) and Bailey Willis (1936) working in Africa. In this hypothesis the normal step faults at the surface were considered to be secondary. The main faults at depth were postulated to be reverse (at 45°) and a consequence of lateral compression, the two opposing overthrust faults holding down the block in between. The overhangs of the thrusts adjusted by gravity giving a series of inslipping normal faults at the surface and obscuring the major faults beneath.

The compression hypothesis found temporary support in the interpretation of the gravity deficiency over the rift valleys of East Africa. In a monumental paper, Bullard (1936) gave the results of 56 gravity measurements made with pendulums in 1933 and 1934 and included the work of Professor Kolschütter's 1899-1900 expedition. The suggestion of a wedge shaped block forced down between reversed faults provided the light matter to explain the gravity deficiency. At this time, the large volumes of light sediments and light volcanics infilling the rifts were unknown, and this interpretation was abondoned when it was recognised that the light sediment could account for a large portion of the negative Bouguer anomaly.

The exploration of the oceans was severely limited by the apparatus and techniques available. In the 1920's, the development of the sonic echo sounder greatly increased the amount of sounding data although matching of the transceiver with an accurately timed recorder was not achieved until the 1950's. Professor F.A. Vening Meinesz developed his pendulum apparatus for use in submarines leading to more accurate gravity measurements and there was little else. It is perhaps hard to realise that seismic work at sea was extremely experimental. Encouraged by Dick Field and William Bowie, Maurice Ewing made the first trials in 1935 and Teddy Bullard succeeded in obtaining reflexions and refractions using two Brixham trawlers in the English channel in that last bright summer before the war.

The only data available were therefore gravity measurements and ocean depths. These reinforced the conclusion of Wegener (1924) that the crust beneath the oceans is quantitatively different from that beneath the continents. This seems obvious now but since the turn of the century the oceans had been considered to be foundered continents (Haug 1908, Suess 1904) and it was hard to abandon this view. Even prior to 1950, many considered that only the Pacific Ocean is truely oceanic and lacking a sialic layer, the Atlantic and Indian Oceans having such a layer thinned by plastic flow (Umbgrove 1947, Gutenberg, 1939). Although stretching and thinning through plastic flow were suggested, there was no concensus on the nature of the mechanism.

As previously mentioned Wiseman and Sewell (1937) noted the connexion and similarity of the Carlsberg and Murray ridges (considered to be upper Tertiary) in the Indian ocean to the East African rift system observing that the former are almost a mirror image of the latter. The distribution of earthquakes (Heck, 1938) strengthened the comparison, implying that simlar rifting processes might be taking place in the Indian ocean as in Africa.

Some progress towards the recognition of rifting on the scale of ocean basins had also been made. For example, Du Toit (1937) identified many coastlines as faultline coasts from their topographic expression and suggested rifting and listric faulting as the mechanism by which the continental margins were created and the sialic crust thinned. He envisaged the ocean basins to be "merely rift valleys of unusual width" and floored by sialic crust. Some words of Umbgrove (1947) probably reflect the thinking at the outbreak of war viz: "an examination of the Atlantic and Indian Oceans, and a comparison with the available geological and seismic data, shows that the bottom probably consists of sialic material. Internal forces caused similar events to occur in this sialic layer and the surrounding continents (e.g. basins and ridges were formed and eventually rift valleys)".

1940 to 1970

During the second world war, studies of the rifts came virtually to a standstill. There was one noteable exception: 1944 saw the publication of the first edition of Arthur Holmes' "Principles of Physical Geology". In it, Holmes describes the concept of sea floor spreading which was to have such a profound influence on rift studies in the 1960's. The germs of the idea can be found in his earlier papers of 1928-9 and 1933 when he was searching for a mechanism for continental drift and invoking large scale mantle convection. There has been some debate as to whether full credit should be given to Arthur Holmes for the idea of sea floor spreading. In the 1944 edition of his book he updates his earlier diagrams and shows the evolution of "new ocean" as the continents move apart: the figure caption describes "ocean floor development on the site of the gap" and leaves little doubt that he understood fully the concept of ocean floor spreading which was destined to become so important.

At about this time, Bernauer (1943) was interested in explaining the young tectonic activity of Iceland having been a member of a German expedition there in 1938. He was a follower of the mantle convection ideas of Holmes, Griggs (1939) and Vening Meinesz (1934) and envisaged Iceland to be formed of new crust consisting of "sima": the new crust forming over an uprising mantle convection current and continuing to develop by the process of dilation, rupture and sideways transport as long as the convection continues.

After the war, the so called "tension hypothesis" for rift formation became generally accepted and the "compression hypothesis" abandoned. Goguel (1949) and Vening Meinesz (1950) suggested that the negative Bouguer gravity anomalies found over the floors of the continental rift valleys could be explained if the valleys have normal faults. With the development of electronic computing techniques in the late 1950's it became possible to quickly compute models to satisfy the gravity and to establish whether the faults become reverse at depth and hence to distinguish between the two hypotheses (Girdler 1964). For normal faulted valleys the gradients quickly diminish to zero over the sides whilst for reverse faulted valleys, the gradients should be quite large 40 to

50 Km from the valley. The latter are not observed and hence the more obvious idea that the faults continue as normal faults at depth was supported. The conclusion later found further support from earthquake mechanism studies (e.g. Fairhead and Girdler 1971, Sykes 1968) the present day motion along the faults being a combination of dip-slip and strike slip.

In the post war period, work on the rifts in the ocean floors expanded at a much more rapid pace than work on the continental rifts. There were two main reasons for this: first, partly as a result of the war, many new instruments became available for exploring the oceans and secondly, the general growth in scientific research meant more financial support and opportunities for visiting unexplored parts.

The new techniques which became available included markedly improved echo sounders with more accurate timing devices and expanded scale recorders: radio sono buoys and hydrophones for seismic refraction work; new energy sources such as airguns for seismic reflexion work; magnetometers for towing behind ships; continuous recording gravimeters and stable platforms: probes for measuring heat flow and later greatly improved positioning of survey ships with the introduction of satellite navigation systems. With all these new techniques and the boost given to research by the International Geophysical year (1957-8) a wealth of new data became available with which to test old and new ideas on the origin of rifts and the evolution of ocean basins.

The sea seismic refraction experiments started just before the war were continued by Maurice Ewing of Columbia University and Maurice Hill of Cambridge University. Maurice Hill developed a system in which charges were fired from a ship and the shock waves were recorded by hydrophones suspended from buoys and the signals transmitted by radio to the ship. The first results in deep water were obtained from a weather ship 500 Km west of Ireland and were most successful. For the first time the now standard section for the crust beneath the oceans of layers 1, 2 and 3 was obtained (Hill and Willmore 1947, Hill and Swallow 1950, Hill 1952). Early in 1949, Maurice Ewing managed to get two ships in the hopes of finding the depth to the Mohorovicic discontinuity (the Moho). A reversed line 56 Km long was shot and the Moho found 5 Km beneath the ocean floor. (Ewing et al 1949, 1950). These results showed that the structure of the ocean floor is completely different from that of the continents as predicted many years previously by Wegener.

The introduction of precision echo sounders (Luskin et al 1954) coupled with the enthusiasm of Maurice Ewing for exploring the oceans and collecting as much data as possible led to the construction of a series of magnificent physiographic maps of the ocean floors by Marie Tharp and Bruce Heezen in the 1950's and 1960's. At last there were maps of the sea floor which could be compared with maps of the continents. Early on, it was noticed that the valley at the crest of the mid-Atlantic ridge is a continuous feature (Ewing and Heezen 1956, 1960, Hill 1960). At a meeting about the floor of the Atlantic Ocean held at the Royal Society in February 1953, J.P. Rothé discussed the distribution of earthquake epicentres in the Atlantic and Indian oceans and showed that the line of epicentres following the mid-Atlantic Ridge continues around the Cape of Good Hope to join with the epicentres marking the central ridge of the Indian Ocean. It became apparent that the earthquakes occurred in the central valley and their distribution could be used to trace the course of the ridge and its central valley where there were no echo sounding records. In their 1956 paper Ewing and Heezen took the logical step and extended this on a world wide scale. The world wide extent of the rift system was gradually confirmed (Ewing and Heezen 1960). This was a remarkable discovery. The recognition that the rifts of Africa and western North America are parts of a world wide system largely in the floors of the oceans has been central to the development of subsequent tectonic theories.

A further technical advance of the greatest importance was the development of the marine magnetometer. At first, the airborne fluxgate magnetometer developed during the war by Victor Vacquier for submarine detection was converted and towed behind a ship. The first trans-Atlantic profile of total magnetic intensity was obtained from Dakar to Barbados aboard the R.V. Atlantis in 1948. An account of the work and the historic profile are given in Heezen, Ewing and Miller (1953). The instrument was somewhat troublesome and cumbersome and was replaced by the proton precession magnetometer (Hill, 1959) which is free from drift and has no moving parts. It became easy to continuously measure the total intensity of the Earth's magnetic field to an accuracy of 1 part in 10^5.

Crossings of the mid-Atlantic rift revealed a large magnetic anomaly (Heezen et al 1953, 1959, Hill 1960, Talwani et al 1961, Girdler 1964) often in excess of 1000 nT. A similarly large anomaly was found on a flight across the Gulf of Aden by W.B. Agocs (Girdler 1958) and large anomalies were observed over the axial rift of the Red Sea (Drake and Girdler 1964). It soon became apparent that a large magnetic anomaly is a characteristic feature of the axial rift zones (U.S. Navy Hydrographic office 1962, British Admiralty Hydrographic office 1963, Keen 1963, Ostenso 1965) and this has been used together with bathymetry and the distribution of epicentres to map the rifts.

There are also many smaller anomalies on either side of the large anomaly associated with the axial rifts. The next interesting discovery was made by contouring these anomalies using many profiles. This was done for a survey conducted from the USCGS vessel Pioneer in the north east Pacific (Mason 1958) which revealed that the anomalies are remarkably linear, the lineations extending for thousands of kilometers. Similar magnetic lineations were found over the western Gulf of Aden (Girdler and Peter 1960), the Red Sea (Drake and Girdler 1964) and the mid-Atlantic (Heirtzler and Le Pichon 1965).

At first, these anomalies proved difficult to interpret. Apart from the large axial anomaly there is no obvious relationship to the bathymetry and the usual practice of assuming the direction of magnetization to be parallel to that of the Earth's present field frequently seemed to fail. A very clear example was seen in the western Gulf of Aden where a simple minimum is expected for a body striking west-east. Large maxima as well as large minima are observed and this led Girdler and Peter (1960) to infer the presence of reversely magnetizated dykes suggesting that remanent magnetization is very important when interpreting oceanic magnetic anomalies. The complete interpretation had to await the work of Vine and Matthews (1963) when they associated their cause with the idea of sea floor spreading which had been revitalised by Hess (1962) and Dietz (1961).

Vine and Matthews argued that if new oceanic crust is formed along the rift zones, it will be magnetized in the current direction of the Earth's magnetic field. It is a corollary of sea floor spreading and geomagnetic reversals that blocks of normally and reversely magnetised material would be carried away from and parallel to the centres of the rifts thus explaining the magnetic lineations. On land, Cox, Doell and Dalrymple (1963, 1964, 1968) had been painstakingly establishing the times of the reversals using the palaeomagnetic method and radiometric (potassium - argon) dating. In a masterly paper, Vine (1966) applied this to the interpretation of the marine magnetic anomalies near the rifts. Knowing the distances from the anomalies and the times from the dating of reversals it was possible to estimate the speeds of formation of the rift zones. The rifts were found to be evolving at rates from 2 to 11 cm/yr. It was obviously important to check the association of these ideas and this has been achieved mainly by examining the ages and magnetic properties of the sediments overlying the magnetized oceanic crust. The sediments have been drilled in many places by the Deep Sea Drilling Project (e.g. : Maxwell et al 1970) and the ages of the oldest sediments agree in a most gratifying way with the ages predicted from the interpretation of magnetic anomalies. It therefore became necessary to adjust to the idea that the rift zones can evolve at remarkably rapid rates on a geological timescale.

The discovery of the magnetic lineations has enabled the oceanic character of the Gulf of Aden and Red Sea to be confirmed. From the early pendulum measurements of Von Triulzi (1898, 1901) Vening Meinesz (1934) and Girdler and Harrison (1957), the Red Sea was known to have a positive Bouguer gravity anomaly in contrast to the negative anomalies found over the rifts in East Africa. This indicated that the Red Sea is underlain by dense material with the physical properties of basalt (Girdler 1958). Later work by the research vessels Atlantis and Vema supported this interpretation with the discovery of high seismic velocities at shallow depths beneath the axial trough (Drake and Girdler 1964). It was suggested that the dyke like bodies with high density, strong magnetization and high seismic velocity were formed as Africa and Arabia moved apart.

This work was complementary to that along the Dead Sea Rift. Quennell (1958, 1959) found further support for the shear proposed by Lartet (1869), Dubertret (1932) and Wellings (1938) ingeniously suggesting that the Dead Sea and Lake Tiberias are rhomboidal gaps in between en echelon faults due to the northward movement of Arabia with respect to Sinai which also opened the Red Sea. In modern terminology the Dead Sea rift is a transform fault transforming the Red Sea in the south to the Taurus - Zagros mountains in the north. Quennell (1959) further noticed that the wrench faults (as they were then known) must follow either a great circle on the Earth's surface or a small circle. The faults forming the Dead Sea rift zone were described as arcs, all arcs having centres at approximately 33°N., 24°E. The movement along the Dead Sea rift was supposed to have taken place in two stages, the first stage being 62 Km corresponding to a rotation of 3° and the second stage 45 Km corresponding to a rotation of $2\frac{1}{2}^{\circ}$. This seems to be the first use of the concept of poles of rotation and rotation angles, a concept which was to become so important in the late 1960's (e.g. Bullard et al 1965, McKenzie and Parker 1967). Further support for the shear along the Dead Sea rift came with the work of Freund (1965) and Freund et al (1970).

To many, this horizontal displacement of more than 100 Km seemed outrageous but meanwhile even larger displacements were being proposed for the ocean floor. The patterns of magnetic lineations previously described were found to have several interruptions and these were related to the fracture zones of Menard (1955). The fracture zones were interpreted as strike - slip faults with displacements of the order of 1000 Km (Vacquier et al 1961) and curiously with an apparent change of magnitude and sense of the displacements along the faults (Menard, 1960). Similar faults were found in the equatorial Atlantic displacing the mid-Atlantic rift (Heezen and Tharp 1961) and these were presumed to be old as they are not seismically active along their lengths (Heezen 1962).

The problem of the relationships between the rifts and the fracture zones was tackled by Wilson (1965) who generalised the lucrative concept of "transform" faults. He developed the idea that the fracture zones were not strike slip faults but faults which evolved simultaneously with the rifting and spreading processes. The concept predicts motion along the fault opposite to that expected for transcurrent faults. This prediction was confirmed by Sykes (1967, 1968) using first motion studies for earthquake mechanisms. The relationships between rifts and offsets (transforms) became clear. Further, careful mapping of the transforms gave the motions between the continents and these together with the timing from sea floor spreading enabled the successive stages of continental drift to be ascertained. Rifting and drifting were now fitted into a unified concept. With this understanding and the quantitative methods of computing poles of rotation (Bullard et al 1965, McKenzie and Parker 1967, Morgan 1968) it became possible to predict motions and hence to plan surveys for studying more fully the processes of rifting and the evolution of the oceans.

The development of shipborne gravimeters and stable platforms (Harrison 1960) enabled continuous profiles of the gravity field over the oceanic rift zones to be recorded. These showed that the rift and associated ridges are in isostatic equilibrium. A new technique of combining gravity and seismic refraction data using an empirical relation between P wave velocities and densities (Nafe and Drake 1963) provided a powerful means of elucidating the structure of the crust and uppermost mantle (Talwani et al 1959).

With this technique, Talwani et al (1965) using early refraction profiles (Ewing and Ewing 1959, Raitt 1956, and Le Pichon et al 1965) computed models for the structure of the mid-ocean ridges. These showed large regions of anomalously low density upper mantle beneath the central parts of the ridges. It was suggested that the regions of "anomalous mantle" might be due to a phase change and the associated expansion might produce the uplift of the ridge and result in rifting. Alternatively, it was thought the "anomalous mantle" might be a mixture of crust and mantle (Ewing and Ewing 1959) or might be due to serpentinization (Hess 1954, 1955).

These studies stimulated more work on land. Ironically, the new instruments had made it easier to work at sea. For continental rift studies the LaCoste Romberg Model G gravimeter was a great advance as it is almost drift-free but the measurements of altitudes is still very tedious and time consuming. Nevertheless, gravity profiles were obtained across the rifts of East Africa (Girdler et al 1969) for comparison with those over the oceanic rift zones. The profiles were found to be similar with a long wavelength negative Bouguer anomaly extending for 1000 Km across the whole of the rift zone. A remarkable discovery was that of a smaller positive Bouguer anomaly over the rift floor in Kenya which is superimposed on the long wavelength negative Bouguer anomaly. The anomaly was first noticed by McCall (1967) and mapped in considerable detail by Searle (1970). The anomaly has a width of 40 to 80 Km and an amplitude of 60 mGal. Interpretation (Searle 1970) suggested the presence of a dense intrusive zone less than 3 Km beneath the floor of the rift. This was confirmed by a seismic line (Griffiths et al 1971) which showed a high refracting layer with velocity 6.4 Km/s at a shallow depth of 2.8 ± 0.5 Km. It was thus proposed that the gravity anomalies indicated thinning beneath the floor of the rift (Girdler et al 1969, Searle 1970). Previously it was traditional to interpret the negative anomaly over the plateau as due to thickening of the crust. At first, the new interpretation was greeted with some sceptism but it now seems generally accepted that the lithosphere thins beneath the whole of the African plateau and extreme thinning occurs as the rifts with local positive anomalies (which have associated volcanism, hot springs and grid faulting) are approached.

The development of probes for the measurements of the temperature gradients in the sediments at the bottoms of the oceans made it relatively easy to obtain heat flow measurements at sea. In 1949, Maxwell and Bullard constructed such a probe containing thermocouples, the probe being attached to a pressure type vessel containing a recording galvanometer. Later, Maxwell replaced the thermocouples by thermistors and the galvanometer by a potentiometer recorder so making it less delicate. The first measurements in the Pacific Ocean were made in 1950 by Revelle and Maxwell (1952) and in the Atlantic in 1952 by Bullard (1954). Then in 1956 measurements in the central valley of the mid-Atlantic revealed exceptionally high values. As more measurements were made, the rifts were found to be characterised by high but variable heat flow confined to narrow zones (Bullard et al 1956 , von Herzen 1959, von Herzen and Uyeda 1963, Langseth et al 1966). Further, Vacquier and von Herzen (1964) showed that in the equatorial Atlantic, the narrow zone of high heat flow is offset by the large transform faults. This suggested that the cause of the high heat flow is a heat leak between the evolving rigid plates and not related in any simple way to an uprising mantle convection pattern.

The variation of heat flow as a function of distance from the rifts was established and McKenzie (1967) showed that to a first order, the heat flow distribution is consistent with sea floor spreading. For comparison with the sea floor spreading models, Sclater and Francheteau (1970) plotted the heat flow as a function of age of the oceanic crust and suggested that this may be a powerful tool in understanding the history of the oceans. It was also recognised that the elevation of the rift-ridge zones is related to the heat flow, the elevation being sensitive to the coefficient of thermal expansion, the basal temperature, the chemical composition and the pressure-temperature stability fields of the subsolidus mineral assemblages. The problems of the elevation of the ridges were investigated in more detail by Sclater et al (1971) who showed that all ridges appeared to show a uniform subsidence rate near the ridge crests and that this may be accounted for by thermal contraction as the cooling lithosphere moves away from the active spreading centres which form the rift zones.

These inferences were made by averaging the heat flow values over various age zones of the ocean floors. It was recognised that the large scatter in the heat flow measurements observed near the rift zones and crests of the ridges may contain valuable information on processes taking place in the rift zones. For example, Palmason (1967) demonstrated the presence of hydrothermal convection in the rift zones of Iceland by studying the temperatures from boreholes, thus supplementing the ideas of Einarsson (1942) for the Tertiary flank zones. In the same paper, Palmason went on to suggest that if the permeability conditions in the upper crust of the mid-Atlantic Ridge are similar to those in Iceland, then free convection of sea water could explain both the low and high values of heat flow observed over the axial ridge zone.

For some time, rocks had been dredged from the rift zones and these were found to include basalts (pillow lavas), gabbros and ultrabasics which were usually serpentinised (Wiseman and Sewell 1937, Shand 1949, Quon and Ehlers 1959, Engel and Engel 1964a, b, Muir et al 1964, Matthews et al 1965, Poldervaart and Green 1965). Bullard (1961) remarked that the decayed state of the rocks dredged by Hill and Matthews in 1958 came as something of a surprise. These rocks were broken and manganese minerals were found deposited in the cracks: if the rocks were shattered and infiltrated by sea water while still hot, the rocks to some depth might be broken up and decayed.

As early as 1871-1875, metalliferous sediments had been found on the East Pacific Rise during the expedition of HMS Challenger. Subsequent studies (Revelle et al 1955, Stornyakova 1964, Arrhenius and Bonatti 1965, Bostrom and Peterson 1966) suggested that these were related to hydrothermal processes associated with the tectonic processes taking place along the Rise.

The Red Sea became a region for the study of hydrothermal processes (Degens and Ross 1969). Here they were recognised in the axial trough and deeps in 1963 to 1965 but anomalously high bottom temperatures had been found in 1948 by the R/V Albatros from Sweden. In 1958 an attempt to measure heat flow from the R/V Vema failed and it was subsequently realised that this was due to the exceptionally high temperatures causing the instrument to go off-scale. Temperatures exceeding 56°C were found in the stable bottom waters, the stability being due to the high salinity (as high as 257‰). The salt water was found to be enriched in many elements which appeared to be due to sea water circulation through the rocks to some depth (Craig 1966,1969). The sediments underlying the brines were found to be enriched in metals.

It thus became realised that the rift zones might be the sites of enrichment in metals of the bottom waters and sediments (Corliss 1971), due to hydrothermal processes, i.e. sea water circulating through the hot underlying basaltic rocks forming in the initial stages of sea floor spreading.

Summary and Conclusions

We have attempted to give a history of rift studies to the start of the Geodynamics Project in 1970. Inevitably, much has been omitted and it is regretted that it is in many ways incomplete.

We have seen that the emphasis moved from land to sea after the second world war. This was largely due to the invention of many new instruments and techniques and the general expansion of science. As more than 70% of the Earth's surface is covered with water this was bound to lead to great discoveries and rich rewards. One of the greatest steps forward was the discovery that the rifts are world wide and connected. This would have gladdened the heart of J.W. Gregory who died in 1932; with his enthusiasm for the enormous extent of the East African rift system "extending for one-sixth the circumference of the Earth" it would probably have come as no surprise! Arthur Holmes died in 1965 and lived to see the results of palaeomagnetism, the recognition of continennental drift and the rebirth of sea floor spreading. The initiation of continental rifting followed by the drifting of the continents and the evolution of the ocean basins has been the central theme of post-war studies which would have delighted Alfred Wegener.

For the continental rifts the concept that the lithosphere thins has taken over from the concept of thickening the crust isostatically to explain the gravity anomalies. The discovery of positive Bouguer anomalies over the floor of some of the rifts has lead to the recognition of extreme thinning in these regions and the suggestion that these are the sites of subsequent continent break up. The exact mechanism for this is not fully understood. Once the continents have rifted and broken apart, we have a much clearer understanding of the way in which the sea floor evolves. This has come largely from the studies of magnetic anomalies and reversals coupled to the theory of sea floor spreading. It is most remarkable that we have quantitative estimates of the speed at which oceanic rift zones evolve.

We have said little about the driving forces responsible for rifting and drifting. The concept of some kind of mantle convection has come to be almost generally accepted but little progress has been made in understanding these processes. It is ironical that one of the principal barriers to the acceptance of continental drift until twenty years ago remains with us.

Acknowledgments

We are greatly indebted to Drs. Kristjan Saemundsson and Gudmundur Palmason for providing information on early studies of the Icelandic rifts.

References

Arrhenius, G., and E. Bonatti, Neptunism and volcanism in the Ocean, Prog. Ocean., 3, p. 7, 1965.

Bernauer, F., Junge Tektonik auf Island und ihre Uhrsachen. In: Spalten auf Island (O. Niemczyk, Ed.) pp 14-64, (Stuttgart, Verlag von Konrad Witter), 1943.

Bostrom, K., and M. N. A. Peterson, Precipitates from Hydrothermal Exhalations on the East Pacific Rise, Econ. Geol., 61, 1258, 1966.

British Admiralty, Hydrographic Office, Bathymetric Magnetic and Gravity Investigations, HMS Owen, 1961-62, 1962.

Bullard, E. C., Gravity Measurements in East Africa, Phil. Trans. Roy. Soc. Lond., A235, 445-531, 1936.

Bullard, E. C., The Flow of Heat through the Floor of the Atlantic Ocean, Proc. Roy. Soc. Lond., A222, 408-429, 1954.

Bullard, E. C., Forces and Processes at Work in Ocean Basins in Oceanography, Mary Sears, ed., Amer. Assoc. Adv. Sci., 39-50, 1961.

Bullard, E. C., Concluding Remarks. Symposium on Continental Drift, Phil. Trans. Roy. Soc., A, 258, 322-323, 1965.

Bullard, E. C., J. E. Everett and A. G. Smith, The Fit of the Continents around the Atlantic, Phil. Trans. Roy. Soc., A258, 41-51, 1965.

Bullard, E. C., A. E. Maxwell and R. Revelle, Heat Flow through the Deep Sea Floor. Adv. Geophys. 3, 153-181, 1956.

Cloos, H. "Hebung - Spaltung - Vulkanismus" Geologische Rundschau, 1939.

Corliss, J. B., The Origin of Metal Bearing Submarine Hydrothermal Solutions, Jour. Geophys. Res., 76, 8128, 1971.

Cox, A., R. R. Doell and G. B. Dalrymple, Geomagnetic Polarity Epochs and Pleistocene Geochronology, Nature, 198, 1049-1051, 1963.

Cox, A., R. R. Doell and G. B. Dalrymple, Reversals of the Earth's Magnetic Field, Science, 144, 1537-1543, 1964.

Cox, A., R. R. Doell and G. B. Dalrymple, Time Scale for Geomagnetic Reversals, in The History of the Earth's Crust, R. A. Phinney, ed., Princeton University Press, p. 101-108, 1968.

Craig, H., Isotopic Composition and Origin of the Red Sea and Salton Sea Geothermal Brines, Science, 134, 1544, 1966.

Craig, H., Geochemistry and Origin of the Red Sea Brines, in Hot Brines and Recent Heavy Metal Deposits in the Red Sea, E. T. Degens and D. A. Ross, eds., Springer-Verlag, New York, p. 208-242, 1969.

Dietz, R. S., Continent and Ocean Basin Evolution by Spreading of the Sea Floor, Nature, 190, 854-857, 1961.

Douville, H., Examen des Fossiles Rapportes du Choa par M. Aubry, Bull. Soc. Geol. France, (3) 14, 223-241, 1886.

Drake, C. L. and R. W. Girdler, A Geophysical Study of the Red Sea, Geophys. Jour. Roy. Astron. Soc. Lon. 8, 473-495, 1964.

Dubertret, L. Les formes structurales de la Syrie et de la Palestine: leur origine, C. R. Lebd. Seane. Acad. Sci. Paris, 195, p. 65-67, 1932.

DuToit, A., Our Wandering Continents, Oliver and Boyd, London, p. 366., 1937.

Einarsson, T., Ueber das Wesen der Heissen Quellen Islands, Vis Isl. (Soc. Sci. Isl.) 26, 1-91, 1942.

Engel, A. E. J., and C. G. Engel, Composition of Basalt from the Mid-Atlantic Ridge, Science, 144, 1330-1331, 1964a.

Engel, A. E. J., and C. G. Engel, Igneous Rocks of the East Pacific Rise, Science, 146, 477--485, 1964b.

Ewing, J. I., and M. Ewing, Seismic Refraction Measurements in the Atlantic Ocean Basins, in the Mediterranean Sea, on the Mid-Atlantic Ridge, and in the Norwegian Sea, Bull. Geol. Soc. Amer., 67, 1623-1640, 1959.

Ewing, M., and B. C. Heezen. Some Problems of Antarctic Submarine Geology, Geophys. Mon. No. 1, Amer. Geophys. Union, p. 75-81, 1956.

Ewing, M., and B. C. Heezen, Continuity of the Mid-Ocean Ridge and Rift Valley in the South Western Indian Ocean Confirmed, Science, N.Y. 131, 1677-1679, 1960.

Ewing, M., J. Hirshman and B. C. Heezen, Magnetic Anomalies of the Mid-Ocean Rift, Reprints Int. Ocean. Cong., N.Y., p. 24-25, 1959.

Ewing, M., J. L. Worzel, J. B. Hersey, F. Press and G. R. Hamilton, (No title), Bull. Geol. Soc. Amer. 60, 1303, 1949.

Ewing, M., J. L. Worzel, J. B. Hersey, F. Press and G. R. Hamilton, Seismic Refraction Measurements in the Atlantic Ocean Basin, Bull. Seism. Soc. Amer. 40, 233-242, 1950.

Fairhead, J. D., and R. W. Girdler, The Seismicity of Africa, Geophys. J.R.A.S. 24, 271--301, 1971.

Freund, R., A Model of the Structural Development of Israel and Adjacent Areas since Upper Cretaceous Times, Geol. Mag. 102, 189-205, 1965.

Freund, R., Z. Garfunkel, I. Zak, M. Goldberg, T. Weissbrod and B. Derin, The Shear Along the Dead Sea Rift, Phil. Trans. Roy. Soc. A267, 107-130, 1970.

Girdler, R. W., The Relationship of the Red Sea to the East African Rift System, Quart. Jour. Geol. Soc. Lon., 114, 79., 1958.

Girdler, R.W., Geophysical Studies of Rift Valleys, Phys, Chem. of the Earth, 5, 121-156, 1964.

Girdler, R. W., J. D. Fairhead, R. C. Searle and W. T. C. Sowerbutts, The Evolution of Rifting in Africa, Nature, Lond., 229, 66-71, 1969.

Girdler, R. W., and J. C. Harrison, Submarine Gravity Measurements in the Atlantic Ocean, Indian Ocean, Red Sea and Mediterranean Sea, Proc. Roy. Soc., Series A, 239, 202-213, 1957.
Girdler, R. W., and G. Peter. An Example of the Importance of Natural Remanent Magnetization in the Interpretation of Magnetic Anomalies, Geophys. Prosp., 8, 474, 1960.
Goguel, J., La Structure des Fossés Africains et la Gravimétrie, Ann. Geophys. Fr. 5, 174-175, 1949.
Gregory, J. W., "The Great Rift Valley", (Murray, London), 1896.
Gregory, J. W., "The Rift Valleys and Geology of East Africa", (Seeley, London), 1921.
Griffiths, D. H., R. F. King, M. A. Khan and D. J. Blundell, Seismic Refraction Line in the Gregory Rift, Nature, Lond., 229, 66-71, 1971.
Griggs, D., A theory of mountain building. Am. J. Sci., 237, 611-650, 1939.
Gutenberg, B., Structure of the Earth's Crust and the Spreading of the Continents, Bull. Geol. Soc. Amer., 47, 1587-1610, 1939.
Harrison, J.C., The Measurement of Gravity at Sea, in "Methods and Techniques in Geophysics", Ed. S. K. Runcorn, I, 211-229 (Interscience), 1960.
Haug, E., Traite du Geologie, 3 Vols., Librairie Armand Colin, Paris, 1908.
Heck, N. H., The Role of Earthquakes and the Seismic Method in Submarine Geology, Proc. Amer. Phil. Soc., 79, 97-108, 1938.
Heezen, B. C., La Topographie et la Geologie des Profoundeurs Oceaniques, Coll. Inter. Dv. CNRS, 83, 295-304, 1959.
Heezen, B. C., The Deep Sea Floor, in Continental Drift, S. K. Runcorn, editor, Int. Geophys. Sci., 3, Academic Press, N.Y., p. 235-288, 1962.
Heezen, B. C., M. Ewing and E. T. Miller, Trans-Atlantic Profile of Total Magnetic Intensity and Topography, Dakar to Barbados. Deep Sea Res., 1, 25-33, 1953.
Heezen, B. C., M. Tharp and M. Ewing, The Floors of the Oceans, I, The North Atlantic, Geol. Soc. Amer. Spec. Pap. 65, 1959.
Heezen, B. C., and M. Ewing, The Mid-Ocean Ridge and its Extension through the Arctic Basin, in Geology of the Arctic, Univ. Toronto Press, 622-642, 1961.
Heezen, B.C., and M. Tharp, Physiographic Diagram of the South Atlantic, the Caribbean, the Scotia Sea, and the Eastern Margin of the South Pacific Ocean, Geol. Soc. Amer., N.Y., 1961.
Heirtzler, J.R., Magnetic Anomalies over the Reykjanes Ridge Deep Sea Res., 13, 427-443, 1966.
Heirtzler, J.R., Evidence for Ocean Floor Spreading Across the Ocean Basins, in the History of the Earth's Crust, R. A. Phinney, ed., Princeton University Press, p. 90-100, 1968.
Heirtzler, J.R., and X. LePichon, Crustal Structure of the Mid-Ocean Ridges, 3, Magnetic Anomalies over the Mid-Atlantic Ridge, Jour. Geophys. Res., 70, 4013-4034, 1965.
Hess, H. H., Geological Hypothesis and the Earth's Crust Under the Oceans, Proc. Roy. Soc., Series A., 222, 341-348, 1954.
Hess, H. H., Serpentines, Orogeny and Epeirogeny, Geol. Soc. Amer. Spec. Pap. 62, 391-408, 1955.
Hess, H. H., History of the Ocean Basins, in Petrological Studies, Buddington Volume, A. E. J. Engel et al., eds, Geol. Soc. Amer. 1962.
Hill, M. N., Seismic Refraction Shooting in an Area of the Eastern Atlantic, Phil. Trans. Roy. Soc., A 244, 561-596, 1952.
Hill, M. N., A Ship-Borne Nuclear Spin Magnetometer, Deep Sea Res., 5, 309-311, 1959.
Hill, M. N., A Median Valley of the Mid-Atlantic Ridge, Deep Sea Res., 6, 193-205, 1960.
Hill, M. N. and J. C. Swallow, Seismic Experiments in the Atlantic, Nature, Lond., 165, 193, 1950.
Hill, M. N. and P. L. Willmore, Marine Seismic Prospecting, Nature, Lond., 159, 707, 1947.
Holmes, A., Radioactivity and Earth Movements, Trans. Geol. Soc. Glasgow, 18, (3), 559-606, 1928-1929.
Holmes, A., The Thermal History of the Earth, J. Wash. Acad. Sci. 23, 169-195, 1933.
Holmes, A., Principles of Physical Geology, London: Nelson, xii + 532 pp., 1944.
Hopkins, W., "Researches in Physical Geology - Third Series", Phil. Trans. Roy. Soc. Lond., Pt. 1, 43-55, 1842.
Keen, M.H., Magnetic Anomalies over the Mid-Atlantic Ridge, Nature, 197, 888-890, 1963.
Langseth, M. G., X. LePichon and M. Ewing. Crustal Structure of the Mid-Ocean Ridges, 5, Heat Flow through the Atlantic Ocean Floor and Convection Currents, Jour, Geophys. Res., 71, 5321-5355, 1966.
Lartet, L. La Geologie de la Palestine, Ann. Sci. Geol. Paris, 1, 1869.
LeConte, J., On the Origin of Normal Faults and of the Structure of the Basin Region, Amer. J. Sci. 38, 257-263, 1889.
LePichon, X., R. E. Houtz, C. L. Drake and J. E. Nafe, Crustal Structure of the Mid-Ocean Ridges, 1, Seismic Refraction Measurements, Jour. Geophys. Res. 70, 319-339, 1965.
Luskin, B., B. C. Heezen, M. Ewing and M. Landisman, Precision Measurement of Ocean Depth, Deep Sea Research, 1, 131-140, 1954.
Mason, R. G., A Magnetic Survey off the West Coast of the United States (Lat. 32°-36° N: Long. 121°-128°W). Geophys. Jour. 1, 320-329, 1958.
Maxwell, A. E., R. P. Von Herzen, K. J. Hsu, J. Andrews, T. Saito, S. F. Percival, E. D. Melow and R. E. Boyce, Deep Sea Drilling in the South Atlantic, Science 168, 1047-1059, 1970.
McCall, G. J. H. Geology of the Nakuru - Thomson's Falls - Lake Hannington area, Geol. Surv. Kenya Rep. 78, 122 pp., 1967.

McKenzie, D. P., Some Remarks on Heat Flow and Gravity Anomalies, Jour. Geophys. Res. 72, 6261-6271, 1967.

McKenzie, D. P. and R. L. Parker, The North Pacific: An Example of Tectonics on a Sphere, Nature, Lond., 216, 1276-1280, 1967.

Menard, H. W., Deformation of the Northeastern Pacific Basin and the West Coast of North America, Bull. Geol. Soc. Amer. 66, 1149-1198, 1955.

Menard, H.W., The East Pacific Rise. Science, 132, 1737-1746, 1960.

Morgan, W. J., Rises, Trenches, Great Faults and Crustal Blocks, J. Geophys. Res. 73, 1959-1982, 1968.

Muir, I. D., C. E. Tilley and J. H. Scoon, Basalts from the Northern Part of the Rift Zone of the Mid-Atlantic Ridge, Jour. Pet., 5, 409-434, 1964.

Nafe, J. E. and C. L. Drake, Physical Properties of Marine Sediments in "The Sea" 3, 794-815, Ed. M. N. Hill, (Interscience), 1963.

Nafe, J. E., J. F. Hennion and G. Peter, Geophysical Measurements in the Gulf of Aden, Preprints of Int. Ocean. Cong., N.Y., p. 42, 1959.

Nicholls, G. D., N. J. Nalwack and E. E. Hays, The Nature and Composition of Rock Samples Dredged from the Mid-Atlantic Ridge between 22°N. and 52°N. Mar. Geol. 1, 333-343, 1964.

Nielsen, N., Islands Tektonik og Wegener-Theorien Rep. 18th Skand, Nat. Congr. Copenhagen, 26-31 Aug, 1929, 4 pp., 1929.

Nielsen, N., Tektonik und Vulkanismus Islands unter Berücksichtigung der Wegener-Hypothese, Geol. Rundschau, 21, 347-349, 1930.

Nielsen, N., Contributions to the physiography of Iceland, with particular reference to the highlands west of Vatnajokull. D. Kgl. Danske Vidensk. Selsk. Skrifter, Naturv. og Matem. Afd., 9, Raekke, IV, 5. 105 pp. 32 plates, 9 maps Copenhagen, 1933.

Ostenso, N. A. Aeromagnetic Evidence for a Trans-Arctic Extension of the Mid-Atlantic Ridge, (Abs.) Trans. Amer. Geophys. Un., 46, 107, 1965.

Palmason, G., On Heat Flow in Iceland in relation to the Mid-Atlantic Ridge, in Iceland and Mid-Ocean Ridges, RIT Soc. Sci. Islandica 111-127, 1967.

Poldervaart, A. and J. Green, Chemical Analysis of Submarine Basalts, Amer. Min. 50, 1723-1727, 1965.

Quennell, A. M. The Structural and Geomorphic Evolution of the Dead Sea Rift, Quart. J. Geol. Soc. Lond. 114, 1-24, 1958.

Quennell, A. M. Tectonics of the Dead Sea Rift, Congr. Geol. International, Mexico 1956, XX Session, 385-405, 1959.

Quon, S. H. and E. G. Ehlers, Rocks of the Northern Part of the Mid-Atlantic Ridge, Bull. Geol. Soc. Amer., 74, 1-7, 1959.

Raitt, R. W., Seismic Refraction Studies of the Pacific Ocean Basin, 1., Crustal Structure of the Central Equatorial Pacific, Bull. Geol. Soc. Amer., 67, 1623-1640, 1956.

Revelle, R., M. Bramlette, G. Arrhenius and E. Goldberg, Pelagic Sediments of the Pacific, Geol. Soc. Amer. Spec. Pap. 62, A. Poldervaard, ed., p. 221, 1955.

Revelle, R. and Maxwell, A., Heat Flow through the Ocean Floor, Nature, Lond., 170, 199-200, 1952.

Rothé, J. P., La Zone Seismique Mediane Indo-Atlantique, Proc. Roy. Soc., Series A, 387, 1954.

Rusnak, G. A. and R. L. Fisher, Structural History and Evolution of the Gulf of California in Marine Geology of the Gulf of California, Tj. H. van Andel and G. G. Shor, eds., Amer. Assoc. Pet. Geol. Mem. 3, p. 144-156, 1964.

Sclater, J. G., and J. Francheteau, The Implications of Terrestrial Heat Flow Observations on Current Tectonic and Geochemical Models of the Crust and Upper Mantle of the Earth, Geophys. Jour. Roy. Astr. Soc. 20, 509-542, 1970.

Sclater, J. G., R. N. Anderson and M. L. Bell, Elevation of Ridges and Evolution of the Central Eastern Pacific, J. Geophys. Res. 76, 7888-7915, 1971.

Searle, R., Evidence from Gravity Anomalies for Thinning of the Lithosphere beneath the Rift Valley in Kenya, Geophys. J. Roy. Astron. Soc. 21, 13-31, 1970.

Shand, S. J., Rocks of the Mid-Atlantic Ridge, Jour. Geol. 7, 89, 1949.

Sieberg, A., Geologische, physikalische und angewandte Erdbebenkunde (Fischer, Jena), 1923.

Skornyakova, I. S., Dispersed Iron and Manganese in Pacific Ocean Sediments, Int. Geol. Rev., 7, 2161, 1964.

Suess, E., The Face of the Earth (Das Antlitz Der Erde), Translated by H. B. C. Sollas, 5 Vols, Clarendon Press, Oxford, 1904.

Sykes, L. R., Mechanism of Earthquakes and Nature of Faulting on the Mid-Ocean Ridges, J. Geophys, Res. 72, 2131-2153, 1967.

Sykes, L. R., Seismological Evidence for Transform Faults, Sea Floor Spreading and Continental Drift, in The History of the Earth's Crust, R. A. Phinney, ed., Princeton Univ. Press, p. 120-150, 1968.

Talwani, M., B. C. Heezen and J. L. Worzel, Gravity Anomalies, Physiography and Crustal Structure of the Mid-Atlantic Ridge, Pub. Bur. Cir. Seism. Int., Series A., Travaux Scientifique, Fasc. 22, 81-111, 1961.

Talwani, M., X. LePichon and M. Ewing, Crustal Structure of the Mid-Ocean Ridges; 2, Computed Model from Gravity and Seismic Refraction Data, Jour. Geophys. Res., 70, 341-352, 1965.

Talwani, M., J. L. Worzel and M. Landisman, Rapid Gravity Computations for Two Dimensional

Bodies with Application to the Mendocino submarine fracture zone, J. Geophys. Res. 64, 49-59, 1959.
Thoroddsen, Th., Islandske Fjorde og Bugter, Geografisk Tidsskrift, 16, 58-82, 1901.
Thoroddsen, Th., Lysing Islands, Icelandic Literary Soc., 1, 1908.
Triulzi, A. E. Von, Expedition S. M. Schiff Pola in das Rothe Meer, nördliche Halfte, Wissenschaftliche Ergebnisse, II, Relative Schwerebestimmungen, Denkschr. K. Akad. Wiss. Math. Nat. Classe, 65, 131-206, 1898.
Triulzi, A. E. Von, Expedition S. M. Schiff Pola in das Rothe Meer, südliche Hälfte Wissenschaftliche Ergebnisse, XII, Relative Schwerebestimmung, Denkschr. K. Akad. Wiss. Math. Nat. Classe, 69, 143-219, 1901.
Umbgrove, J. H. F., The Pulse of the Earth, 2nd Edition, Martinus Nnhoff, The Hague, p. 358, 1949.
U.S. Navy Hydrographic Office, Operation Deep Freeze 61, 1960-61, Marine Geophysical Investigations (TR-105), 1962.
Vacquier, V. and R.P. Von Herzen, Evidence for Connection between Heat Flow and the Mid-Atlantic Ridge Magnetic Anomaly, J. Geophys. Res., 69, 1093-1101, 1964.
Vacquier, V., A.D. Raff and R. G. Mason, Horizontal Displacements in the Floor of the N.E. Pacific Ocean, Bull. Geol. Soc. Amer. 72, 1251-1258, 1961.
Vening Meinesz, F.A. Gravity Expeditions at Sea II, 1923-32, Delft, 1934.
Vening Meinesz, F.A. Les Grabens Africains, Resultat de Compression ou de Tension dans la Croute Terrestre? Bull. des Seances Inst. Roy. Colonial. Belge. 21, 539-552, 1960.
Vine, F. J., Spreading of the Ocean Floor: New Evidence, Science, 154, 1405-15, 1966.
Vine, F. J., Magnetic Anomalies Associated with Mid-Ocean Ridges, in the History of the Earth's Crust, R. A. Phinney, ed., Princeton Univ. Press, p. 73-89, 1968.
Vine, F. J., and D. H. Matthews, Magnetic Anomalies over Ocean Ridges, Nature, 199, 947-949, 1963.
Von Herzen, R. P., Heat Flow Values from the Southeastern Pacific, Nature, 183, 882-883, 1959.
Von Herzen, R. P. and S. Uyeda, Heat Flow through the Eastern Pacific Floor, Jour. Geophys. Res., 68, 4219-4250, 1963.
Wandel, C. F., Rejseberetning, in Den Danske Ingolf-Expedition 1895-1896 Vol. 1, 1-19, (Bianco Lunos Kgl., Kjobenhavn), 1898.
Wayland, E. J., Rift Valleys and Lake Victoria, Compte Rendu XV Int. Geol. Congr. S. Afr. (Pretoria 1930) 2, 323-353, 1929.
Wegener, A., The Origin of Continents and Oceans, 3rd edition, (English translation by J. G. A. Skerl), Methuen, London, p. 212, 1924.
Wellings, F. E., in "Wellings Observations of Dead Sea Structure" by Bailey Willis, Bull. Geol. Soc. Amer. 49, 1938.
Willis, Bailey, East African Plateaus and Rift Valleys, Carnegie Inst. Washington, Pub. 470, 3588 pp., 1936.
Wilson, J. Tuzo, New Class of Faults and their Bearing on Continental Drift, Nature, 207, 343-347, 1965.
Wiseman, J. D. H. and R. B. S. Sewell, The Floor of the Arabian Sea. Geol. Mag., 74, 219, 1937.

RECENT GROUND DEFORMATION IN CONTINENTAL AND OCEANIC RIFT ZONES

Eysteinn Tryggvason

Nordic Volcanological Institute, University of Iceland, Reykjavik, Iceland

Abstract. The active rift zones of the world form a continuous chain of volcanically and seismically active belts in the world oceans, the mid-ocean ridges, with a few extensions into the continents. Furthermore, active rift zones of limited length exist at several locations on the continents, such as the Baikal zone, the Rhine graben and the Rio Grande graben. Ground deformation in four segments of the world rift zones, as deduced from geodetic observations are discussed. These zone segments are: The North Atlantic rift zone where it crosses Iceland, the East African rift zone in Ethiopia and Djibouti, the Baikal rift zone in southern Siberia and the Rhine graben in Germany. The results indicate that the existing topographic relief is increasing, low areas are subsiding and high regions are rising. A significant part of the deformation in Iceland and Afar seems to be associated with discrete rifting events which occur at intervals of decades or centuries and are closely related to volcanic activity.

Introduction

The mid-ocean rift system, roughly 60.000 km in length, is presently recognized by numerous earth scientists as accreting plate boundaries, where the crust on opposite sides moves away from each other, and new crust is formed. This view is supported by a multitude of observations, such as parallel stripes of magnetic anomalies, earthquake focal mechanism, sediment thickness and age of basement rocks.

The surface deformation associated with this relative displacement of the oceanic crust cannot be observed in detail due to the water cover, except where the crest of the mid-ocean ridges is above sea level, as in Iceland, or where the rift system enters the continents, as in Afar.

Rift systems do also occur in the continents, such as the East-African rift zone, the Baikal rift zone, the Rhine graben, rifts in western North America, and the Jordan Valley rift. These rift zones are characterized by recent volcanism, seismic activity and normal faulting, and their structure indicates that recent tensional processes in the continental rift zones are similar to those of the mid-ocean ridges.

The present review deals with the most recent deformation of a few segments of the rift zones based on precise geodetic measurements. The precise observations needed to detect ground deformation in the rift zone were rarely made before about 1960 but increasing effort has been placed on such observations during the last two decades. This short period of precise observation puts severe limitations to the conclusions which can be drawn from them, especially with regard to variation with time in the deforming processes.

Four regions, where some information has been obtained on recent deformation, are here selected as examples of continental and oceanic rift zones.

Iceland is a portion of the Mid-Atlantic ridge. Geodetic measurements aimed at deformation studies were initiated in 1938. Increasing effort has been placed on these measurements since 1965.

The north end of the East Africa rift system in Ethiopia and Djibouti is a border case between oceanic and continental rift zones. Geodetic measurements for deformation purposes were started in 1969.

The Baikal rift zone lies in central Asia and has no connection with the mid-ocean ridges. Extensive remeasurements of level lines during the last decades show the vertical component of ground deformation, but no data on measured horizontal displacements are known to the present author.

The Rhine graben lies in southern Germany and has no obvious connection with the mid-ocean rift system. Great effort has been made to determine the present ground deformation with geodetic means over several decades.

Most other rift zones of the world have not been investigated in such a detail that a clear picture of their present ground deformation have been obtained. The principal exception is the Rio Grande rift in New Mexico, U.S.A., but a comprehensive treatment of the present state of knowledge of the structure and processes at work related to this rift zone has been published recently (Riecker, 1979).

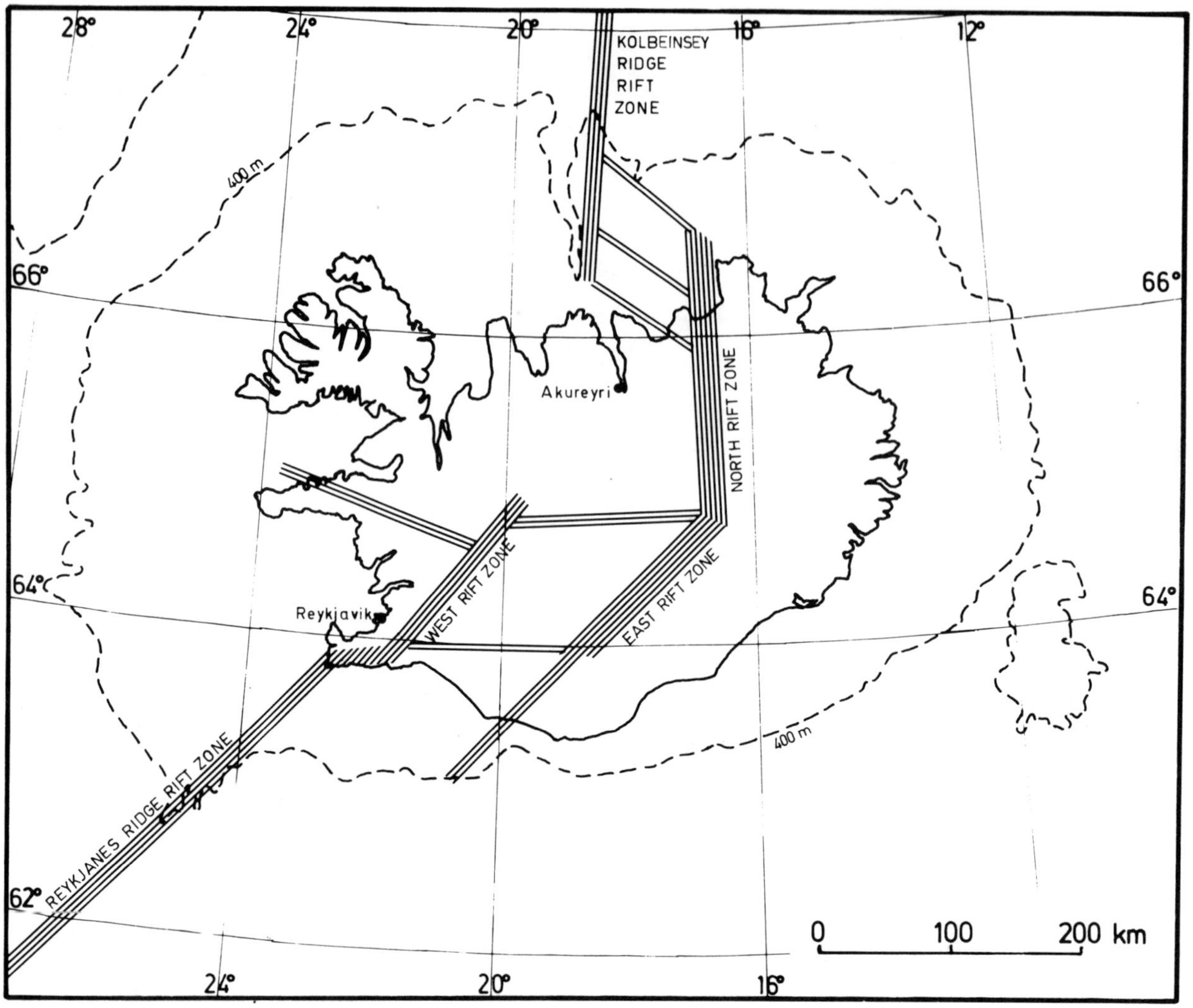

Fig. 1. The rift system of Iceland slightly simplified. The principal rift zones are shown with four or five parallel lines and other volcanic zones with three parallel lines. Zones of seismic activity connecting the rift zones (transform faults) are shown with two parallel lines. The boundary of the Iceland platform is indicated by the 400 m depth contour.

The Iceland Rift System

The rift system in Iceland is a direct continuation of the Mid-Atlantic ridge to the south and the Kolbeinsey ridge to the north, although branchings and offsets (transform faults?) occur in South Iceland and off the north coast (Fig.1).

The open fissures and graben structures in the Iceland rift zone were early recognized as a sign of yield to a regional east-west tensional stress (Nielsen, 1930), and an estimate of the rate of tensional movement in recent times in North Iceland, based on open fissures in postglacial lavas, gave a value of 3.56 m per km per 1000 years (Bernauer, 1943). This value was obtained by measurements in the Krafla fissure swarm in North Iceland, which is some 3 to 5 km wide and Bernauers result indicates an average widening of this fissure swarm of some 10 to 15 meters per 1000 years during the last 5000 to 10000 years.

In light of these rather obvious signs of recent tensional movement, and the rather crude estimates of the rate of movement, an effort to observe this rate by direct measurements was started in 1938 by establishing a network of precisely measured bench marks across the rift zone in North Iceland (Niemczyk and Emschermann, 1943). Remeasurements of this network were delayed by the second world war but since 1965 it has been remeasured several times and extended greatly both by adding bench marks and introducing new and more precise measuring techniques (Gerke, 1969; Gerke, 1974; Spickernagel, 1966; Schleusener and Torge, 1971). Further measurements of crustal movement in the Iceland rift zone have been performed since 1966 (Tryggvason,

1968; Decker et al., 1971; Brander et al., 1976).

A dramatic rifting event in North Iceland which started in 1975 has caused great increase in the effort aimed at understanding the tectonic processes in Iceland (Björnsson et al., 1977). Much of this effort has been towards measuring ground movement and crustal deformation.

Measured Horizontal Deformation in Southwest Iceland

Precise distance measurements on the southwest tip of the Reykjanes peninsula (Fig.2) in 1968 to 1972 show some significant but small length changes (Brander et al., 1976). The network consists of some 40 bench marks and the measured lines are usually about one km. These measurements support the hypothesis of a left lateral transform fault on the Reykjanes peninsula (Tryggvason, 1968) although the measured displacements are not parallel to the active zone, but a significant tensional component is indicated. Thus the active zone of the Reykjanes peninsula can be termed as a leaky transform fault where the opposite sides are moving at an average rate of 9 mm per year relative to each other (Brander et al., 1976).

The Thingvellir area has been the subject of a great effort to determine horizontal deformations. Measurements by a group from Imperial College in London 1968 to 1972 across the Thingvellir graben show significant length increases of 11 and 14 millimeters on two lines while all other lines in the nearly 30-line measuring program showed no significant length changes. This result indicates that the Thingvellir graben is widening at an average rate of about 3 mm/year (Brander et al., 1976).

Another series of distance measurements in the Thingvellir area was made by a German team in 1967 and 1971 (Gerke, 1974). The first interpretation indicated that no horizontal displacement of bench marks could as yet be established in

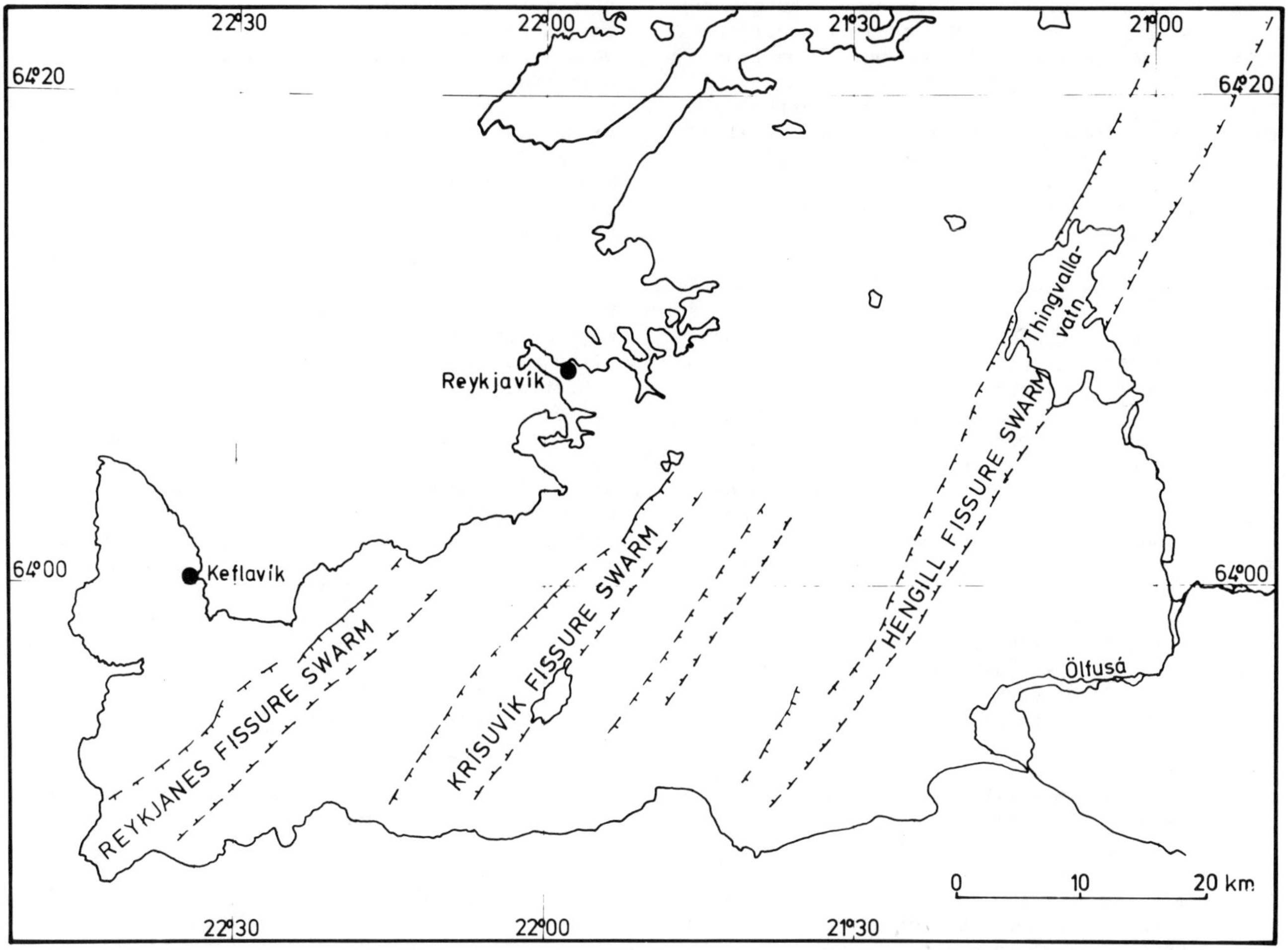

Fig. 2. Details of the West Rift Zone in Iceland with en echelon fissure swarms. Each fissure swarm consists of a graben with numerous open fissures while few open fissures are found between the swarms.

geodetic control network covering an area of 20 km in N-S direction by 30 km in E-W direction. A much smaller quadrilateral across the Thingvellir graben showed a general increase of line length of about 10 ppm on the average, which is interpreted as a widening of the graben of about 1 cm per year (Gerke, 1974).

A third series of distance measurements across the Thingvellir area was made by a team from the U.S.A. in 1967, 1970 and 1973 (Decker et al., 1971; Decker et al., 1976). The first remeasurement showed no significant length changes in the Thingvellir area, while the second remeasurement in 1973 showed some significant length changes of 1 to 3 centimeters. The whole Thingvellir graben seems to have widened by about 2.0 centimeter from 1967 to 1973 or about 3 mm per year on the average, in good agreement to the values obtained by the Imperial College.

It is of interest to compare results of the U.S.A. team and the German team, which partly used the same bench marks. The distance from point 5001 on the west side of the Thingvellir graben to point 3212 in the center of the graben increased by 1.2 cm from 1967 to 1971 according to the German measurements, and by 2.8 cm from 1967 to 1973 according to the U.S. measurements. The distance from 5001 to 539 on the east side of the graben increased by 2.3 cm from 1967 to 1970 according to the U.S. measurements, by 3.4 cm from 1967 to 1971 according to the German measurements and by 3.0 cm from 1967 to 1973 according to the U.S. measurements.

Vertical Deformation in Southwest Iceland

Precise leveling along several selected lines in the rift zone in southwest Iceland in 1966 to 1971 show definite vertical displacements (Tryggvason, 1974). On the Reykjanes peninsula, near Vogar, a tilt towards south or southeast is observed, at a rate of approximately 0.4 microradians per year, in addition to some fault displacements. The observed displacement of about 7 mm on the fault Hrafnagjá and indication of minor displacements (about 1 mm) on two other faults are supposedly caused by an earthquake swarm in 1967 located 20 to 25 km southwest of the leveling line (Tryggvason, 1970a).

A leveling line across the Thingvellir graben shows a subsidence centered in the eastern part of the graben, but no fault displacements. Tilt rates of about 0.26 microradians per year are observed near the major faults on either side of the graben, and this tilt rate seems to be maintained towards the ends of the 10 km leveling line. Thus the eastern part of the Thingvellir graben is subsiding at a rate of about one millimeter per year relative to the west end of the leveling profile about 5 km northwest of the area of greatest observed subsidence. Apparently, the subsidence bowl extends outside the line of precise leveling, so the total subsidence may be considerably more than one millimeter per year (Tryggvason, 1974).

Deformation in the East Rift Zone

Precision leveling across the lava shield of the new volcanic island Surtsey was performed several times from 1967 to 1970 (Tryggvason,1970b and 1972). These levelings show that the whole lava shield was subsiding, and the rate of subsidence decreased approximately by a factor 2 each year. Maximum subsidence of about 30 cm/year was observed at the highest elevation in 1967-1968 but in 1969-1970 this had decreased to about 10 cm/year. Most of this subsidence is interpreted as due to cooling of the volcanic pile, which in places exceeds 200 m in thickness.

Distance measurements across the volcanic fissure of Heimaey during the eruption of 1973 (Brander and Wadge, 1973) showed a progressive right lateral shear on the eruption fissure from early February to April 1973, but the eruption started on January 23, 1973. The measured rate of deformation decreased nearly exponentially with time and the time constant was about 7.5 days. The calculated total strain along the principal strain axes was: ε_1 = 36 ppm (compression) and ε_2 = -420 ppm (dilatation), and the direction of the axis of maximum compression was N45°E (Brander and Wadge, 1973).

Precision levelings along short leveling lines in the vicinity of the volcano Katla, in the Mýrdalsjökull glacier in 1967 to 1971 showed no progressive ground deformation, but significant response to snow accumulation on the glacier. This was interpreted as an indicator of low viscosity material below 6.5 to 8.5 km thick elastic crust (Tryggvason, 1973).

Distance measurements across the East rift zone north of the volcano Hekla in 1967, 1970 and 1973 show 6 to 7 cm (±3 cm) widening of a 10 km wide zone about 25 km northeast of Hekla between 1967 and 1970, which may have occurred during the Hekla eruption in May to June 1970 (Decker et al. 1971). However, the same section of the measuring line showed some contraction, although barely significant, between 1971 and 1973. The whole rift zone shows no significant extension or contraction for the period 1967-1973 (Decker et al., 1976).

Measured Deformation in the North Rift Zone Before December 1975

An extensive geodetic network was established in 1938 across the North rift zone in Iceland (Fig.3) in order to observe slow ground deformation (Niemczyk and Emschermann, 1943). Remeasurement of this network in 1965 showed westward displacements of up to 4.8 m for bench marks near the center of the network, when two endpoints about 115 km apart were considered stable. Reexamination of the result indicated systematic

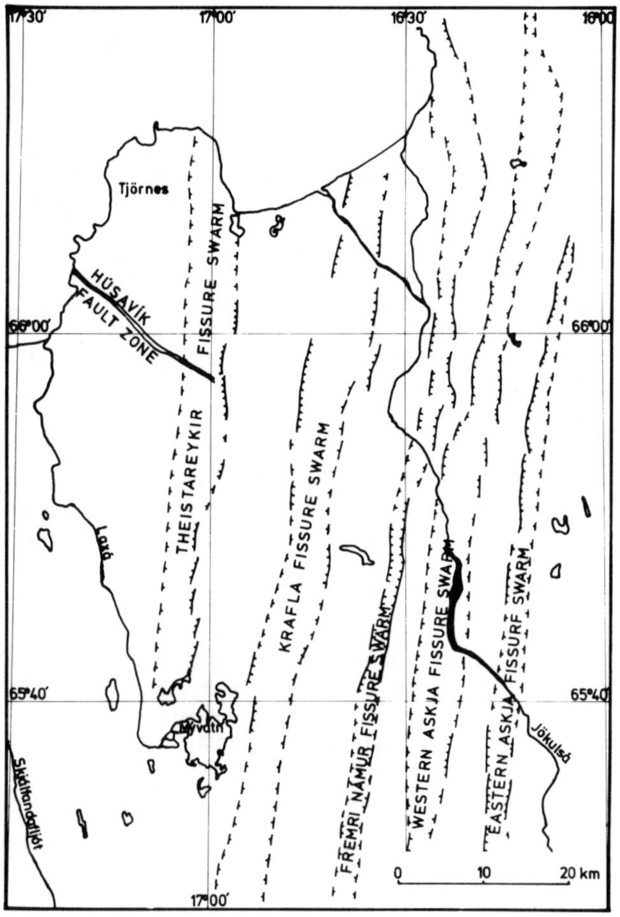

Fig. 3. Details of the northern part of the North Rift Zone in Iceland. A rifting episode which started in 1975 has caused widening of the Krafla fissure swarm of about 5 m. The area of repeated inflation and deflation during the rifting episode is centered about 12 km northeast of lake Mývatn.

errors caused by different scales in the western and the eastern part of the network in the 1938 measurements. Adjustments for this apparent scale error reduced the computed displacement of the bench mark to be within the confidence limits of the measurements, resulting in no significant horizontal deformation between 1938 and 1965 (Gerke, 1969).

Further remeasurements in 1971 and 1975 with increasing accuracy showed a zone of disturbance, coinciding roughly with the N-S trending Krafla fissure swarm and stable blocks on either side. During the period 1965 to 1971 did the stable blocks move towards each other up to 50 cm in the southern part of the area of measurements and about 10 cm in its northern part while bench marks within each stable block did not move relative to each other. Bench marks within the disturbed zone were displaced northwards up to about 20 cm. During the period 1971 to 1975 the same zone of disturbance appeared, and the stable areas on either side moved away from each other by some 30 cm in the southern part of the network and 10 cm in its northern parts. Two points within the disturbed zone moved significantly southwards (Gerke et al., 1978).

A small profile (3 km long) across the Krafla fissure swarm near Hrútafjöll (65°50'N) was measured in 1938 and again in 1965, 1967 and 1971. According to these measurements, the fissure swarm was contracted some 54 cm between 1938 and 1965, expanded about 5 cm from 1965 to 1967 and did not change in width between 1967 and 1971. The measurements of 1938 were not as accurate as later measurements and should be judged with caution (Gerke, 1974). Another source reports contraction of 45.4 cm from 1938 to 1965 and a further contraction of 2.4 cm from 1965 to 1967 (Gerke, 1969).

Vertical control of this line in 1938, 1965 and 1967 (in 1938 partly triangulation, otherwise precision leveling) indicated a subsidence of the 3 km wide graben of the Krafla fissure swarm of about 17 cm relative to the west flank of the graben between 1938 and 1965 (Spickernagel,1966). The east flank of the graben was uplifted 11 cm relative to the west flank during the same interval. From 1965 to 1967 no movement was indicated on the graben faults but a general tilt towards west of roughly 7 microradians resulting in an uplift of the east flank of the graben by about 20 millimeters relative to the west flank (Gerke, 1969).

A small network across the Fremri Námur fissure swarm (Klaustur) was measured in 1965, 1968 and 1971. An extension of the 2.5 km wide fissure swarm of about 6.5 cm is indicated from 1965 to 1971 or roughly 1 cm per year (Gerke, 1974). Precision leveling showed some barely significant vertical displacements.

Precision leveling along a 3.3 km east-west profile across the Theistareykir fissure swarm at latitude 65°58'N was performed in 1966, 1968, 1970 and 1972. The result showed progressive tilt of the ground towards east of 0.3 microradians per year on the average. The west end of the profile tilted about 0.7 microradians per year while the central and east part tilted only 0.2 microradians per year on the average (Tryggvason, 1974).

Short leveling profiles 4 km to the east and 5 km to the west of the above profile were leveled in 1970 and 1972. The eastern profile indicated westward tilt of 0.4 ± 0.2 microradians per year while the western profile showed a tilt of about 0.7 microradians per year towards east. Three short profiles in the Laxá area, wholly to the west of the North Iceland rift zone were measured in 1970 and 1973. They all showed eastward tilt of 0.3 to 1.0 microradians per year (Tryggvason, 1974).

Measured Deformation in North Iceland After December 20, 1975

An episode of repeated rifting, faulting and volcanic eruptions started in the North Iceland rift zone on December 20, 1975 (Björnsson et al., 1977). A variety of measurements made within the rift zone since 1975 show a continuous procession of events of rifting, vertical and horizontal ground movement, earthquake swarms and volcanic activity. All this activity has been in the Krafla fissure swarm, one of several fissure swarms in the North Iceland rift zone (Fig.3). Similar events of rifting, faulting and volcanic activity are reported in the Askja fissure swarm in 1874 to 1875, and the Krafla fissure swarm in 1724 to 1729, and opening of fissures is reported in the Theistareykir fissure swarm in 1618 (Björnsson et al., 1977). Thus it seems that major tectonic episodes occur in the North Iceland rift zone approximately once in 100 to 150 years, and each episode is confined to only one of the several fissure swarms making up the rift zone (Björnsson et al., 1979).

The measurements and observations which have been applied to monitor the tectonic episode which started in 1975 in North Iceland include precision levelings (Björnsson, 1976), tilt measurements (Tryggvason, 1978a; Sindrason and Ólafsson, 1978), gravity measurements (Torge and Drewes, 1977; Torge and Kanngieser, 1978), measurements of width of fissures (Björnsson et al., 1979), change of river courses and lake shores (Tryggvason, 1976) and distance measurements with geodimeter (Tryggvason, 1978b; Gerke et al., 1978).

The ground deformation in the North Iceland rift zone since 1975 has been characterized by successive uplift and subsidence of the area of the Krafla central volcano (Krafla caldera). Each period of uplift has lasted from one to seven months while the subsidence lasts from about 1 to 20 days. During the subsidence events, certain portion of the fissure swarm, outside the Krafla caldera, is widened through fissuring and faulting. Sometimes the same portion of the fissure swarm is affected by several subsidence events while other parts of the swarm are affected by only one subsidence event. The maximum widening of the fissure swarm in one event may exceed 2 meter (Björnsson et al.,1979). The total widening of the fissure zone has been measured to exceed 3 meter in places (Tryggvason, 1980). The widening seem to take place in only about one km wide zone along the central axis of the fissure swarm. The segment of the swarm which is active in each subsidence event may exceed 20 km in length while about 80 km of the fissure swarm has been active in one or more of the 10 subsidence events which occurred in 1975 through 1978 (Tryggvason, 1980). During the widening of the fissure zone, open fissures are formed and the central part of the fissure swarm subsides relative to its flanks by an amount similar to or slightly less than the widening. The flanks of the fissure swarm are at the same time uplifted relative to areas farther away from the swarm. This uplift may be roughly 20 per cent of the widening (Björnsson et al., 1979; Tryggvason, 1980). The area immediately outside the zone of widening is simultaneously contracted in east-west direction, the contractional strain being greatest (up to 1.8×10^{-4}) nearest to the zone of active fissuring but diminishing with distance. At 40 km distance from the fissure zone, the horizontal displacement is only some 15 per cent of the displacement at the flanks of the fissure swarm (Björnsson et al., 1979; Tryggvason, 1980; Gerke et al., 1978).

Summary of the Iceland Rift Zone deformation

The rift zone and its surrounding is continuously stretched in east-west direction and at the same time the central part of the rift zone subsides. This stretching and subsidence affects a zone some 80 km wide and the subsidence rate appears to be about 1 cm/year at its center (Tryggvason, 1974), but the rate of widening has not been established by direct measurements. When tensional stress has reached some critical value, the crust fails along the central axis of the rift zone, as happened in North Iceland in 1975, and magma intrudes the fissure at depth, but above the magma the fissure collapses forming a narrow zone of intense faulting and subsidence (Tryggvason, 1980). The flanks of the fractured zone are contracted to release the tensional stress that has been built up since previous fracturing. This contraction results in uplift of the flanks to approximately the same level as they had immediately after previous rifting episode. The rifting episode may be multiple as has been observed in North Iceland from 1975 to 1979 and in 1874-1875. The 1975 to 1979 rifting episode in North Iceland has resulted in a total widening of some 5 m (the measured amount of 3 to 4 m plus some rifting in 1975 and 1976 before extensive measuring program started), and about 80 km segment of the rift zone was active. The total uplift of the flanks of the fissured zone is poorly determined as 50 to 100 cm and the subsidence of the 1 to 2 km wide zone of intense fracturing may be about two meter on the average. The contraction of the flanks of the fissured zone amounts to nearly 2×10^{-4} and the widening of a 80 km wide zone centered in the zone of fissuring is probably 0.5 to 1.0 m. This means that some contraction has taken place at distances greater than 40 km from the fissure zone, if we assume constant rate of movement of the main body of the crustal plates.

The total widening of the fissure zone of roughly 5 m may be considered to represent the stretching of the crustal plates since 1730, at the end of the Mývatn fires volcano-tectonic episode which affected the same area as the present episode, or in 250 years. This represents

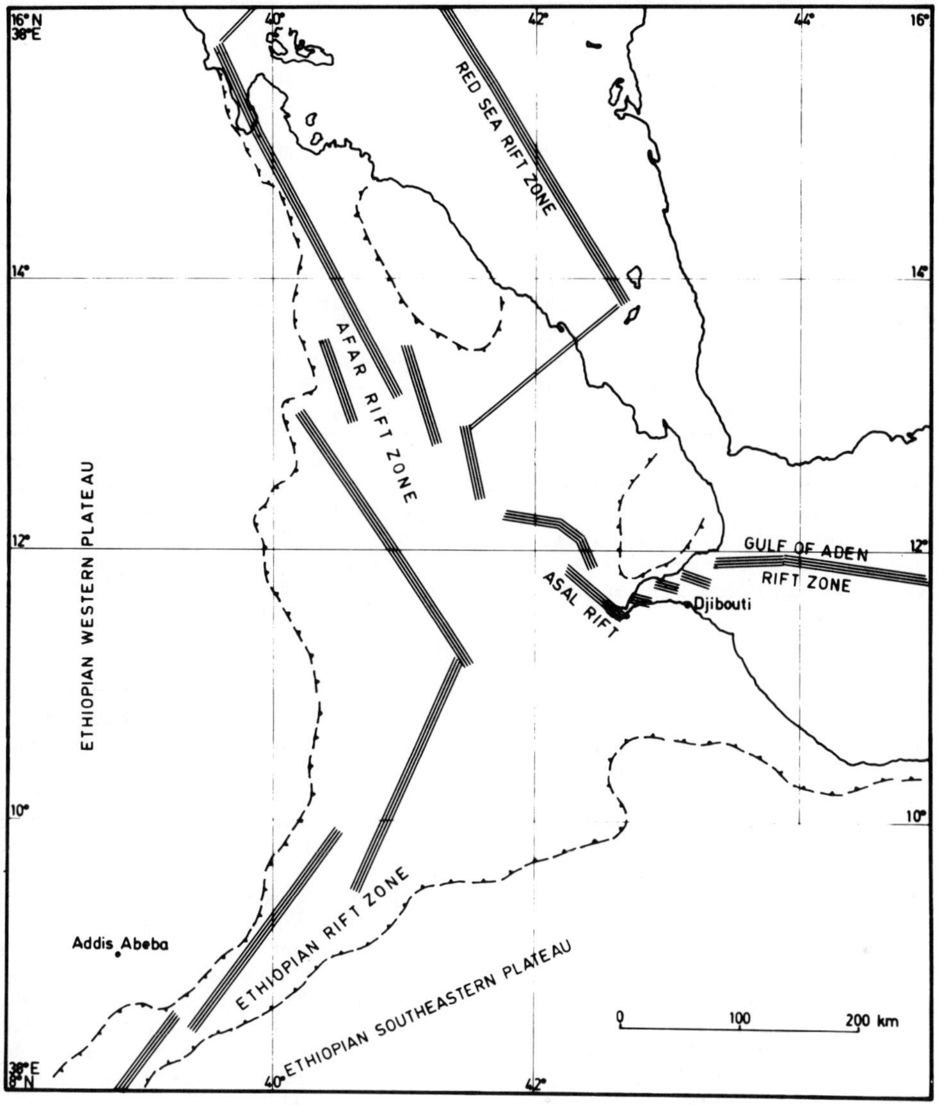

Fig. 4. The Afar rift system showing approximate location of individual rift zones (four parallel lines) and possible fracture zones (two parallel lines). Approximate boundaries of the rifted areas are shown by broken hachured lines. The Asal rift zone became active in Nov. 1978. (Mostly after Kronberg et al., 1975).

2 cm per year as the average widening of the rift zone, if no creep or displacements caused by earthquakes has occurred during this period.

The East African Rift Zone in Ethiopia and Djibouti

The East African rift system is one branch of the rift systems extending from the triple junction in the Afar area (Fig.4). The other two branches are the Red Sea - Dead Sea rifts and the Gulf of Aden rifts and its continuation in the Carlsberg ridge (Laughton, 1966). This whole rift system is commonly considered as a continent-ward extension of the mid-ocean ridge system, which has not as yet formed oceanic area due to its low age and/or slow plate movements. The age of the East African rift and the associated volcanism has been determined as roughly 25 million years (Rogers, 1966; Logatchev et al., 1972; Williams, 1972).

The seismicity of the East African rift system clearly shows that it is presently tectonically active and focal plane solutions indicate a tensional stress field in south-easterly direction in East Africa while in the Gulf of Aden and the Red Sea the rifting is related to a tensional stress field in north-easterly direction (Fairhed and Girdler, 1972).

Direct measurements of ground deformation in

the East African rift system include tidal gauge observations in the Red Sea and Gulf of Aden, and geodetic measurements in the Afar region and the rift valley of Ethiopia

In Aden, tide gauge shows a gradual rise of the sea level by about 2.7 mm/year from 1937-1967, indicating a subsidence of the station of the order of one mm/year, when eustatic rise of sea level is subtracted from the observed rise (Faure, 1975).

A network of 22 geodetic stations was established in the republic of Djibouti in 1972 and 1973 by the Institut Geographique National in the area of the Asal-Ghoubbet graben. This network was measured with geodimeter and triangulation in order to detect ground deformation. A precision leveling line along about 100 km of the road crossing the graben was also established and leveled.

A volcanic fissure eruption occurred in the Asal-Ghoubbet graben in November 1978, accompanied by an intense earthquake swarm and rifting of the ground. Remeasurement of the 1972-1973 network in November 1978 and March 1979 showed the following ground deformation:

A three km wide zone along the Asal-Ghoubbet graben was widened up to 2.4 m. Greatest widening was observed across the Ghoubbet bay and the total length of the widened zone is about 25 km.

The flanks of the widened zone were compressed in direction perpendicular to the zone by up to 0.9×10^{-4} or even somewhat more.

The zone of widening was heavily faulted and it subsided up to 70 cm while its flanks were uplifted as much as 18 cm. This uplift decreases with distance from the faulted zone and seems to disappear in less than 5 km (Allard et al., 1979; Tarantola et al., 1979; Kasser et al., 1979; Ruegg et al., 1979).

The average spreading rate of the Asal rift has been estimated as 1.5 cm/year (Delibrias et al., 1975; Mohr, 1978), although measurements of fault displacements, dikes and fissures indicate a slower spreading rate of 0.3 to 0.6 cm/year on the average (Schaefer, 1975). This indicates that the rifting of 1978 has resulted from accumulation of tensional stress over a period of a few centuries.

An extensive work of repeated distance measurements in the Ethiopian rift valley has been conducted since 1969 (Mohr et al., 1975, 1978). The principal result of the first five years of measurements were as follows:

Several of the lines of measurements show significant increase in length, although this is near the error limit of the measurements, but no measured lines show significant shortening. The average lengthening has been interpreted as tensional strain rate of $6\text{-}16 \times 10^{-7}$/year and widening of the rift zone of 3-5 mm/year. Small right lateral shear movement is indicated (Mohr et al., 1978). The direction of the lines, which show significant increase in length from 1969 to 1974, is approximately SE-NW while all lines, which have other orientations, show no significant length changes (Mohr et al., 1978, Fig.7). Thus the tensional strain axis in the Ethiopian rift zone near 8°30′N is approximately SE-NW and strain rate in this direction, within the rift zone is roughly $10 \pm 5 \times 10^{-7}$ per year.

The Baikal Rift System

The Baikal rift system (Fig.5) extends over a distance of 2500 to 3000 km in SW-NE direction from northwestern Mongolia to South Yakutia near the southern edge of the Precambrian Siberian platform (Florensov, 1966; Logatchev and Florensov, 1978). This rift system has no evident structural connections with other active rift systems, and lies entirely within the Asiatic continent. The rift system is characterized by deep sedimentary basins and the deep fresh water basin of Lake Baikal along the elongated Baikal uplift, high degree of seismicity and low degree of volcanic activity (Kiselev et al., 1978; Solonenko, 1978).

The earliest volcanic activity associated with the Baikal rift zone may be as old as late Cretaceous but the greater bulk of the fissure eruptions are of Miocene or early Pliocene age (Kiselev et al., 1978).

The evolution of the rift system and its sedimentary basins appears to have come in two major stages. The earlier stage of Eocene to early Pliocene age is characterized by slow plastic deformation of the basement, while the later stage, from middle Pliocene to Recent, was dominated by rapid deformation and faulting (Logatchev and Florensov, 1978).

The high seismicity of the rift offers excellent opportunity to study the stress field from solution of earthquake focal mechanism. A great number of fault plane solutions show that the tensional axis is predominantly horizontal and perpendicular to the direction of the rift zone, except at both ends of the zone, where the tensional axis is near vertical (Golenetsky and Misharina, 1978).

Repeated levelings by the Institute of Geology and Geophysics of the U.S.S.R. Academy of Sciences, indicate that the whole Baikal area is being uplifted. A very striking aspect of this uplift is the high correlation between the land elevation and rate of uplift. The rift zone and its immediate surroundings show the highest rate of crustal movements, where 100 m increase in elevation corresponds to one mm/year increase in the rate of uplift, and the maximum rate of uplift is about 26 mm/year (Kolmogorov and Kolmogorova, 1978).

The present information on the vertical component of crustal deformation in the Baikal area seem to indicate that a large area centered near Lake Baikal is being uplifted. This dome of uplift has a diameter of more than 1000 km, and its central part is uplifted at a rate of 2 to 3 cm/year.

The volume of the uplift per unit time can be crudely estimated from the existing information

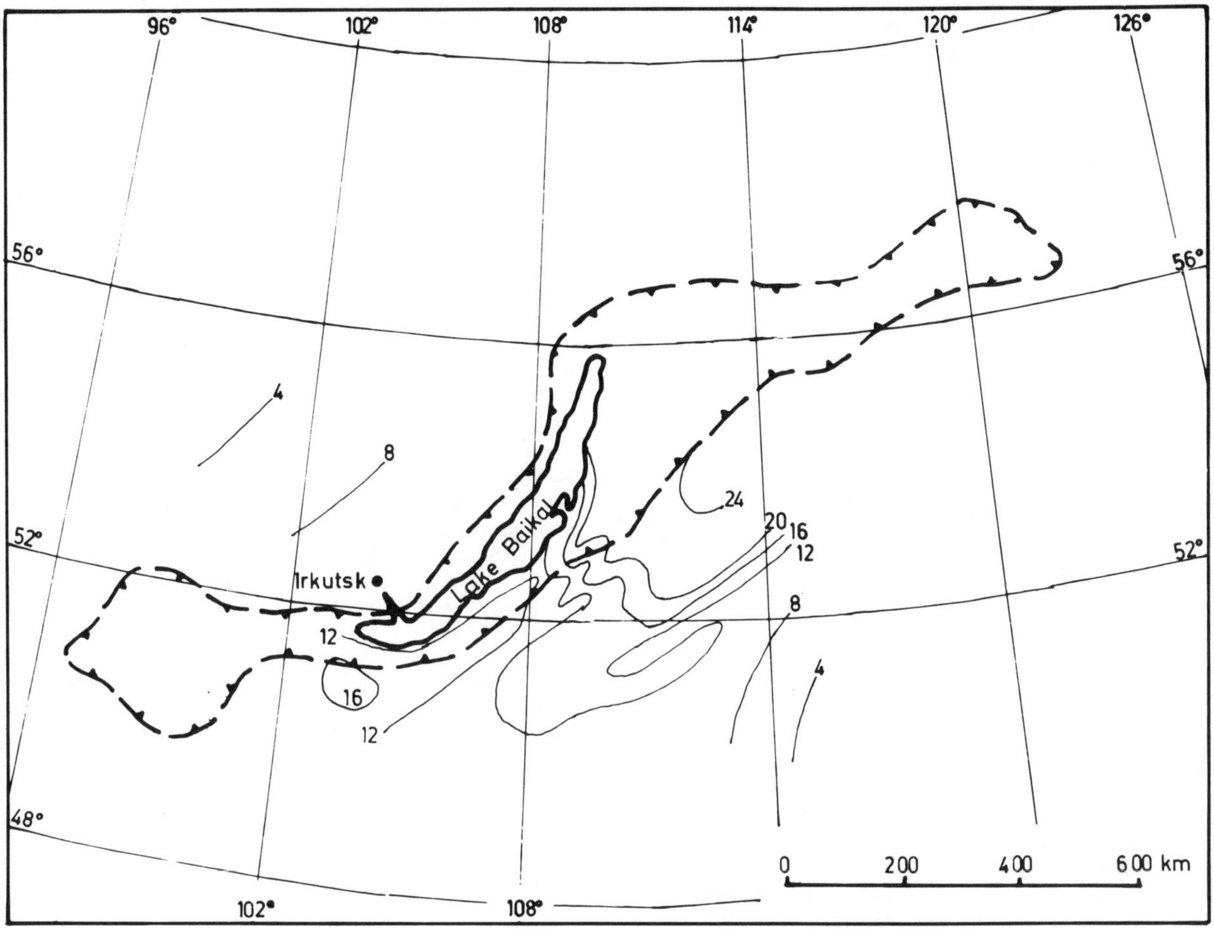

Fig. 5. The Baikal rift system. Broken hachured line shows the boundary of the rift system (Solonenko, 1978) and thin lines show the observed rate of uplift in mm/year (Kolmogorov and Kolmogorova, 1978).

(Kolmogorov and Kolmogorova, 1978). Taking the area of uplift as 10^6 km^2 and the average uplift 10 mm/year, the volume of uplift becomes 10 km^3/year.

No measurements show the horizontal component of crustal deformation in the Baikal area, but focal mechanism solution of earthquakes indicate tension perpendicular to the rift zone (Golenetsky and Misharina, 1978). The magnitude of this tensional deformation is very much in doubt. Most of the earthquakes in the rift zone are associated with normal faulting and the epicenters are concentrated along elongated belts, parallel to the rift system (Golonetsky and Misharina, 1978) but correlation between these belts of high seismic activity and the belts of steep gradient in vertical crustal movement (Kolmogorov and Kolmogorova, 1978) is not clear.

The Rhine Graben

The Rhine graben (Fig.6) is a part of a discontinuous system of rifts, grabens and seismically active zones, which probably extend from the Atlantic-Arctic mid-ocean ridge off northern Norway to the Mediterranean in Italy. The best known sections of this system are the Oslo graben in Norway where faulting started in Permian times and the Rhine graben where vertical movement started in lower Oligocene and is continuing at present (Bederke, 1966). The Rhine graben is about 300 km long and 30-40 km wide and its direction is about N21°E.

The present day stress field in Central Europe as deduced from focal plane solutions of earthquakes and in-situ stress measurements shows the mean direction of the horizontal component of maximum compressive stress is 142° ± 20° (Ahorner, 1975; Greiner, 1975). The focal plane solutions for earthquakes in and around the Rhine graben show strike-slip mechanism to dominate, and if slip-plane is subparallel to the graben, left-lateral displacement is indicated (Ahorner, 1975).

Extensive geodetic observations of the Rhine graben area over several decades have been

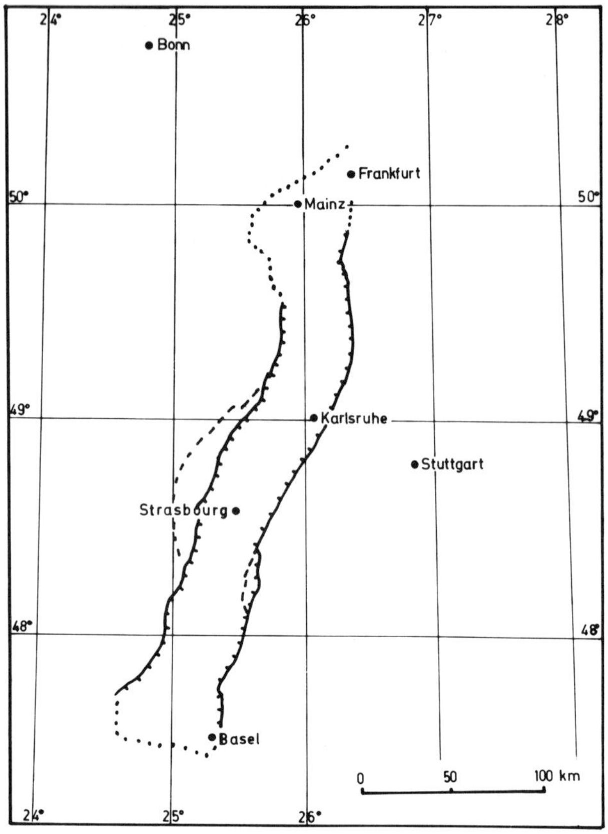

Fig. 6. The outlines of the Rhinegraben (hachured lines) and connected fault systems (dashed lines). The approximate termination of the graben towards north and south is shown by dotted lines. (Largely from Illies, 1975).

treated in order to detect crustal deformation. These include gravity and gravity gradient observations, tilt observations, precision leveling, triangulation, and distance measurements (Mälzer and Schlemmer, 1975; Groten et al., 1979; Mälzer et al., 1979). It appears that the geodetic observations to date fail to show significant horizontal component of deformation while the vertical component is convincingly demonstrated. Inside the Rhine graben subsidence dominates. The average rate of subsidence in the northern part of the Rhine graben near Ludwigshafen and Worms is 0.3 to 0.7 mm/year (Groten et al., 1979) and similar subsidence values are reported by Mälzer et al. (1979) using a different treatment of the data from northern part of the Rhine graben. More irregular subsidence is indicated in the southern part of the graben (Mälzer and Schlemmer, 1975). Subsidence of approximately 1.0 mm/year is found near Breisach and near Kehl although subsidence of 0.2 to 0.4 mm/year seems to be more common. Even uplift of up to 0.4 mm/year is indicated in the Rhine graben between Rastatt and Bruchsal.

The observed rate of vertical ground movement correlates well with the thickness of Pliocene and Pleistocene sediments. The maximum thickness of Pliocene sediments in the northern part of the Rhine graben is about 760 m and that of Pleistocene sediments about 380 m. This shows that the present rate of vertical ground movement is roughly 10 times greater than the average rate since the beginning of Pliocene (Illies et al., 1979).

In the areas immediately outside the Rhine graben, uplift of 0 to 0.2 mm/year is indicated in the Renish Massif in the north while slight subsidence is indicated in the Black Forest area (Mälzer and Schlemmer, 1975; Mälzer et al., 1979).

The horizontal component of deformation in the Rhine graben area, although not as yet clearly demonstrated by direct measurements, can be inferred from seismicity, stress and strain relief measurements and the vertical movement. Focal mechanism of earthquakes shows that some widening of the Rhine graben occurs, but the major part of the horizontal deformation is left-lateral shear (Ahorner, 1975). The seismotectonic slip rate has been estimated as 0.05 mm/year in the upper Rhine graben, while the geodetically inferred slip rate is 0.2 to 0.7 mm/year (Ahorner, 1975). This indicates that most of the geodetic slip rate occurs as a seismic creep, or else, that it occurs in few large earthquakes, none of which has occurred since 1700, in the time interval considered by Ahorner.

Although the present tectonics of the Rhine graben area is primarily left-lateral shear, indications are, that earlier in the history of the rift system, tension perpendicular to the graben direction prevailed (Illies, 1975).

Concluding Remarks

The present rate and nature of ground deformation in the rift zones of the world is known to a limited degree on a few short segments of these zones, and still fewer segments are considered here as illustrative examples. The present state of knowledge is insufficient to form an accurate model of the rift zone deformation, but a few aspects are emerging.

The tensional strain rate perpendicular to the rift zones is in the oceanic rift in Iceland and Ethiopia of the order of 10^{-6} per year during periods of low activity. In Iceland, and probably also in Afar, major rifting episodes occur at intervals of roughly one century. During these rifting episodes, contractional strain of roughly 10^{-4} occur on the flanks of the rift zone, releasing all the tensional strain which has been built up during the roughly 100 years of quiescence. Similarily the tilt rate towards the rift zone is observed to be 0.4 to 0.6 x 10^{-6} during quiet periods in Iceland, while tilt of about 100 x 10^{-6} away from the rift zone occurs during a rifting episode (Tryggvason, 1980). Another interpretation, equally true, is that the flanks of the rift zone, together with the rift zone itself, subsides continuously during quiet periods but the flanks are uplifted, probably by nearly the same amount, during the brief rifting episodes. The rift zone itself subsides further during the rifting episodes. Thus the land elevation immediately outside the rift zone oscillate up and down without any indicated net vertical displacement while the narrow rift zone subsides continuously at a rate of a few millimeters per year between rifting episodes and subsides further some tens of centimeters during these episodes.

Rifting episodes in both Iceland and Afar are associated with magmatic activity (Björnsson et al., 1977; Kasser et al., 1979). The interpretation of this association in Iceland is that rifting does not occur unless molten magma is available to fill the fissure in a form of a dike up to a shallow depth. The subsidence of the rift zone is then due to collapse of the near-surface formation into the fissure above the dike. The brittle-elastic model proposed to explain observed deformation in Afar (Tarantola et al., 1979) seems not to require molten magma to intrude the formed fissure, although it occurred during the observed rifting event in 1978.

The deformation within and near the two continental rift zones, the Rhine graben, and the Baikal rifts, seems to differ significantly from that of the oceanic rift zone in Iceland, and also in Afar, which may be termed as oceanic. The rate of deformation in the Rhine graben is quite slow, at least one order of magnitude slower than in Iceland and Afar. The deformation of the Baikal rift zone appears to be dominated by large scale uplift of a vast area. This has been correlated to a large volume of the upper mantle with abnormally low seismic velocity and probably higher temperature than normal upper mantle (Zorin and Flovensov, 1979). Then the uplift is due to isostatic adjustment and the rifting due to tensional stress above the slowly rising high temperature mantle material.

References

Ahorner, L., Present-day stress field and seismotectonic block movements along major fault zones in central Europe, Tectonophysics, 29, 233-249, 1975.

Allard, P., H. Tazieff and D. Dajlevic, Observations of seaflor spreading in Afar during the November 1978 fissure eruption, Nature, 279, 30-33, 1979.

Bederke, E., The development of European rifts, pp. 213-219, The World Rift Systems, T.N. Irvine (ed.), Geological Survey of Canada, paper 66-14, 1966.

Bernauer, F., Junge Tektonik auf Island und ihre Ursachen, pp. 14-64, Spalten auf Island, O. Niemczyk (ed.), Verlag von Konrad Wittwer, Stuttgart, 1943.

Björnsson, A., Jardhraeringar vid Kröflu (Rifting and volcanism in the Krafla area 1975-1977), Natturufraedingurinn, 46, 177-198, 1976.

Björnsson, A., K. Saemundsson, P. Einarsson, E. Tryggvason and K. Grönvold, Current rifting episode in north Iceland, Nature, 266, 318-323, 1977.

Björnsson, A., G. Johnsen, S. Sigurdsson, G.Thorbergsson and E. Tryggvason, Rifting of the plate boundary in north Iceland 1975-1978, J. Geophys. Res., 84, 3029-3038, 1979.

Brander, J. and G. Wadge, Distance measurements across the Heimaey eruptive fissure, Nature, 244, 496-498, 1973.

Brander, J., R.G. Mason and R.W. Calvert, Precise distance measurements in Iceland, Tectonophysics, 31, 193-206, 1976.

Decker, R.W., P. Einarsson and P.A. Mohr, Rifting in Iceland: New geodetic data, Science, 173, 530-532, 1971.

Decker, R.W., P. Einarsson and R. Plumb, Rifting in Iceland: Measuring horizontal movements, Greinar, V, 61-71, Soc. Scient. Isl., 1976.

Delibrias, G., G. Marinelli and L. Stieljes, Spreading rate of the Asal rift: A geological approach, pp 214-221, Afar Depression of Ethiopia, Vol. I, A. Pilger and A. Rösler (eds.), E. Schweizerbart'sche Verlagsbuchandlung, Stuttgart, 1975.

Fairhead, J.D. and R.W. Girdler, The seismicity of the east African rift system, Tectonophysics, 15, 115-122, 1972.

Faure, H., Recent crustal movements along the Read Sea and Gulf of Aden coasts in Afar (Ethiopia and T.F.A.I.), Tectonophysics, 29, 479-486, 1975.

Florensov, N.A., The Baikal rift zone, pp. 173-180, The World Rift System, T.N. Irvine (ed.), Geological Survey of Canada, paper 66-14, 1966.

Gerke, K., Measurement of recent crustal movements in Iceland, pp. 237-246, Problems of Recent Crustal Movements, J.D. Boulanger et al. (eds.), U.S.S.R. Academy of Sciences, Moscow, 1969.

Gerke, K., Crustal movements in the Myvatn- and in the Thingvallavatn-area, both horizontal and vertical, pp. 263-275, Geodynamics of Iceland and the North Atlantic Area, L. Kristjánsson (ed.), D. Reidel Publ. Comp., Dordrecht, 1974.

Gerke, K., D. Möller and B. Ritter, Geodätische Lagemessungen zur Bestimmung horizontaler Krustenbewegungen in Nordost-Island, pp. 23-33, Festschrift für Walter Höpcke zum 70. Geburtstag, Hannover, 1978.

Golenetsky, S.I. and L.A. Misharina, Seismicity and earthquake focal mechanism in the Baikal rift zone, Tectonophysics, 45, 71-85, 1978.

Greiner, G., In-situ stress measurements in southwest Germany, Tectonophysics, 29, 265-274, 1975.

Groten, E., C. Gerstenecker and G. Hein, Geodetic high-precision measurements in active tectonic areas; example: The Rhinegraben, Tectonophysics, 52, 157-165, 1979.

Illies, J.H., Recent and paleo-intraplate tecto-

nics in stable Europe and the Rhinegraben rift system, Tectonophysics, 29, 251-264, 1975.
Illies, J.H., P. Claus, H.-U. Schmincke and A. Semmel, The quaternary uplift of the Rhenish shield in Germany, Tectonophysics, 61, 197-225, 1979.
Kasser, M., J.C. Lépine and J.C. Ruegg, Mise en évidence de mouvements d'extension sur la dorsale émergée d'Asal-Ghoubbet en République de Djibouti, Comptes Rendus Acad. Scient. Paris, Serie D, 288, 469-472, 1979.
Kiselev, A.I., H.A. Golovko and M.E. Medvedev, Petrochemistry of Cenozoic basalts and associated rocks in the Baikal rift zone, Tectonophysics, 45, 49-59, 1978.
Kolmogorov, V.G. and P.P. Kolmogorova, Some results from studying recent crustal movements in the Baikal rift zone, Tectonophysics, 45, 101-105, 1978.
Kronberg, P., M. Schönfeld, R. Günther and P. Tsombos, ERTS 1-data on the geology and tectonics of the Afar/Ethiopian and adjacent regions, pp. 19-27, Afar Depression of Ethiopia, Vol. I, A. Pilger and A. Rösler (eds.), E. Schweizerbart'sche Verlagsbuchhandlung, Stuttgart, 1975.
Laughton, A.S., The Gulf of Aden, in relation to the Red Sea and the Afar depression of Ethiopia, pp. 78-97, The World Rift System, T.N. Irvine (ed.), Geological Survey of Canada, paper 66-14, 1966.
Logatchev, N.A., V.V. Beloussov and E.E. Milanovsky, East African rift development, Tectonophysics, 15, 71-81, 1972.
Logatchev, N.A. and N.A. Florensov, The Baikal system of rift valleys, Tectonophysics, 45, 1-13, 1978.
Mälzer, H. and H. Schlemmer, Geodetic measurements and recent crustal movements in the southern upper Rhinegraben, Tectonophysics, 29, 275-282, 1975.
Mälzer, H., G. Schmitt and K. Zippelt, Recent vertical movements and their determination in the Rhenish massif, Tectonophysics, 52, 167-176, 1979.
Mohr, P., Afar, pp. 145-172, Annual Review of Earth and Planetary Sciences, Vol. 6, F.A. Donath (ed.), Palo Alto, 1978.
Mohr, P.A., A. Girnius, J.R. Chermiack, E.M. Gaposchkin and J. Latmier, Recent crustal deformation in the Ethiopian rift valley, Tectonophysics, 29, 461-469, 1975.
Mohr, P., A. Girnius and J. Rolff, Present-day strain rates at the northern end of the Ethiopian rift valley, Tectonophysics, 44, 141-160, 1978.
Nielsen, N., Tektonik und Vulkanismus Islands unter Berücksichtigung der Wegener-Hypothese, Geologische Rundschau, 21, 347-349, 1930.
Niemczyk, O. and E. Emschermann, Sonderdreiecksmessung auf Island zur Feststellung feinster Erdkrustenbewegungen, pp. 80-113, Spalten auf Island, O. Niemczyk (ed.), Verlag von Konrad Wittwer, Stuttgart, 1943.
Riecker, R.E., (editor), Rio Grande Rift. Tectonics and magmatism, 438 pp., American Geophysical Union, Washington D.C., 1979.
Rogers, A.S., Some comments on the rift in Ethiopia, p. 98, The World Rift System, T.N. Irvine (ed.), Geological Survey of Canada, paper 66-14, 1966.
Ruegg, J.C., J.C. Lépine and A. Tarantola, Geodetic measurements of rifting associated with a seismo-volcanic crisis in Afar, Geophysical Research Letters, 11, 817-820, 1979.
Schaefer, H.U., Investigations on crustal spreading in southern and central Afar (Ethiopia), pp. 289-296, Afar Depression of Ethiopia, Vol. I, A. Pilger and A. Rösler (eds.), E.Schweizerbart'sche Verlagsbuchhandlung, Stuttgart, 1975.
Schleusener, A. and W. Torge, Investigations of secular gravity variations in Iceland, Zeitschrift für Geophysik, 37, 679-701, 1971.
Sindrason, S. and H. Ólafsson, A magnetoresistor geotiltmeter for monitoring ground movement, Nordic Volcanological Institute, report 7806, 16 pp., Reykjavik, 1978.
Solonenko, V.P., Seismotectonics of the Baikal rift zone, Tectonophysics, 45, 61-69, 1978.
Spickernagel, H., Höhenmessungen in Nord-Island, Mitteilungen aus dem Markscheidewesen, 73, 139-152, 1966.
Tarantola, A., J.C. Ruegg and J.C. Lépine, Geodetic evidence for rifting in Afar: A brittle-elastic model of the behaviour of the lithosphere, Earth Plan. Sci. Lett., 45, 435-444, 1979.
Torge, W. and H. Drewes, Gravity changes in connection with the volcanic and earthquake activity in northern Iceland 1975-1976, Jökull, 27, 60-70, 1977.
Torge, W. and E. Kanngieser, Local gravity variations in northern Iceland connected with earthquake and volcanic activity, Presented to the 8th meeting of the International Gravity Commission, Paris, 1978.
Tryggvason, E., Measurements of surface deformation in Iceland by precision leveling, J. Geophys. Res., 73, 7039-7050, 1968.
Tryggvason, E., Surface deformation and fault displacement associated with an earthquake swarm in Iceland, J. Geophys. Res., 75, 4407-4422, 1970a.
Tryggvason, E., Precision levelling on Surtsey in 1968, Surtsey Research Progress Report, V, 113-116, 1970b.
Tryggvason, E., Precision levelling in Surtsey, Surtsey Research Progress Report, VI, 158-162, 1972.
Tryggvaosn, E., Surface deformation and crustal structure in the Myrdalsjökull area of south Iceland, J. Geophys. Res., 78, 2488-2497, 1973.
Tryggvason, E., Vertical crustal movements in Iceland, pp. 241-262, Geodynamics of Iceland and the North Atlantic area, L. Kristjánsson (ed.), D. Reidel Publ. Comp., Dordrecht, 1974.
Tryggvason, E., Landslagsbreytingar samfara jardskjalftunum 1975-1976 (Ground movement in north

Iceland during the earthquake swarm of 1975-1976), Natturufraedingurinn, 46, 124-128, 1976.
Tryggvason, E., Tilt observations in the Krafla-Myvatn area 1976-1977, Nordic Volcanological Institute, report 7802, 45 pp., Reykjavik, 1978a.
Tryggvason, E., Distance measurements in 1977 in the Krafla-Myvatn area and observed ground movements, Nordic Volcanological Institute, report 7810, 47 pp., Reykjavik, 1978b.
Tryggvason, E., Subsidence Events in the Krafla area, north Iceland, 1975-1979, J. Geophys. 47, 141-153, 1980.
Williams, L.A.J., the Kenya rift volcanics: A note on volumes and chemical composition, Tectonophysics, 15, 83-96, 1972.
Zorin, Y.A. and N.A. Florensov, On geodynamics of Cenozoic uplifts in central Asia, Tectonophysics, 61, 271-283, 1979.

IN MEMORIAM

RAPHAEL FREUND

Dr. Raphael Freund, an associate of the ICG Working Group 4 died on 24 February 1980 at the early age of 46.

Dr. Freund (Raphy, as he was known to all his friends and scientific colleagues) first came to the International scene at the symposium on "The World Rift System" in Ottawa, Canada, 1965. His lecture was entitled simply "Rift Valleys" and he brought together a wealth of detailed field evidence and the results of laboratory model experiments in his efforts to explain the origin and evolution of rift structures. He was an advocate of large-scale horizontal movements at a time when they were still unfashionable but his lecturing technique was superb and he carried his audience with his infectious enthusiasm.

He was one of the most lively speakers at any meeting. Many will recall his fine exposition at the Royal Society Discussion Meeting on the evolution of the Red Sea in London in 1969 when he became master of the viewgraph. With translucent plastic sheets of different colours he demonstrated to an enthralled audience how the shear he observed along the Dead Sea rift was accompanied by the opening of the Red Sea to the south and the crumpling of the Toros-Zagros mountains to the north. The trouble he took over his lectures was rewarded by spontaneous and enthusiastic applause.

It gave him very great pleasure that in 1979 he was able to organise an international meeting in his own country, the Dead Sea Rift Symposium which was held in Jerusalem under the co-sponsorship of the Inter-Union Commission on Geodynamics. In a three-day post-symposium excursion to the northern part of the rift system, Raphy explained the complexities of this part of the World Rift System, emphasizing at the same time the problems that still remained to be solved.

Raphy Freund was born in Breslau, Germany and after six months his family emigrated to Israel. After graduating with distinction from high school in Haifa he went on to the Hebrew University of Jerusalem to read geology obtaining his M.Sc. in 1959 and his Ph.D. in 1963. He was to remain based at the Hebrew University for the rest of his life becoming successively Instructor (1959), Lecturer (1963), Senior Lecturer (1966), Associate Professor (1972), Full Professor (1977) and first Director of the Institute of Earth-Sciences.

As he lived almost on top of the Dead Sea rift, it is not surprising that this became his main object of study. He searched for and found an enormous amount of geological data for elucidating the timing of the shear movements along the Dead Sea rift zone. He was an enthusiastic field geologist and it was fortunate that he was still well enough to be able to show a lot of the evidence for horizontal shear to the international gathering last year. Further afield, he made contributions to the studies of strike-slip faults in New-Zealand and in his last year worked on dynamic models of subduction zones.

Raphy had a great zest for life. He loved music, literature and sport and he shared with the late Sydney Chapman an inexaustible enthusiasm for cycling. He will be greatly missed at home and abroad.

RONALD GIRDLER
GUDMUNDUR PALMASON

Footnote: The review article "The Role of Shear in Rifting" was received shortly before he died. He was already very ill when it was written and was never able to revise it. As this is an invited contribution, the editors feel it should be published. Editorial corrections have been made but it is hoped that the scientific content remains as he intended.

THE ROLE OF SHEAR IN RIFTING

R. Freund

Department of Geology, Hebrew University Jerusalem, Israel.

Abstract. Horizontal shear on transform faults is an integral feature of oceanic rift systems. These faults are usually perpendicular to the rifts. Many attempts have been made to explain their origin, the most promising involving accretional growth of the faults and spreading orientated inhomogeneities in the ocean floor. Small strike-slip faults occur associated with continental rifts: these are probably related to the rifting because they usually trend obliquely to the rift in the direction of the maximum effective shear stress which coincides with the stress responsible for the origin of the rift.

Introduction

The term shear here refers to horizontal shear, i.e. the strike-slip component. This is because normal faults which are the main structures of rift valleys comprise by definition both vertical (dip slip) and horizontal (heave) components. An ideal normal fault does not have a strike-slip component, the subject of this review. Indeed, in theory (Anderson, 1951) the formation of rifts, grabens and normal faults does not involve or require any strike-slip component. Nevertheless, strike-slip faults are commonly associated with oceanic and continental rifts. In the oceans they are mostly perpendicular to the rifts and on the land they are oblique to them. In this review, several features are described and attempts to explain them.

Ridge-Ridge Transform Faults in the Oceans

The importance of horizontal shear in the ocean rift zones is indicated by the abundance (Heezen and Tharp, 1965) of ridge-ridge transform faults (Wilson, 1965), by the strike slip motions derived from earthquake fault plane solutions (Sykes, 1967) and from direct observations (Arcyana, 1975). Yet despite many attempts to explain the origin of these faults, we still do not know how and why they come into existence. The ideas put forth include complex stresses between edges of tensile fractures (Orowan, 1969, Oldenburg and Brune, 1975), various asymmetric growth patterns of the ocean floor (Ball and Harrison, 1970; Weissel and Hayes, 1971; Oldenburg and Brune, 1972; O'Bryan and others, 1975), considerations of least energy dissipation (Lachenbruch and Thompson, 1972; Froideveaux, 1974), serpentine protruding along pre-existing fractures (Bonatti, 1973), thermal contraction (Turcotte, 1974; Collette, 1974), flow lines and shear weakening (Vroman, 1976) and mechanical anisotropy (Freund and Merzer, 1976a). These ideas emerge from phenomena observed in the ocean floor and from ridge transform patterns produced in wax models (Oldenburg and Brune, 1972, 1975: O'Bryan and others, 1975; Freund and Merzer, 1976a), supported also by similar features observed on the surface of the Mauna Ulu lava lake (Duffield, 1972). We shall review these features and discuss the explanations relating to them.

Being parallel to the spreading direction (Le Pichon, 1968), the orientation of the transform faults coincides with one of the principal normal stresses (tensile if the spreading is due to pulling, compressional if it is due to pushing), so that the shear stress on these faults is nil. If the orientation of the stress was the same when these faults were formed, they could not be the result of pure mechanical failure as suggested by Orowan (1969) and Oldenburg and Brune (1975) unless the shear strength of the ocean floor is very small in the direction of spreading as it is in the anisotropic wax film (Freund and Merzer, 1976a). This is not so, as the seismic travel times do not show mechanical anisotropy of the ocean floor, except perhaps one case (Christensen, 1972a) where it is weaker along, not across the ridge.

From the absence of strain across the transform faults, it may be deduced that at the present stage of fully developed faults, no compressive normal stress occurs across them, so that the friction is nil and no energy is invested on the movement along them. Nor is the useful work (i.e. deformation) accomplished by the motion on the faults. The earthquakes on the transform faults are probably created on small irregularities. The transform faults serve, however, to reduce the length of oblique ridges and at the same time increase their width, so that if the ridge spreading is an energy consuming process, then the creation of the faults reduces the energy dissipation of the system (Lachenbruch and Thompson, 1972; Froideveaux, 1974). This may be a factor in pushing the system towards the ridge-transform pattern from an oblique ridge configuration: yet it cannot be the dominant cause for the creation of transform faults because they appear also on ridges whose original trend is perpendicular to the spreading both on the ocean floor south of Australia and in wax models (O'Bryan and others, 1975).

Indeed, the original obliquity of the ridge seems to be related to the origin of several major transform faults such as the Vema and Romanch fracture zones in the Atlantic and the Alula-Fartak trench in the Gulf of Aden, as these faults extend through their fracture zones to sharp bends in the coast lines. Two explanations have been suggested for this relationship in addition to the one already mentioned. First, Bonatti (1973) regarded these bends as ancient transverse fractures along which protrusions of stagnant serpentine bodies rising from the ridge produced the transform faults. Secondly, Vroman (1976) suggested that the spreading between these bends lead to shear weakening of the ocean floor, and that these weakened zones fractured subsequently to transform faults. Again this cannot be the single or the dominant factor in the creation of transform faults, because most transform faults do not extend to bends in the coast lines, and moreover transforms also occur on ridges which originated in the ocean, such as the present ridges in the Indian Ocean and East Pacific.

It is very unlikely that thermal contraction of the ocean floor initiated the transform fault as proposed by Turcotte (1974) and Collette (1974), because the thermal cracks should be perpendicular to the ridge, the contraction parallel to the ridge being taken up by the rifting. Also in oblique ridges these cracks would be oblique again and not in the direction of spreading.

Another fact about the ridge-ridge transforms is that the movement on them exceeds (or at least will exceed in time) the length of the active fault (Freund, 1974). In other words the fault is as a rule older than the rocks lining it on both sides. At each end of the transform fault new rock solidifies on the ridge side against the older rock on the opposite side of the existing fault. At present the transform faults are being constructed or accreted rather than disrupted or fractured, although the latter process must also take place between the newly solidified rock and the opposite wall or near this place.

Since symmetric spreading on the ridge segments at the two ends of the transform faults maintains their length, it may be suggested that the transform faults originate and grow by asymmetric spreading of ridge segments. Transform faults may grow where the spreading rate increases abruptly provided one side is fixed (Ball and Harrison, 1970, Weissel and Hayes, 1971). In wax models, transform faults have been observed to grow (Oldenburg and Brune, 1972; O'Bryan and others, 1975) between ridge segments on which the spreading was asymmetric and alternated from one side to the other. In similar experiments, it has been observed (Freund and Merzer, 1976) that the same process in the opposite direction has lead to the diminution of existing transforms. It is not known which of these processes leads to the creation of the transform faults in the oceans as the resolution of the magnetic anomaly pattern for the early stages of spreading is not sufficiently good.

O'Bryan and others (1975) proposed another accretion process of changing a zigzag spreading center into a ridge-transform pattern by the quicker heat loss to the walls in the narrower oblique parts of the former. This process has neither been observed in the wax models nor in the magnetic anomalies of the ocean floor.

The petrology of the transform faults is a cardinal feature in their existence and may even be the cause of their origin. Dredging and seismological considerations (V_p/V_s ratio, Poisson ratio) show that they are lined by serpentines while the rest of the ocean floor layer 3 is composed of metabasites (Miyashiro and others, 1969, 1970; Melson and Thompson, 1970, 1971; Christensen, 1972 b; Bonatti, 1973 ; Christensen and Salisbury, 1975). The fault zones may be serpentinised because they provide channels for the percolation of sea water to great depth, where temperatures of 300° to 350°C at 0.5 to 1 Kbar for sufficiently long time provide the conditions for the hydration of Mg-Fe silicates to serpentines (Johannes, 1968). Normal ocean floor is not serpentinised because the lava rises rapidly at the ridge axes and cools before this process can take place. As the serpentines show a marked weakening at temperatures of 300° to 600° (Raleigh and Paterson, 1965), the mechanical weakness of the fault zone is amplified by the serpentinisation. The intrusion of the lava in the spreading ridge terminates at the transform fault because it seems to be easier for the weak serpentine to yield by shear than the propagation of the tensile crack across it. The serpentinite-weakening of the transform faults is probably the reason for the stability of the ridge-transform pattern.

If this is the case, at the present advanced stage of this set-up, it is tempting to speculate that selective serpentinisation of certain parts of the ocean floor is the cause of the origin of the transform faults. Conditions for serpentinisation may occur also without existing faults. Lines of partially serpentinised rocks may form at the wake of every spot on the ridge where the lava fails to rise fast enough to create new ocean floor at the average pace of the spreading. Such a retardation may occur in oblique segments which provide narrower channels for the rise of the lava, or above points where the temperature of the mantle is lower, or where the water content is smaller, or where the minerals are more refractive. In these cases lines of depressed topography will form in

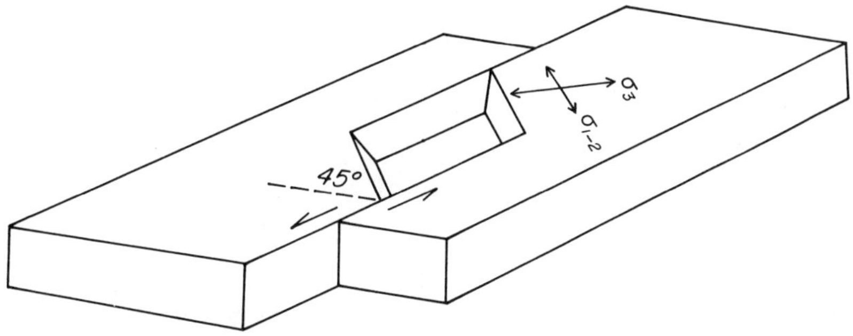

Fig. 1. A schematic model of a rhomb graben at the left-slip fault. The vertical strike slip fault becomes an inclined normal fault along the graben whose length exceeds the displacement, and the connecting faults are normal and their strike deviates by c. 45° from the main strike slip, normal to the least principal stress coaxial with the shear stress.

the direction of spreading on both sides of the ridge whose length and degree of serpentinisation depend on the intensity, size and duration of lava retardation.

Wherever the lava is rising intermittently and the ridge freezes between the pulses, the tensile crack through which the lava rises extend perpendicular to the spreading, in the plane containing the maximum and intermediate principal normal stresses. As suggested above, these cracks may terminate at a serpentinised belt, because its shear resistance may be smaller than its resistance to the propagation of the tensile crack, particularly where two tensile cracks approach the serpentinised belt from both sides at an offset and combine effort in a shear couple. In brief, though at an advanced stage, the transform faults and the serpentine lining them are related as chicken and eggs, it seems that the serpentine belts can develop independant of transform faults and hence lead to the creation of the latter, whereas transform faults cannot form independently.

Horizontal Shear in Continental Rifts

The phenomenon of horizontal shear in continental rifts is less common and more variable than in the oceanic rifts, and the evidence for its existence is sometimes circumstantial. On the other hand, it seems that the associations of strike slip faults and normal faults is less of a problem, as they usually conform to the empirical rules of mechanical shear failure. The phenomena can be classified into three groups according to the dominant direction of the motion: (a) along the rift, (b) oblique to the rift and (c) across the rift. The first and the last group exhibit systematic fault patterns which are common to both groups.

(a) Rifts or Grabens along Strike-Slip Faults

Although strike-slip faults are not genuine rifts, the secondary grabens along them are included in this review because unfortunately the entire features have often been called rifts (e.g. the "Dead Sea rift" which exhibits associations of normal faults and strike-slip faults which resemble features observed in "genuine" extensional rifts).

Small and large grabens occur along the Dead Sea Rift (Quennell 1956, 1958; Freund, 1965; Zak and Freund, 1966; Freund and others, 1968; Garfunkel, 1970), along the Alpine fault system in New Zealand (Kingma, 1958; Lensen, 1958; De Ridder and Lensen, 1960; Clayton, 1966; Freund, 1971), along the San Andreas fault system (Gastil, 1968; Moore and Baffington, 1968; Lomnitz and others, 1970; Garfunkel, 1972; Cromwell, 1974), and along smaller faults in Sinai (Garfunkel, 1970; Bartov, 1974) and in Iran (Tchalenko and Ambraseys, 1970). They are reported on faults whose horizontal displacements range between 100 m to 100 km, and the length of these grabens varies accordingly, though as a rule they are somewhat longer than the displacement.

Two types of grabens have been recognised and defined: rhomb grabens and wedge grabens. The first have also been called "openings" by Clayton (1966) and "pull aparts" by Cromwell (1974).

(i) Rhomb grabens. These appear where two long segments of strike slip fault are offset (Figure 1) in the same sense as the horizontal displacement of the strike slip fault (e.g. offset to the left when viewed along a left hand strike slip fault). These two long segments might or might not have been connected by a small oblique fault before the movement on the strike slip fault, but after the movement normal faults surround the entire rhomb graben. The vertical strike slip fault segment turns to the inclined normal fault gradually near the end of the rhomb graben. The oblique border faults and other longitudinal or oblique faults which occur frequently in these grabens are all normal. The orientation of the oblique faults concerning the two long segments is always about 45° to the long faults, so that they are perpendicular to the least principal stress which is coaxial with the shear stress of the major strike-slip faults.They are parallel also to the oblique ocean ridge segments which occur at the

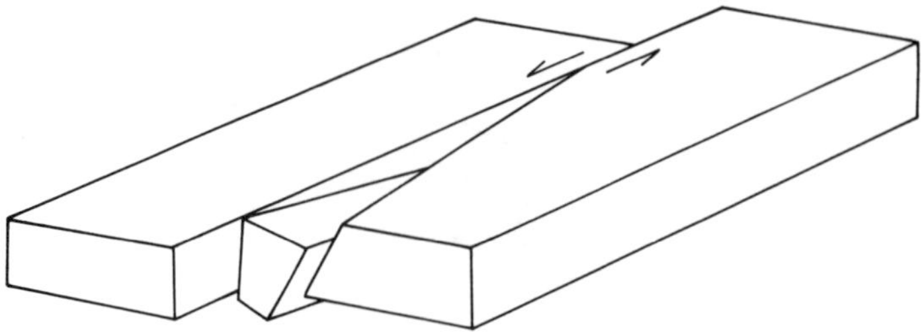

Fig. 2. A schematic model of wedge graben which occurs between a strike slip fault and a branch fault which deviates from the main fault by an acute angle against the direction of shear. The branch fault is normal, but the dip of the main fault remains unaltered. The deepest part occurs at the tip.

termination of the strike slip fault, such that the Red Sea at the end of the Dead Sea Rift and the Salton Trough at the Southern end of the San Andreas are probably the largest and most widely known rhomb grabens.

(ii) Wedge grabens. These occur where a small splay fault leaves the main one at a very acute angle (c. 10°) against the direction of shear (Figure 2). The deepest part of the wedge grabens occurs at their tip. The splay fault is normal but the other side is either vertical or even somewhat reversed depending on the nature of the major strike slip fault.

In addition to these well defined structures, various depressions and sag points occur along strike slip faults but little can be said about their nature and origin. Many strike slip faults run along narrow deep valleys (hence their being called rifts) whose origin is probably not tensional as the two sides dip steeply toward the fault. The origin of these V-shaped fault structures is not known so far.

(b) Rift Valleys with Oblique Motion

The evidence for the "transtensional" structures (Harland, 1971) varies in kind and in degree of confidence, and they exhibit very little in common.

It may be suggested that the straight, coast to coast fracture zones in the Gulf of Aden (Laughton, 1966) indicate that the Red Sea (as already speculated by Lartet as long ago as 1869) and the Gulf of Aden originated as oblique rifts. The oblique shear zones and faults observed on both sides of the Red Sea (Abdel Gawad, 1968) and Gulf of Aden (Laughton, 1966) could thus be regarded as "Riedel Shears" of the longitudinal component of the movement at the early stages of extension.

Right lateral motion along the Ethiopia-Kenya rift associated with clockwise rotation of the Somali block relative to Africa may be necessary to account for the discrepancy between the restorations of the Red Sea and the Gulf of Aden (Baker, 1965) and to conform with other features in Afar (Mohr, 1968). However, there is no direct record of horizontal offsets or strike slip faults in these rifts, and a detailed pole study of this problem (Girdler and Daracott, 1972) turned out to be inconclusive on this point.

Right hand strike slip movement occurs on a fault crossing the Ruwenzori Mountains (western rift) from Lake Albert to Lake Edward (McConnell, 1959, 1972). Earthquake first arrivals (Sutton and

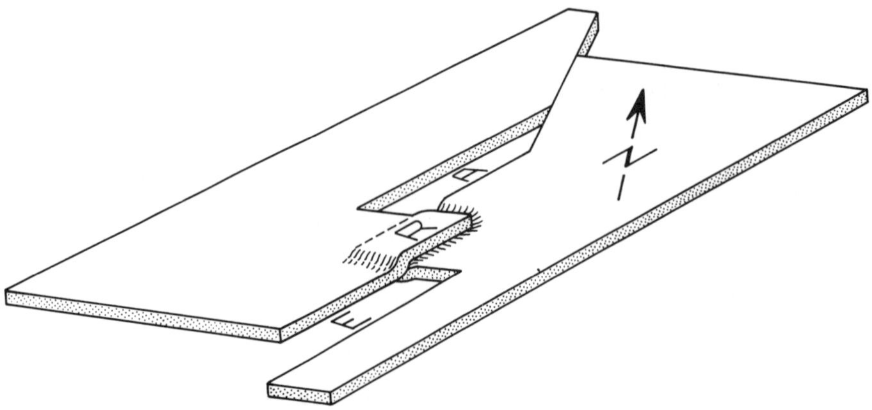

Fig. 3. A model proposed to explain the contemporaneous creation of the Ruwenzori (R) uplift with the rifting of the Albert (A) and Edward (E) grabens by oblique motion across a step-like fault configuration.

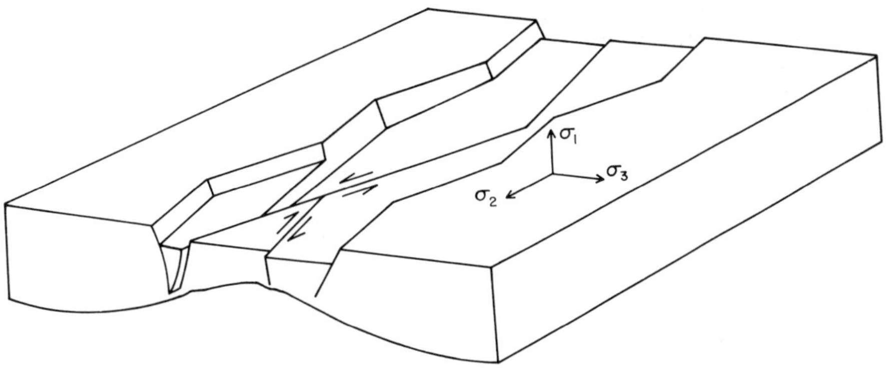

Fig. 4. Zigzag pattern of normal faults in rift valleys and rare strike slip faults associated with the rifting as they appear in clay models (Freund and Merzer, 1976b) and in several rifts.

Berg 1958) suggest both right lateral shear and normal faulting with orientations in accord with the faults. If this right lateral movement is contemporaneous with the Tertiary to Recent rifting, of the Western Rift, the Ruwenzori uplift may be due to compression between two rhomb grabens due to oblique motion (Figure 3) acting of an offset fault configuration (Freund, 1966).

Illies and Greiner (1976) suggest that changes in stress orientation show that the Rhine Graben experienced two stages of motion. In mid-Eocene to early Moicene, the motion was extensional as indicated by normal faulting and subsidence (Illies, 1977). Since mid-Pliocene, the motion has been left lateral as indicated by in situ stress measurements which give compressive axis oblique to the rift. The presence of young reverse faults support this.

(c) Horizontal shear in genuine extensional continental rifts.

Recently, reports of strike-slip faults in rifts have appeared. Yairi and Mitsutani (1969) suggest that the N-S system of normal faults and the E-W system of right hand faults in the Lake Tanganyika rift are probably contemporaneous, and that the three en echelon deep basins of this lake are probably offset by the right slip faults during the creation of these basins by the normal faults.

In New Mexico, the 030° trending left-slip Tijeras fault crosses the Sandia uplift on the east side of the Albuquerque - Bellen basin of the Rio Grande rift. The 010° trending right-slip Piciris fault extends along the Sangre de Cristo uplift farther north, and a small 320°-trending right slip crosses the Ladron uplift on the southwestern corner of this basin (Kelley and Northrop, 1975; Kelley, 1979). These investigators relate the strike slip faults to old, Precambrian to Laramide deformation rather than to the Terriary rifting. Chapin (1979), and Muelberger (1979) expresses the opinion that faults (lineaments) parallel to the above extend between the en echelon basins of the Rio Grande rift, and that they were active in the process of rifting through scissor-like torque. This appears at the surface as opposing the sense of normal faulting reflecting transverse shear at depth.

In Norway, Ramberg and Spjeldnaes (1978) report NE-trending left slip fault offsetting the rhomb-porphyry Tyveholmen dyke on the west side of the N-S Oslo graben. These faults are consistent with transversal tension of the rift and compression along the rift, and the writers enquire whether the strike slip faults are related to the rifting. That the answer may be positive is demonstrated by clay model experiments (Freund and Merzer, 1976b) where two sets of conjugate strike slip faults appear sometimes as an integral part of the process of rifting with the narrow bisectrix of two sets, indicating the compression trends along the rift. The character of these faults in the experiments, (applicable to the Rio Grande and Oslo grabens) is shown schematically in Figure 4.

Far more common than strike slips is the zigzag pattern of normal faults in rift valleys, recognized by Krenkel (1925) in his pioneering study of the African rifts. A similar angular pattern with lozenge-shaped criss-cross faulting occurs in extensive tensional terrains such as the Basin and Range (Donath, 1962) and the Afar depression (CNR-CNRS Afar Team, 1973). In these cases there are four sets of normal faults arranged in pairs with common strike and opposite dip. The angle between the two pairs varies between 15° to 90°, with 30° to 60° being the most frequent.

This pattern has been attributed to lines of original weakness in the country rocks, to successive stages of faulting under different stress orientations (Anderson, 1951), to a strike slip component along the rift at some stage in its tectonic evolution (Bederke, 1966) and to mutual interference of faulting (Oertel, 1965). The mechanical analysis following Freund and Merzer's (1976b) clay experiments show that both the common lozenge and zigzag patterns and the rare strike slip faults are the result of the horizontal attenuation of the crust along the rift which is much the same as the vertical attenuation which is the cause of the depression of the rift, though the horizontal attenuation is far less effective because it has to overcome the horizontal resistance of the sub-crust to the horizontal motions. Both vertical and horizontal attenuations are in fact the three dimensional response of Poisson effect to the crustal extension across the rift.

Finally, several writers (e.g. Illies, 1974; Bergerat and others, 1979) suggest that individual rifts is Europe and Africa are connected by "continental transform zones" extending along transtensional rifts or manifested by zones of Riedel shears.

References

Abdel Gawad, M., New evidence of transcurrent movements in the Red Sea area and petroleum geology, Amer. Assoc. Petrol. Geol. Bull. 53, 1466-1479, 1969.

Anderson, E. M., The Dynamics of Faulting and Dyke Formation, Oliver and Boyd, London, 209 pp., 1951.

Arcyana, Transform faults and rift valleys from bathyscap and diving saucer, Science, 190, 108-116, 1975.

Baker, B. H., The structural pattern of the Afro-Arabian rift system in relation to plate tectonics, Phil. Trans. Roy. Soc. London, A267, 383-391, 1965.

Ball, M. M. and C. G. A. Harrison, Crustal plates in central Atlantic, Science, 167, 1128-1129, 1970.

Bartov, J., A Structural and Palaeogeographical Study of the Central Sinai Faults and Domes, Ph.D. Thesis, The Hebrew University, Jerusalem, 143 pp. (in Hebrew with an English Abstract), 1974.

Bederke, E., The development of European rifts, Geol. Surv. Canada Paper 66-14, 213-219, 1966.

Bergerat, F., Chorowicz, J., Mukonki, M. na Bantu, Different types of transform faults and zones in European and in East African rift systems, Abst. Dead Sea Rift Symposium, Jerusalem, 1979.

Bonatti, E., Origin of offsets in the mid-Atlantic ridge in fracture zones, J. Geol., 81, 144-156, 1973.

Chapin, C. E., Evolution of the Rio Grande rift, a summary in R. E. Riecker (ed.), Rio Grande Rift, Tectonics and Magmatism, Am. Geophys. Union, 1-5, 1979.

Christensen, N. I., Seismic anisotropy in the lower oceanic crust, Nature, 237, 450-451, 1972a.

Christensen, N. I., The abundance of serpentines in the oceanic crust, J. Geol., 80, 709-719, 1972b.

Christensen, N. I., and M. H. Salisbury, Structure and constitution of the lower oceanic crust, Rev. Geophys. Space Phys. 13, 57-86, 1975.

Clayton, L., Tectonic depressions along the Hope fault, a transcurrent fault in north Canterbury, New Zealand, N.Z.J. Geol. Geophys. 9, 95-104, 1966.

CNR-CNRS Afar Team, Geology of northern Afar (Ethiopia), Rev. Geog. Phys. Geol. Dyn., 15, 443-490, 1973.

Collette, B. J., Thermal contraction joints in a spreading sea floor as origin of fracture zones, Nature, 251, 299-300, 1974.

Crowell, J. C., Origin of Late Cenozoic basins in southern California, in: W. R. Dickinson (Ed.), Tectonics and Sedimentation, SEPM, Sp. Pub. 22, 190-204, 1974.

DeRidder, N. A., and G. J. Lensen, Indirect evidence for transcurrent faulting and some examples from New Zealand and the Netherlands, Tech. Bull., 15, 1-13, 1960.

Donath, F. A., Analysis of Basin Range structure, south central Oregon, Geol. Soc. Amer. Bull. 73, 1-16, 1962.

Duffield, W. A., A naturally occurring model of global plate tectonics, J. Geophys. Res., 77, 2543-2555, 1972.

Freund, R., A model of the structural development of Israel and adjacent areas since Upper Cretaceous times, Geol. Mag. 102, 79-100, 1965.

Freund, R., Rift Valleys, Geol. Surv. Canada, Paper, 66-14, 330-44, 1966.

Freund, R., The Hope fault, a strike slip fault in New Zealand, N.Z. Geol. Surv. Bull. n.s. 86, 46 pp., 1971.

Freund, R., Kinematics of transform and transcurrent faults, Tectonophysics, 21, 93-134, 1974.

Freund, R., and A. M. Merzer, Anisotropic origin of transform faults, Science 192, 137-138, 1976a.

Freund, R., and A. M. Merzer, The formation of rift valleys and their zigzag fault pattern, Geol. Mag. 113, 561-568, 1976b.

Freund, R., Zak, I. and Z. Garfunkel, Age and rate of sinistral movement along the Dead Sea Rift, Nature, 220, 253-255, 1968.

Froideveaux, C., Energy dissipation and the geometric structure at spreading plate boundaries, Earth Planet. Sci. Letters 20, 417-424, 1974.

Garfunkel, Z., The tectonics of the Western Margins of the Southern Arava, Ph.D. Thesis, Hebrew University, Jerusalem, 204, pp. (in Hebrew with an old English abstract), 1970.

Garfunkel, Z., The tectonics of the Salton trough, southern California and Mexico, and their relation to plate tectonics, in: R. W. Rex (ed.) Cooperative Investigation of Geothermal Resources in the Imperial Valley, and their potential value for Desalting of water and other purposes, University of California at Riverside IGPP-UCR, 7233 pp. H1-H33, 1972.

Gastil, R. G., Fault systems in northern Baja California and their relations to the origin of the Gulf of California, in: R. W. Dickinson and A. Grantz (eds.), Proc. Conference on Geologic Problems of San Andreas Fault System, Stanford University Publications Geological Series 11, 283-286, 1968.

Girdler, R. W. and B. W. Daracott, African poles of rotation, Comments Earth Sci., 2, 131-138, 1972.

Harland, W. B., Tectonic transpression in Caledonian Spitzbergen, Geol. Mag. 108, 27-42, 1971.

Heezen, B. C. and M. Tharp, Tectonic fabric of the Atlantic and Indian oceans and continental drift, Phil. Trans. Roy. Soc. London, A258, 90-106, 1965.

Hess, H. H., Seismic anisotropy of the uppermost mantle under the oceans, Nature, 203, 629-631, 1964.

Illies, J. H., Taphrogenesis and plate tectonics, in: Approaches to Taphrogenesis, IUCG Report, 8, 433-460, 1974.

Illies, J. H., Ancient and recent rifting in the Rhine graben, Geol. Minjb. 56, 329-350, 1977.

Illies, J. H. and G. Greiner, Rhinegraben and the Alpine System, Geol. Soc. Amer. Bull. 89, 770-782, 1976.

Johannes, W., Experimental investigations of the reaction Forsterite +H_2O Serpentine + Brucite. Contr. Mineral. Petrol. 19, 309-15, 1968.

Kelley, V. C., Tectonics, middle Rio Grande Rift, New Mexico, in: R. E. Riecker (ed.); Rio Grande Rift, Tectonics and Magmatism, Amer. Geophys. Union, pp. 57-70, 1979.

Kelley, V. C. and S. A. Northrop, Geology of Sandia Mountains and vicinity, New Mex. Bur. Mines Min. Res. Mem. 29, 136 pp., 1975.

Kingma, J. T., Possible origin of piercement structures, local unconformities and secondary basins in the eastern geosyncline, New-Zealand, N.Z.J. Geol. Geophys. 1, 269-274, 1958.

Lachenbruch, A. H. and G. A. Thompson, Oceanic ridges and transform faults; their intersection angle and resistance to plate motion, Earth Planet. Sci. Letters, 15, 116-122, 1972.

Lartet, L., La Geologie de la Palestine, Paris, 1869.

Laughton, A. S., The Gulf of Aden, Phil. Trans. Roy. Soc. Lond. A259, 150-171, 1966.

Lensen, G. J., A method of horst and graben formation, J. Geol. 66, 597-586, 1958.

LePichon, X., Sea floor spreading and continental drift, J. Geophys. Res. 73, 3661-3679, 1968.

Lomnitz, C., Mooser, F., Allen, C. R., Brune, J. N., and Thatcher, W., Seismicity and tectonics of the northern Gulf of California region, Mexico, Preliminary results, Geofisica Internacional 10, 37-48, 1970.

McConnell,R.B., Outline of the geology of the Ruwenzori Mountains, Overseas Min. Resour. 7, 245-268, 1959.

McConnell, R. B., Geological development of the rift system of Eastern Africa, Geol. Soc. Amer. Bull. 83, 2549-2572, 1972.

Melson, W. G. and G. Thompson, Layered basic complex in ocean crust, Romanche, equatorial Atlantic Ocean, Science 168, 817-820, 1970.

Melson, W. G. and G. Thompson, Petrology of a transform fault zone and adjacent ridge segments, Phil. Trans. Roy. Soc. London, A268, 423-441, 1971.

Miyashiro, A., Shido, F. and M. Ewing, Composition and origin of serpentinities from the mid-Atlantic ridge near 24° and 30° north latitude, Contr. Miner. Petrol. 23, 117-127, 1969.

Miyashiro, A., Shido, F., and M. Ewing, Petrologic model for the mid-Atlantic ridge, Deep Sea Res., 17, 109-123, 1970.

Mohr, P. A., Transcurrent faulting in the Ethiopian rift system, Nature, 218, 938-941, 1968.

Moore, D. G. and E. C. Buffington, Transform faulting and growth of the Gulf of California since the late Pliocene, Science, 160, 1238-1241, 1968.

Muelberger, W. R., The Embuds Fault between Pilar and Arroyo Hondo, New Mexico: an active intercontinental transform fault, New Mexico Geol. Soc. Guidebook, 30th Field Conf. Santa Fe County, pp. 77-82, 1979.

O'Bryan, J. W., Cohen, R., and W.N. Gilliland, Experimental origin of transform faults and straight spreading center segments, Geol. Soc. Amer. Bull. 86, 193-796, 1975.

Oertel, G., The mechanism of faulting in clay experiments, Tectonophysics, 2, 343-393, 1965.

Oldenburg, R. W. and J. N. Brune, Ridge-transform fault spreading pattern in freezing wax, Science, 178, 301-304, 1972.

Oldenburg, R. W. and J. N. Brune, An explanation for the orthogonality of ocean ridges and transform faults., J. Geophys. Res. 80, 2575-2585, 1975.

Orowan, E., The origin of oceanic ridges, Scient. Amer. 221, 5, 102-119, 1969.

Picard, L., On Afro-Arabian graben tectonics, Geol. Rundschau 59, 337-381, 1970.

Quennell, A. M., Tectonics of the Dead Sea rift, 20th Inter. Geol. Congr., Mexico, 385-405, 1956.

Quennell, A. M., A structural and geomorphic evolution of the Dead Sea rift, Quart. J. Geol. Soc. Lond., 114, 1-24, 1958.

Raleigh, C. B., Experimental deformation in ultramafic rocks and minerals, in: P. T. Wyllie (ed.), Ultramafic and Related Rocks, J. Wiley and Sons, N.Y. pp. 191-199, 1967.

Raleigh, C. B. and M. S. Patterson, Experimental deformation of serpentinites and its tectonic implications, J. Geophys. Res. 70, 3965-3985, 1965.

Ramberg, I. B. and N. Spjeldnaes, The tectonic history of the Oslo region, in: I. B. Ramberg and E. R. Neumann (eds.), Tectonics and Geophysics of Contiental Rifts, D. Reidel Publishing Co., Holland, 167-194, 1978.

Sutton, G. H., and E. Berg, Seismological studies of the western rift valley of Africa, Trans. Amer. Geophys. Union 39, 474-481, 1958.

Sykes, L. R., Mechanism of earthquake and nature of faulting in the mid-ocean ridges, J. Geophys. Res. 72, 2131-2153, 1967.

Tchalenko, J. S. and N. N. Ambraseys, Structural analysis of Dashte Bayaz (Iran) earthquake features, Geol. Soc. Amer. Bull. 81, 41-60, 1970.

Turcotte, D. L., Are transform faults thermal contraction cracks? J. Geophys. Res. 79, 2573-2577, 1974.

Vroman, A. J., A tentative explanation of the origin of "Transform Faults" (with critical remarks on the term). Tectonophys. 30, T 11 - T16, 1976.

Weissel, J. K. and D. E. Hayes, Asymmetric seafloor spreading south of Australia, Nature 231, 518-521, 1971.

Wilson, J. T., A new class of faults and their bearing on continental drift, Science 207, 343-347, 1965.

Yairi, K. and S. Mitzutani, Fault system of Tanganyika rift at the Kigoma area, western Tanzania, J. Earth. Sci., Nagoya Univ. 17, 71-96, 1969.

Zak, I. and R. Freund, Recent strike slip faults along the Dead Sea rift, Israel J. Earth Sci., 15, 33-37, 1966.

THE SEISMICITY OF THE EAST AFRICAN RIFT SYSTEM AND COMPARISON WITH OTHER CONTINENTAL RIFTS

J.D. Fairhead & G.W. Stuart
Department of Earth Sciences, The University of Leeds, Leeds LS2 9JT. (U.K.)

Summary. This contribution reviews the seismicity associated with the East African, Baikal, Rhine and Rio Grande Rift Systems.

Of these, the East African Rift System is considered by many to be the best example of extensional tectonics relating to an incipient plate boundary. On a regional basis, the East African Rift System follows a broad, NNE-SSW trending gravity lineament that decreases in importance as does the surface expression of rifting, southwards. Locally rifting and its associated seismicity are influenced by the structural trends of the Precambrian basement geology and follow the mobile belts that surround the older, more stable continental nuclei (cratons). This has resulted in a complex, inter-connecting Rift System, the northernmost part of which is the most mature and represents an advanced stage of rift development. Although the seismicity of this northern part is poorly monitored, compared with farther south, evidence from detailed micro earthquake studies within the Kenya Rift indicates that a major part of the tensile stress is being released by microearthquakes along the axial zone of the rift. The Western Rift and rifting further south, in contrast, do not appear to have such well developed zones of lithospheric weakness, thus resulting in tensional stresses being released by larger magnitude earthquakes occurring over a zone several times broader than the rift valley itself. In southern Africa, incipient rifting has been successfully delineated by its associated seismicity which follows the regional trend of the rift system and parallels the structural trends associated with the Damara and Katanga mobile belts. Focal mechanism studies, including seven new solutions, indicate that the Rift System is the result of horizontal tensional stresses orientated normal to the overall strike of the Rift System. The lack of strike slip mechanisms, compared with the Oceanic Rift System, probably relate to the slow tectonic processes that control the initial break up of a continental lithosphere. Away from the Rift System strike slip, compressional and tensional focal mechanism solutions are presented. The relation of these mechanisms to the stress field associated with the Rift System is unclear.

Faulting and associated seismicity of the Baikal, Rhine and Rio Grande Rifts are controlled to a large extent by pre-existing structural weaknesses in the basement geology. However, unlike the East African Rift System, these Rifts are unconnected with the Oceanic Rift System and, in general, lie close to zones of compressional tectonics to suggest that they represent secondary effects of mountain building events.

1. Introduction

The best example of continental rifting is considered by many to be the East African Rift System since it has a clear structural link with the World-wide Oceanic Rift System via the Afar depression of Ethiopia and the Gulf of Aden. As such, the East African Rift System probably reflects the initial processes leading to future continental disruption and provides a unique area to investigate the interaction of crustal and upper mantle rifting processes at an incipient plate boundary and the transitional tectonics from oceanic to continental rifting.

Other examples of active continental rifts are the Baikal, Rhine and Rio Grande Rifts. Although these rift systems are extensional in character, they differ from the East African Rift System in that they do not have any clear structural connection to constructive plate margins. Instead they tend to lie on the convex side of regions experiencing compressional tectonics associated with destructive plate margins suggesting that they may represent secondary effects of a compressional stress regime (Burke, 1978; Tapponier & Molner, 1979).

The major part of this article concentrates on an updated review of the sesmicity associated with the East African Rift System while other rift structures are only briefly reviewed.

2. The African Rift System

Seismicity studies by Rothé (1954), Gutenberg & Richter (1954), Sykes and Landisman (1964) and Fairhead & Girdler (1970, 1971) show that the narrow zone of shallow seismicity associated with the Indian Ocean branch of the Oceanic Rift System (Lilwall, this volume) continues through the Gulf of Aden and connects with the shallow seismicity associated with the continental rift system of East Africa. This seismicity connection, together with the structural connection and the extensional nature of the rift system, deduced from focal mechanism studies (Sykes, 1967; Fairhead & Girdler, 1971), provide convincing evidence that the East African Rift System is a developing constructive plate boundary. This boundary is complex and forms a series of inter-connecting rifts through eastern and southern Africa. Rifting (and its associated seismicity) appear to be controlled by the Precambrian basement geology such that it tends to follow the structural trends of the mobile belts which surround the older continental nuclei (cratons). The overall trend of the rift system is SSW and closely follows the axis of the regional negative Bouguer anomaly which has been interpreted by Girdler (1975) and Fairhead & Reeves (1977) as a zone of thin lithosphere.

(a) Red Sea and Gulf of Aden Rifts. McKenzie et al (1970) and others have shown that the axes of the Red Sea and Gulf of Aden Rifts are seismically active. These rifts have formed by relative motion of Arabia away from the Nubia and Somalia crustal blocks (Fig. 2) such that their floors are now oceanic crust, (see Falcon, Gass, Girdler & Laughton, 1970). Since these features are considered to be embryonic oceans they are not reviewed here except for the structures that lie adjacent to the Afar depression where the oceanic and continental rift systems meet.

(b) Gulf of Suez and Arava-Dead Sea Rifts

Sea floor spreading in the Red Sea is terminated at its northern end by the Gulf of Suez and Arava-Dead Sea Rifts (Levant fracture zone). Historic seismicity for this region (Ben Menahem et al., 1976) indicates that the Levant fracture zone and its associated branch faults are active. Recent sesmicity (Fig. 1) has been restricted to the mouth of the Gulf of Suez and an isolated event NE of Cairo. Freund (1965) has shown from field evidence that a total of 110 km of left lateral strike slip motion has taken place along the Arava-Dead Sea rift. This contrasts with at least 150km of opening of the northern Red Sea rift (Girdler, 1965). The difference was considered to be taken up by normal faulting within the Gulf of Suez graben (Picard, 1966; McKenzie et al., 1970). However, focal mechanism studies by Ben-Menahem et al. (1976) and Pearce (1977b, 1980) support Abdel-Gawad (1969) view that there is left lateral motion along the Gulf of Suez. The sense of this strike slip motion is consistent with the interpretation of folds found in the Suez area by Freund (1965). Ben Menahem et al. (1976) have shown that if 25 km of strike slip motion has taken place then

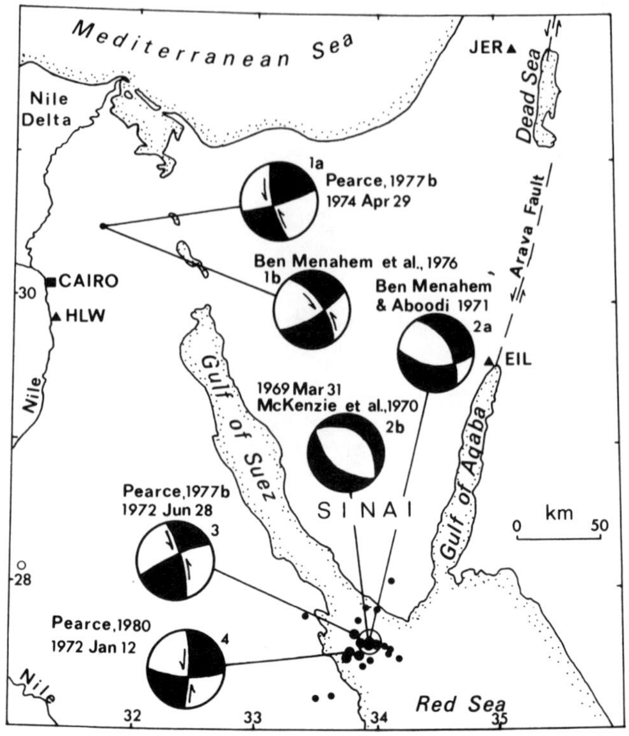

Fig. 1. Recent seismicity of the Gulf of Suez and Arava-Dead Sea Rifts. Solid triangles represent seismograph stations and dots - earthquake epicentres(see Fig. 7 for details).Focal mechanism solutions 1a, 3 & 4 after Pearce (1980), 1b - Ben Menahem et al. (1976), 2a - Ben Menahem & Aboodi (1971) and 2b - McKenzie et al. (1970).

this removes the discrepancy between the movement along the Avara-Dead Sea Rift and the opening of the Red Sea.

Ben Menahem et al. (1976) have further shown that the relative slip along the Levant fracture zone, to the north of the Dead Sea, diffuses sideways along branch faults such that the edges of the plate lose their coherency as they approach the Amanus-Taurus regions of continental collision. The slip along the Gulf of Suez Rift probably dies out in a similar manner such that the Sinai region should not be considered as a separate plate, but rather a splinter of the African plate (Ben Menahem et al., 1976).

(c) Afar and the Ethiopian Rift The seismicity of this region is shown in Figure 2 and represents instrumentally detected events reported by the ISC and USGS for the period Jan. 1950 to June 1979. The detectability of earthquakes for this region is generally poor with only events with body wave magnitude greater than about m_b = 4.8 being fully recorded. Prior to 1972, the only seismograph station in the region was Addis Ababa (AAE). Since 1972 a network of stations has been established in and about the Gulf of Tadjura (Fig. 3). These stations have successfully recorded the seismicity of the western end of the Gulf of Aden. However, the response of this network is shaped to record local earthquakes and only the largest events in Fig. 2 have been reported to the ISC. Detectability of this region should be improved with the installation in Feb. 1980 of 3 seismograph stations in Sudan, two of which are shown in Fig. 2 (open triangles).

(i) Afar The seismicity of eastern Afar during the period 1972-1979 is shown in Figure 3. The seismicity is closely associated with the oceanic ridge (plate boundary) of the Gulf of Aden and Gulf of Tadjura and enters Afar along a series of NE-SW en-echelon transform faults which progressively off-set the ridge to the SW (Lépine & Ruegg, 1973, 1976; Abdallah et al. 1979).

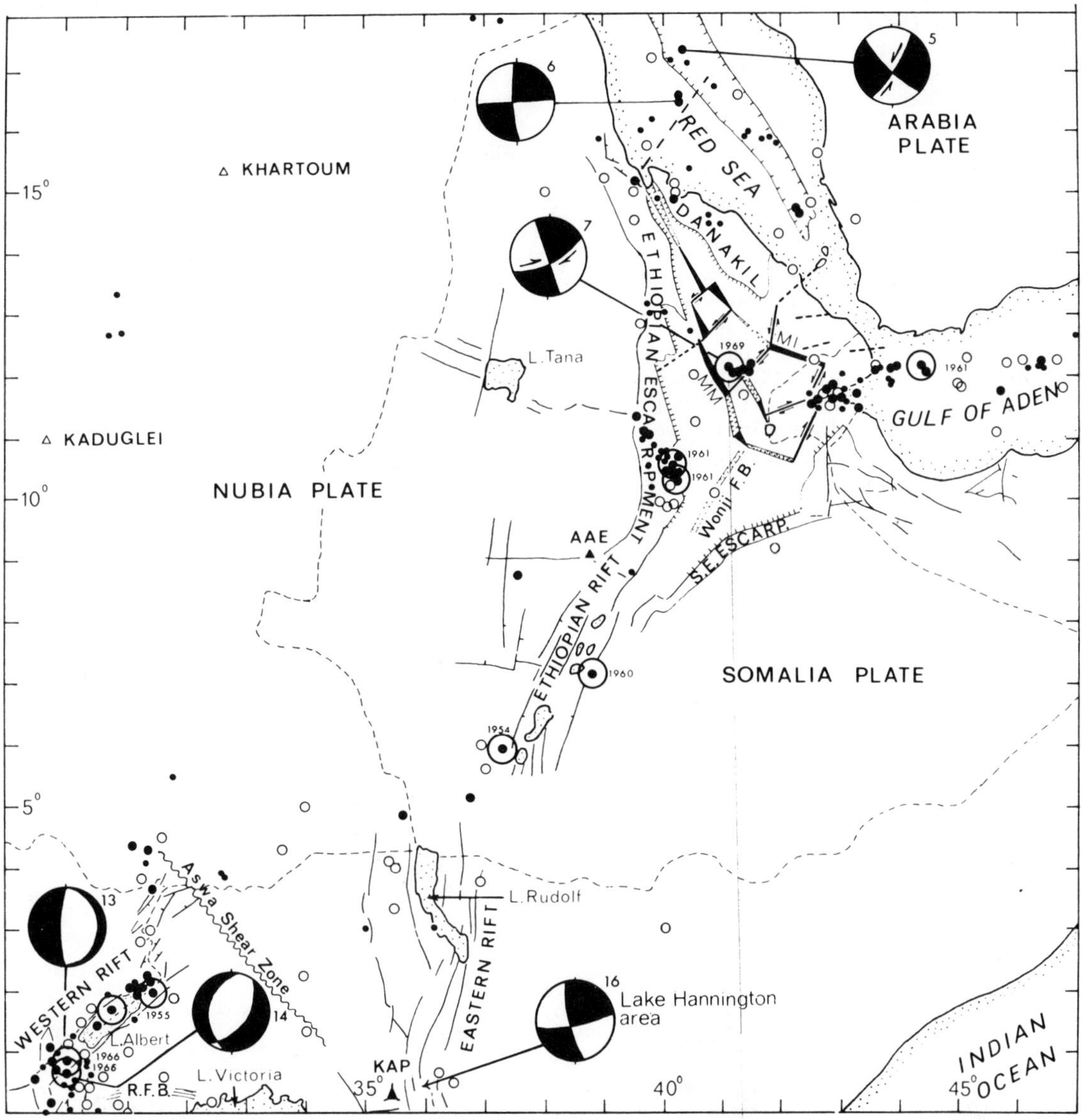

Fig. 2. Seismicity of the Afar region and the East African Rift System north of the equator. Open triangles represent new seismograph stations (see text), dots - earthquake epicentres (see Fig. 7 for details) and open circles - earthquakes with $m_b \leqslant 4.0$ or no reported magnitude. The plate tectonic model for Afar is after Barberi & Varet (1975) with solid zones representing axial volcanism connected by transform faults (motion given) and thick dashed lines structural/ volcanic lineations. MM represents Manda Mararo and MI Manda Inakir volcanic axes. The thin dashed line represents National boundaries and RFB - Ruwenzori fault belt. Focal mechanism solution 5 after Sykes (1970), 6 - this study, 7 - McKenzie et al (1970), 13 - Sykes (1967), 14 - Banghar & Sykes (1969) and 16 - Hamilton et al. (1973).

Composite focal mechanism solutions have been reported for the Asal rift and the Arta and Maskali transforms (solutions 8-10, Fig. 3, Abdallah et al., 1979) and are consistent with the sense of opening as deduced by plate tectonics. Two further solutions are reported here for the 1978 Nov 07 event (solution 11) and the 1973 April 01 event (solution 12) and are shown in Figures3 and 8. The 1978 Nov 07 event gives a strike slip mechanism with nodal planes striking at about 045° to the axis of spreading in the Asal Rift. The tensional axis for this solution is normal to the strike of the Rift. The 1973 April 01 event is also a strike slip mechanism with either left lateral movement along a NW-SE nodal plane or right lateral movement along a NE-SW nodal plane. The right lateral motion is consistent with the sense of motion along the transforms off setting the spreading axis in the Gulf of Tadjura.

The seismicity for the remainder of Afar is less well known and as a result the seismicity appears 'patchy' with the main events associated with the Serdo earthquakes of 1969 and the Kara Kore earthquakes of 1961. Gouin (1976, 1979) using felt tremors and instrumental reports compiled a historic seismicity map of the Ethiopian region and found that the main zone of earthquake activity is along the Ethiopian escarpment. A similar pattern of activity is found in Fig. 2.

Focal mechanism solutions determined by McKenzie et al. (1970) on the two largest Serdo events (solution 7) give identical strike slip motion with nodal planes striking 063° and 333°. The 333° nodal plane is considered by McKenzie et al. (1970) to be the fault plane representing relative motion between Nubia and Somalia (i.e. northern end of the East African Rift System). However, recent tectonic studies of Afar by Barberi & Varet (1975) prefer the nodal plane 063° to be the fault plane since it fits the direction and sense of motion of the proposed transform linking the Manda Mararo volcanic axis to the Manda Inakir volcanic axis (Fig. 2). These volcanic axes are considered by Barberi & Varet (1975) to be two of a series of discrete spreading centres connected by transform faults. A plate tectonic model, after Barberi & Varet, representing plate motion for the last 2 Ma is shown in Fig. 2 and is the product of extensive geological, geophysical and isotopic studies (see Pilger & Rosler, 1975). The plate boundaries in Afar tend to reflect the result of high spreading rates of the Red Sea and Gulf of Aden opening (1.0 to 1.5 cm a^{-1}) rather than the low spreading rate of 3 to 5 cm a^{-1} found for the Ethiopian Rift (Mohr et al., 1978).

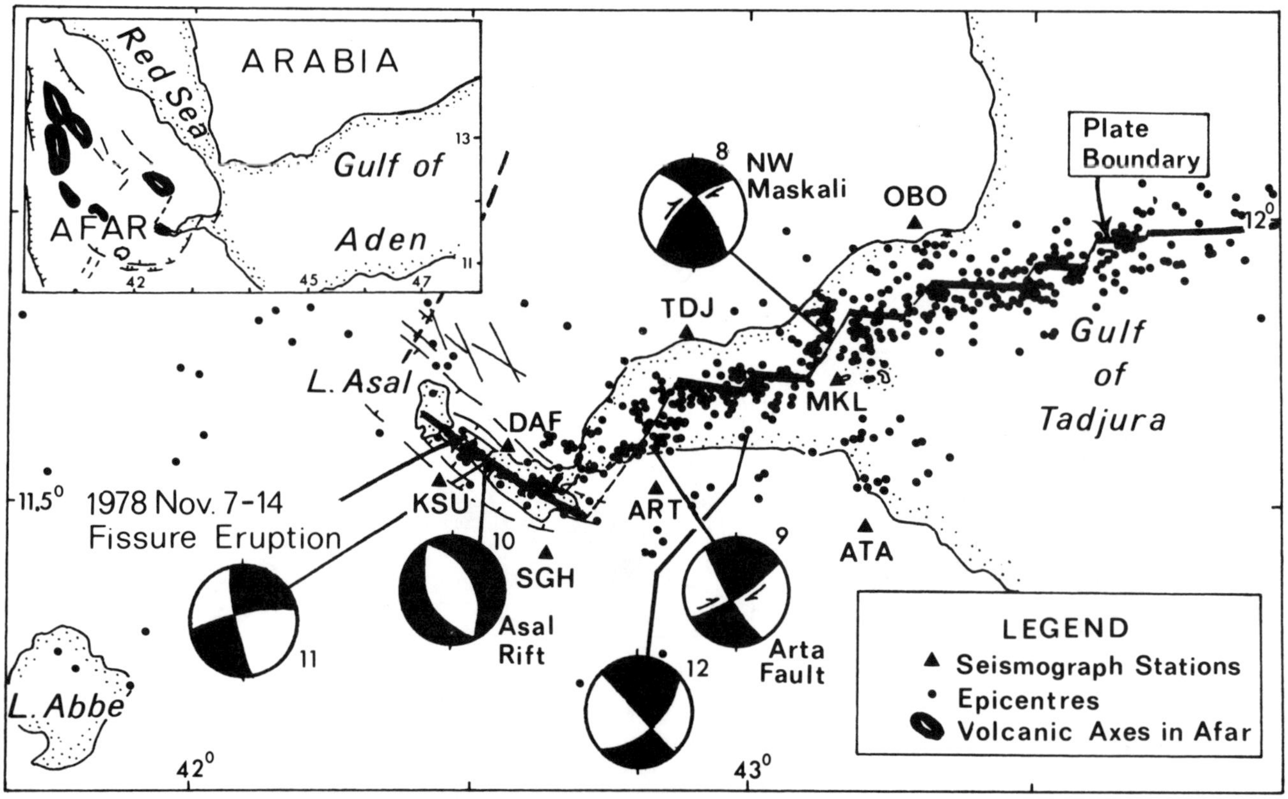

Fig. 3. The seismicity of the Gulf of Tadjura after Lépine & Ruegg (1973, 1976) and Abdallah et al. (1979) for the period 1972-1979. Epicentres not differentiated by magnitude. Focal mechanism solutions 8-10 after Abdallah et al (1979) and 11 & 12 this study.

A new focal mechanism solution (solution 6) for the 1978 Jan. 17 southern Red Sea event is shown in Figs. 2 and 8. Although the nodal planes are poorly constrained by the data, the solution does suggest strike slip faulting in the region connecting northern Afar to the axial trough of the Red Sea.

(ii) Ethiopian Rift Recent deformation within the Ethiopian Rift is essentially restricted to an 8 km wide zone of intense faulting (Wonji fault belt) which follows the axis of the Rift in an *en echelon* pattern. This zone coincides with the youngest volcanic centres and with a positive gravity anomaly that Searle & Gouin (1972) consider to represent the effect of emplacement of dense axial intrusions to within a few kilometres of the rift floor. Large earthquakes have occurred along the length of the Ethiopian Rift during historical times (Gouin, 1976, 1979). Micro-seismicity studies along the rift floor (Molnar et al., 1970) find a surprisingly low level of activity compared with that found in the Kenya Rift.

(d) East Africa The Ethiopian Rift structures do not pass directly into either the Eastern or the Western Rifts but are offset. The precise nature of the structural link between these rifts remains unclear, although Fig. 2 suggests an E-W seismicity link with the Western Rift.

(i) Western Rift The Western Rift is a well defined rift valley extending in an arc from 4°N to 9°S swith its floor occupied by numerous lakes of which Lake Tanganyika in the largest (Fig. 4). Its seismicity has been monitored since the 1950's by a network of seismograph stations established by the Institut de Recherche Scientifique du Zaire (IRS) formally known as Institut pour la Recherche Scientifique en Afrique Centrale (IRSAC) (De Bremaecker 1955). Seismicity studies using this network (De Bremaecker, 1959; Sutton & Berg, 1958; Wohlenberg, 1969; Ndontoni, 1976) indicate that the earthquakes with $m_b \geq 4.2$ are being well recorded and occur within a zone of width of about 100 km along the whole length of the Western Rift. At the northern end of the Rift (i.e. north of Lake Albert) both faulting and seismicity die out along a series of *en echelon* faults which terminate against the Precambrian Aswa shear zone (Fig. 2).

Earthquakes do occur outside this 100 km wide zone containing the Western Rift and form well defined zones extending away from the Rift in areas with no surface expression of faulting. The best example of this is the 600 km long zone extending WSW into Zaire from the Rift at 1°S (Wohlenberg, 1969, Fairhead & Henderson, 1977). Earthquakes along this Missisi-Wasongola zone are well located and may be related, on the other side of the Rift, to the off rift seismic zone following the E-W folds of the Ruwenzori fold belt (RFB) into the northern part of Lake Victoria (Maasha, 1975a) (Fig. 2).

Micro-earthquake studies using temporary seismograph networks have been made in the Lake George-Ruwenzori-Lake Albert region by Rykounov et al. (1972) and Maasha (1975b). These studies reveal a high level of seismicity with focal depths ranging down to 25 km except for the Ruwenzori host block where depths extend to 40 km. Composite focal mechanism solutions (solutions 15a-i, upper focal hemisphere projections) were made by Maasha (1975b) for 9 localities surrounding Ruwenzori (Fig. 5). The general tectonic picture to emerge is an extensional axis trending E-W to ESE-WNW. Fault mechanisms mainly indicate normal faulting giving a relative upward movement to the Ruwenzori host block compared to the surrounding country. The epicentre locations for the focal mechanism solutions for the 1966 March 20 event (solution 14, Sykes, 1967) and the 1966 May 17 (solution 13, Banghar & Sykes, 1969) have been relocated using the method of Joint Epicentre Determination (Fairhead & Girdler, 1971) giving locations shown in Fig. 6 which agree better with the known surface effects (Wohlenberg et al. 1966).

Inspection of Figs. 4, 6 and 7 clearly indicate that the seismicity of the Western Rift follows the western edge of the Tanganyika Shield with rift structures lying within and paralleling the trends of the younger Precambrian basement rocks of the Kibaride belt (to the north) and the Ubendian belt (to the south) (McConnell, 1967, 1972; Cahen, 1970). Volcanism is restricted to only small sections of the rift between Lakes Albert and Tanganyika (Fig. 4) with Nyiragongo and Nyamuragira volcanic centres currently being the most active. McConnell (1972) considers the location of this volcanism to be associated with structural nodes i.e. junctions of mobile belts.

(ii) Eastern Rift Tectonically, the Eastern (Gregory) Rift can be divided into northern and southern parts with the division being the E-W line of volcanic centres of Northern Tanzania (i.e. at 3°S) which terminates the Kenya Rift Valley. North of 3°S, the Kenya Rift Valley is 60 to 80 km wide and about 1 km deep traversing the Kenya domal uplift. The Rift floor has undergone extensive fracturing (grid faulting) with recent volcanism, geothermal activity and a positive gravity anomaly coinciding with its axis (Baker et al., 1972; Fairhead, 1976).

Earthquake detection for the Kenya Rift is poor compared to the Western Rift with only one seismograph station Nairobi (NAI) close to the Rift. Thus detectability is probably similar to that of the Ethiopian Rift. Figs. 2 & 5 clearly indicate the absence of teleseismically recorded activity within the rift since 1950. Prior to 1950, the only events to be recorded were the two events close to Lake Hannington of 1928 (Richter, 1958; McCall, 1967). The seismicity for the rift reported by Wohlenberg

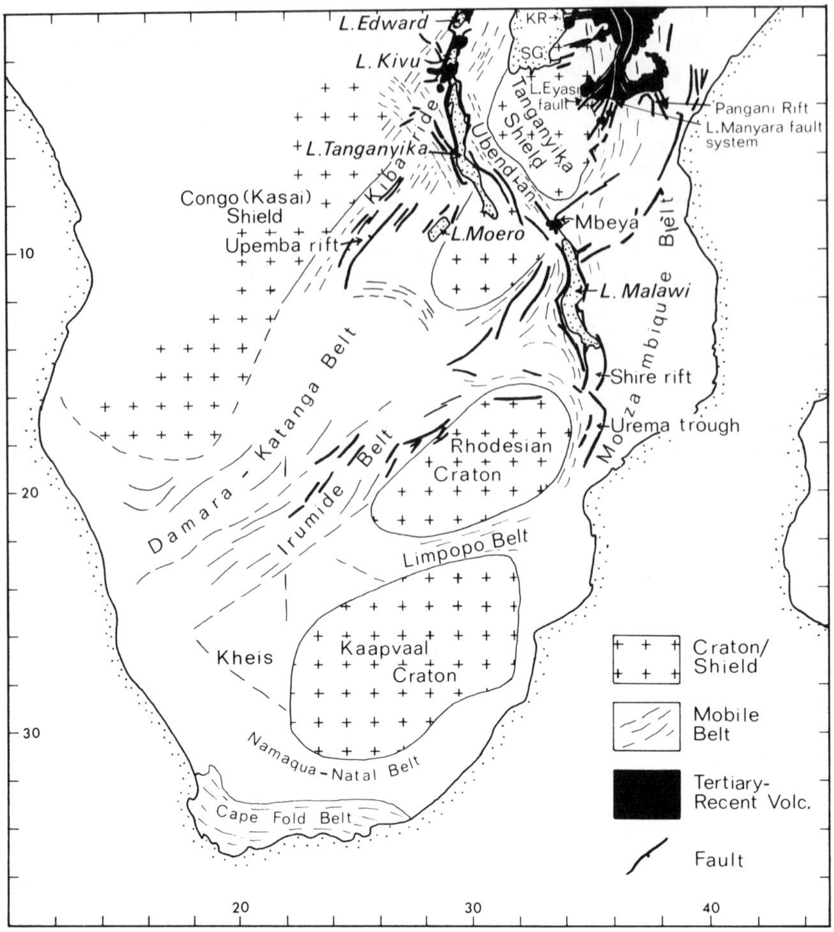

Fig. 4. Basement geology for Southern Africa showing the relation of the East African Rift System with respect to the mobile belts and the older continental nuclei (cratons). SG represents Speke Gulf and KR Kavirondo Rift.

(1969) and Băth (1975) using IRS data also indicate it to be aseismic. The Speke Gulf and Kavirondo Rift structures located to the west of the Kenya Rift are shown by them to be seismically active.

Micro-earthquake studies (Tobin et al., 1969; Molner & Aggarwal, 1971; Hamilton et al., 1973) on the other hand, show the floor of the Kenya Rift to be very active. Hamilton et al. (1973) used an 8 station radio telemetry network found that during a total of 7 months of recording at 3 sites (see below) there was a very close spatial correlation between seismicity and known geothermal areas.

In the Lake Naivasha area (1°S) 41 events were recorded and were restricted to a N-S elongate zone 8 km by 3 km with focal depths ranging from near surface to 4 km, in an area SW of the lake, sub-paralleling the main structural features of the rift valley, band of craters, faults and steam jets. A composite focal mechanism solution (solution 18, Fig. 7) indicates an E-W tensional stress axis and supports the interpretation that the earthquakes are caused by regional stresses and delineate a 'through going' fault zone.

In the Eburru area, NW of Lake Naivasha, 24 events were located of which 14 occurred beneath the geothermal area on Eburro volcano. A composite focal mechanism (not shown) suggests E-W compressional stress. However, the absence of epicentre trend and swarm nature of this activity suggests the activity relates to volcanic processes and not to the regional stress field.

In the Lake Hannington area (0.25^{o}N) 104 events were located and found to be more broadly distributed, extending from the eastern edge of the rift valley to the south of the lake where the main geothermal activity occurs. Focal depths range from 6-10 km and a composite focal mechanism (solution 16) indicates right lateral motion on an ENE-WSW fault plane (Fig. 2).

A temporary installed seismic array at

Kaptagat (KAP) (Fig. 2) also found the axis of the rift to be extremely seismically active (Maguire & Long, 1976).

The lack of teleseismically recorded events is considered by Fairhead & Girdler (1971) to be due to the low tensile strength of the lithosphere due to the presence of hot igneous material at very shallow depth along the axis of the rift (Searle, 1970; Fairhead, 1976) such that tensional stresses are being adequately released by the observed micro-seismic activity.

South of 3°S, the faulting is not confined to a narrow rift valley but spread out in a fan arrangement of block faults. This tectonic change is reflected in both the seismic and geothermal activity and has led Fairhead (1980) to suggest the E-W line of volcanoes delineate a leaky transform fault. The area containing the Manyara and Eyasi faults have in the last 20 years been very seismically active. The spatial distribution of earthquakes (Figs. 6 & 7) and energy release (Båth, 1975) demonstrate that two trends exist in Tanzania, the first follows the southward continuation of the Manyara fault into the Mbeya region and the second strikes SE, from the volcanic chain, paralleling the Pangani fault structures towards the coast of Tanzania from where the seismicity follows the coast line (off-shore) of southern Tanzania and northern Mozambique. McConnell (1972) considers the faulting in Tanzania, and consequently the seismicity, to be controlled by the presence of the Tanganyika Shield which has prevented the propogation of fractures through it to any great extent (Fig. 4). Thus the faulting (and seismicity) follows its boundary with the Mozambique belt. In Kenya this boundary lies to the west of the rift valley.

One focal mechanism (solution 19) has been determined for this area and is shown in Fig. 7. The mechanism has either NW-SE left lateral strike slip motion (Fairhead & Girdler, 1971) or near vertical motion along a fault striking E-W (Maasha & Molnar, 1972). The main difference between these solutions is the poor control of the NE-SW striking nodal plane.

(e) Seismicity of Southern Africa Earthquake detection for southern Africa is good, such that the present seismograph network records the majority of events with magnitudes greater than m_b = 4.0. To obtain a more uniform picture of the seismicity for southern Africa all events in Fig. 6 with magnitudes less than m_b = 4.0 have been removed. This magnitude cut off has been suggested by Fairhead & Henderson (1977) to differentiate between 'spurious' and tectonic events. Since in most cases the locational accuracy of these 'spurious' events is good they can be considered to represent either background seismicity of the continent or secondary effects associated with active tectonic zones. The

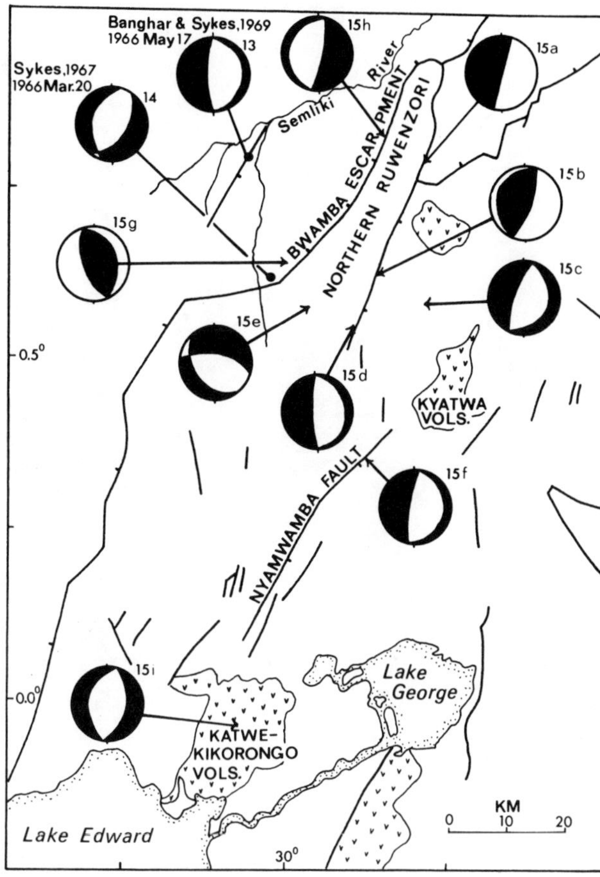

Fig. 5. Focal mechanism solutions for the Ruwenzori section of the Western Rift (seismicity not shown). Focal mechanism solution 13 after Banghar and Sykes (1969), 14 - Sykes (1967) and solutions 15a-i after Maasha (1975b).

resulting epicentre distribution for events from 1950 to June 1979, is plotted together with all available focal mechanism data in Fig. 7.

Lake Malawi-Urema Rift Figs. 6 and 7 clearly show that the high level of seismicity associated with the Western Rift continues south along the Lake Malawi-Shire Rifts in Malawi into the Urema Trough of central Mozambique. South of 22°S the seismicity decreases and the seismic trend becomes uncertain. The decrease in seismicity may either represent a real decrease or the failure by the seismograph network to detect this seismicity. Comparison of Figs. 6 and 7 reveal that the majority of the activity associated with the Urema Trough is of low magnitude. The Urema Trough lies within the Mozambique belt and borders the Rhodesian craton. Inspection of Figs. 4 and 6 indicate that the edges of the Rhodesian craton are seismically active but within the craton very little seismicity has been reported although it is the best monitored craton in Africa.

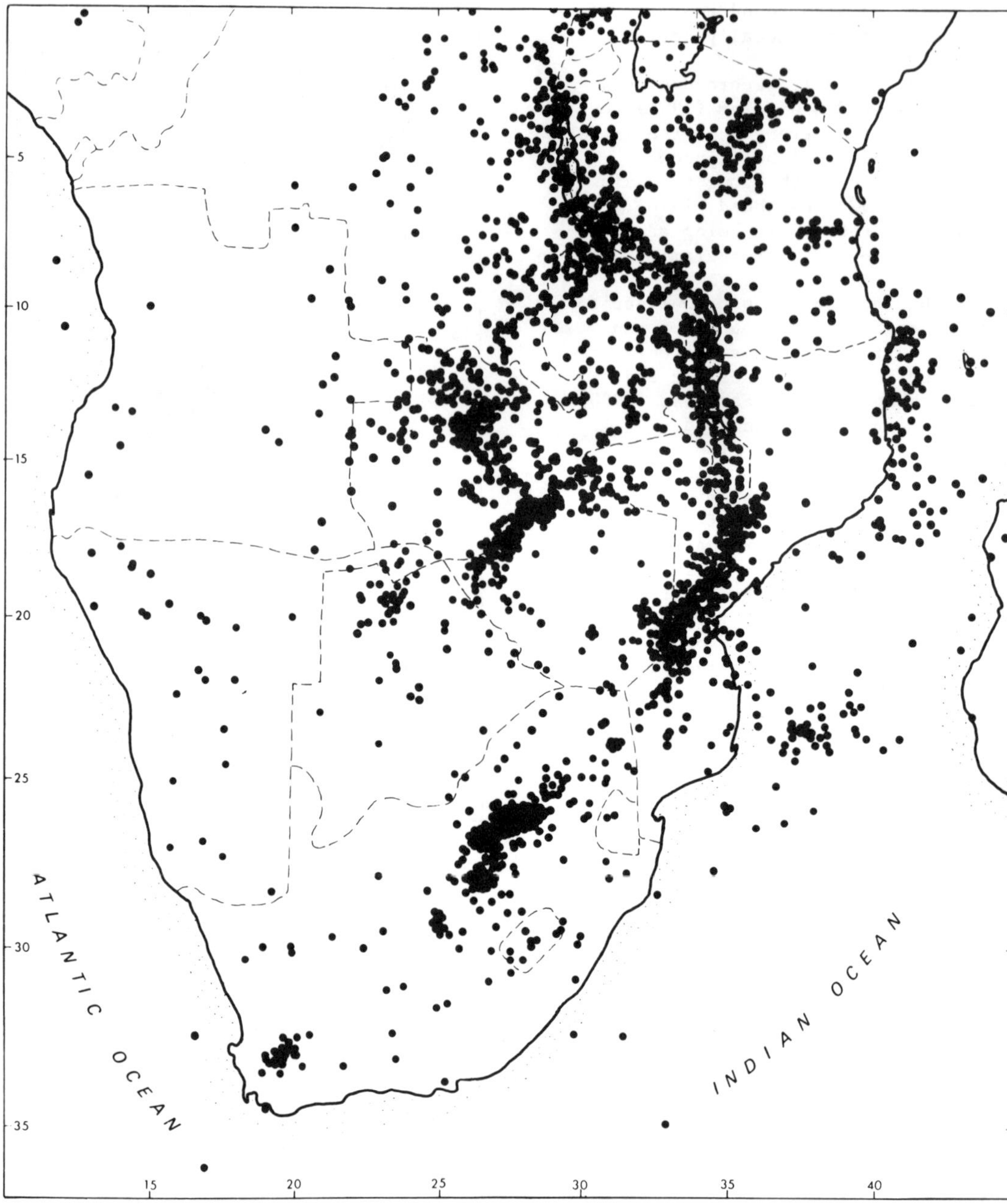

Fig. 6. The seismicity of Africa, south of the equator, for all events reported by the ISC & USGS for the period 1950-1979. Dash lines represent National boundaries.

Two focal mechanisms have been determined for this rift by Maasha & Molnar (1972) and the present authors (Figs. 7 and 8). Both solutions (solutions 20, 21) indicate normal faulting with fault planes striking approximately N-S indicating crustal extension normal to the trend of the rift.

Zambia-Botswana seismic zones The seismicity so far reviewed has been related to rifting. Detailed seismicity studies by Bram (1972) in the Katanga province of southern Zaire, Reeves (1972) and Scholz et al. (1976) in the Okavango region of northern Botswana and regional seismicity studies by Fairhead and Henderson (1977) show that there exists NE-SW trending seismicity zones that are not associated with major rift structures. One seismic zone strikes SW from the southern end of Lake Tanganyika through Lake Moero following the border region of Zaire and Zambia and parallels the Upemba graben before

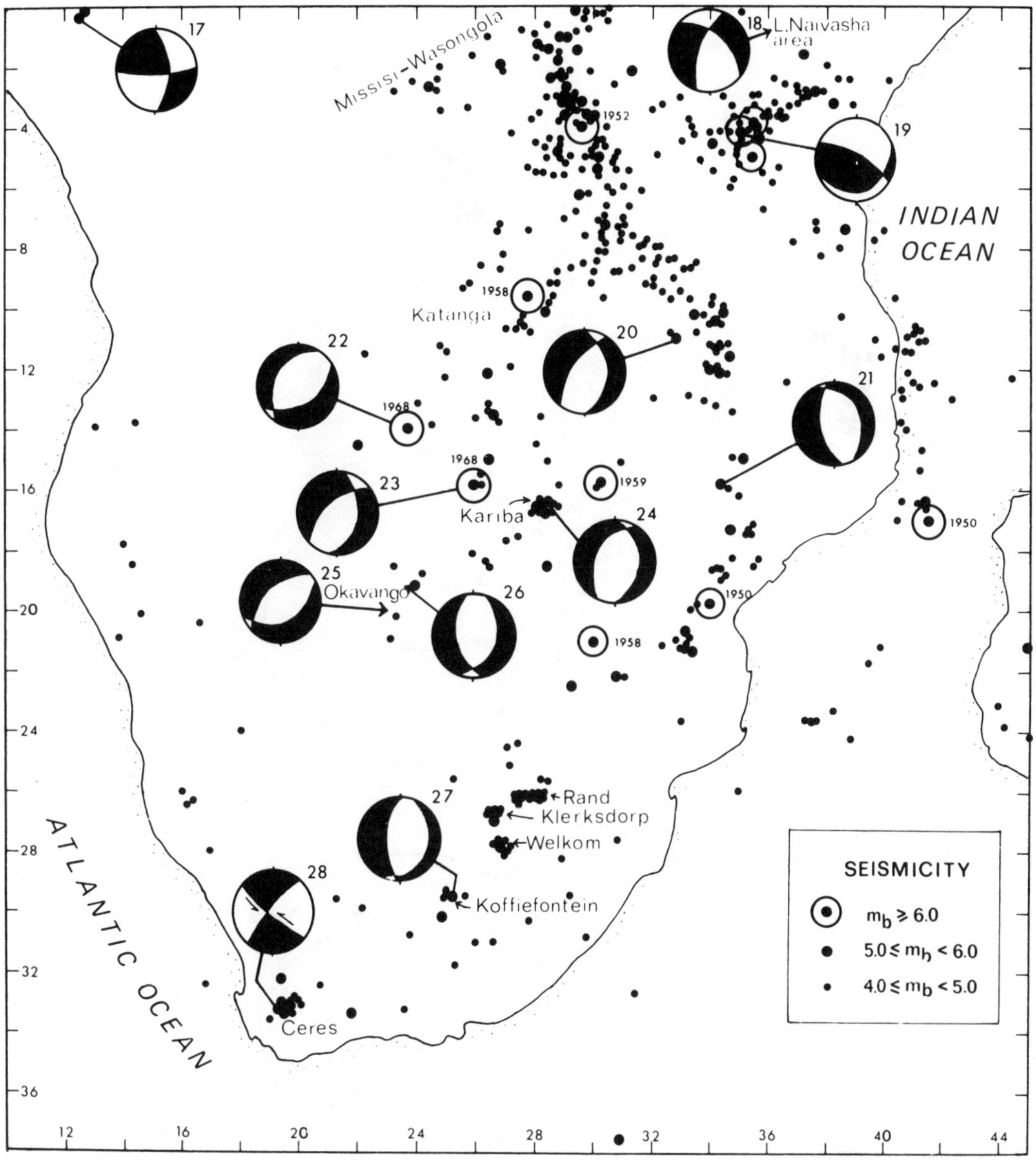

Fig. 7. The seismicity of Africa, south of the equator, for events with magnitudes $m_b \geqslant 4.0$. Focal mechanism solutions 17, 20 and 27 - this study, 18 - Hamilton et al. (1973), 19 and 21 - Maasha & Molnar (1972), 22, 23 & 28 - Fairhead & Girdler (1971), 24 - Sykes (1967), 25 & 26 - Scholz et al. (1976).

terminating at about longitude 22°E. The other zone parallels the first and is located along the Zambia-Rhodesia border striking SW from about 15°S, 30°E through Kariba into northern Botswana before again terminating at about longitude 22°E. The two zones are connected by an area of diffuse seismicity NW of Lake Kariba in central Zambia. These zones lie within and parallel the structures of the Katanga and Damara orogenic belts and possibly delineate the southern and northern margins of the Congo (Kasai Shield) and Rhodesia cratons.

The seismicity associated with Lake Kariba is mainly induced and correlated with infilling the dam in 1963 (Fairhead & Girdler, 1969; Gough and Gough, 1970). Subsequent to infilling the seismicity has continued at a high level and cannot be entirely reconciled with transient loading caused by seasonal fluctuations in Lake level. A suggestion by Scholz et al. (1976) supported by Fairhead & Henderson (1977) is that Kariba lies on a NE-SW seismo-tectonic

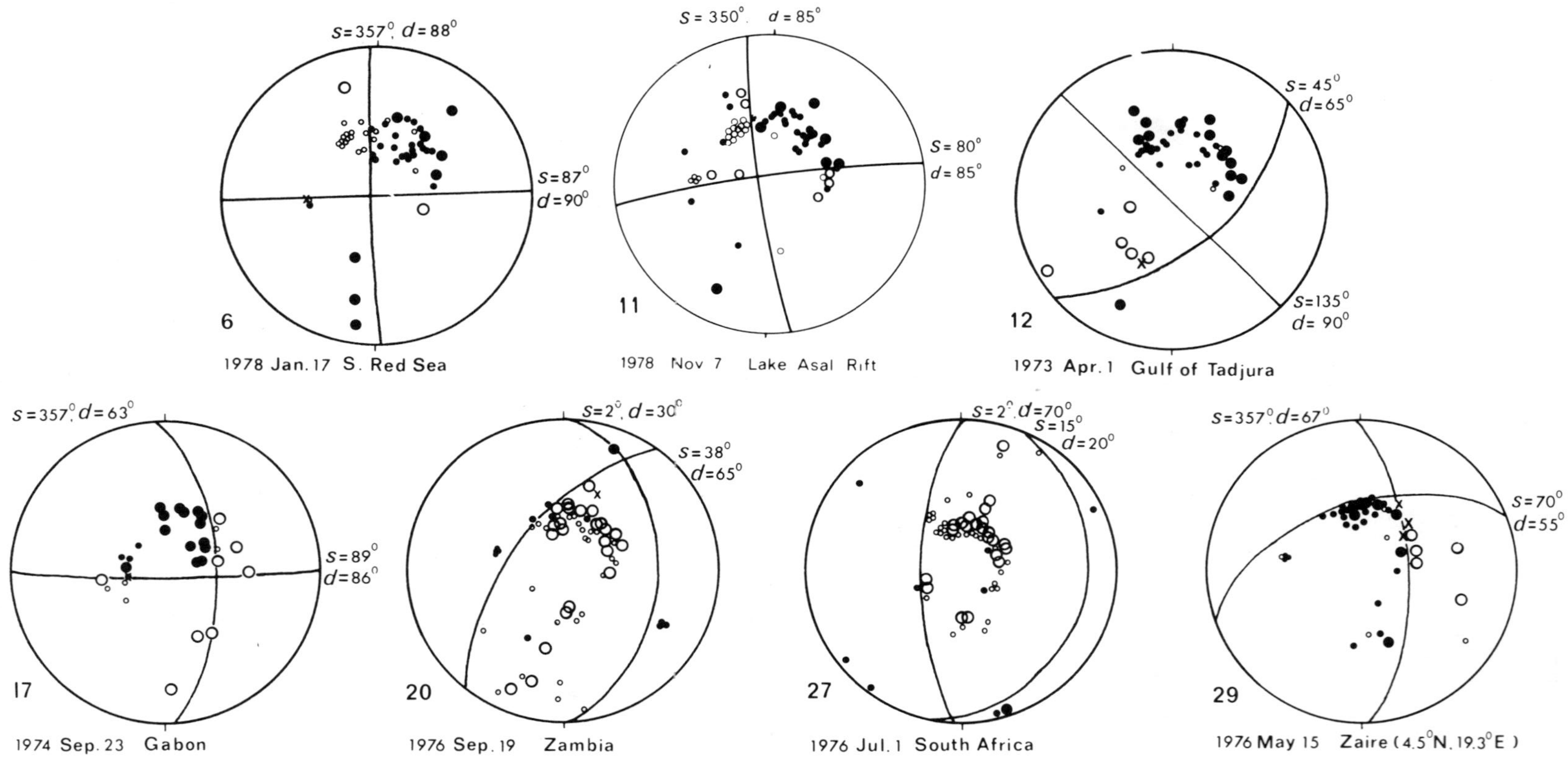

Fig. 8. Focal mechanism solutions using equal area projections of the lower hemisphere of the focal sphere. Event numbers relate to Figs. 3 and 7. Dots represent compressions, open circles dilations, and crosses wave character suggesting station near to nodal plane. Large symbols represent first motion data from long period (LP) records and small symbols represent less certain LP data and/or bulletin short-period data. 's' and 'd' represent the strike and dip of the nodal planes.

zone such that seasonal loading of the dam provides an additional factor affecting stress release.

Focal mechanism solutions (solutions 22-26, Fig. 7) all indicate normal faulting with fault planes having a general strike direction of NE-SW.

Maasha & Molnar (1972) investigated the stress drop and seismic moment of 8 earthquakes of which 4 events (22, 23, 24 & 28) were considered to be outside the rift system and 3 events (7, 13 & 14) lie within the rift system. The 4 events outside the rift system and event 19 were shown by them to have the highest stress drop. This result is consistent with the idea that the well developed Rift System, particularly its northern part, contains narrow zones of crustal/lithospheric weakness and that the leading edge (southernmost parts) of the areas of incipient rifting there exists on plate boundary or developed zone of weakness.

(f) Other Seismic Activity

(i) Kaapvaal Craton Earthquakes The historical seismicity of South Africa for the period 1620 to 1970 has been reviewed by Fernandez and Guzman (1979) and reveals that the Kaapvaal craton has a long history of earthquakes. Figures 5 & 7 clearly indicate that the main area of seismicity is centred on the gold mining areas of South Africa. This seismicity is mainly induced and results from rock bursts in the East, West and Far West Rand, Kelerksdorp and Welkom gold mines (areas indicated in Fig. 7). Heunis (1976) using data from a 3D seismic network in the Western Deep levels mine (now the world's deepest), collected since the early 1960's, reveals the spatial relation of the seismicity to the working face. The seismicity tends to concentrate in front of and above the working face. Heunis (1976) finds that the severity of rockbursts appears to be influenced by the geological joint systems of which the dominant strike direct is NNE-SSW. Earth tremors resulting from mining activities can be large (Fig. 7) e.g. the Welkom event of 1976 Dec 8 had m_b = 5.2 and killed four people when a block of flats collapsed in the centre of Welkom (Fernandez & Labuschagne, 1979).

The NNE structural trend found in the Western Deep levels mine may explain the apparent NE-SW trend in the seismicity across the Kaapvaal craton. The trend is supported by the focal mechanism solution (solution 27, Figs. 7 & 8) for the Koffienfontein earthquake of 1976 July 1 which give normal faulting with nodal planes striking approximately NS. No surface faulting was observed (per. comm. Fernandez, 1979) and the event is unrelated to mining activities.

The SE edge of the Kaapvaal craton is shown in Fig. 7 to be moderately seismically active.

(ii) Ceres Earthquakes These earthquakes are located about 100 km NE of Cape Town and began in 1969 in an area where earthquakes were uncommon. Focal mechanism studies of the 1969 Sept. 29 event (solution 28) give strike slip motion with nodal planes striking NW & NE (Fairhead & Girdler, 1971; Green & Bloch, 1971). Detailed monitoring of the after-shocks and surface effects clearly indicate the NW-SE nodal plane to be the fault plane with left lateral movement (see Seismologiese Reeks 4, 1974).

(iii) Gabon-Zaire Earthquakes Three earthquakes have occurred since 1950 to the west of the Rift System, close to the equator. These events together with the historic events plotted by Sykes (1978, his Fig. 5) form a NE-SW seismic trend that parallels the Tertiary to Recent Cameroon volcanic line. The 1974 Sept. 23 Gabon event has been studied by Sykes (1978) who derived a focal mechanism solution indicating thrust-faulting with a compressional axis paralleling the Cameroon volcanic line. This event is considered complex (Pearce, per comm. 1979) since the P phase on the short period records tends to be emergent with many SP records having opposite P wave polarity to that found using LP records. Because of this, the solution has been reworked using only first motion data from LP records and gives a strike slip mechanism (event 17, Fig. 7 & 8). The azimuthal distribution of first motions is such that the nodal planes are not well determined.

The 1976 May 15 Zaire event (solution 29, Fig. 8) is also poorly constrained but suggests predominantly a thrust fault mechanism in a NW-SE direction that is normal to the Cameroon volcanic line. Compression tectonics are well documented in this part of Africa (Freeth, 1979).

(g) Regional Stress Field for the East African Rift System Many of the published focal mechanism solutions for Africa as well as the 7 new solutions presented in Fig. 8, have poor azimuthal distibution of first motion data such that the strikes and dips of the nodal planes can easily be in error by as much as $\pm$ 20°. This uncertainty does not prevent these solutions from being used to investigate the regional stress field association with the Rift System but is sufficient to prevent the differentiation between relative plate motions that probably exist between the Red Sea and Gulf of Aden axes. The accuracy of focal mechanism solutions can be improved by a new computer method developed by Pearce (1977a, 1980). This method uses the polarity and relative amplitudes of P, pP and sP phases for shallow focus earthquakes and the seismogram modelling techniques of Hudson (1969a, b) and Douglas et al. (1972). This method is ideal for small magnitude events (e.g. solutions 1, 3 & 4, Fig. 1) and can provide confidence limits for the direction and dip of the nodal planes.

Focal mechanism solutions associated with the East African Rift System and its continuation into southern Africa indicate a predominance of normal faulting with the tensional stress axis orientated WNW-ESE which is normal to the SSW to SW trend of the Rift System through Africa. The most striking difference between the East African Rift System and the Oceanic Rift System is the lack of strike slip focal mechanisms associated with Rifting in Africa. This suggests that strike slip displacements are taking place aseismically and/or are being taken up by en echelon faults. The primary source for tensional stress within the African lithosphere is considered by the authors to result from regional uplift caused by the isostatic response of the crust to the presence of a developing and rising low density, upper mantle body beneath the Rift System. Girdler (1975) and Fairhead & Reeves (1977) consider this low density body to be the asthenosphere which has thinned the continental lithosphere beneath the Rift System by a process that may eventually lead to the break-up of Africa.

Outside the Rift System, strike slip and compressional mechanisms predominate. Their relation to the regional stress field is unclear. Only one tensional mechanism (solution 27, Fig. 7) has been found and may relate to regional uplift of the Kaapvaal craton. Its E-W tensional axis is broadly consistent with the in situ stress measurements reported by Gay (1977).

(h) 'b' values Earthquakes follow a distribution of size of the form log (N) = a - bM, where a and b are constants for a set of earthquakes, N, the number of earthquakes with magnitude greater than M (see Lilwall, this volume for details). Table 1 summarises the 'b' values for the different parts of the East African Rift System.

3. The Baikal Rift

At first sight the seismicity of the Baikal Rift zone appears to form a northern branch of the Central Asiatic seismic belt. However, seismicity of an extensional nature is confined to the Baikal Rift zone and is isolated from the World Rift System by neighbouring areas of compressional tectonics.

Rifting in the Baikal region is considered to have started in the Eocone, beginning around South Baikal and propagating in westerly and north-easterly directions (Logatchev and Florensov, 1978). The graben formation was predominantly controlled by pre-existing deep basement faults which followed the boundary between the Pre-Cambrian Siberian craton and the Saya-Baikal fold belt (Sherman, 1978), Logatchev and Florensov (1978) also point out that the Bolnai lineament, an ancient shear zone, appears to impede the southern extension of the rift (Fig. 9). Rifting in the Baikal region has not resulted in simple grabens but in numerous assymmetrical fault-bounded depressions which are considered by Zorin and Rogozhina (1978) to be in different stages of their development.

The present seismograph network detection is such that over 2000 earthquakes of magnitude greater than 1.5 - 2.0 are recorded annually.

TABLE 1

Location	Constant 'b'	Reference
Gulf of Suez	1.99	Fairhead & Girdler, 1970
Central Afar (Serdo)	0.76	Fairhead & Girdler, 1971
Western Rift	0.65 to 0.94	Wohlenberg 1969; Ndontoni 1976
Sub-regions of Rift	0.81 to 1.05	Ndontoni 1976; Maasha 1975a; Fairhead & Girdler, 1971, Lahr & Pomeroy, 1970.
Micro earthquakes (Ruwenzori)	0.62 ± 0.01 (dry season) 0.43 ± 0.05 (wet season)	Maasha, 1975b.
Virunga Volcanic Region	1.20 ± 0.07	Ndontoni, 1976.
Eastern Rift	0.79 to 1.35	Wohlenberg, 1969; Ndontoni, 1976; Bath, 1975; Fairhead & Girdler, 1971
Upemba-Moero Rift	0.72 to 0.87	Bram, 1972; Ndontoni 1976.
Kariba Dam events	1.45	Gough & Gough, 1970

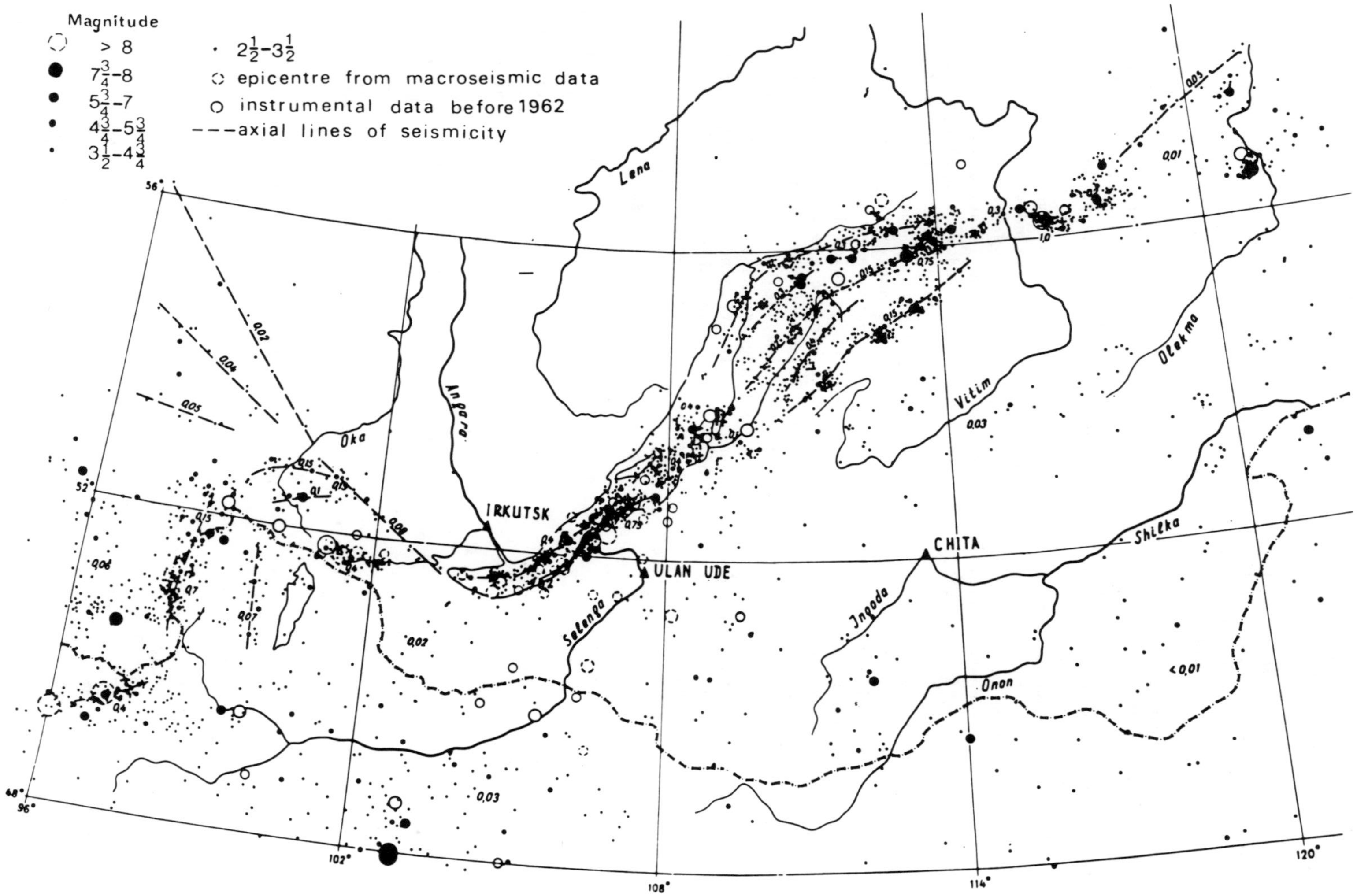

Fig. 9. Earthquake epicentres in the Baikal seismic zone after Golenetsky and Misharina (1978). The numerical values indicate seismic activity A_{10}, where A_{10} is the annual number of energy class 10_2 Rautian scale earthquakes in an area of 1000 km^2.

Earthquakes occur both as individual shocks and as swarms. Furthermore, Golentsky and Misharine (1978) note that the large earthquakes show no apparent temporal or spatial sequences indicative of progressive breaking of the crust in the rift zone. The foci of the small earthquakes are concentrated in the upper crust, with epicentres often grouped in linear belts, usually striking parallel to the main rift structures (Fig. 9). The seismicity is associated with vertical fault movement both in and along the edges of the assymetric depressions and adjacent horsts. Field evidence for significant horizontal movements has not been firmly established (Solonenko, 1978). The geographical distribution of recent large earthquakes and the presence of small rift-like depressions of Pleistocene age suggest that the rift zone is both lengthening and widening along its eastern boundary (Solenenko, 1978; Golenetsky and Misharino, 1978).

However, the seismic activity is not always simply connected with the mapped fault zones. Focal mechanism studies sometimes give solutions that are unrelated to surface faulting suggesting that crustal inhomogeneity is evident in these regions with different focal mechanisms confined to different layers of the crust (Golenetsky and Misharina, 1978).

Solonenko (1978) noted that seismotectonic observations indicate that crustal movements in the Baikal Rift zone have not been due to slow continuous crustal deformation but have been episodic with outbursts of assumed seismic activity associated with volcanic eruptions during the Lower Pleistocene, Interglacial and Middle Holocene epochs. This is also reflected in the present day seismicity which consists of alternating periods of renewed activity and quiescent periods.

The orientation of the tensional axes determined from earthquake foci studies (Fig. 10) indicates that the central section of the Baikal Rift zone is being subjected to horizontal tension perpendicular to the strike of the fault structures. Earthquakes at the northern and southern extremities of the rift zone, however, show a variable pattern of stress axis orientation consistent with the greater stress inhomogeneity in these areas (Golentsky and Misharina, 1978). Only a small group of earthquakes have been studied which may represent transverse tectonism.

It is the general opinion of Soviet scientists, who have been engaged in detailed studies of the Rift system (Logatchev & Florenso, 1978; Puzyrev et al., 1978), that the extension, causing rifting, results from processes within the upper mantle. This is in disagreement with Molnar and Tapponier (1975) and Tapponier & Molnar (1979) who suggest that the rifting might have resulted from the transmission of stress due to the collision of the Indian and Eurasian plates.

4. The Rhine Graben

The earthquake activity associated with the Rhine graben system is the most conspicuous feature on the seismicity map of Western Europe north of the Alpide belt (Fig. 11). This graben system forms a triple junction centred on the Mainz area, with branches consisting of the Upper Rhine (to the S), Lower Rhine (to the NW) and the Hessian grabens (to the NE) as shown in figure 11. With the exception of the Hessian graben, which has been notably inactive since Tertiary times, these grabens are marked by recent tectonic activity of an extensional character (Schneider 1967; Ahorner, 1970).

The grabens began to develop in the Middle Eocene and are now filled with up to 4km of sediment in places. Malzer (1967) calculated that the Upper Rhine is still subsiding at a rate of 0.5mm a^{-1} and it has been noted that not all the grabens have a symmetric form.

The southern part of the Rhine graben is seismically the most active (Fig. 11). From south to north, the seismicity lies on the eastern side of the Upper Rhine from Basle to Strasbourg before switching to the west side then back to the east between Karlsruhe and Heidelberg. North of Mainz, at the southern border of the Hercynian Rhenish Massif, the earthquake zone changes strike to a north-westerly direction. This seismic zone, which is associated with the Lower Rhine graben between Mainz and Koln, divides between Koln and Aarchen. An east-west branch continues through Belgium to the Channel at Ostend and is known as the Brabant earthquake zone, while the NW trending seismicity continues from Koln along the Maas valley and joins with the disseminated zone of seismicity associated with the North Sea region (Hysebye et al, 1975). Throughout, the seismicity consists of a patchy accumulation of epicentres.

Ahorner (1975) has calculated that the rate of seismic activity (earthquake frequency per magnitude class per unit area) in the Rhine graben system is at least 5 to 10 times less than in the East African Rift system. Epicentres are confined to the upper 25km of the crust (Hiller et al 1967; Schneider, 1967). Mueller (1970) correlated the depth of maximum earthquake activity at about 10km with the roof of the so-called sialic low velocity zone determined from seismic refraction work in the region.

Focal mechanism studies (Ahorner 1975), the results from which are illustrated in Fig. 12, indicate that within the Upper Rhine strike slip mechanisms dominate by about 3 to 1 over dip slip mechanisms. Left lateral motion is seen along fault planes parallel to the graben axis and right lateral motion on those perpendicular to the axis. These mechanisms are in agreement with a regional compressional axis trending in a NW-SE direction (see Fig. 13). However there are a few dip slip

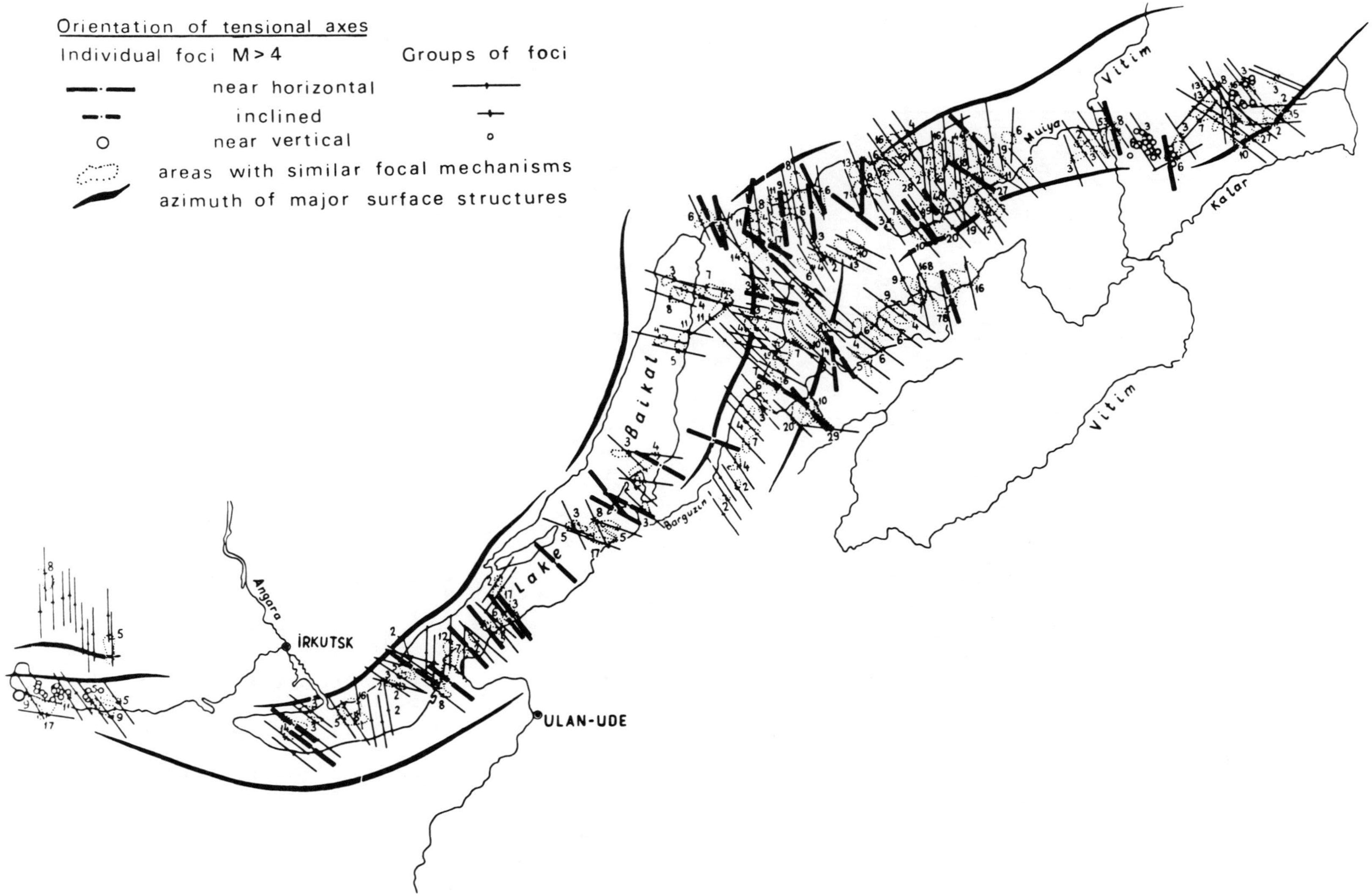

Fig. 10. Orientation of the tensional axes in earthquake foci in the Baikal Rift zone after Golenetsky and Misharina (1978). Near horizontal is defined as the zenith angle $60^{\circ} \leq \delta \leq 90^{\circ}$, inclined as the zenith angle $35^{\circ} \leq \delta \leq 55^{\circ}$ near vertical as the zenith angle $0^{\circ} \leq \delta \leq 30^{\circ}$. The numbers represent the number of shocks analysed to determine the group focal mechanisms.

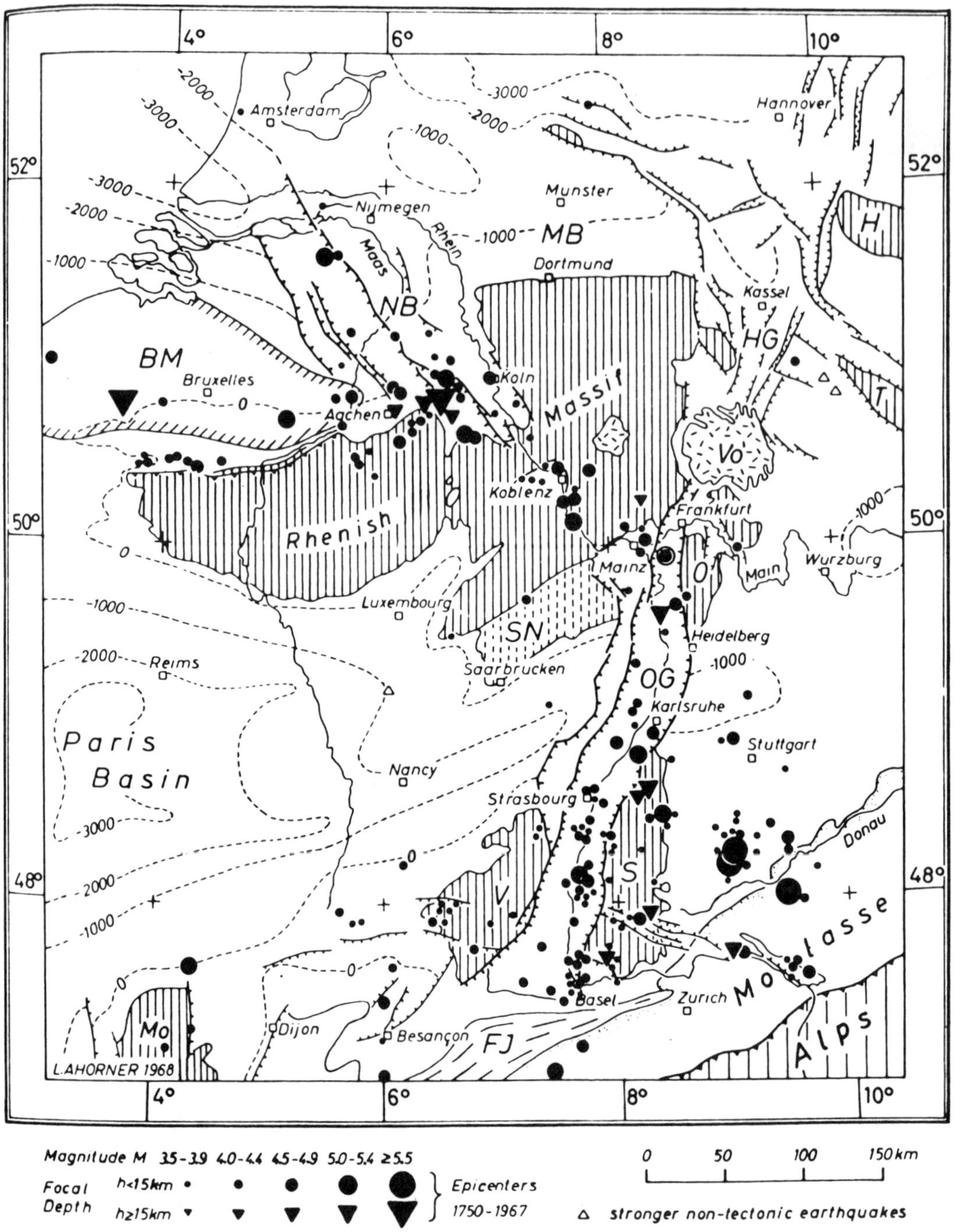

Fig. 11. Seismicity map of the Rhine Graben system after Ahorner 1970. Epicentres from 1750 to 1967 with magnitudes M > 3 are illustrated. OG = Upper Rhine graben, NB = Lower Rhine graben, HG = Hessian graben, Vo = Vogelsberg volcano. Paleozoic massifs are hatched. Broken lines give contours of Paleozoic basement covered by later formations.

mechanisms on the fault planes parallel to the graben which disagree with this overall trend (Fig. 12).

The Lower Rhine graben has a somewhat different seismotectonic regime with both strike slip and dip slip mechanisms more frequent. The dip slip mechanisms belong to the tensional faults striking NE-SW sub-paralleling the axis of the graben. The strike slip mechanisms are associated with fault planes trending WNW or NNE, obl-

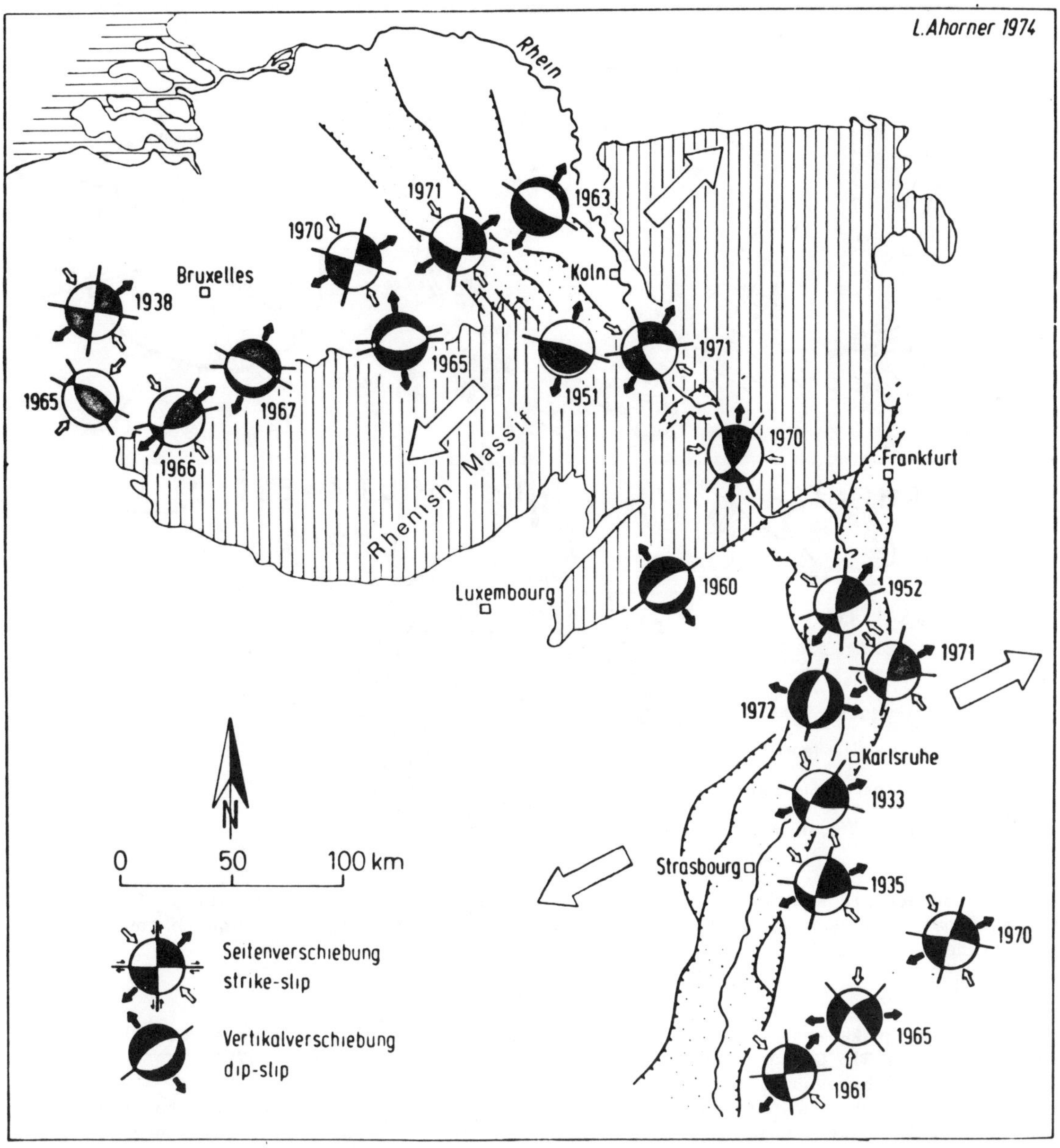

Fig. 12. Focal mechanisms of earthquakes in and around the Rhine graben, after Ahorner (1975). The projections are on to the lower hemisphere with the black quadrants denoting compressional P-wave first motions.

ique to the graben trend (Fig. 12).

Ahorner (1975) calculated the average amount of seismotectonic block movement along the fracture zones using the historical earthquake method of Brune (1968) to be about 0.05mm a^{-1} in the Upper Rhine, 0.23mm a^{-1} in the Lower Rhine and 0.07 mm a^{-1} in the Brabant earthquake zone. However, geodetically observed slip rates are roughly 5-10 times higher. This implies that a large proportion of the accumulated strain is released as aseismic creep movements. Ahorner (1970) has therefore suggested that vertical block movements along normal faults in the Lower Rhine may take place aseismically.

By integrating all the focal mechanism studies, and in-situ stress measurements, Ahorner (1970) inferred that there is a mean compressional axis in this part of Europe with a direction of $142^\circ \pm 20^\circ$ consistent with the transmission of stress, due to the collision of the African and Eurasian plates. Under the influence of this regional stress field the diagonally arranged fracture zone of the Upper Rhine is affected by horizontal shear forces as are the WNW-ESE

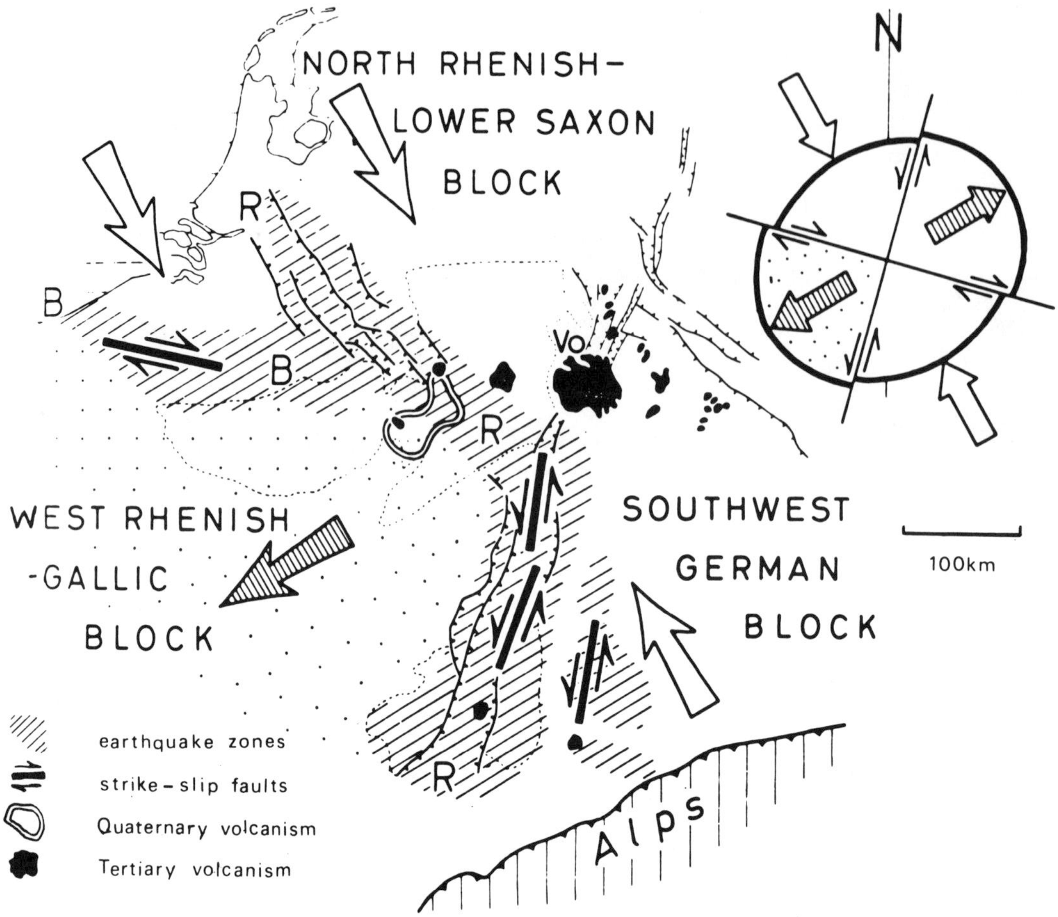

Fig. 13. Seismotectonic scheme of the Rhine graben system with present-day stress field and possible horizontal block movements after Ahorner (1970). R = Rhenish earthquake zone, B = Belgian earthquake zone.

striking Brabant earthquake zone (Fig. 13).

5. Rio Grande Rift The Rio Grande rift is a tensional structure which began to form in the mid Tertiary along ancient zones of weakness (Chapin and Seager, 1975) at the same time as the Rocky Mountains. Situated in New Mexico, it consists of a series of north-south trending fault bounded basins arranged *en echelon.*

The rift is still active as shown by fault scarps cutting alluvial fans (Chapin 1971), earthquake and micro-earthquake activity (Sanford et al, 1972; Eeilinger and York, 1979; Hoffman, 1975) and contemporary vertical movements (Reilinger and Oliver, 1976). Hoffman (1975) documents the historic activity in the region, which has included earthquakes up to intensity VIII MM, but implies the rift is relatively quiet at present. He also notes that there is a complete lack of correlation between rift structure and zones of micro-earthquake activity. In the northern part of the rift, north-east of the Jemez Mountains, the seismicity forms a pronounced NNE trending pattern across the rift (Reilinger and York, 1979).

Acknowledgements We wish to thank M. Coulibaly (ORSTOM Senegal), L.M. Fernandez (Geol. Survey., Rep. South Africa), H. Jarosch (Weizmann Inst., Israel) and J.L. Tournier (ORSTOM Ivory Coast) for supplying first motion data, R.G. Pearce for helping with the focal mechanism studies and making available unpublished data and N.B. Henderson for supplying epicentre data from the Rhodesian seismological network.

References

Abdallah, A., Courtillot, V., Kasser, M., LeDain, A.Y., Lépine, J.C., Robineau, B., Ruegg, J.C., Tappionier, P. & Tarantolla, A. 1979. Relevance of Afar seismicity and volcanism to the mechanics of accreting plate boundaries. Nature, 282: 17-23.

Abdel-Gawad, M. 1969. New evidence of transcurrent movements in Red Sea areas and petroleum implications. Am. Assoc. Pet. Geol. Bull., 55: 1466-1479.

Ahorner, L. 1970. Seismotectonic relations between the graben zones of the Upper and Lower Rhine valley. In: H.J. Illies and S. Mueller (Eds). Graben Problems. Schweizerbart, Stuttgart, pp. 155-156.

Ahorner, L. 1975. Present day stress field and seismotectonic block movements along major fault zones in Central Europe. Tectonophysics, 29, 233-249.

Baker, B.H., Mohr, P.A. & Williams, L.A.J. 1972. Geology of the eastern rift system of E. Africa. Geol. Soc. Am., Sp. Pap., 136: 67pp.

Banghar, A.R. & Sykes, L.R. 1969. Focal mechanisms of earthquakes in the Indian Ocean and surrounding areas. Geophys. J.R. astr. Soc. 3: 433-440.

Barberi, F. & Varet, J. 1975. Recent volcanic units of Afar and their structural significance. In: A. Pilger & A. Rosler (editors) Afar dperession of Ethiopia. Stuttgart (Schweizerbart): 174-178.

Båth, M. 1975. Seismicity of the Tanzania region. Tectonophysics, 27: 353-379.

Ben Menahem, A. & Aboodi E. 1971. Tectonic patterns in the northern Red Sea region. J. Geophys. Res., 76: 2674-2689.

Ben Menahem, A., Nur, A. & Vered, M. 1976. Tectonics, seismicity and structure of the Afro-Eurasian junction - the breaking of an incoherent plate. Phys. Earth and Planet Inter., 12: 1-50.

Bram, K. 1972. Seismicity of Katanga and western Zambia, southwest of East Africa Rift Systems, from 1960-1971. Bull. Seismol. Soc. Am., 62: 1211-1216.

Brune, J. 1968. Seismic moment, seismicity and rate of slip along major fault zones. J. Geophys. Res., 73, 777-784.

Burke, K. 1978. Evolution of continental rift systems in the light of plate tectonics. In: Ramberg, I.B. and Newman, E.R. (Eds.) Tectonics and Geophysics of Continental Rifts. D. Reidel Publ. Co., Holland. p. 1-9.

Cahen, L. 1970. Igneous activity and mineralisation episodes in the evolution of the Kibaride and Katangide orogenic belts of central Africa. In: T.N. Clifford and I.G. Gass (editors), African magmatism and Tectonics. Oliver & Boyd, Edinburgh: 97-117.

Chapin, C.E. 1971. The Rio Grande rift. Part I Modifications and additions: New Mexico Geological Society, 22nd Field Conference Guidebook, p. 191-201.

Chapin, C.E. and Seager, W.R. 1975. Evaluation of the Rio Grande rift in the Socorro and Las Crurces areas: New Mexico Geological Society, 26th Field Conference Guildebook, p. 297-321.

DeBremaecker, J.C. 1955. Realisation et programme de l'IRSAC en seismologie. Bull Acad. Roy. Sci. Coloniales, Brussels, 1(4):643-664.

DeBremaecker, J.C. 1959. Seismicity of the West African rift valley. J. Geophys. Res. 64: 1961-1966.

Douglas, A., Hudson, J.A. & Blamey, C. 1972. A quantitative evaluation of seismic signals at teleseismic distances-III computed P and Rayleigh-wave seismograms. Geophys. J.R. astr. Soc., 28: 385-410.

Fairhead, J.D. 1976. The structure of the lithosphere beneath the Eastern Rift, East Africa, deduced from gravity studies. Tectonophysics, 30: 269-298.

Fairhead, J.D. 1980. The structure of the cross-cutting volcanic chain of Northern Tanzania and its relation to the East African Rift System. Tectonophysics 65: 193-208.

Fairhead, J.D. & Girdler, R.W. 1969. How far does the Rift System extend through Africa? Nature, 221: 1018-1020.

Fairhead, J.D. & Girdler, R.W. 1970. The seismicity of the Red Sea, Gulf of Aden and Afar triangle. Phil. Trans. R. Soc. London. A267: 49-74.

Fairhead, J.D. & Girdler, R.W. 1971. The seismicity of Africa. Geophys. J.R. astr. Soc., 271-301.

Fairhead, J.D. & Henderson, N.E. 1977. The seismicity of southern African and incipient rifting. Tectonophysics, 41: T19-T26.

Fairhead, J.D. & Reeves, C.V. 1977. Teleseismic delay times, Bouguer anomalies and inferred thickness of the African lithosphere. Earth Planet. Sci. Lett. 36: 63-76.

Falcon, N.L., Gass, I.G., Girdler, R.W. & Laughton, A.S. (editors). 1970. A discussion on the structure and evolution of the Red Sea and nature of the Red Sea, Gulf of Aden and Ethiopia Rift Junction. Phil. Trans. Roy. Soc. Lond. A267: pp.417.

Fernandez, L.M. & Guzman, A.J. 1979. Seismic history of southern African (south of 17°S). Seismol. Ser. geol. Surv. S. Afr., 9: pp.38.

Fernandez, L.M. & Labuschagne, P.G.A. 1979. Catalogue of earhtquakes in Southern Africa and surrounding areas for 1976. Seismol. Ser. geol. Surv. S. Afr., 8: pp.29.

Freeth, 1979. Deformation of the African plate. Earth Planet Sci. Lett., 45: 93-104.

Freund, R. 1965. A model of the structural development of Israel and adjacent areas since upper Cretaceous times. Geol. Mag., 102: 189-205.

Gay, N.C. 1977. Principal horizontal stresses in southern Africa. Pure Appl. Geophys. 115: 3-10.

Girdler, R.W. 1965. The role of translational and rotational movements in the formation of the Red Sea and the Gulf of Aden. In: T.N. Irvine (editor), The World Rift System, U.M.C. Symp., Ottawa, Ont., September 6-8, 1965: 65-95.

Girdler, R.W. 1975. The great negative Bouguer anomaly over Africa. EOS, Trans. Am. Geophys. Union 56(8): 516-519.

Gough, D.I. & Gough, W.I. 1970. Load induced earthquakes at Lake Kariba. Geophys. J.R. astr. Soc., 21: 79-101.

Golenetsy, S.I. and Misharina, L.A. 1978. Seismicity and earthquake focal mechanisms in the Baikal Rift zone. Tectonophysics, 45 71-85.

Gouin, P. 1976. Seismic zoning in Ethiopia. Bull. Geophys. Obs. Ethiopia, 17: pp.46.

Gouin, P. 1979. Earthquake history of Ethiopia and the Horn of Africa. Ottawa, Ont. I.D.R.C. 259p.

Green, R.W.E. & Bloch, S. 1971. The Ceres, South Africa earthquake of September 29, 1969. Bull. seism. Soc. Am., 61: 851-859.

Gutenberg, B. & Richter, C.F. 1949, 1954. Seismicity of the Earth and associated phenomena, 1st and 2nd editions, Princeton University Press.

Hamilton, R.M., Smith, B.E. & Knapp, F. 1973. Earthquakes in geothermal areas near Lakes Naivasha and Hannington, Kenya (unpublished report).

Heunis, R. 1976. Rock bursts and the search for an early warning system. Optima 1976 (1): 36-42.

Hiller, W., Rothé, J.P-. & Schneider, G. 1967. La séismicité du Fosse Rhenan. -Abh. geol. Landesant Baden Wurttemberg, 6, 98-100.

Hoffmann, J.P. 1975. The seismic history of the Rio Grande rift. Earthquake Information Bulletin, 7, No. 3, p.8-13.

McKenzie, D.P., Davies, D. & Molnar, P. 1970. Plate tectonics of the Red Sea and East Africa. Nature, 226: 242-248.

Mohr, P., Birnius, A. & Rolff, J. 1978. Present-day strain rates at the northern end of the Ethiopian Rift Valley. Tectonophysics,44: 141-160.

Molnar, P. & Aggarwal, Y.P. 1971. A microearthquake survey in Kenya. Bull seism. Soc. Am., 61: 195-201.

Molnar, P., Fitch, J. & Asfaw, L.M. 1970. A micro-earthquake survey in the Ethiopian rift. E.S.S.A., Earthquake Notes, 41(2): 37-44.

Molnar, P. & Tapponnier, P. 1975. Tectonics of Asia: consequences and implications of a continental collision. Science, 189: 419-426.

Mueller, St. 1970. Geophysical aspects of graben formation in continental rift systems. In: H.J. Illies and S. Mueller (editors), Graben Problems. Schweizerbart, Stuttgart, pp. 27-37.

Ndontoni, Z. 1976. Seismicity of the Western African Rift Valley. Individual studies by participants at the International Institute of siemology and earthquake engineering, 12: 121-140.

Pearce, R.G. 1977a. Fault plane solutions using relative amplitudes of P and pP. Geophys. J.R. astr. Soc., 50: 381-394.

Pearce, R.G. 1977b. A study of earthquake modelling with applications to the East African Rift System. Ph.D. Thesis, University of Newcastle upon Tyne.

Pearce, R.G. 1980. Fault plane solutions using relative amplitudes of P and surface reflections: Further studies. Geophys. J.R. astr. Soc., 60: 459-487.

Picard, L. 1966. Thoughts on the graben system of the Near East. Geol. Surv. Can., Pap., 66-44: 22-34.

Pilger, A. & Rosler, A. (editors). 1975. Afar Depression of Ethiopia. Stuttgart, (Schweizerbart), pp.416.

Puzyrev, N.N., Mandelbaum, M.M., Krylov, S.V., Mishenkin, B.P., Petrik, G.V. and Krupskaya, G.V. 1978. Deep structure of the Baikal and other continental rift zones from seismic data. Tectonophysics, 45: 15-22.

Reeves, C.V. 1972. Rifting in the Kalahari? Nature, 237: 95-96.

Reilinger, R.E. and Oliver, J.E. 1976. Modern uplift associated with a proposed magma body in the vicinity of Socorro, New Mexico: Geology, 4: 583-586.

Reilinger, R.E. and York, J.E. 1979. Relative crustal subsidance from levelling data in a seismically active part of the Rio Grande rift, New Mexico. Geology, 7: 139-143.

Richter, C.F. 1958. Elementary Seismology. Freeman & Co., San Francisco & London. pp.768.

Rothe, J.P. 1954. La zone seismique mediane Indo-Atlantique. Proc. R. Soc. Lond., A222: 387-397.

Rykounov, L.N. Sedov, V.V., Savrina, L.A. & Bourmin, V.Ju. 1972. Study of micro-earthquakes in the rift zones of East Africa. Tectonophysics, 15: 123-130.

Sanford, A.R., Alptekin, O.S. & Toppozada, T.R. 1973. Use of reflection phases on microearthquake seismograms to map an unusual discontinuity beneath the Rio Grande rift. Bull. Seism. Soc. Am., 63 2021-2034.

Schneider, G. 1967. Erdbeben und tektonik in Sudwest Deutschland. Tectonophysics, 5: 459-511.

Scholz, C.H., Koczynski, T.A. & Hutchins, D.G. 1976. Evidence for incipient rifting in southern Africa. Geophys. J.R. astr. Soc., 44: 135-144.

Searle, R.C. 1970. Evidence from gravity anomalies for thinning of the lithosphere beneath the rift valley in Kenya. Geophys. J.R. astr. Soc., 21: 13-31.

Searle, R.C. & Gouin, P. 1972. A gravity survey of the central part of the Ethiopian Rift Valley. Tectonophysics, 15: 41-52.

Seismologiese, Reeks A. 1974. Die Aardbewin van 29 September 1969 in die Suidwestelike Kaapprovinsie, Suid-Afrika seismol. Ser. Geol. Surv. S. Afr., 4: pp.48.

Sherman, S.I. 1978. Faults of the Baikal Rift. Tectonophysics, 45: 31-39.

Solonenko, V.P. 1978. Seismotectonics of the Baikal Rift Zone. Tectonophysics, 45: 61-69.

Sutton, G.H. & Berg, F. 1958. Seismological studies of the western rift valley of Africa. Trans. Am. Geophys. Union, 39: 474-481.

Sykes, L.R. 1967. Mechanism of earthquakes and nature of faulting on the mid-ocean ridges. J. Geophys. Res., 72: 2131-2153.
Sykes, L.R. 1970. Focal mechanism solutions for earthquakes along the World Rift System. Bull. Seism. Soc. Am., 60(5): 1749-1752.
Sykes, L.R. 1978. Intraplate seismicity, reactivation of pre-existing zones of weakness, alkaline magmatism and other tectonism post-dating continental fragmentation. Rev. Geophys Space Phys. 16(4): 621-688.
Sykes, L.R. & Landisman, M. 1964. The seismicity of East Africa, the Gulf of Aden and the Arabian and Red Seas. Bull Seism. Soc. Am., 54: 1927-1940.
Tapponier, P. & Molnar, P. 1979. Active faulting and Cenozoic tectonics of the Tien Shan, Mongolia, and Baikal regions. J. Geophys. Res. 84: 3425-3459.
Tobin, D.G., Ward, P.L. & Drake, C.L. 1969. Micro-earthquakes in the rift valley of Kenya. Bull. seism. Soc. Am., 80: 2043-2046.
Wohlenberg, J. 1969. Remarks on the seismicity of East Africa between 4^oN-12^oS and 23^oE-40^oE. Tectonophysics, 8: 567-577.
Wohlenberg, J., Janssen, T. & Loupekine, I.S. 1966. Earthquake reconnaisance Mission Uganda Earthquake, UNESCO, Paris.
Zorin, Yu.A. & Rogozhina, V.A. 1978. Mechanism of rifting and some features of the deep-seated structure of the Baikal Rift Zone. Tectonophysics, 45: 23-30.

SEISMICITY OF THE OCEANIC RIFTS

R C Lilwall

Institute of Oceanographic Sciences, Blacknest, Brimpton, Nr Reading, Berkshire

Abstract Seismicity along the mid-oceanic rifts appears as a narrow zone within the confines of the median valley. Microearthquake surveys show an active zone as little as 5 km across. A few reliable depth of focus determinations of both micro and large earthquakes are all less than 10 km below the sea bed, probably indicating the lower limit of the brittle crust. Fault plane solutions show that the ridge crest morphology is dominantly the result of "normal" dip-slip movement, there being little evidence of reversed faulting. Earthquake swarms, usually of limited spatial extent, are frequent on the ridge crest but not in the centre of fracture zones. The association of swarming with volcanism has led to faulting models based on both magma intrusion and withdrawal, but at present there is no direct evidence for a large scale high level magma chamber beneath slow-spreading ridges. Maximum seismic moments of ridge crest and fracture zone earthquakes may be determined by the thickness of the brittle crust which is related to an isotherm marking the transition between seismic to aseismic slip. Higher b values on ridge crests compared with fracture zones are believed to be caused by lower stress states on the former but convincing direct evidence on this is not yet available.

Introduction

Studies of the seismicity of mid-oceanic ridges have made an important contribution to the concept of plate tectonics. The seismically active mid-ocean ridge system, which represents over half of the earth's major plate boundaries, is largely inaccessible to direct observation and remote from seismograph stations. In spite of this, epicentre and focal mechanism determinations have revealed both plate boundaries and their relative motion.

This review is concerned with seismicity in relation to the processes taking place along the plate boundary itself and in particular to summarise the seismological data in the context of geological and morphological information becoming available about the ridge crest. In some instances this requires a resolution which approaches or exceeds that possible from available data and processing techniques, these limitations are stated and discussed where appropriate.

Although we are concerned primarily with the seismicity of the rifted part of the oceanic ridge system a strict division between this and the fast-spreading non-rifted and transform sections is inappropriate. Where necessary, discussion includes work on these regions.

Spatial Distribution of Ocean Rift Seismicity

Accurate locations of earthquakes provide a direct link between seismicity and resulting morphological features. Discussion is complicated because until recently, with the notable exception of Iceland, the remoteness of much of the ocean ridge system has resulted in earthquake locations relying almost exclusively on data from distant (teleseismic) stations. The advent of sonobuoy and ocean bottom recordings is now leading to a small but steady flow of papers devoted to specific areas but with variable seismological content. Since the coverage, nature and limitations of these two data sources are different they are described separately below. Observations on Iceland are also separated as they do not fall into either of the above categories and cannot be typical of the ridge system generally.

World Distribution

An association of seismic activity with the mid-oceanic ridges has been known for many years (Gutenburg & Richter, 1954). However, the establishment of the World Wide Standard Station Network and the routine determination of hypocentres, initially by the United States Coast and Geodetic Survey, enabled Barazangi & Dorman (1969) to publish maps which show world seismicity with a clarity not previously attained. A striking feature of these maps is the excellent definition of the mid-oceanic ridges. Fig (1) shows a similar map to that published by Barazangi & Dorman but including only epicentres along the mid-oceanic ridge system and with no distinction between ridge crest and associated fracture zones. The power of epicentre distri-

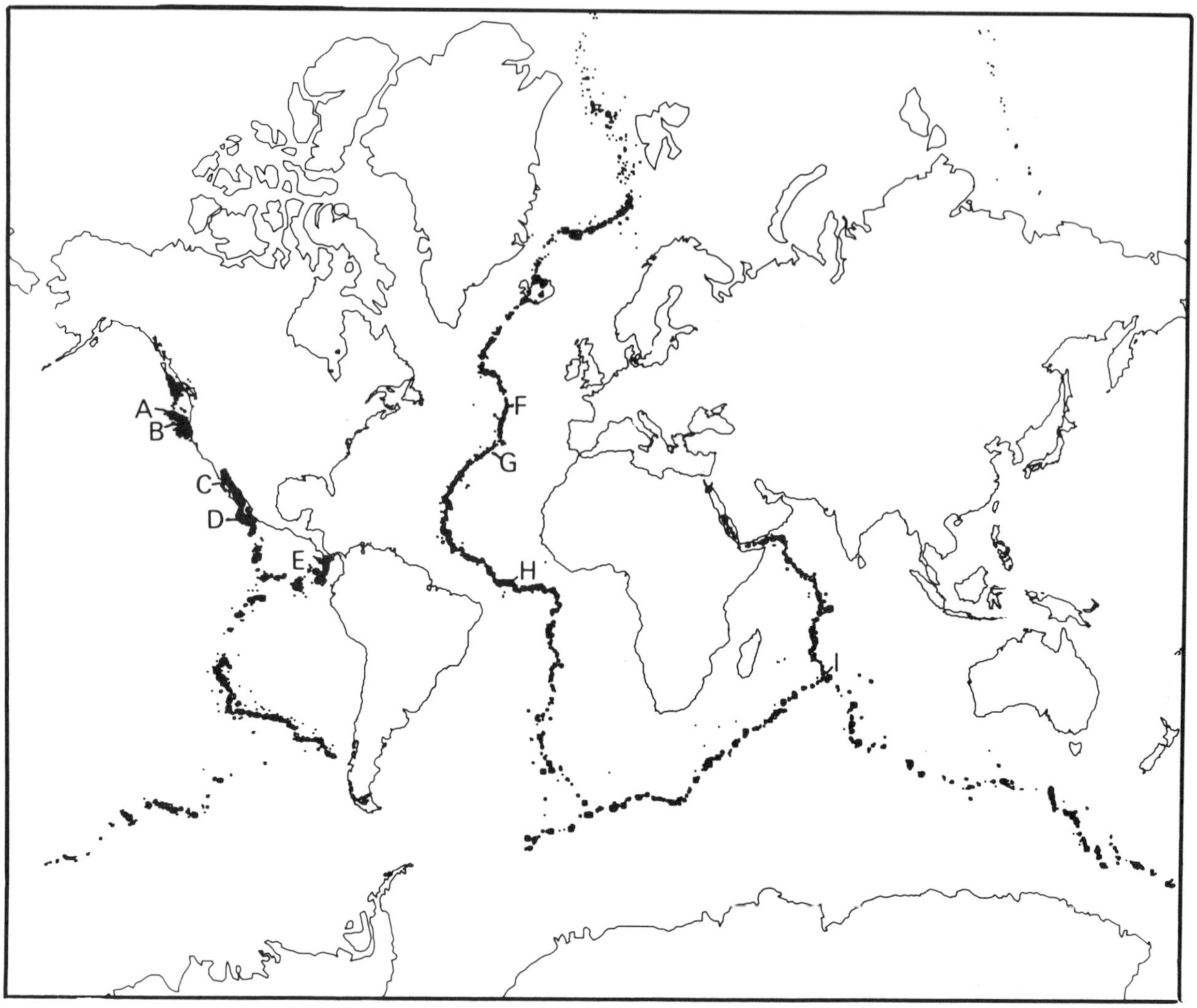

Fig 1. Epicentres along the world's major oceanic spreading centres for the period 1964-77. Letters denote areas having local surveys mentioned in this review; A = Blanco FZ, B = Gorda Rise, C = Gulf of California, D = Rivera FZ, E = Galapagos Rise, F = Mid-Atlantic ridge 45°N, G = Mid-Atlantic ridge 37°N, H = East St. Pauls FZ, I = Indian Ocean Triple Junction.

butions in delineating plate boundaries in otherwise remote regions is clearly shown in this map. Several papers have been written on this subject which divide the ridge system regionally: eg see Sykes (1965), Husebye et al (1975), Bungum & Husebye (1977)*, Einarsson (1979)*, for the Arctic and North Atlantic; Stover (1966), Forsyth (1975)*, for the South Atlantic; Stover (1968), Sykes (1970a)*, for the Indian Ocean; Sykes & Landisman (1964), McKenzie, Davies & Molnar (1970)*, Fairhead & Girdler (1970), for the Red Sea and Gulf of Aden; Tobin & Sykes (1968), Forsyth (1972)*, Anderson et al (1974)*, Stover (1973) for the Pacific Ocean. These papers, many of which include focal mechanism studies (marked *), provide substantial evidence towards our understanding of plate tectonics.

Little is said about the detailed processes occurring along the plate boundaries, that is at the ridge crests and fracture zones. The value of such studies in this context is ultimately dependant on the accuracy of the hypocentre determinations themselves. For shallow hypocentres located exclusively using distant stations the problems of epicentre and depth are separate.

Several of the studies mentioned above claim epicentral accuracies of 10 km , but this is optimistic for the majority of locations. Husebye et al (1975), in a study on the accuracy of epicentres in the Arctic region come to the conclusion that a figure of 20 km should be put on the lower limit of features resolvable by studying the better recorded (observed by greater than 10 stations) earthquakes. In another study on the same region Bungum & Husebye (1977) indicate that the accuracy of the most well recorded events may approach 10 km but that events recorded by less than 15 stations should be disregarded in this context. This applies to the Arctic region, unfortunately for the South Atlantic, Indian Ocean and much of the Pacific Ocean the seismic station distribution is poor, and increased scatter is visible even on world maps (Barazangi & Dorman, 1969). Taking 30 km as a distance typical of that between the crestal mountains which flank the rift, it seems that most epicentres at present located using distant stations would fall within this zone but are of little use in resolving smaller features. Furthermore errors in travel time curve used for the locations lead to systematic mislocation of even the well recorded events by up to 15 km, depending on the region (Lilwall & Underwood, 1970). Commentary on apparent bias of epicentres to one side or other of the median valley (Johnson & Vogt, 1973; McGregor & Rona, 1975) must therefore be viewed with caution.

Depth determinations based purely on the first arrival P phase times from distant stations are poorly controlled and systematic errors of up to 200 km are possible (Lilwall & Underwood, 1970). Use of the surface reflections pP and sP can resolve this, but for shallow events (less than about 50 km) they are difficult to identify routinely and hence are not reported or are misread. The general lack of depth phases in bulletins for mid-oceanic ridge events is perhaps diagnostic of shallow focus and the large range of depths published by the International Seismological Centre (see Table 1) do not represent the real distribution but are largely the result of the problems outlined above. Reliable depth determinations of large mid-ocean ridge events are very rare but all give shallow sources. Weidner and Aki (1973) have studied the Rayleigh wave amplitude and phase spectra of two fracture zone (transform) and two ridge crest ("normal" dip slip) earthquakes, all from the Mid-Atlantic Ridge. Correction for transmission effects is important and makes the technique a lengthy process. All these events were found to be less than 10 kms depth below the sea bed, the ridge crest events being shallower (note that all depths quoted in this paper are below sea bed). Duschenes and Solomon (1977) have confirmed depths of less than 5 km for the ridge crest events by identifying the seabed and surface reflected phases (denoted pP and pwP) in the P coda. They point out however that it is not always possible to identify such phases unequivocally in the presence of reflections off horizontal discontinuities other than sea bed or surface. Pearce (1980) however has demonstrated a shallow depth (5.4 km) for a dip slip earthquake in the Gulf of Aden by computing synthetic seismograms which show reflections off likely interfaces as well as those off seabed and sea surface. The agreement between observed and synthetic shown in Fig (2) is a convincing confirmation of the shallow origin of this source.

It is unlikely that a significant improvement will be made on routine epicentre determinations but very good depth determinations are possible with due effort for individual cases. Such effort would seem worthwhile because although ocean-bottom instruments will give increasing numbers of precise depth determinations in the future, the events located from short term surveys may not be typical of the larger ones seen by the world network.

Sonobuoy and Ocean Bottom Observations

The presence of seismic stations at short ranges has the advantages of a low magnitude detection threshold and the potential to obtain accurate hypocentres. Until recently observations from local stations situated on or near the ridge crest have been restricted to regions which are atypical, Iceland being notable. The early 1970's saw increasing numbers of deployments of sea surface hydrophones (sonobuoys) and ocean bottom seismographs (OBS) along the mid-ocean ridges with the object of observing seismicity. The locations of these studies are shown in Fig (1) which indicates a sparse and uneven coverage at present along the world's ridge system.

Publications to date in this field have variable seismological content in terms of their contribution towards our understanding of ridge-crest seismicity. This arises primarily from the difficulty in obtaining data from arrays of sufficient number of stations to enable location of both epicentre and depth. A knowledge of the location capability of such arrays is important when assessing the value of their results.

Sonobuoys have advantages over ocean bottom instruments of relative cheapness and ease of deployment. They have the limitations of short operation time, inability to directly detect S waves and the problems of drift and correction for seabed topography. Johnson and Jones (1978) have investigated the location capability of sonobuoy arrays. They found that events at distances from the centre of more than about $1\frac{1}{2}$ times the array radius (10-20 km) within the array were often unlocatable and even the favoured events within the array could lie in a circle 2 km radius. These conclusions applied even without the possible adverse effect of incorrect seismic structure or incorrect choice of focal depth. Lilwall and Francis (1978) have studied the location ability of small OBS arrays of 3 or 4 instruments with 10 km spacing. Given an event with well read P and S arrivals it can be located to better than 2 km in both epicentre and depth within a region of about 30 km radius. Francis et al (1978) investigated the effect of errors in the seismic structure assumed

TABLE 1. Depth of Focus Determinations of Oceanic Ridge Events

Source/Reference	Data/Method	Location	Depth km
ISC Bulletins 1964-1977	Mostly Teleseismic P times	All oceanic rifts and Fracture Zones	0-298
Weidner & Aki (1973)	Rayleigh Wave amplitude and phase spectra	Mid Atlantic Ridge 2 ridge crest events 2 fracture zone events	 3 ± 2 6 ± 3
Duschenes & Solomon (1977)	Identification of pP,sP	Mid Atlantic Ridge 2 ridge crest events	4 ± 2 2 ± 1
Pearce (1980)	Modelling short period waveform	Gulf of Aden "normal" dip slip	5.4 ± 1.5
Thatcher & Brune (1971)	Pg-Pn times	Gulf of California	7
Reichle & Read (1977)	Sonobuoy array	Gulf of California	0-2
Klein et al 1977	Local network	Reykjanes Peninsula Swarm	0-7 Peak 2-3
Ward & Bjornsson (1971)	Local networks	Iceland Geothermal areas	2-6
Einarsson & Brandsdottir (1979) Brandsdottir & Einarsson (1978)	Local network	Krafla NE Iceland swarm	0-6 Peak 2.5-4
Jones & Johnson (1978)	Sonobuoy array	Gorda Rise (3 events) Gorda Rise-Blanco FZ intersection (2 events)	2.5-6.5 6.5-10.0
Prothero et al (1976)	OBS array	Rivera fracture zone	<10
Francis et al (1978)	OBS array	Eastern end of St Paul's Fracture Zone	0-11 Peaks at 0 and 7
Lilwall et al (1978)	OBS array	Mid Atlantic Ridge at 45°N	0-11 Peak 8

for locating events. The likely errors were typical of the differences found in refraction surveys. Errors in hypocentre are small ($\angle 0.5$ km) for events near the stations but systematic errors greater than 2 km arose for events well outside the array. In all locations where depth of focus cannot be determined the choice of an incorrectly assumed depth may lead to systematic epicentre errors, especially for events outside the array. In spite of these problems local networks can give locations up to an order of magnitude more accurate than possible using only distant stations, for a region a few tens of kilometres across.

An encouraging result of most Sonobuoy and OBS surveys on oceanic rifts so far is that large numbers of microearthquakes are usually recorded during relatively short time intervals. Microearthquake activity appears relatively high and continuous along the ridges, but surveys of longer duration are needed to establish the continuity.

Successful sonobuoy surveys involving three or more instruments have been reported on the Blanco fracture zone, Gorda Rise, Gulf of California and the Mid-Atlantic Ridge near 37°N.

Johnson & Jones (1978) describe the results of two short deployments (12 hour and 14 hour) of tripartite sonobuoy arrays on the Blanco fracture zone. Epicentres indicate a close association with the bathymetric expression of the fracture zone, unlike locations relying on distant observations which appear systematically mislocated by about 30 km. Although the events are all within 5 km of the major bathymetric features, detailed interpretation is not possible

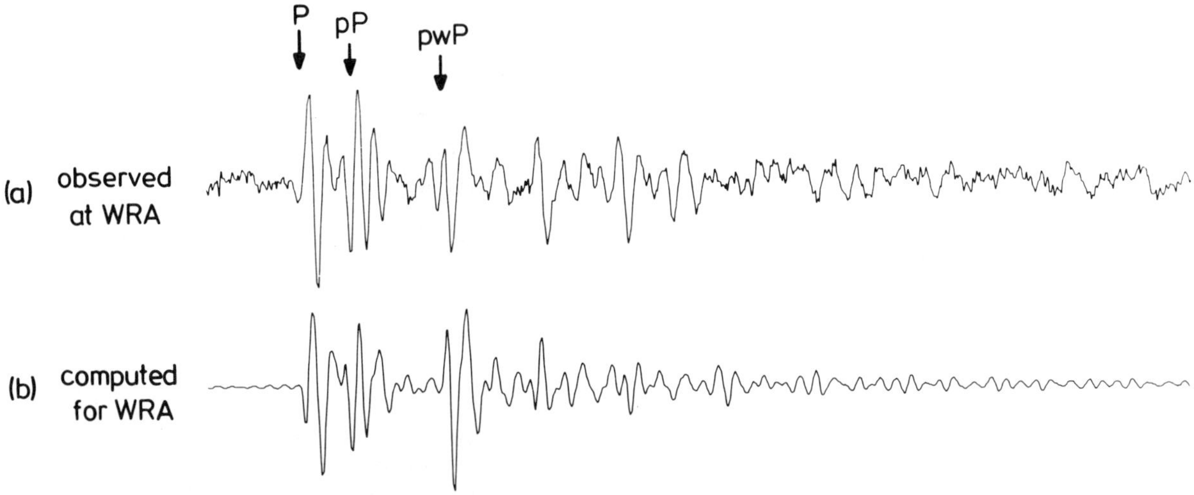

Fig 2. Observed and synthetic seismograms, after Pearce (1980), for the earthquake of April 15 1971 in the Eastern Gulf of Aden recorded by the station WRA in Australia. The agreement between the sea-bed and sea surface reflected phases pP and pwP demonstrate the shallow focus (5.4 ± 1.5 km) of this event. Time separation between P and Pwp is 7 seconds.

because of the depth assumption (here 0.1 km) necessary with three instruments giving P readings only.

On the Gorda rise, a slow spreading ridge with marked rifting (Atwater & Mudie, 1968), quadripartite sonobuoy deployments (Jones & Johnson, 1978) show activity within the rift and not systematically displaced as indicated by epicentres globally determined. Hypocentres could only be determined for 5 events which were all shallow (see Table 1) with those at the intersection of the ridge and Blanco fracture zone deepest.

In the Gulf of California, transform faults along its length are offset by short spreading axes. They are relatively fast spreading (half rate 3 cm/year) and at the northern end are seismically active with frequent swarms. Sonobuoy observations (Reid et al, 1973; Reichle & Reid, 1977) show most activity at the spreading centres and within the associated graben type basins. Depths of the swarm events are put at 2 km but no confidence bounds given. In contrast Thatcher and Brune (1971) put larger events seen in the area at 7 km depth. The high activity seen at the northern end of the Gulf compared with that at the southern end, and on East Pacific Rise, may be the result of high sedimentation rates in the north (Reichle & Reid, 1977).

On the Mid-Atlantic Ridge near latitude 37°N two sonobuoy microearthquake surveys were made as part of project FAMOUS (Reid & MacDonald, 1973; Spindel et al, 1974). Epicentre locations were based on a simple half space (V_p=5 km/sec) velocity model beneath the sea bed and an assumed depth of focus at the sea bed. No details as to how sea bed topography corrections, if any, were made and it is likely that the quoted epicentre accuracies (both better than 2 km) in terms of absolute locations are optimistic. The study area consists of a section of rift terminated at each end by short fracture zones. Activity along both fracture zones appears restricted to a narrow zone less than 5 km wide, terminating at the intersections (Reid & MacDonald, 1973). Epicentres along the rift section occupy a similar narrow zone but along the eastern wall (Spindel et al 1974). Incorrect assumed depth, over-simplified velocity structure together with the small size of the array (1.5-2.0 km) make systematic errors likely. Locations using two OBS deployed in the same area have put activity along the western wall (Francis, Porter & McGrath, 1977) but in this study there is an additional problem of epicentre ambiguity.

Successful OBS deployments of three or more instruments have been reported from the East Pacific Rise and Rivera fracture zone, eastern end of St Paul's fracture zone, and the Mid-Atlantic Ridge near latitude 45°N.

Prothero et al (1976) have described tripartite OBS array deployments on the Rivera fracture zone and East Pacific Rise. In the vicinity of the OBS array, the fracture zone consists of a bathymetric trough flanked by ridges, within which observed epicentres were located. Two small offsets in the otherwise parallel trend suggest small spreading centres as in the Gulf of California. Reliable depth locations are all less than 10 km on the fracture zone. Near the junction with the East Pacific Rise, depths are less than 5 km but are poorly constrained. No activity was detected on the extension of the fracture zone beyond

TABLE 2. b Values Determined from Microearthquake Surveys: S, OBH, OBS denotes Instrument Type as Sonobuoy, Ocean Bottom Hydrophone and Seismometer Respectively

Source/Reference	Location	b Value
Ward et al 1969	Krafla region Iceland	0.83
Brandsdottir & Einarsson 1978	Krafla NE Iceland	0.6 - 3.0
Klein et al (1977)	Reykjanes Peninsula	0.87
Thatcher & Brune (1971)	Gulf California	0.9
Reid et al (1973)	Gulf California	1.51(S)
MacDonald & Mudie (1973)	Galapagos Rise	3.0(S)
Johnson & Jones (1978)	Blanco fracture zone	1.5(S)
Solomon et al (1977)	Mid Indian plate triple Junction	1.0 (OBH)
Francis & Porter (1973)	Mid Atlantic Ridge 45°N	0.83 (OBS)
Francis et al (1977)	Mid Atlantic Ridge 37°N	1.44 (OBS) 1.53 (OBH)
Lilwall et al (1977)	Mid Atlantic Ridge 45°N	1.04 (OBS)
Francis et al (1978)	Eastern end St Paul's Fracture Zone	0.59 (OBS)

its junction with the East Pacific Rise, or on the Rise itself.

Epicentres located at the eastern end of St Paul's Fracture Zone using a quadripartite OBS array (Francis, Porter & Lilwall, 1978) occupy a triangular zone filling the corner between the spreading axis of the Mid-Atlantic Ridge and the transform part of the fracture zone Fig (3). Events are seen at ranges from the array centre of up to 58 km along the fracture zone but only 20 km along the ridge. Epicentres within the triangle have an uneven distribution and several trends may exist, but the clearest is parallel to the median valley of the Mid-Atlantic ridge and along the central axis. Focal depth distributions of well located events are shown in Fig (4 bottom). The peak at less than 1 km depth consists of very small events. The peak at 7 km are larger events situated on or near the ridge axis. The residual activity between 0-5 km depth is made up mostly of off axis events.

The Mid-Atlantic Ridge at latitude 45°N has been the site of two OBS surveys. The ridge crest here has been well surveyed and shows a well developed median rift and is free from major fracture zones. In a preliminary OBS survey Francis and Porter (1973) demonstrated that activity was concentrated within the median valley. A later survey consisted of a tripartite array which located many epicentres within a narrow linear zone only 5 km wide (Lilwall, Francis and Porter, 1977). These epicentres, which are for an assumed depth of 4 km , are shown in Fig (5). The slight bias to the east of the axis illustrates the effect of an incorrect assumed depth. Use of only well controlled solutions gives depths peaking at 8 km and epicentres centrally placed beneath the median valley floor (Lilwall, Francis & Porter, 1978).

Observations in Iceland

Icelandic seismicity has been the subject of much research and its possible relevance to our understanding of the mid-ocean ridges is often stated, eg Trygvason (1973). Although it is atypical of the Mid-Atlantic Ridge as a whole the absence of the sea makes observations, which may give clues towards the nature of seismicity elsewhere, relatively easy.

Microearthquake surveys in Iceland (Ward, Palmason & Drake, 1969; Ward & Bjornsson, 1971) show isolated areas within the diffuse volcanic zone which have steady activity. These areas are typically 5 km radius and near geothermal fields which exhibit ground fissure systems. In

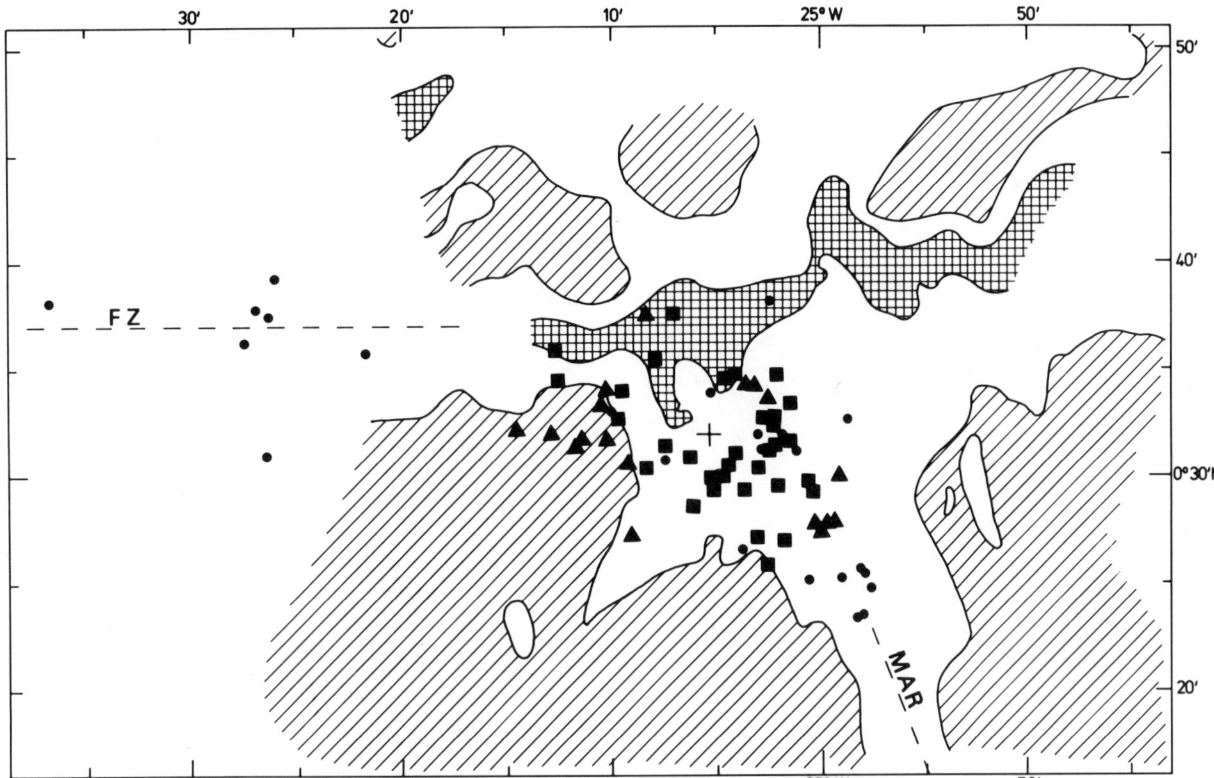

Fig 3. Microearthquakes located at the junction of the Eastern end of St Paul's fracture zone (FZ) and the Mid-Atlantic ridge (MAR), after Francis et al 1978. Circles mark events with poor depth determinations. Squares mark events with depth greater than 5 km. Shallow events are denoted by triangles. Sea depth shallower than 1900 fm hatched diagonally, greater than 2300 fm cross hatched.

addition to this steady activity swarms periodically occur, several of which have been well monitored recently. One at Borgarfjordur, to the west of the neo-volcanic zone, and therefore to some extent intra-plate is described by Einarsson, Klein & Bjornsson (1977). Most events were shallower than 7 km and none below 10 km.

In North Iceland, what is described as a major rifting episode started in 1975 involving repeated inflation followed by deflation of the Krafla volcano (Bjornsson et al, 1977). The deflation is associated with magma movement into a rifted zone to the north and south of the Krafla caldera and is accompanied by intensive earthquake swarms. Altogether an 80 km section of the plate boundary is involved (Bjornsson et al, 1977), but the width of the earthquake zone is less than 5 km and marked at the surface by a faulted belt. Focal depths are mostly less than 6 km with most between 2.5-4.0 km (Einarsson & Brandsdottir, 1980, Brandsdottir & Einarsson, 1979). Considerable widening and relative subsidence of the fault zone has occurred during this activity. From this and similar historical accounts it appears that rifting is an episodic rather than continuous process in Iceland (Bjornsson et al, 1977).

Another earthquake swarm that was well observed instrumentally occurred on the Reykjanes peninsula during September 1972 (Klein et al, 1977). A large number of hypocentres were located to an accuracy of 0.5 km. The epicentres are shown in Fig (6) and again fall within a very narrow zone. Fig (4 top) shows that the depth distribution peaks between 2-3 km .

Discussion

Harrison & Stieljes (1977) describe faulting models which have been hypothesised to explain the morphology of the median valley. Three models have been suggested which can maintain the median valley as a steady state feature. Model 1 involves normal faulting on the inner faults of the valley progressively changing to reverse faulting along the same planes on the outer, thereby cancelling out the throws. Model 2 again involves normal faulting on the inner faults with more normal faults further out but with fault planes dipping away from the median

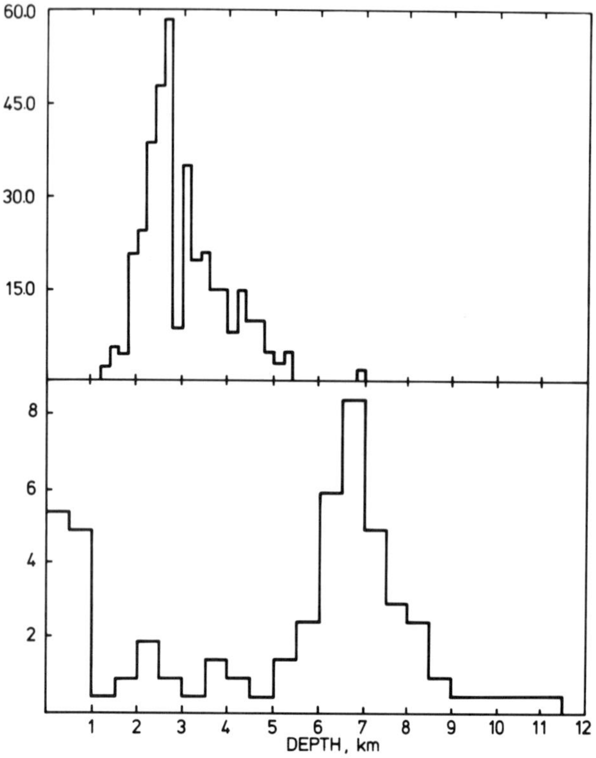

Fig 4. Depth distributions found for two microearthquake surveys. Top: the September 1972 Reykjanes peninsula swarm (after Klein et al 1977). Bottom: for the Eastern end of St. Paul's fracture zone (after Francis et al 1978).

valley. In model 3 normal faulting occurs in the central part only with the crust undergoing a rotation further out.

In all three models seismicity is restricted to within the median valley. The narrowness of the seismic zone indicated by epicentres of the larger events confirms this to be true, but further detail cannot be resolved. Microearthquake observations both at sea and for swarms in Iceland show the active zone to be less than 5 km across and probably centrally placed. If so movement is on the innermost faults, favouring model 3. It can be argued however that microearthquake activity observed on a few relatively short surveys is unrepresentative of the larger events observed world wide. This is supported by the slightly greater depth of the microearthquakes compared with the few depths so far determined for larger events.

Earthquakes on the ridge crest so far located occur at less than 10 km depth. Fracture zone events may be slightly deeper than ridge crest events. Microearthquakes show a peak of activity between 5 and 10 km on slow spreading ridges (Mid-Atlantic Ridge and possibly the Gorda rise). Activity may peak at shallower depths in Iceland and beneath fast spreading centres in the Gulf of California. These summary statements are based on few observations with limited regional sampling and so interpretation can only be tentative (see Table 1).

An explanation of the limited depth of focus is that it reflects the depth down to which brittle rock extends, ie is a measure of lithospheric thickness. On slow spreading ridges it is relatively deep (7-8 km), on fast spreading ridges and Iceland shallower, presumably because of higher heat flow. Observations on the maximum size of earthquakes observed (discussed later) corroborate this. An explanation of the peak in the microearthquake distribution has been suggested by Francis et al (1978). Activity is restricted to a zone of competent rock, sandwiched between ductile rock below and highly cracked (therefore unable to support stress) rock above. Above and below this zone the rock deforms aseismically. The limited distance within the median valley over which microearthquakes are seen (Francis & Porter, 1973, Francis et al, 1978) and the waveforms of P arrivals from these events (Lilwall, 1980) both confirm the existence of the lower zone. Some shallow off-axis events described by Francis et al, (1978) are explained by the progressive increase in the strength of the rocks with age as a result of mineralisation caused by hydrothermal circulation in the highly-fractured crust. Larger earthquakes, if they are slightly off axis, can involve a greater thickness of competent rock at shallower depth. A depth of 7-8 km to the ductile zone even on slow spreading ridges however is inconsistent with heat flow models (Sleep, 1975, Kuznir & Bott, 1976) but this could be explained if hydrothermal circulation penetrates as deep as 6 km as suggested by de Wit and Stern (1976).

Focal Mechanism Studies

Our present understanding of the mechanisms of mid-oceanic ridge earthquakes is essentially the same as described by Sykes (1967). Fracture zone earthquakes are strike-slip with the sense of motion consistent with the transform hypothesis of Wilson (1965). Ridge crest events, including those on continental extensions of the ridge, have predominantly normal faulting, an observation which fits well with the morphology of the rifted median valley. Papers on mid-oceanic ridge seismicity already mentioned, as well as papers by; Anderson et al (1974), Banghar & Sykes (1969), Bungum (1977), Sykes (1970a, 1970b), Thatcher (1972) all confirm the original observations. A survey by Sykes and Sbar (1973) shows a predominance of normal faulting even for intra plate earthquakes in crust away from the ridge crest up to 20 million years old. There is little evidence of reversed faulting as expected in model 1 described by Harrison & Stieltjes (1977).

Occasional "anomalous" mechanisms are seen however (Einarsson, 1979), and many mechanism

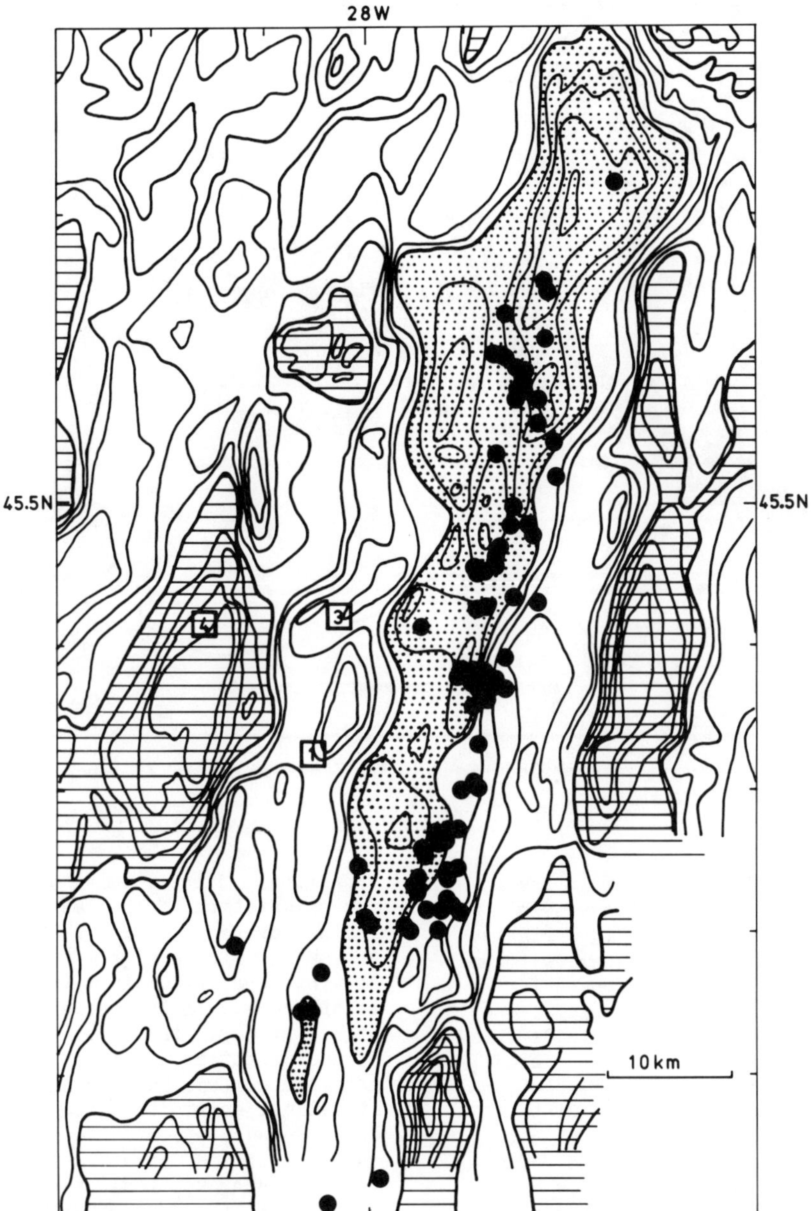

Fig 5. Microearthquakes located on the Mid-Atlantic ridge near 45°N (after Lilwall et al 1977). Numbered squares show OBS locations. Sea depths less than 1000 fm hatched greater than 1500 fm dotted, contour interval 100 fm.

solutions exhibit non-orthogonality of the nodal planes as in Fig (7), an observation first noted by Sykes (1967). A survey of the ways that those anomalous mechanisms are thought to arise is given by Einarsson (1979). Briefly, they may be the result of deviations from the double couple source such as a superimposed explosive component or an extension failure mechanism (Robson et al, 1968). Alternatively they may be an artefact of the projection used in mapping the observations on the focal sphere. Solomon & Julian (1974) assume the latter mechanism and have devised a velocity distribution beneath the events which changes the raypaths, giving a corrected projection. Refraction surveys however do not substantiate the postulated velocity structure (Steinmetz, Whitmarsh & Moreira, 1977) but admittedly they do not sample the structure directly beneath the axis.

The reviewer has some misgivings about

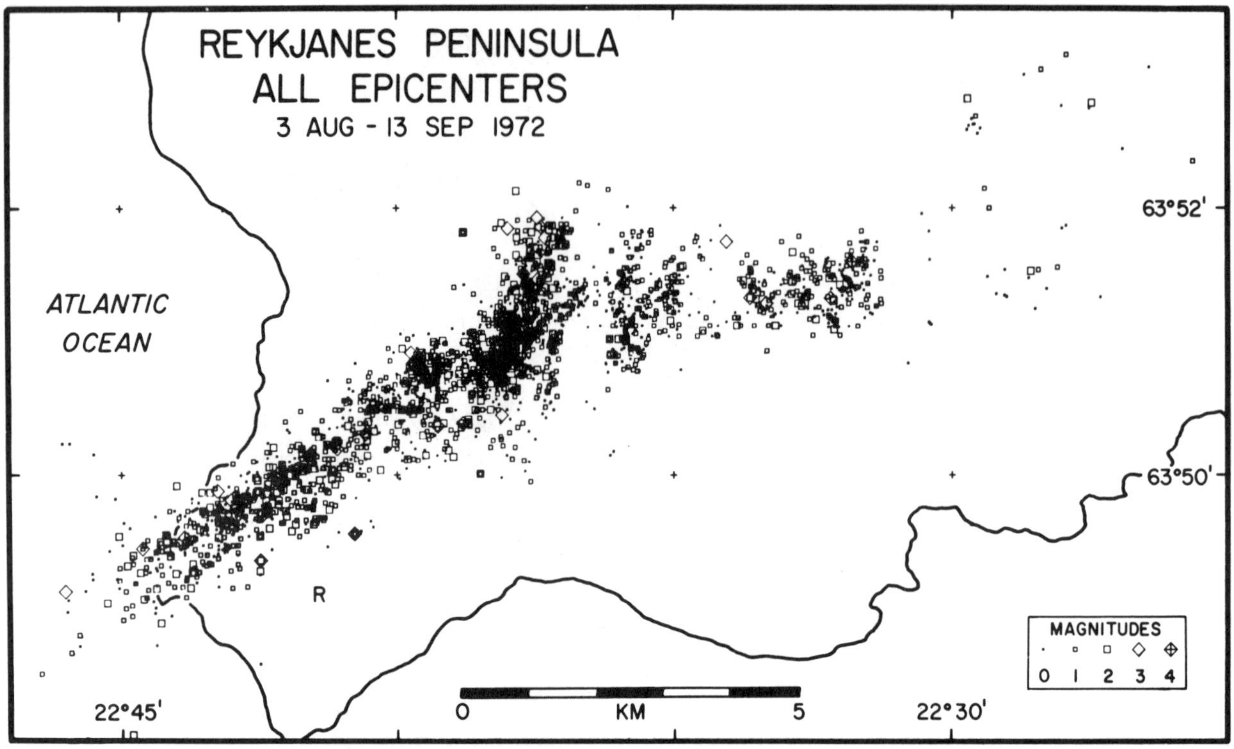

Fig 6. Reykjanes peninsula swarm locations (from Klein et al 1977). Depth distribution of these events is in Fig. 4 (top).

detailed interpretation of anomalies in the fault plane mechanism solutions. Many solutions are often poorly constrained (see Fig 7 right). Ritsema (1967) has shown that for shallow focus events only about 5% of the focal sphere is covered by observations in the "seismic window" ranges between 30 and 90 degrees. Outside this range the seismogram onset may be complicated by multi-pathing. Nevertheless a few solutions are sufficiently constrained to demonstrate nonorthogonality. It is likely however that this is the result of interference from reflections off interfaces above these shallow events. These reflections can interfere with the direct wave, especially at long periods, with the result that first motions do not have the accepted quadrant distribution. Warnings about this problem can be found in the literature (Keylis-Borok, 1957, Burridge, Lapwood and Knopoff, 1964), and recently several workers have demonstrated convincingly that this results in nonorthoganality (Hart, 1978; Nabelek & Trehu, 1980; Trehu et al, 1980).

Local station networks rarely give sufficient constraint on focal mechanism solutions. An exception is the study of the September 1972 Reykjanes peninsular swarm described by Klein et al (1977). Normal, "oblique", and strike slip solutions were found, as might be expected at the junction of a spreading axis and suspected transform. All mechanism solutions have a common orientation of the least compressive stress in the horizontal and NW. Near the centre of the swarm, strike slip events occur below events showing normal faulting, an effect interpreted as the result of the increased lithostatic stress at depth thus changing the maximum stress direction. Sonobuoy or OBS observations have not yet yielded first motion solutions for individual events but Jones & Johnson (1978) have produced a composite solution for three events from the Gorda Rise. Their chosen fault plane is near vertical, reversed and strikes parallel to the median valley. Such a mechanism is evidence of "unfaulting" (model 1) but the accuracy of the solution depends ultimately on the hypocentre location, which may be incorrect in a sonobuoy survey. Waveforms of swarm microearthquakes recorded by OBS on the Mid-Atlantic Ridge near 45°N have been interpreted by Lilwall (1980) in terms of normal faulting but with small dips. If these events, which are at 8 km depth, are on continuations of the fault scarps seen on the sea bed within the median valley, then the fault surfaces must be curved. Curved fault surfaces could be another cause of anomalous first motion distributions. But the possibility remains that the microearthquakes are not directly linked with the larger fault movements.

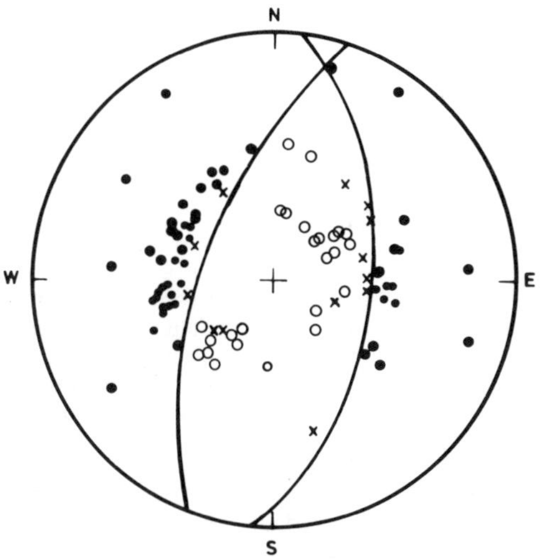

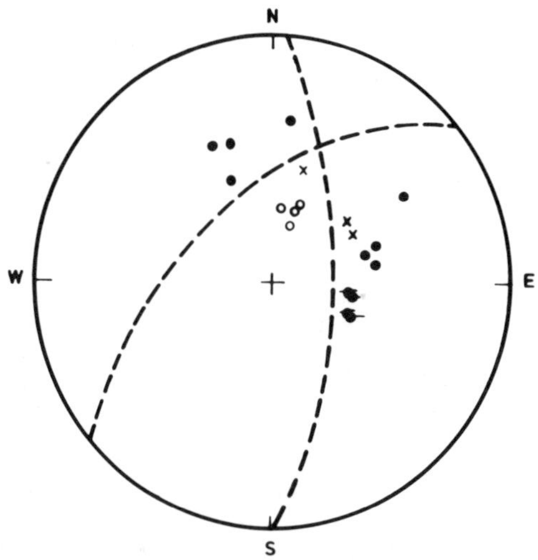

Fig 7. Two focal mechanisms of ridge crest earthquakes with a strong dip slip "normal" component, both of which are said to exhibit non-orthagonality of the nodal planes; left: Mid-Atlantic Ridge event of 20 September 1969 (after Solomon & Julian 1974); right: Gulf of California event 20 March 1969 (after Thatcher & Brune 1971). Dots indicate compressions, circles dilations, crosses "nodal character".

Temporal Behaviour - Swarming

Earthquakes often cluster both in space and time. Mogi (1967) has classified earthquake sequences into three types: type 1 are main-shock-aftershock sequences; type 2 has the addition of foreshocks but still with an obvious main event; type 3 are termed swarms which have no obvious mainshock. This classification of earthquake sequences is probably an oversimplification (see Gibowicz, 1973) but adequate for the present description.

Earthquake sequences, clustering closely in both space and time are often seen on long period records from stations bordering the Atlantic Ocean. In a survey of these sequences, mainly from the Mid-Atlantic Ridge, Sykes (1970c) found that the great majority can be classified as type 3. Such swarm activity is confined to the ridge crests, rarely from fracture zones. It is wrong however to assume that swarms are associated exclusively with "normal" faulting, Tathem & Savino (1974) have described swarms in the Gulf of California and the Queen Charlotte Islands which are dominantly strike slip. Nevertheless both examples are in close proximity to the junction of the fracture zone with a ridge crest.

Francis and Porter (1971) have made a statistical study of Mid-Atlantic Ridge earthquakes reported by the world network. In general 25% of events reported belong to clusters rising to 40% in regions where the world detection capability is best. Clusters are short lived, 90% of events occur within 10 hours of each other. They also come from small regions, 50% occur within 20 km of each other. With an assumed relative location accuracy of 10 km for the larger events nearly all the apparent size of such clusters can be attributed to location error (Francis & Porter, 1971). A similar observation has been made for a swarm of over 70 events with magnitude (M_b) greater than 4.0 from the Gulf of California (Thatcher & Brune, 1971). The world network produced epicentres over a region 50 km across, local observations put them in a region one tenth this size. Most activity was over within six hours.

Earthquake swarms have been observed by sonobuoys in association with the small spreading centres in the Gulf of California (Reichle & Reid (1977)). On the Mid-Atlantic Ridge crest OBS surveys have observed clusters of a few events to several hundred, with locations coming from a zone as little as 2 km across (Lilwall, Francis & Porter, 1978), and many showing identical records but differing widely in amplitude (Francis & Porter, 1973; Francis, Porter & McGrath, 1977). The latter events must have the same mechanism within a region only a few hundred metres in extent and are perhaps repeat movements of one fault.

Swarming is clearly a feature of ridge crest seismicity and suggests a link with volcanism. Sykes (1970c) has reviewed much of the literature on swarms and concludes that swarms frequently occur in volcanic regions, although many apparently do not. Mogi (1967) has examined the conditions favourable to swarming when microfractures are generated during the deformation of rock specimens in the laboratory. Elastic shocks accompanying microfracturing behave like earthquake sequences of type 1 in homogeneous materials and/or uniformly applied stress whilst type 3 swarm like behaviour occurs in heterogeneous materials and concentrated stress sources. These results strengthen the possible link with volcanic areas which are usually heterogeneous with magma a potential concentrated stress source.

The causal connection between swarming and volcanism remains obscure however because most volcanism, in the broad sense of magma movement cannot be confirmed by direct visual observation, especially at the ridge crest. Observations of swarm sequences in Iceland however give strong evidence of magmatism. Migration of activity with the September 1972 Reykjanes peninsula swarm (Klein et al, 1977) suggests fault creep or fluid flow. There is no direct proof that magma was involved but gaps in the epicentre distribution shown in Fig (6) may delineate magma bodies. Observations of the activity associated with the Krafla volcano in NE Iceland show a direct connection of both rifting and magma movement with earthquake swarms. As already described the swarm activity occurs during deflation of the central volcano. Brandsdottir & Einarsson (1978) show that the activity propagates away from its source (at 0.5 ms^{-1}) and in one instance glowing pumice was ejected at the surface when the swarm reached a borehole. Even with such clear evidence the causal connection between the rifting and earthquake activity with the magmatism still remains uncertain.

A qualitative model explaining the connection between magmatism and swarming has been devised by Hill (1977). In this model extension of dykes to accommodate magma is taken up by fault movements between adjacent dyke tips. The type of faulting (strike slip, normal, or reversed) and its orientation can be shown to depend on the principal stresses acting on the region. On spreading centres the least principle stress is the tension acting normal to the ridge, the greatest, at least at depth, will be lithostatic loading. According to Hill this leads to normal faulting striking parallel to the ridge, as observed.

Another hypothesis intended to explain the seismicity and morphology observed in the median valley has been developed by Francis (1974). This is based on observations of swarm activity during the collapse in 1968 of the Fernandina caldera in the Galapagos islands (Simkin & Howard, 1970; Filson, Simkin & Lei-Kuang Leu, 1973). According to Francis a similar collapse of the median valley floor (model 3 of Harrison & Stieljes) assumed to be above a series of very elongated magma chambers, would account for the seismic activity and rifting. Whitmarsh & Laughton (1975) have interpreted curvature seen on side scan sonar, of median valley scarps as they approach, fracture zones as the result of the rounded end of a magma chamber. The analogy with caldera collapse as seen at Fernandina requires events of low stress drop (Francis 1974) and a relatively shallow magma chamber about 2 km below the surface. Direct evidence for both of these, as we will see in this and the subsequent section, is at present not available.

Both the above models associate median valley seismicity, especially swarm activity, with magma movement. In a sense they represent opposite extremes of magma injection forming new crust or magma withdrawal resulting in collapse. Both could be separate episodic parts of a large process. The presence of some form of magma reservoir beneath the axis seems to be a common factor. Unfortunately on slow spreading ridges, which in general have the most well defined rifting, the evidence on theoretical observational grounds is lacking. Heat flow models (Sleep, 1975, Kuznir & Bott, 1976) indicate a permanent magma chamber probable on fast spreading ridges but possibly not on slow spreading ones. Seismic refraction (Orcott, Kennet & Dorman, 1975) and reflection (Herron et al, 1978) experiments have both provided proof for a magma chamber at shallow depths (2 km) beneath the fast spreading East Pacific Rise. On the Mid-Atlantic Ridge, seismic refraction experiments have yet to reveal any sizeable molten zone in the top 7 km (Nisbet & Fowler, 1978). If however such chambers are transient they may have been missed with the sparse sampling of refraction surveys so far undertaken. Microearthquake observations on the Mid-Atlantic Ridge (Lilwall, Francis & Porter, 1978) show little attenuation of S waves which propagate steeply upwards at the ridge from sources near 8 km depth, further evidence against a high level molten zone. But analysis of the seismogram waveforms of the same sources (Lilwall 1980), the depth distribution of these microearthquakes, and their limited range of seismic observation all make the existence of a deeper (depth 7-8 km) molten or partially molten zone likely. Detection of such a zone beneath the median valley structure and topography, will be very difficult using seismic refraction or reflection methods. Absence of a permanent shallow magma chamber but a partially molten zone at depth is incorporated in a model proposed by Nisbet & Fowler (1978) to account for the formation of oceanic crust along slow spreading ridges. In this model rising melt is trapped beneath the rigid crust but some may rise to form dykes which occasionally might lead to the development of a temporary high level chamber.

Other Observations

Magnitudes and Seismic Moments

Some generalisations can be made about the maximum sizes of mid-ocean ridge earthquakes. Maximum magnitudes on transforms are larger than those on adjacent ridge crests (Francis 1968a). Maximum magnitudes and/or seismicity is generally greater on slow spreading than fast spreading ridges (Isacks et al, 1968, MacDonald & Mudie,

1974). An explanation of this is that maximum size is limited by the thickness of the brittle zone, which because of differences in heat flow may be greater on slow spreading than fast spreading ridges, and greatest on transforms.

Seismic moment (denoted Mo) is a better measurement of earthquake size than magnitude as it can be related to the physically realisable quantities of average fault displacement ($\bar{D}$), fault area (S) and rigidity of the rocks (u) by the relationship:

$$Mo = u\bar{D}S \qquad (1)$$

In rare cases Mo can be measured directly using the above equation but more frequently it can be estimated from magnitude provided the latter is measured at periods much greater than the duration of the source.

Burr and Solomon (1978) have studied seismic moment data for oceanic transform events. Fig (8) shows convincingly that maximum moments of oceanic transform events decrease with increasing spreading rate. From this and other source parameter data on transform events Burr and Solomon conclude that the data supports the hypothesis that earthquake size in each environment is limited by the depth to an isotherm which marks the transition between seismic and aseismic slip. Incorporation of ridge crest earthquakes into this model is more difficult and data comparable with Fig (8) are at present too few to be convincing. Nevertheless Solomon & Burr (1979), assuming the thermal model of Sleep (1975), show that the available data is consistent with the limiting isotherm hypothesis provided the events are located 10-20 km off axis. This implies activity restricted to faults on the median valley walls but as already discussed the locations of larger events cannot resolve this in detail. Reliable microearthquake locations appear to contradict the model by placing activity axially and at depth, but as mentioned previously they may represent the same process as the large scale activity.

<u>b values</u>

It is well known that magnitudes of earthquakes follow a distribution of size of the form:

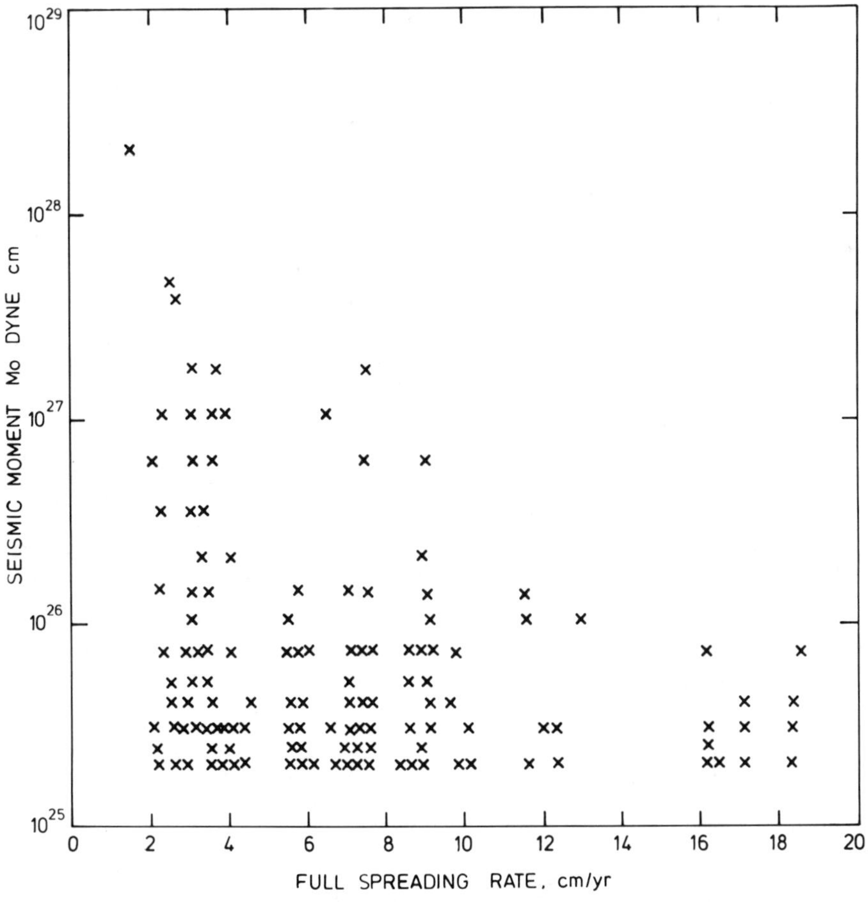

Fig 8. Seismic moments of oceanic transform earthquakes plotted against spreading rates of adjacent ridge crests (after Burr & Solomon 1978).

$$\mathrm{Log}(N) = a\text{-}bM \qquad (2)$$

with a and b constant for a set of earthquakes, and N the number of earthquakes with magnitude M or alternatively the number with magnitude greater than M. Either definition does not alter the value of b, which although typically about 1.0, has been found to vary to some degree in various regions of the world. In particular Francis (1968a,b) has shown that it differs between rift and transform events observed by the world network. For rift and transform events respectively the b values are 1.72 and 0.99 based on short period body wave magnitudes (m_b), and 1.3 and 0.65 based on the long period surface wave magnitude (M_s). The difference found between magnitude scales is evidently a possible cause of variation in b. The origin of the difference here is well known and arises because the m_b scale progressively underestimates the true size of larger events as a result of the fall off of seismic energy radiated at short periods (circa 1s). This roll-off does not occur except for the largest events at the 20 second period used on the surface wave M_s scale, and over much of its range this magnitude can be scaled linearly to seismic moment Mo.

Although the world data reveal an indisputable difference between transform and ridge crest, b values based on locally recorded events are confused. There seems little doubt that a major cause of this are differences in the instruments employed and magnitude scales used. Table 2 gives some recent b values from mid ocean ridge surveys. Relatively high values found using sonobuoys may result partly from the enhanced sensitivity of these instruments at high frequencies. The recording and playback system used by MacDonald and Mudie (1974) had a peak sensitivity at 40 Hertz. Most of the remainder are for ridge crests with b values over too great a range to be interpreted. A relatively low value for the eastern end of St Paul's fracture zone agrees with the findings for large events.

Interpretation of b values has been influenced by laboratory studies on microfracturing already mentioned in the context of swarming, (Mogi,1967). High b values (and swarm type 3 sequences) are found to be associated with heterogeneity and concentrated stress whilst low b values (and type 1 sequences) are associated with uniformity in structure and stress. Scholz (1968) found that stress level was the overriding factor, higher b values occurring at lower stress. Deviations from the simple linear law were found with regular structures (Mogi, 1967) or where the specimen size limits larger shocks (Scholz,1968). Scholz (1968) also found that cataclastic deformation in marble was also accompanied by shocks with a high b value.

To account for the difference between ridge crest and fracture zone earthquakes Francis (1968b) originally proposed that elevated temperatures in the crust at the axis lead to cataclastic deformation and hence high b values. On fracture zones temperatures will tend to be lower in the crust which would behave in a brittle manner giving lower b values. In a more recent paper Francis (1974) showed that events which were probably along the caldera wall during the Fernandina collapse had a high b value and low stress drop whilst those caused by subsidiary fracturing on the caldera floor had a lower b value. If, as Francis suggests, ridge crest events are the result of caldera collapse then this would also explain their relatively high b value. An objection to this interpretation is that the resulting magnitude frequence curve should be concave upwards and not convex as observed. The convexity may be caused by a lack of large earthquakes because of a concentrated stress source or fault size limited to that of the caldera. (Personal communication, Einarsson, 1980).

Stress Drop Measurements

Measurements of stress drops of earthquakes would resolve some of the problems posed in the previous section on b values.

Stress drops measured for events from inferred spreading centres in the Gulf of California appear to be lower than events from adjacent Baja California (Thatcher & Brune, 1971, Thatcher, 1972). Stress drops measured for events on the Rivera fracture zone observed by OBS are also low compared with events observed on land in California (Prothero et al, 1976). To enable inter comparisons Solomon & Burr (1979) have compiled the available source parameter data on mid-ocean ridge earthquakes and also made some further determinations. Their summary of results is contained in Fig (9), a plot of seismic moment against fault length. Seismic moment Mo, fault length L, width W and stress drop s are related as follows:

$$Mo = CLW^2s \qquad (3)$$

where C depends on the fault geometry (Knopoff, 1958, Aki, 1966). The data in Fig (9) are consistent with $Mo \propto L^3$ which is taken by Solomon and Burr as supporting the rule of constant stress drop s and geometrical similarity. From this, and in reviewing other work on apparent stress of ridge, transform and intra plate earthquakes, Solomon and Burr conclude that there is no evidence that the general state of stress varies between rift and transform plate boundaries.

It would be wrong to assume that the available data proves that regional variations in stress drop do not exist. Firstly systematic variations in fault geometry (CLW^2) and stress drop (s) may exist which would invalidate conclusions drawn from Fig (9). More important is that the inversion of stress drops involves the determination of seismic moment and fault dimensions. The former involves the correction

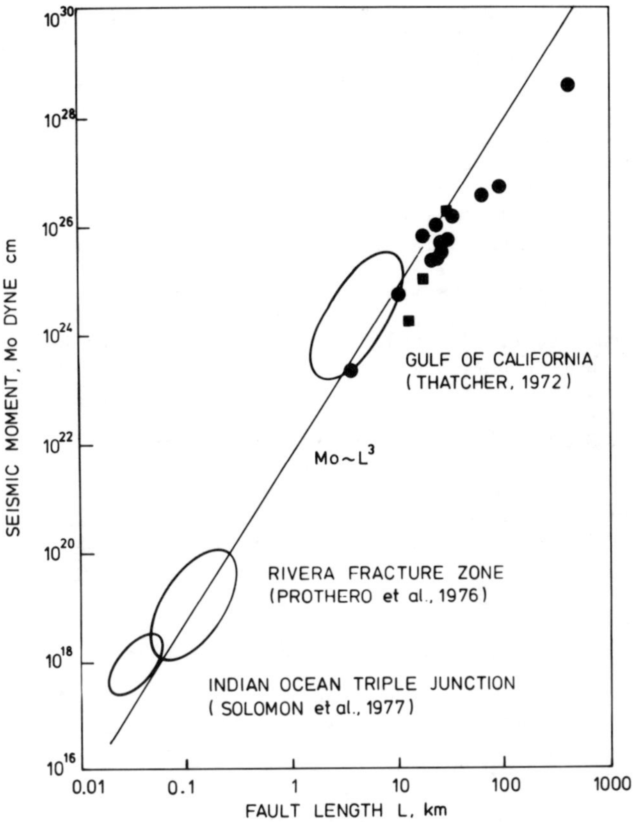

Fig 9. Survey of seismic moment and fault length data for ridge crest and transform earthquakes (after Solomon & Burr 1979). Dots and Crosses represent additional determinations made by Solomon & Burr for large transform and ridge crest events respectively.

of amplitude data for the radiation pattern and both attenuation and propagation between source and receiver. This is a very difficult task for the smaller local events but which has received little attention in many of the results shown in Fig (9). Estimation of fault dimensions from seismological data involves the a priori assumption of fault model and geometry which if incorrect can lead to large errors in stress drop (Knopoff & Mouton, 1975, Madariaga, 1977).

Summary

1. Globally determined earthquake epicentres show that most seismic activity is between the crests of the median valley, but further detail cannot be resolved. A few reliable depth determinations are all less than 10 km.
2. A small number of microearthquake studies show an active zone as little as 5 km across. Depths are less than 10 km peaking between 5 and 10 km on the Mid-Atlantic Ridge and at less than 5 km in Iceland. The maximum depths probably represent the lower limit of the brittle zone.
3. Focal mechanisms published for events over most of the mid ocean ridges show a predominance of normal faulting on the crests. Many solutions show non orthogonal nodal planes for which several explanations have been proposed but not confirmed. No significant reverse faulting has been observed.
4. Earthquake swarms are frequent on the ridge crest and near intersections with fracture zones. They occur over the entire range of earthquake magnitudes observable and within regions usually less than 10 km across. The possible connection with magmatism has resulted in models of faulting involving intrusion (dyke emplacement) and magma withdrawal (caldera collapse). At present there is no evidence for a permanent high level magma chamber on slow spreading ridges.
5. Maximum size of both ridge crest and transform earthquakes appears determined by the depth down to which brittle fracture is possible. In any environment this is determined by the depth of an isotherm which marks the change from seismic to aseismic slip.
6. Ridge crest and transform earthquakes have different b values. This has not been confirmed for microearthquakes possibly because of differences in measuring local magnitudes. The current interpretation is that the variation results from lower stress states on the ridge crest.
7. Stress drop estimations apparently show no differences for ridge crest and transform earthquakes. There are however some doubts about their value as at present determined.

Acknowledgements

This review was undertaken as part of the research program of the Institute of Oceanographic Sciences a component body of the Natural Environment Research Council. I thank Dr H I S Thirlaway and his colleagues for the facilities provided at Blacknest. In particular I thank Dr T J G Francis and Dr. P Einarsson for critically reading the manuscript and Mrs P Courtney and Mrs J Bojie for help in its final preparation.

References

Aki, K., Generation and Propagation of G waves from the Niigata earthquake of June 16, 1964. Part 2 - Estimation of earthquake moment, released energy and stress strain drop from the G wave spectrum, Bull. Earthquake Res. Inst, Tokyo Univ, 44, 73-88, 1966.

Anderson, R.N., Forsyth, D.W., Molnar, P., and Mammerickx, J., Fault Plane solutions of earthquakes on the Nazca Plate Boundaries and the Easter Plate, Earth Planet Sci. Let, 24, 188-202, 1974.

Atwater, T.M., and Mudie, J.D., Block Faulting on the Gorda Rise, Science, 159, 729-731, 1968.

Banghar, A.R., and Sykes, L.R., Focal mechanisms of earthquakes in the Indian Ocean and

adjacent regions, J. Geophys. Res, 74, 632-649, 1969.
Barazangi, M., and Dorman, J., World seismicity maps compiled from ESSA, Coast and Geodetic survey, Epicentre Data, 1961-1967, Bull. Seism. Soc. Amer, 59, 369-380, 1969.
Bjornsson, A., Saemundsson, K., Einarsson, P., Tryggvason, E., and Gronvold, K., Current rifting episode in North Iceland, Nature, 266, 318-323, 1977.
Brandsdottir, B., and Einarsson, P., Seismic Activity associated with the September 1977 deflation of the Krafla Central Volcano in NE Iceland, J. Volcanol.Geothermal.Res, 6, 197-212, 1979.
Bungum, H., Two focal-mechanism solutions for earthquakes from Iceland and Svalbard, Tectonophysics, 41, 715-718, 1977.
Bungum, H., and E.S., Husebye, Seismicity of the Norwegian Sea: The Jan Mayen Fracture Zone, Tectonophysics, 40, 351-360, 1977.
Burridge, R., Lapwood, E.R., and Knopoff, L., First motions from seismic sources near a free surface, Bull. Seism. Soc. Amer, 54, 1889-1913, 1964.
Burr, N.C., and Solomon, S.C., The relationship of source parameters of oceanic transform earthquakes to plate velocity and transform length, J. Geophys. Res, 83, 1193-1205, 1978.
de Wit, M.J., and Stern, C.R., A model for ocean floor metamorphism, seismic layering and magnetism, Nature, 264, 615-619, 1976.
Duschenes, J.D., and Solomon, S.C., Shear wave travel time residuals from oceanic earthquakes and the evolution of oceanic lithosphere, J. Geophys. Res, 82, 1985-2000, 1977.
Einarsson, P., Seismicity and earthquake focal mechanisms along the Mid-Atlantic plate boundary between Iceland and the Azores, Tectonophysics, 55, 127-153, 1979.
Einarsson, P., Klein, F.W., and Bjornsson, S., The Borgarfjordur earthquakes of 1974 in West Iceland, Bull. Seism. Soc. Amer, 67, 187-208, 1977.
Einarsson, P., and Brandsdottir, B., Seismological evidence for lateral magma intrusion during the July 1978 deflation of the Krafla volcano in NE Iceland, J. Geophys. Res, In Press, 1980.
Fairhead, J.D., and Girdler, R.W., The seismicity of the Red Sea, Gulf of Aden and Afar Triangle, Phil. Trans. Roy. Soc. London, 267, 49-74, 1970.
Filson, J., Simkin, T., and Lei-Kuang Leu, Seismicity of a Caldera Collapse: Galapagos Islands 1968, J. Geophys.Res, 78, 8591-8622, 1973.
Forsyth, D.W., Mechanisms of earthquakes and plate motions in the East Pacific, Earth Planet. Sci. Lett, 17, 189-193, 1972.
Forsyth, D.W., Fault Plane solutions and Tectonics of the South Atlantic and Scotia Sea, J. Geophys.Res, 80, 1429-1443, 1975.
Francis, T.J.G., The detailed seismicity of mid-oceanic ridges, Earth Planet. Sci. Lett, 4, 39-46, 1968a.
Francis, T.J.G., Seismicity of mid-oceanic ridges and its relation to properties of the upper mantle and crust, Nature, 220, 899-901, 1968b.
Francis, T.J.G., A new interpretation of the 1968 Fernandina Caldera collapse and its implications for the mid-oceanic ridges, Geophys.J. Roy. Astr. Soc, 39, 301-318, 1974.
Francis, T.J.G., and Porter, I.T., A statistical study of mid-Atlantic Ridge earthquakes, Geophys. J. Roy. Astr. Soc, 24, 31-50, 1971.
Francis, T.J.G., and Porter, I.T., Median Valley seismology: the Mid-Atlantic Ridge near 45°N, Geophys. J. Roy. Astr. Soc, 34, 279-311, 1973.
Francis, T.J.G., Porter, I.T., and Lilwall, R.C., Microearthquakes near the eastern end of St. Paul's Fracture Zone, Geophys. J. Roy. Astr. Soc, 53, 201-217, 1978.
Francis, T.J.G., Porter, I.T., and McGrath, J.R., Ocean bottom seismograph observations on the Mid-Atlantic Ridge near lat 37°N, Geol. Soc. Amer. Bull, 88, 664-677, 1977.
Gibowicz, S.J., Patterns of earthquake swarm activity, Seism. Obs. Wellington Bull, 213, 637-658, 1973.
Gutenburg, B., and Richter, C.F., Seismicity of the Earth. 2nd Ed, 310pp, Princeton University Press, Princeton NJ, 1954.
Harrison, C.G.A., and Stieltjes, L., Faulting within the median valley, in Present State of Plate Tectonics, edited by J. Bonnin and R.S. Dietz, Tectonophysics, 38, 137-144, 1977.
Hart, R., Body Wave Observations from the September 1969, North Atlantic Ridge Earthquake, EOS, 326, 1978.
Herron, T.J., Ludwig, W.J., Stoffa, P.L., Kan, T.K., and Buhl, P., Structure of the East Pacific Rise Crest from Multichannel Seismic Reflection Data, J. Geophys. Res, 83, 798-804, 1978.
Hill, D.P., A model for earthquake swarms, J. Geophys. Res, 82, 1347-1352, 1977.
Husebye, E.S., Gjoystdal, H., Bungum, H., and Eldholm, O., The seismicity of the Norwegian and Greenland Seas and adjacent Continental Shelf areas, Tectonophysics, 26, 55-70, 1975.
Isacks, B., Oliver, J., and Sykes, L.R., Seismology and the new global tectonics, J. Geophys. Res, 73, 5855-5899, 1968.
Johnson, S.H., and Jones, P.R., Microearthquakes located on the Blanco Fracture Zone with sonobuoy arrays, J. Geophys. Res, 83, 255-261, 1978.
Johnson, G.L., and Vogt, P.R., Mid-Atlantic Ridge from 47° to 51°N, Geol. Soc. Amer. Bull, 84, 3443-3462, 1973.
Jones, P.R., and Johnson, S.H., Sonobuoy array measurements of active faulting on the Gorda Ridge, J. Geophys. Res, 83, 3435-3440, 1978.
Keylis-Borok, V.I., The study of earthquake

mechanism, Pub. Dom. Obs, 20, 279-294, 1957.
Klein, F.W., Einarsson, P., and Wyss, M., The Reykjanes Peninsula, Iceland, earthquake swarm of September 1972, and its tectonic significance, J. Geophys. Res, 82, 865-888, 1977.
Knopoff, L., Energy Release in Earthquakes, Geophys. J. R. Astr. Soc, 1, 44-52, 1958.
Knopoff, L., and Mouton, J.O., Can one determine seismic focal parameters from far field radiation? Geophys. J. R. astr. Soc, 42, 591-606, 1975.
Kuznir, N.J., and Bott, M.H.P., A thermal study of the formation of oceanic crust, Geophys. J. R. astr. Soc, 47, 83-95, 1976.
Lilwall, R.C., Fault mechanisms and sub-crustal seismic velocities on the Mid-Atlantic Ridge, Geophys. J. R. astr Soc, 60, 245-262, 1980.
Lilwall, R.C., and Francis, T.J.G., Hypocentral resolution of small ocean bottom seismic networks, Geophys. J. Roy. astr. Soc, 54, 721-728, 1978.
Lilwall, R.C., Francis, T.J.G., and Porter, I.T., Ocean-bottom seismograph observations on the Mid-Atlantic Ridge near 45°N, Geophys. J. Roy. astr. Soc, 51, 357-370, 1977.
Lilwall, R.C., Francis, T.J.G., and Porter, I.T., Oceanbottom seismograph observations on the Mid-Atlantic Ridge near 45°N - further results, Geophys. J. Roy. astr. Soc, 55, 255-262, 1978.
Lilwall, R.C., and Underwood, R., Seismic network bias maps, Geophys. J. Roy. astr. Soc, 20, 335-339, 1970.
MacDonald, K.C., and Mudie, J.D., Microearthquakes on the Galapagos spreading centre and the seismicity of fast-spreading ridges, Geophys. J. Roy. astr. Soc, 36, 245-257, 1974.
Maderiaga, R., Implications of stress-drop models of earthquakes for the inversion of stress drop from seismic observations, Pure Appl. Geophys, 115, 301-316, 1977.
McGregor, B.A., and Rona, P.A., Crest of the Mid-Atlantic Ridge at 26°N, J. Geophys. Res, 80, 3307-3314, 1975.
McKenzie, D.P., Davies, D., and Molnar, P., Plate Tectonics of the Red Sea and East Africa, Nature, 226, 243-248, 1970.
Mogi, K., Earthquakes and fractures, Tectonophysics, 5, 35-55, 1967.
Nabelek, J.L., and Trehu, A.M., Shallow Earthquakes and Non-orthogonal Nodal Planes, J. Geophys. Res, submitted, 1980.
Nisbet, E.G., and Fowler, C.M.R., The Mid-Atlantic Ridge at 37° and 45°N: some geophysical and petrological constraints, Geophys. J. R. astr. Soc, 54, 631-660, 1978.
Orcutt, J.A., Kennet, B.L.N., and Dorman, L.M., Structure of the East Pacific Rise from an ocean bottom seismometer survey, Geophys. J. Roy astr. Soc, 45, 305-320, 1975.
Pearce, R.G., Complex P waveforms from a Gulf of Aden earthquake, Geophys. J. Roy. astr. Soc, In press, 1980.
Prothero, W.A., Reid, I., Reichle, M.S., and Brune, J.N., Ocean bottom seismic measurements on the East Pacific Rise and Rivera Fracture Zone, Nature, 262, 121-124, 1976.
Reichle, M., and Reid, I., Detailed study of earthquake swarms from the Gulf of California, Bull. Seism. Soc. Amer, 67, 159-171, 1977.
Reid, I., and MacDonald, K., Microearthquake study of the Mid-Atlantic Ridge near 37°N using sonobuoys, Nature, 246, 88-90, 1973.
Reid, I., Reichle, M., Brune, J., Bradner, H., Microearthquake studies using sonobuoys: Preliminary results from the Gulf of California, Geophys. J. Roy. astr. Soc, 34, 365-379, 1973.
Ritsema, A.R., Mechanisms of European earthquakes, Tectonophysics, 4, 247-259, 1967.
Robson, G.R., Barr, K.G., and Luna, L.C., Extension failure: an earthquake mechanism, Nature, 218, 28-32, 1968.
Scholz, C.H., The frequency-magnitude relation of microfracturing in rock and its relation to earthquakes, Bull. Seism. Soc. Amer, 58, 399-415, 1968.
Simkin, T., and Howard, K.A., Caldera Collapse in the Galapagos Islands, 1968, Science, 169, 429-437, 1970.
Sleep, N.H., Formation of oceanic crust: some thermal constraints, J. Geophys. Res, 80, 4037-4042, 1975.
Solomon, S.C., and Burr, N.C., The Relationship of source parameters of Ridge-Crest and Transform Earthquakes to the Thermal Structure of Oceanic Lithosphere, Tectonophysics, 55, 107-126, 1979.
Solomon, S.C., and Julian, B.R., Seismic constraints on ocean-ridge mantle structure: Anomalous fault-plane solutions from first motions, Geophys. J. Roy. astr. Soc, 38, 265-285, 1974.
Solomon, S.C., Mattaboni, P.J., and Hester, R.L., Microseismicity near the Indian Ocean Triple Junction, Geophys. Res. Lett, 4, 597-600, 1977.
Spindel, R.C., Davis, S.B., MacDonald, K.C., Porter, R.P., and Phillips, J.D., Microearthquake survey of Median Valley of the Mid-Atlantic Ridge at 36°30'N, Nature, 248, 577-579, 1974.
Steinmetz, L., Whitmarsh, R.B., and Moreira, V.S., Upper mantle structure beneath the Mid-Atlantic Ridge north of the Azores based on observations of compressional waves, Geophys. J. Roy. astr. Soc, 50, 353-380, 1977.
Stover, C.W., Seismicity of the South Atlantic Ocean, J. Geophys. Res, 73, 3807-3820, 1968.
Stover, C.W., Seismicity of the Indian Ocean, J. Geophys. Res, 71, 2575-2581, 1966.
Stover, C.W., Seismicity and Tectonics of the East Pacific Ocean, J. Geophys. Res, 78, 5209-5220, 1973.

Sykes, L.R., The seismicity of the Arctic, Bull. Seism. Soc. Amer, 55, 501-518, 1965.
Sykes, L.R., Mechanism of earthquakes and nature of faulting on the Mid-Oceanic Ridges, J. Geophys. Res, 72, 2131-2153, 1967.
Sykes, L.R., Seismicity of the Indian Ocean and a possible Nascent Island arc between Ceylon and Australia, J. Geophys. Res, 75, 5041-5055, 1970a.
Sykes, L.R., Focal mechanism solutions for earthquakes along the world Rift System, Bull. Seism. Soc. Amer, 60, 1749-1752, 1970b.
Sykes, L.R., Earthquake swarms and sea-floor spreading, J. Geophys. Res, 75, 6598-6611, 1970c.
Sykes, L.R., and Landisman, M., Seismicity of East Africa, the Gulf of Aden and the Arabian and Red Seas, Bull. Seism. Soc. Amer, 54, 1927-1940, 1964.
Sykes, L.R., and Sbar, M.L., Intraplate earthquakes, Lithospheric stresses and the Driving Mechanism of Plate Tectonics, Nature, 245, 298-302, 1973.
Tatham, R.H., and Savino, J.M., Faulting mechanisms for two oceanic earthquake swarms, J. Geophys. Res, 79, 2643-2652, 1974.
Thatcher, W., Regional variations of seismic source parameters in the Northern Baja California Area, J. Geophys. Res, 77, 1549-1565, 1972.
Thatcher, W., and Brune, J.N., Seismic study of an oceanic ridge earthquake swarm in the Gulf of California, Geophys. J. Roy. astr. Soc, 22, 473-489, 1971.
Tobin, D.G., and Sykes, L.R., Seismicity and tectonics of the northeast Pacific Ocean, J. Geophys. Res, 73, 3821, 1968.
Trehu, A.M., Nabelek, J.L., and Solomon, S.C., Source Characterisation of Two Reykjanes Ridge Earthquakes: Surface Waves and Moment Tensors; P Waveforms and Non-orthogonal Nodal Plane, J. Geophys. Res, submitted, 1980.
Tryggvason, E., Seismicity, earthquake swarms and plate boundaries in the Iceland region, Bull. Seism. Soc. Amer, 63, 1327-1348, 1973.
Ward, P.L., and Bjornsson, S., Microearthquakes, swarms and the geothermal areas of Iceland, J. Geophys. Res, 76, 3953-3982, 1971.
Ward, P.L., Palmason, G., and Drake, C., Microearthquake survey and the Mid-Atlantic Ridge in Iceland, J. Geophys. Res, 74, 665-684, 1969.
Weidner, J.D., and Aki, K., Focal depth and mechanism of Mid-Ocean Ridge Earthquakes, J. Geophys. Res, 78, 1818-1831, 1973.
Whitmarsh, R.B., and Laughton, A.S., The fault pattern of a slow-spreading ridge near a fracture zone, Nature, 258, 509-510, 1975.
Wilson, J.T., A new class of faults and their bearing on continental drift, Nature, 207, 343-347, 1965.

INVESTIGATION OF THE SUBCRUSTAL LITHOSPHERE AND ASTHENOSPHERE BY CONTROLLED SOURCE SEISMIC EXPERIMENTS ON LONG-RANGE PROFILES

K. Fuchs[1], L. P. Vinnik[2]

[1]Geophysical Institute, University of Karlsruhe, Hertzstr.16
D-7500 Karlsruhe, F.R.G.

[2]Institute of Physics of the Earth, Academy of Sciences of USSR
Bolshaya Gruzinskaya, 10 Moscow 123 242, USSR

Abstract. Seismic refraction experiments on long-range profiles form the most powerful tool to investigate the structure and physical properties of the subcrustal lithosphere. Observations on such profiles have been carried out on continents and oceans. These experiments are compiled and the main results are reviewed. In the proper sense a long-range profile requires numerous mobile stations or shotpoints deployed as densely spaced linear arrays on profiles with observation distances of 500 to 1000 km and more with sufficient crustal control, overlapping and reversed observations. Only a few long-range profiles fulfill these conditions which are required to correct times and velocities for lateral variations. Most of the long-range profiles probe the upper mantle to a depth of 100 km, some even reach below the asthenosphere. They revealed both vertical and lateral variations of structure and mechanical properties in the subcrustal lithsophere, low velocity zones occur already at shallow depth in the continental upper mantle and are embedded in zones of unexpectedly high 'in situ' velocities of up to 8.6 - 8.7 km/s. Similar high velocities have been detected in the oceanic upper mantle. The high velocities and also their increase with depth are in strong contrast to the velocities predicted for an isotropic garnet pyrolite mantle. They should be considered as indirect evidence for the existence of a considerable amount of anisotropic minerals with a preferred orientation of which olivine is the most likely candidate. There is also direct evidence for the presence of anisotropy in the topmost continental mantle as large as observed in the oceanic uppermost mantle. The presence of anisotropy in the subcrustal lithosphere of both oceans and continents may reflect dynamical processes in the lithosphere-asthenosphere system.

Harvard Law:
Under the most rigorously controlled conditions of pressure, temperature, volume, humidity, and other variables, the earth will do as it pleases.

Introduction

Physical properties of the upper rigid layer of the earth - the lithosphere - and the underlying plastic shell - the asthenosphere - are directly related to many dynamic processes observed at the surface of the earth. The most important information about those properties comes from seismic observations on long-range profiles with controlled sources. On continents refracted waves penetrating into the mantle return to the earth's surface at distances of more than 200 km from the source. Thus the profiles providing data on the lower lithosphere and asthenosphere should be at least several hundred kilometres long.

The first detailed observations on a long-range profile were carried out in the early 60's in Soviet Central Asia using the Deep Seismic Sounding technique (DSS) normally applied in the USSR for seismic prospecting of the crust. The record sections were obtained on a 600 km long profile with station spacings of about 100 m which is typical for layered media (Ryaboy, 1966).

In the middle of the 1960's long-range profiling was conducted in North America the largest being the Early Rise experiment. Underwater explosions in Lake Superior were recorded on 9 profiles up to a length of 3500 km with average station spacings of several ten kilometres. Analyses for these observations were published in a number of papers (Roller and Jackson, 1966; Green and Hales, 1968; Mereu and Hunter, 1969; Iyer, Pakiser, Stuart and Warren, 1969; Massé, 1973; Ansorge, 1975). In the eastern regions of North America at depths around 100 km unexpectedly high P-wave velocities (8.4 - 8.5 km/s) were found and pronounced differences in the velocity distributions for the western and the eastern regions of the USA have been documented.

A drawback of the North American experiment, in comparison with the experiments of the following decade, lies in the absence of reversed and overlapping profiles which are necessary for the control of lateral variations in the crust and upper mantle and for more reliable velocity determinations, as well as in the relatively wide station spacings which impeded correlations of secondary phases.

During the last decade studies of the sub-crustal lithosphere-asthenosphere by controlled source experiments were conducted in many other regions. Inspite of considerable difficulties due to unfavourable signal to noise ratio promising studies of the oceanic upper mantle by ocean bottom seismographs were initiated in several locations (for a review see Asada and Shimamura, 1979). On continents the most important field work was carried out in the USSR and in Western Europe.

In Western Europe many observations on long-range profiles were conducted with a station spacing of less than 10 km and crustal control was established by a sequence of reversed crustal profiles (for a review see Fuchs, 1979). Seismic waves were generated by underwater explosions with charges at optimum depth, for the theory see Wielandt (1975) or by large borehole shots. A set of about 150 'MARS 66' stations (Berckhemer, 1970) is available for these experiments in Western Europe. In the USSR observations were conducted on a number of profiles up to 1000 - 2000 or more km long (e.g. Burmakov et al., 1975; Egorkin et al., 1977; Vinnik and Egorkin, 1980). The waves were generated by powerful explosions on land, recording was made by stations of type 'Taiga' and 'Tcherepacha' (Turtle) with station spacings of about 20 km. The systems of observations included reversed and overlapping profiles.

The locations of long-range profiles have been compiled on a world map (Fig. 1). The various experiments differ by station spacings, by depth penetration (i.e. range) and by the arrangements of profiles. The most reliable data are provided by the observations on reversed and overlapping profiles with crustal control and with average distance between the neighbouring stations of less than 10 - 20 km. These requirements are fulfilled only on some profiles with long ranges of observations. Such profiles are drawn with a solid line in Fig. 1; the other profiles are represented by dashed lines. The profiles of the Early Rise experiment, for instance, although extremely long are therefore represented by dashed lines.

In the following sections we discuss the principal results of seismic observations on long-range profiles: the nature of the wave fields, seismic models of the subcrustal lithosphere-asthenosphere, possible geodynamic applications of the results obtained and the directions of future research.

Seismic wave fields observed on long-range profiles

The final result of the observations on a long-range profile is usually a velocity-depth distribution of the upper mantle. The observed wave field is determined in such a way that its theoretically computed wave field should be in satisfactory agreement with the observed one. It remains, however, always more complicated than the calculated one. Furthermore, the velocity-depth distribution depends on the phase correlation across the seismic record section. The phase correlation is a relatively simple and reliable procedure for first arrivals but could become somewhat subjective for later arrivals. During the last decade evidence has accumulated that not only the crust but also the uppermost mantle are laterally inhomogeneous. The refraction method allows only the detection of lateral heterogeneities with a very broad scale length. Seismic wave scattering at heterogeneities with a scale of about 10 km produces large amplitude fluctuations of P-waves as is observed for distant earthquakes recorded by seismic arrays (e.g. Aki, 1973). Waves scattered at these heterogeneities form precursors to the seismic phases PP and PKPPKP (e.g. Vinnik, 1974).

Scattering at lateral velocity heterogeneities may be responsible also for certain features of the wave fields observed on long-range profiles. In order to obtain some rough quantitative estimates of the effect of scattering one can use the acoustic theory of scattering of sound caused by weak fluctuations of the refractive index (Chernov, 1960). We shall assume that the field of fluctuations is isotropic with the spatial autocorrelation function

$$K(\rho) = K(o) \exp(-\rho^2/a^2)$$

where a is the radius of correlation often called the scale of heterogeneity. The decay of wave energy with distance due to scattering can be described by the expression $\exp(-\alpha r)$ where α is the coefficient of scattering and r the distance. If the wavelength is small in comparison with the scale of heterogeneity then the coefficient of scattering is given by

$$\alpha = \sqrt{\pi}\ K(o)\ \kappa^2 a$$

where κ is wave number. We shall assume a = 10 km. $K(o) = 10^{-4}$ close to the values obtained by the methods of earthquake seismology (e.g. Aki, 1973; Vinnik, 1974). A value for K(o) is assumed corresponding to the r.m.s. value of velocity fluctuations of 1 % with respect to the mean value. If it is further assumed that the dominant frequency of oscillation is 3.3 Hz and the velocity is 8 km/s then $\alpha = 0.012$ which means that the energy loss of the primary wave due to scattering is around 50 % per 50 km distance. Most of this energy is scattered in the direction of propagation of the primary wave within the solid angle $^1/\kappa a$ which in our example is 0.04. Since the primary wave according to Fermat's principle propagates along the minimum time path the scattered waves are part of the later arrivals. In the

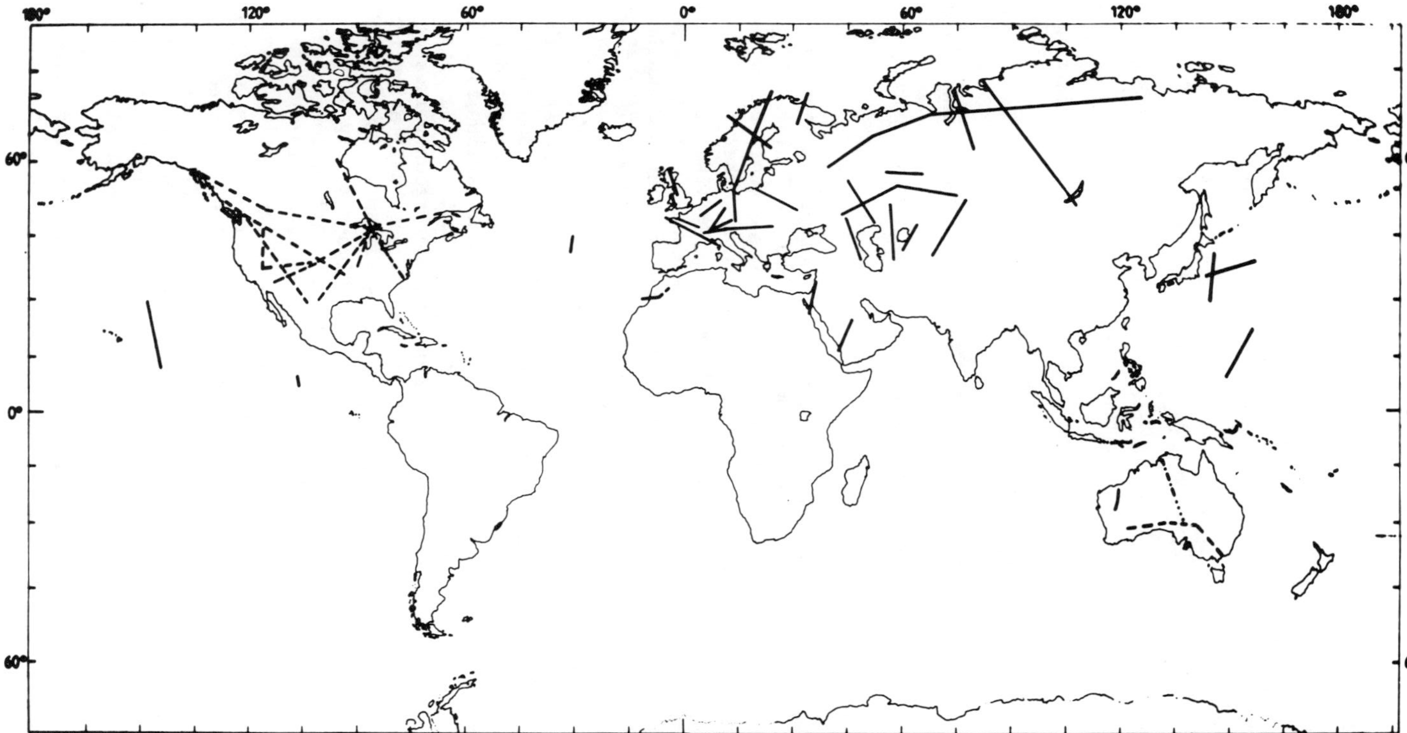

Fig. 1 Position map of long-range profiles. Solid lines: profiles with crustal control and reversed observations. Dashed lines: profiles without crustal control and without reversed observations. Dashed-dotted line in Australia signifies a joint explosion and earthquake experiment. References: USSR: Ryaboy (1966, 1977), Kosminskaya et al.(1972), Burmakov et al. (1975), Egorkin et al. (1977), Vinnik and Ryaboy (1981), Vinnik and Egorkin (1980). Europe: France: Hirn et al. (1973, 1975) Kind (1974), Steinmetz et al. (1974), Ansorge (1975), Hirn (1976), Kennett (1976); Britain: Bamford et al. (1976), Kaminski et al. (1976), Faber (1978), Faber and Bamford (1979); Northern Europe: Hirschleber et al. (1975), Cassell and Fuchs (1979), Working Group of the ESC, Reporter: Lund (1980); Alps: Alpine Explosion Seismology Group (1976, 1977), Miller et al. (1977, 1978); F.R. Germany: Ansorge et al. (1979), Fuchs (1979, 1980), Mechie et al. (1981); Spain: Working Group for Deep Seismic Sounding in Spain (1977). North America: Ryall and Stuart (1963), Hill and Pakiser (1966), Roller and Jackson (1966), Green and Hales (1968), Lewis and Meyer (1968), Warren et al. (1969), Borcherdt et al. (1969), Hales (1969), Iyer et al. (1969), Mereu and Hunter (1969), Gurbuz (1970), Hill (1972), Hales and Nation (1973), Massé (1973), Forsyth et al. (1974), Ansorge (1975), Berry and Forsyth (1975). Near East: Blank et al. (1979), Lamson and Leone (1980a, b), Ginzburg et al. (1979a, b), Perathoner et al. (1981). Australia: Finlayson et al. (1974), Hales and Rynn (1978), Drummond (1979), Hales et al. (1980a, b), Leven (1980), Leven et al. (1981); African-Atlantic Margin: Makris et al. (1981); Atlantic Ocean: Steinmetz et al. (1977), Gebrande et al. (1980), RRISP Working Group (1980); Pacific Ocean: Asada and Shimamura (1976, 1979), Orcutt and Dorman (1977); Gulf of Mexico: Hales et al. (1970); Mediterranean Sea: Hirn et al. (1977).

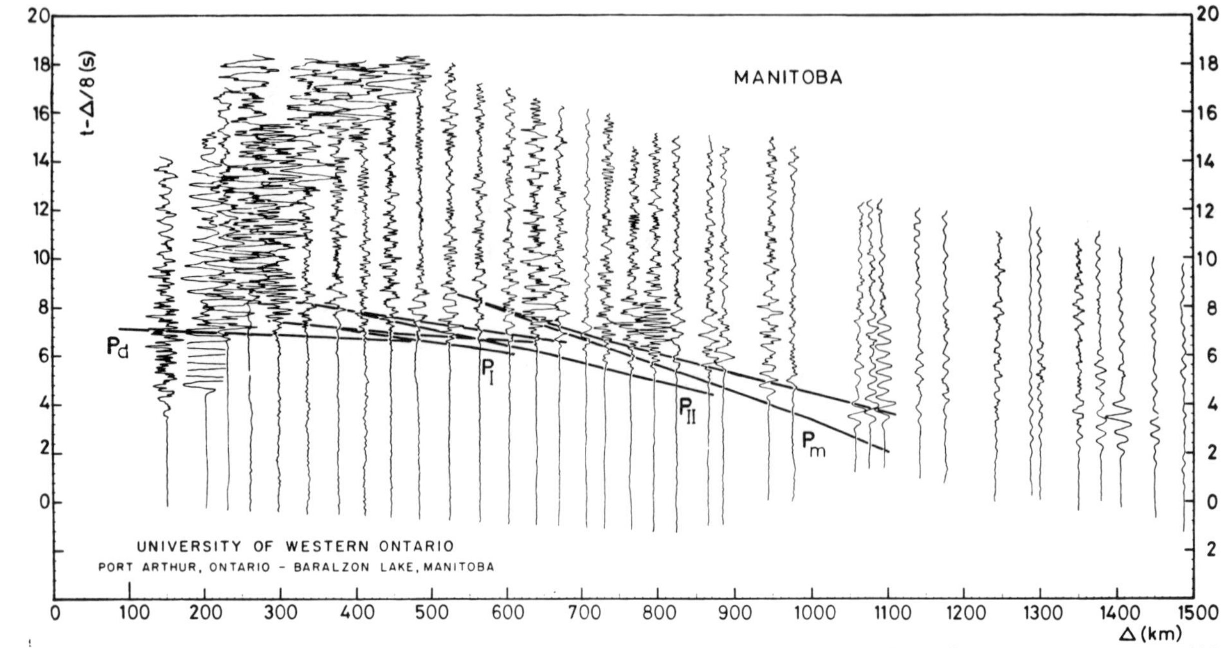

Fig. 2 Seismic record section from Early Rise Experiment, Manitoba Profile (Mereu and Hunter, (1969) with correlations by Ansorge (1975).

shadow zones scattered waves form even first arrivals.

The theory mentioned above is valid only on the assumptiom that the ratio of the scattered energy to the primary energy is much less than 1. This is obviously not the case in the long-range seismic experiments. However, this theory implies that on a long-range profile a considerable part of the wave energy in the later arrivals may consist of scattered waves. Therefore, interpreting later arrivals in terms of regular (reflected or refracted) waves needs much care. Having this in mind we shall further discuss the examples of seismic record sections obtained on long-range profiles in several regions of the earth.

Fig. 2 shows an example of a seismic record section of the Early Rise experiment: the Manitoba profile across the Canadian shield (Ansorge, 1975). Clear features of this section are an apparent velocity of about 8.1 km/s of the first arrivals between 200 and 600 km distance and an increasing velocity up to 8.45 - 8.5 km/s at distances beyond 600 km. The correlation of later phases is more problematic due either to the absence of distinct arrivals or the interference of several wave groups. Similar wave fields were observed on other profiles crossing the Canadian shield (Massé, 1973).

Fig. 3 shows typical examples of seismic record sections from experiments in Western Europe: the profile of 1971 through France and an example of the LISPB experiment of 1974 in Britain. Most of the energy is returned to the surface as the P_mP reflection from the crust-mantle boundary which appears on the left of the sections. The P-wave arrivals returned from the subcrustal lithosphere to the surface show the following characteristics: the P_n-phase from the topmost mantle immediately below the crust-mantle boundary decays rather rapidly with distance and becomes practically unobservable beyond 250 km. At around 300 km a new sequence of P-phases starts with strong amplitudes. The average apparent group velocity of the first arrivals in this group to distances as great as 800 - 1000 km is about 8.1 - 8.2 km/s.

Because of the small station separation the mantle P-phase starting at around 300 km can be resolved by a number of branches produced by overcritical reflections from various first - or second - order discontinuities. The phase velocities of these branches reach values up to 8.5 - 8.6 km/s or even higher, well above the group velocity of 8.1 - 8.2 km/s. Corresponding models consist of thin high-velocity strata embedded in material of 'normal' to low velocities (Hirn et al., 1973, 1975; Faber, 1978). It should be noted that the high-velocity phases are detected mostly in later arrivals. Only in Southern Germany they have been observed as first arrivals on reversed profiles with velocities of up to 8.7 km/s (Ansorge et al., 1979).

As discussed in detail by various authors (e.g. Ansorge 1975; Faber, 1978; Hirn et al., 1975) the correlations presented are not quite unique leading to different models. However, basically all authors agree in the principal physical nature of these phases and the resulting models show the same basic features: a sequence of high- and low-velocity layers in the subcrustal lithosphere.

Fig. 4 shows examples of less known record

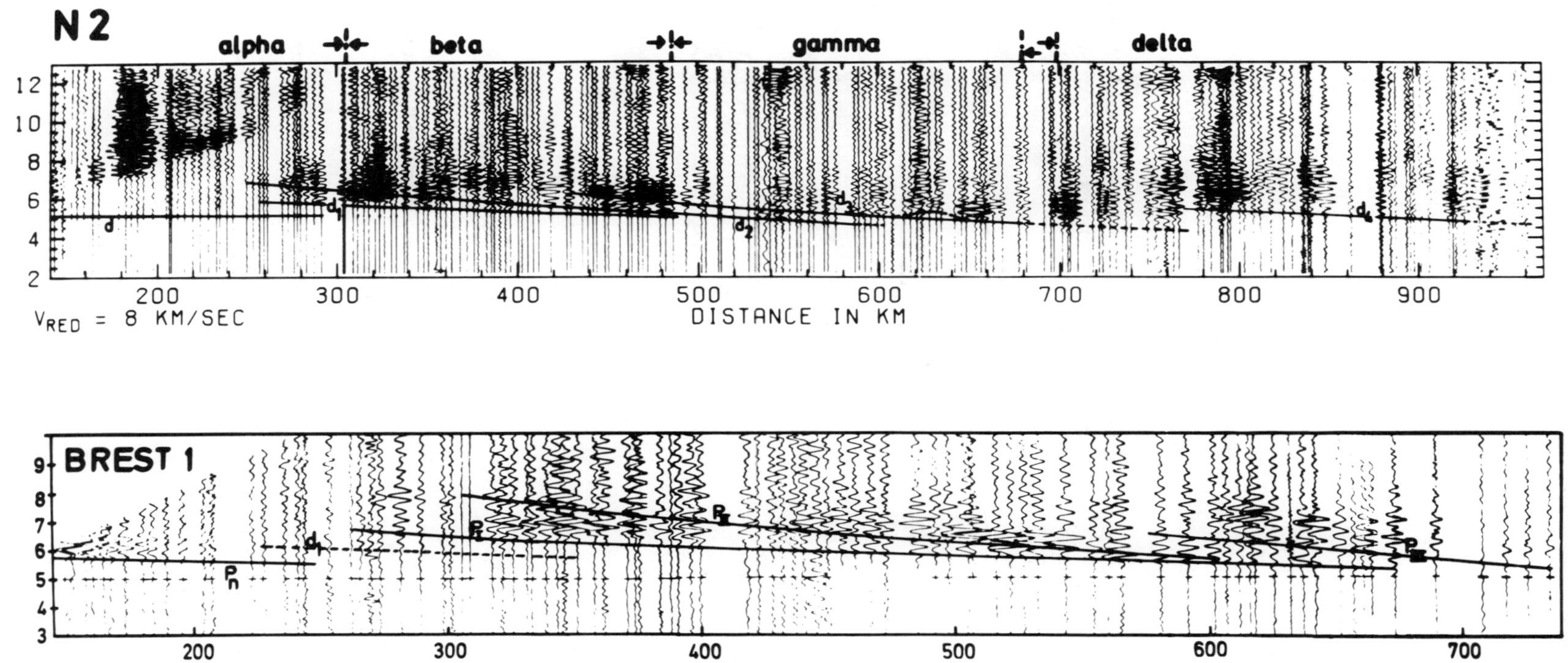

Fig. 3 Two seismic record sections from Western Europe. Reduction velocity: 8 km/s. Top: Britain profile N 2 from LISPB experiment in Britain (Faber, 1978; Faber and Bamford, 1979). - Bottom: France - profile Brest 1 from lithospheric profile experiment in France (Hirn et al., 1973; Kind, 1974; Faber, 1978).

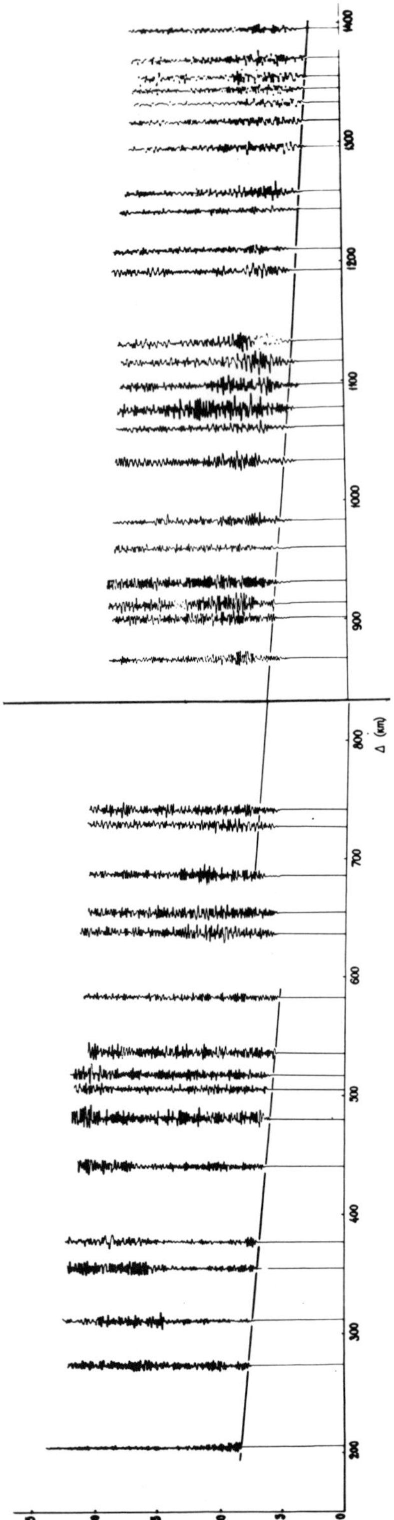

Fig. 4a

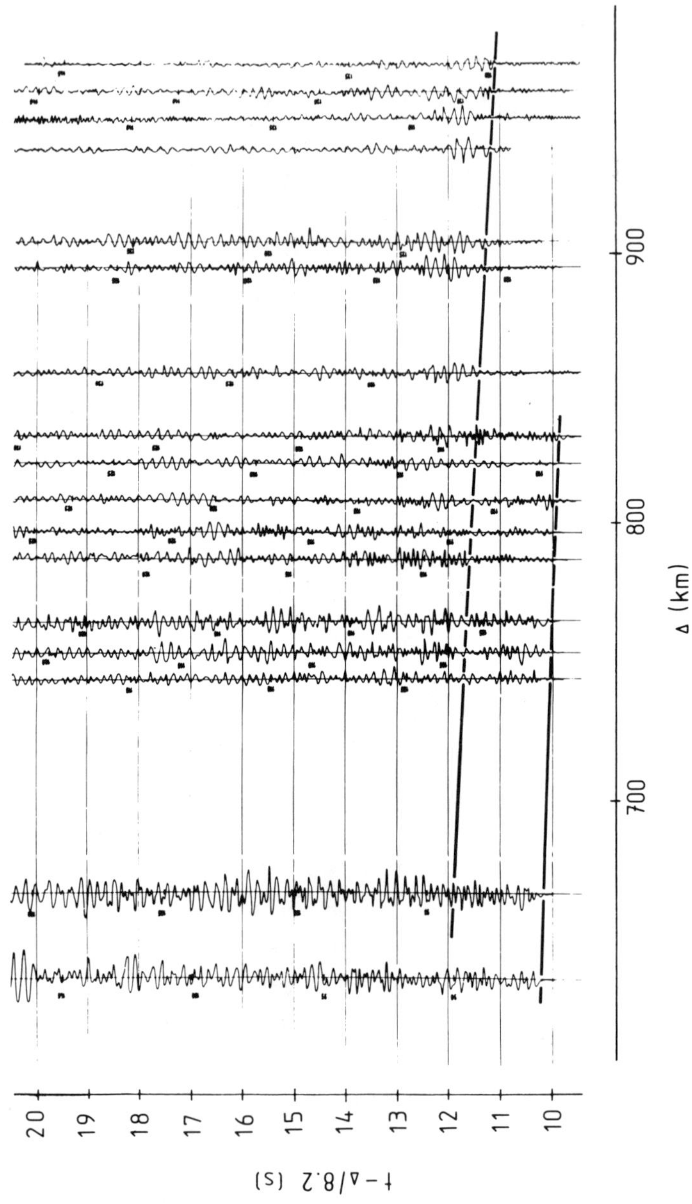

Fig. 4b

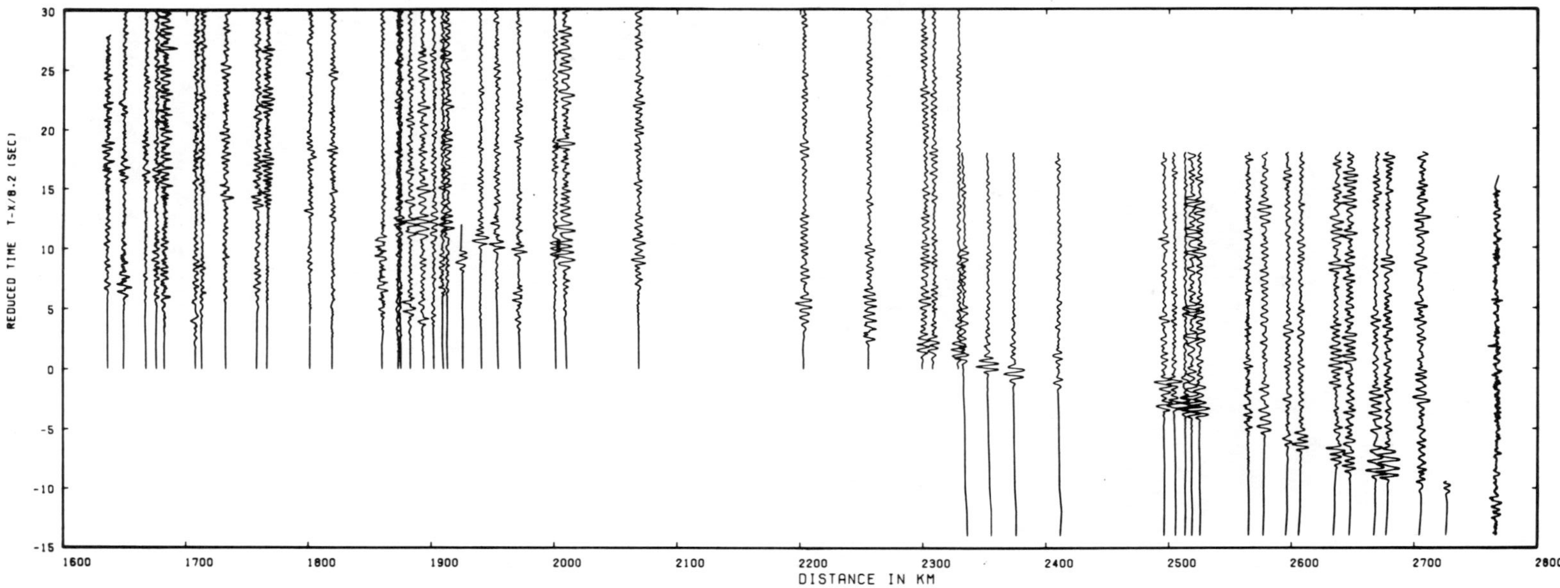

Fig. 4c

Fig. 4 Seismic record sections in the USSR. - a) Profile Kineshma-Vorkuta and b) latitudinal profile in West Siberia (Vinnik and Egorkin, 1980). Reduction velocity: 8 km/s. - c) SW-meridional profile Dikson-Hilok, displayed distance range 1500 to 2750 km (Egorkin and Pavlenkova, 1981). Reduction velocity 8.2 km/s. Amplitudes are normalized with regard to maximum amplitude in every trace.

sections obtained during the last decade in the USSR. The observational system in the north-eastern region of the East-European platform included the profile Kineshma-Vorkuta 1400 km long and a few reversed and overlapping profiles. The record sections of this profile are displayed in Fig. 4a.

In the record section of the main profile two refracted branches are visible. The first with an apparent velocity of 8.55 km/s is observed in a distance range from 200 to 600 km. The second corresponds to the wave-train with large amplitudes emerging in later arrivals at a distance of about 800 km and becoming the first arrivals at a distance of about 1100 km. The largest amplitudes are formed by an overcritical reflection. The dominant periods in this branch are on average longer than in the arrivals of the first refraction line. Both lines are almost parallel, but the second line is late relative to the first one by about 3 s. Offsets of this kind are characteristic of low-velocity layers. A qualitatively similar wave field is observed in the record section of the overlapping profile. It implies that the principal features of this field are generated by vertical rather than by lateral velocity variations.

The apparent velocity of the first arrivals preceding the second refracted branch in a distance range from 600 to 1100 km does not exceed 8.2 km/s. This velocity is too slow to be appropriate for any refracted arrival but is formed most likely by scattered waves.

In Siberia, seismic observations were conducted along the latitudinal profile Vorkuta-Tiksi and along the meridional profile Dikson-Hilok (see Fig. 1). A part of the seismic record section from the latitudinal profile from 600 to 1000 km is displayed in Fig. 4 b. At shorter distances from 1500 to 2200 km a weak phase with an apparent velocity of about 8.7 km/s is recorded as first arrival. This phase is interpreted by Vinnik and Egorkin (1980) as being refracted in a layer at 150 - 200 km depth. A more energetic phase is recorded in later arrivals and becomes the first arrival at distances beyond 220 km. This phase is associated with refraction or reflection near 400 km depth. Relatively long periods of oscillations in this phase are notable.

Fig. 5 shows schematically the wave fields observed on two almost perpendicular profiles across the ancient Siberian platform in East Siberia (see Fig. 1). On the latitudinal profile two reversed travel time curves 1400 and 1800 km long were obtained. The common features of both travel time curves is a shadow zone beginning at epicentral distances around 400 - 700 km. Low velocity arrivals in this zone are formed by scattered waves. Among these waves from a distance of about 900 km a wave with the apparent velocity of about 8.7 km/s is clearly visible in both wave fields. At distances over 1200 km another shadow zone with low apparent velocities of the first arrivals is observed.

On the meridional profile two reversed travel-time curves were obtained along a distance range of more than 2000 km. A part of the record section of the profile to the North is shown in the Fig. 4c. The common features of the wave fields on both reversed profiles are as follows: an apparent velocity of around 8.2 km/s of the first arrivals at distances less than 500 km, appearance of the wave with a higher apparent velocity at a distance around 500 km, a shadow zone at distances around 1000 km and the wave with the apparent velocity of around 8.7 km/s detectable at distances of more than 1400 km.

The travel time data shown in Fig. 5 imply pronounced differences in the wave fields observed on perpendicular profiles within the same region. The differences between the records from reversed profiles although less pronounced are evident as well. In particular the apparent velocities of the wave detected at distances between 500 and 1000 km differ by 0.2 km/s on the reversed meridional profiles and differences between arrival times can exceed 1.0 s.

An example of the record sections of a profile at the Western margin of the Pacific is shown in Fig. 6. At distances less than 500 km a phase with the apparent velocity around 8.2 km/s is seen as the first arrival. At distances around 500 km there is a sharp increase of the apparent velocity. At distances around 1000 km the high-velocity phase is replaced by another phase with 2 s delay. The frequencies in the phase recorded at distances over 1000 km are markedly lower. In general the structure of this wave field looks similar to the wave fields observed in Siberia if differences in absolute travel times are disregarded.

The fundamental property of the wave fields in various regions at distances up to 2000 km is the presence of several mantle phases replacing each other as first arrivals with increasing distance from the source. The reliability of the detection of separate phases and the accuracy of estimates of their parameters depend on the availability of reversed and overlapping observational systems. In almost all of the reviewed experiments the offsets of travel times branches suggest the existence of a sequence of low- and high-velocity layers in the subcrustal lithosphere and asthenosphere.

Seismic models of the subcrustal lithosphere and asthenosphere

Velocity-depth distributions of the subcrustal lithosphere from various parts of the world are compiled in Fig. 7. The models are divided into three groups according to the tectonic settings: old Precambrian platform, young continental platforms and oceans. Since in this review no attempt was made to reinterprete the original record sec-

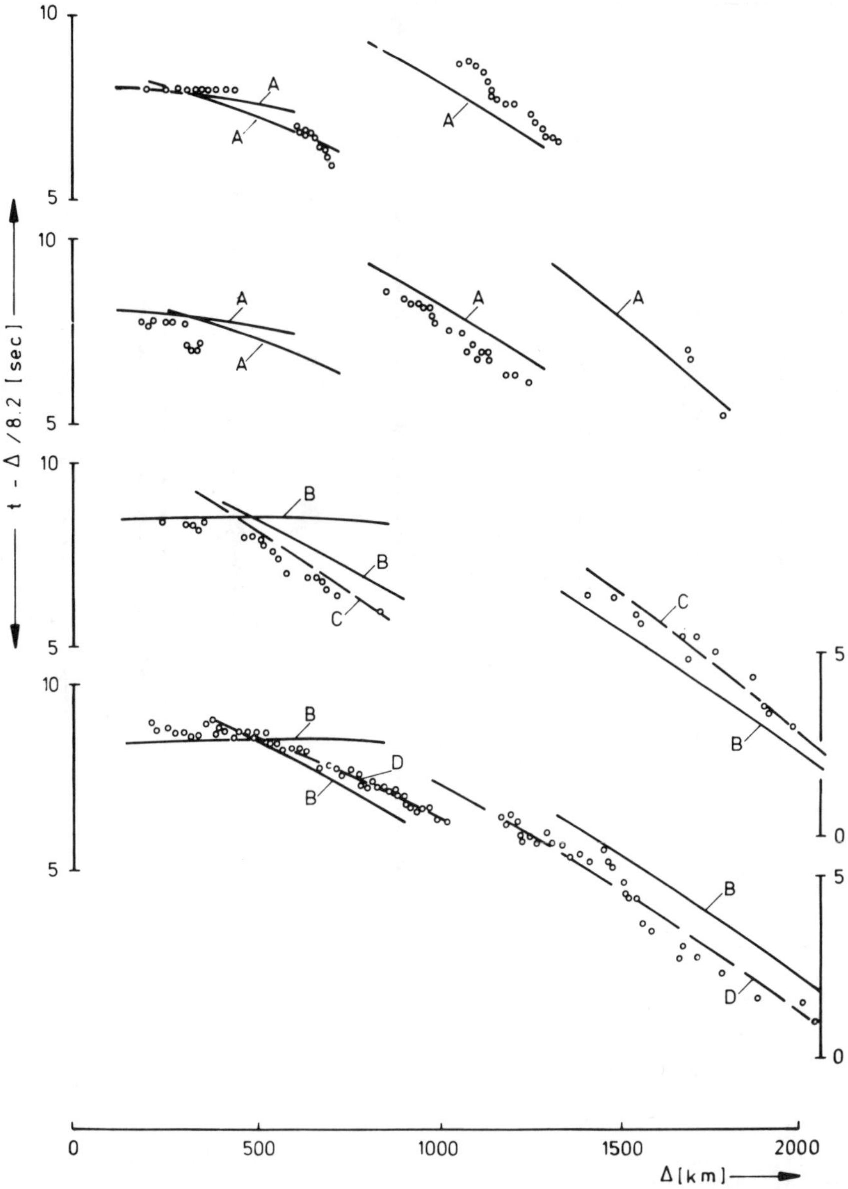

Fig. 5 Traveltime diagrams on two almost perpendicular reversed profiles across the ancient Siberian Platform in East Siberia. Observed times (circles) are compared with theoretical times (solid lines) for the models A to D displayed in Fig. 7a.

tions, the models reflect to a certain degree the personal judgement and experience of the different authors. As pointed out the strongest differences arise by the correlation of weak secondary arrivals, to a smaller degree by the inversion methods. Many velocity-depth distributions were determined from travel times alone and it may be expected that vertical velocity gradients in these models are not always well constrained. The existence of a low-velocity layer is often evident from seismic record sections but it is difficult to derive independently the velocity drop and the layer thickness. Analyses of amplitude combined with that of travel times could reduce this ambiguity.

The most advanced technique of data inversion is the numerical modelling of the wave field and the comparison of the simulated and the observed record sections. The preferred model should explain not only the observed travel times but also the amplitudes and frequencies of the recorded signals. The velocity-depth distributions in

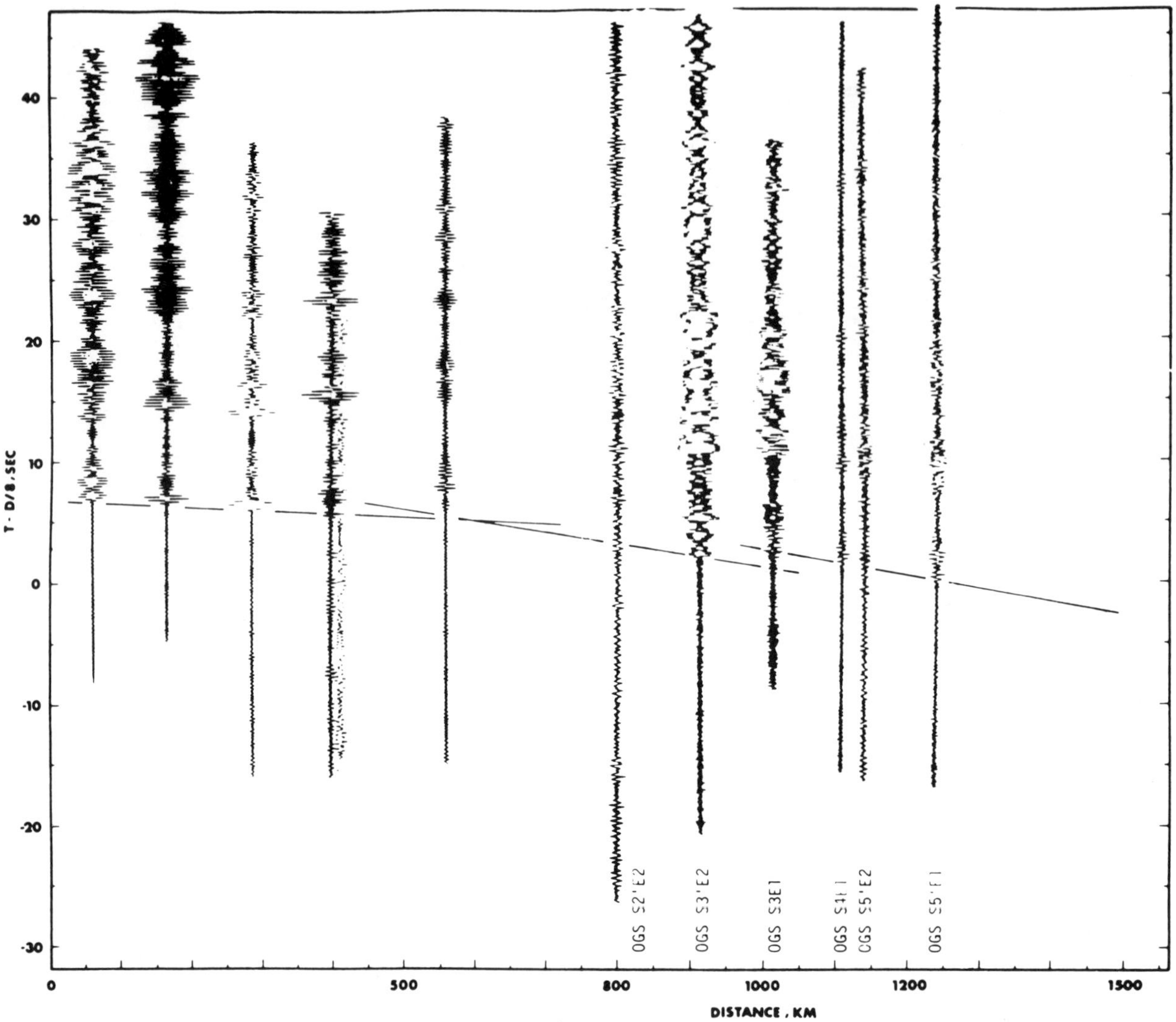

Fig. 6 Seismic record sections from a profile at the western margin of the Pacific (Asada and Shimamura, 1976).

France (Hirn et al., 1973; Kind, 1974) and Britain (Faber and Bamford, 1979) have been derived by this method.

Fig. 7a shows the velocity models of the upper mantle of the old Precambrian platforms. For the Canadian shield the two models by Massé (1973) and by Ansorge (1975) derived from the Early Rise data are presented. The most important features of both models are velocities of about 8 km/s in the topmost mantle, the velocity increases to 8.4 km/s at about 80 km (Massé) and to 8.5 km/s at about 120 km (Ansorge). While Massé's model includes a thin low-velocity zone at a depth of around 100 km, Ansorge's model has two additional shallower low-velocity zones. Velocity determinations for the subcrustal lithosphere have been made also in other parts of the Canadian shield (e.g. Gurbuz, 1970; Mereu and Hunter, 1969; Bates and Hall, 1975). In the depth range between 50 and 90 km an abrupt velocity increase from 8.0 - 8.1 km/s to 8.3 - 8.5 km/s has been detected everywhere. Gurbuz (1970) even attempts to map lateral depth variations of this discontinuity. Low-velocity layers are absent or not very much pronounded in these models.

For the later discussion in every set of models the velocity-depth distribution (Leven et al., 1980) for a garnet pyrolite mantle (Ringwood, 1975) with a geotherm listed in Table 1 is included. It is essentially matching a seismic model of the upper mantle in Northern Australia (Hales et al., 1980a).

For the East-European platform velocity cross-sections of the upper mantle were obtained in the regions of the Russian plate (Vinnik and Ryaboy, 1979) of the North-Caspian coastal depression (Vinnik and Ryaboy, 1979) and of the Fennoscandian shield (Hirschleber et al., 1975). In

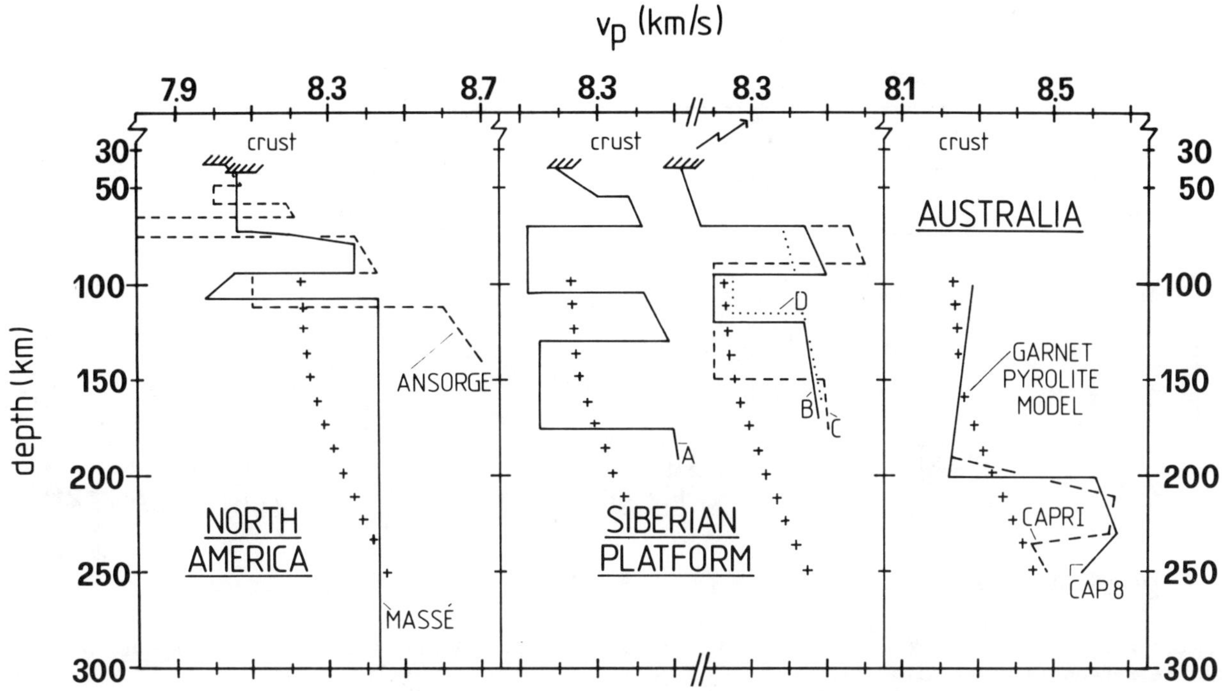

Fig. 7a Depth velocity distribution from long-range profiles on old platforms. Left: North America (Ansorge, 1975; Massé, 1973). Middle: Siberian Platform, models A, B, C, D (Vinnik and Egorkin, 1980) Right: Australia (Hales et al., 1980a; Leven, 1980; Leven et al., 1981). - The crossed distribution is the velocity predicted for an isotropic garnet pyrolite mantle (Leven, 1980; Leven et al., 1981). It is included in all models in Fig. 7a and b for comparison.

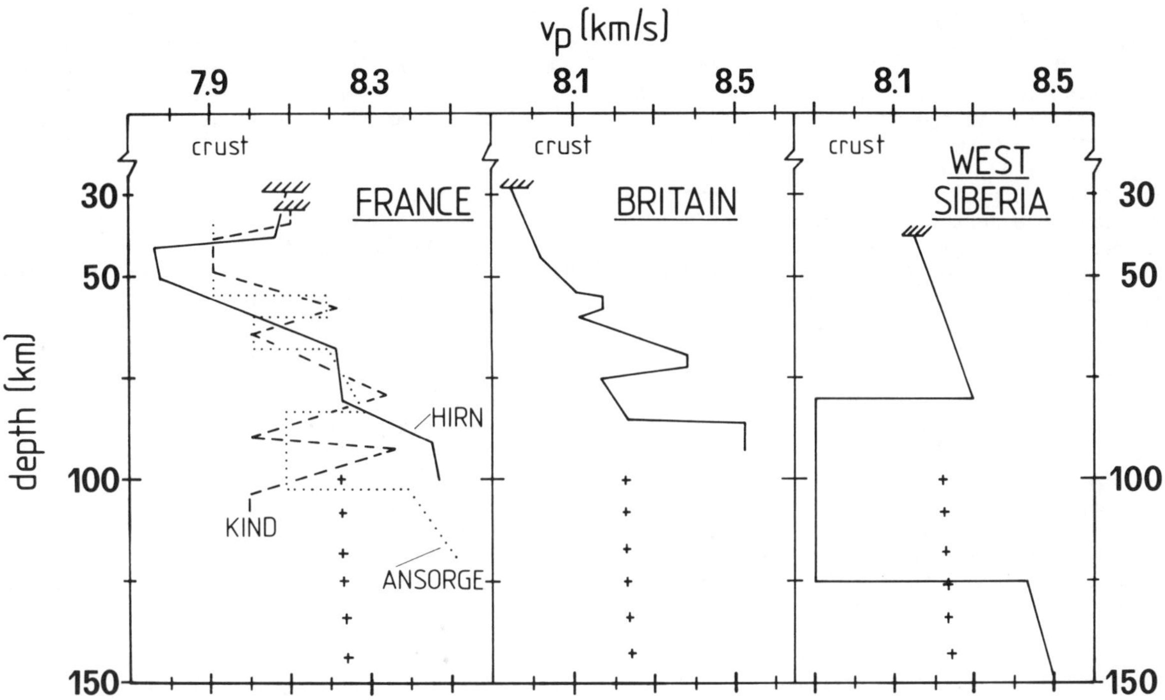

Fig. 7b Depth-velocity distribution from long-range profiles on young continental platforms. Left: France (Hirn et al., 1973; Kind, 1974; Ansorge, 1975). Middle: Britain (Faber, 1978). Right: West Siberia (Vinnik and Egorkin, 1980). For the crossed distribution see Fig. 7a.

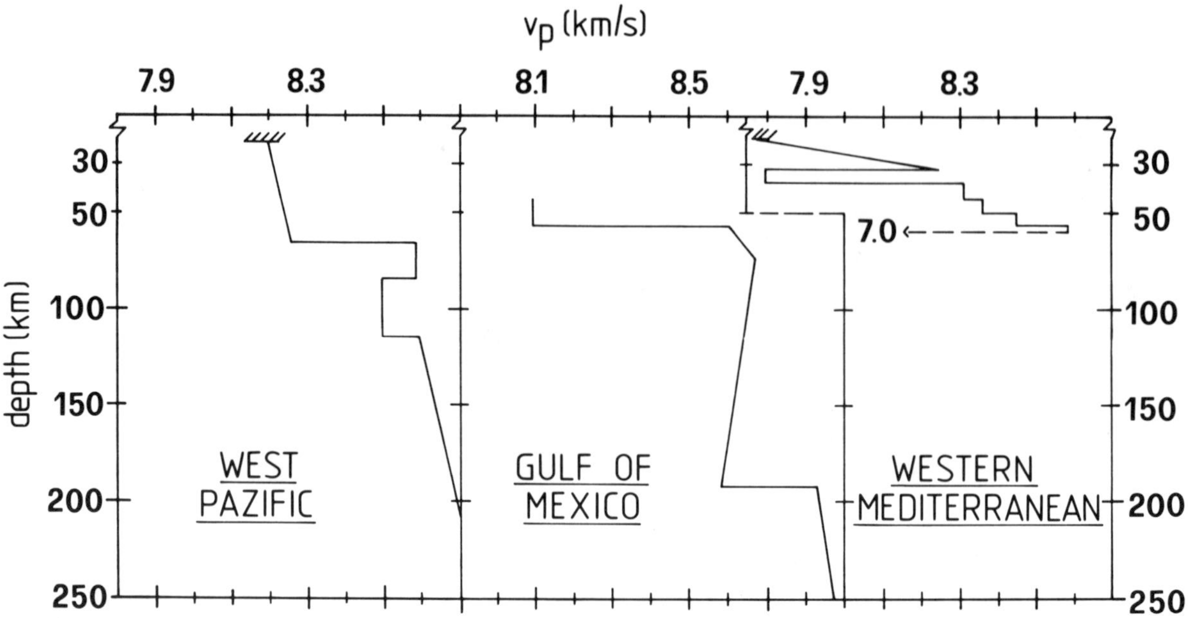

Fig. 7c Depth-velocity distribution from long-range profiles in the oceans. Left: West Pacific (Asada and Shimamura, 1976, 1979). - Middle: Gulf of Mexico (Hales et al., 1970). - Right: Western Mediterranean (Hirn et al., 1977).

the region of the Russian plate a velocity of around 8.4 km/s is observed almost immidiately from below the Moho discontinuity and reaching to a depth of 60 - 70 km. Below this depth there is a well pronounced low velocity-layer. In the North-Caspian coastal depression the P-velocity between the Moho discontinuity and 70 km depth is close to 8.1 km/s. At greater depth it increases to 8.45 km/s. In the region of the Baltic shield the velocity in the uppermost layer of the mantle is around 8.0 km/s, with indications of higher velocities (Cassell and Fuchs, 1979).

For the Siberian platform (Fig. 7a) velocity distributions to a depth of 200 km were derived from the observations along latitudinal and meridional profiles (see Fig. 1). The upper mantle along the latitudinal profile (model A) possesses only small lateral variations as can be recognized from the small travel-time residuals referred to model A on the reversed profiles in Fig. 5 (two top graphs). However, the upper mantle along the meridional profile is laterally heterogeneous. The model B is intermediate between the models C and D that were derived independently from two reversed travel-time curves (Fig. 5, two bottom graphs). Furthermore, the velocity distributions on the meridional and latitudinal profiles differ as is already evident from the differences of arrival times of the corresponding seismic phases (see Fig. 5). In the upper parts of the models there is a similar increase of the velocity from 8.1 -8.2 km/s to 8.4 - 8.5 km/s in the depth range less than 100 km. The difference exists mainly in the number of the low velocity layers.

The velocity distribution under the Precambrian shield of Northern Australia although mainly derived from earthquake observations (Hales et al., 1980a) has been added because the wide depth-range of the hypocenters north of Australia constrains the velocity distribution around a depth of 200 km rather closely. While the model CAP 8 was derived from travel times the model CAPRI (Leven, 1980; Leven et al., 1980) is more refined by the application of synthetic seismograms. The velocity distribution in the top sub-crustal lithosphere above 100 km is less known in this area.

In summary the typical velocity-depth distribution of the old platforms are characterized by the following main features. Between the crust-mantle boundary and a depth of 50 - 60 km the P-velocity rises from 8.0 - 8.2 km/s to 8.4 - 8.5 km/s, in part abruptly; this depth-region is underlain by a low-velocity zone with a top at a depth of around 80 - 100 km. The thickness of the low-velocity zone is in the order of a few

Table 1: Geotherm in Australia and garnet-pyrolite P-velocities (Leven, 1980; Leven et al., 1981)

Depth (km)	100	150	200	250
Temperature (^{o})	770	1050	1180	1320
Velocity (km/s)	8.23	8.25	8.34	8.44

tens of kilometres. Below the bottom of this zone, the velocity is around 8.5 km/s or larger.

There is clear evidence for lateral variations in the subcrustal lithosphere of old platforms. Most interesting is the difference between the elevated parts of the old platforms (shields) and the subsided parts. The velocity distributions of the Canadian shield and those of the Russian and Siberian platforms differ in the development of the low-velocity zone. The reality of this difference is supported already by direct comparison of the corresponding record sections. Analysis of travel times of mantle P-phases in the East-European platform revealed pronounced differences between the Russian plate and the Baltic shield (Vinnik et al., 1978). These differences are mainly due to lateral heterogeneities of the low-velocity zone.

The second type of velocity distribution is that from young continental platforms (Fig. 7b). Three alternative models are presented which are derived from the observation on the long-range profile in France (Hirn et al., 1973). The main phases of the P-wave field from the upper mantle have to be interpreted as supercritical reflections from high-velocity layers with first or second order discontinuities (see previous section and Fig. 3). The observations in France were interpreted by several authors: two alternative models by Kind (1974) and Ansorge (1975) are included in Fig. 7b. In spite of the differences of the details of these models their common main features can be recognized: a thin low-velocity layer at a depth between 40 and 60 km and an unusually strong increase of the velocity with depth below this layer. From additional observations Steinmetz et al. (1974) extended this model to a depth of 200 km with a low-velocity zone between 100 - 200 km. The model derived from the long-range experiment LISPB in Britain (Faber, 1978) by the joint inversion of times and amplitudes is quite similar to that in France with high velocity layers embedded in 'normal' to low-velocity material and a strong increase of velocity with depth.

A well-pronounced low-velocity layer exists also in the velocity distribution for the West-Siberian plate. The corresponding record section is shown in Fig. 4b. It should be remembered that the thickness of the low-velocity layer depends on the value assumed for the velocity in this layer. The presence of the shallow low-velocity layer similar to that in France could not be detected in Siberia due to limitations of the observational system. The models of the upper mantle in France, Britain and West Siberia differ from the models of the old platforms by a somewhat lower average velocity in the upper 30 - 80 km depth range. This result is significant and does not depend on the method of inversion.

The detection of an azimuthal anisotropy of P-velocities in the topmost upper mantle immediately below the Moho in West Germany (Bamford, 1973, 1976, 1977) is one of the important findings of recent studies by controlled source seismology. The magnitude of the anisotropy is 7 - 8 % with a direction of N 20° E of the maximum velocity of 8.3 - 8.4 km/s. A search for a similar phenomenon in Northern Britain and in the Eastern USA was not successful while in the western regions of the USA a weak anisotropy has been found (Bamford et al., 1979). Direction of this anisotropy is close to that observed in the oceanic lithosphere near the shore of California.

Direct evidence of anisotropy in the continental upper mantle at some depth is extremely poor since it is quite difficult to realize a suitable observational system. Nevertheless, Hirn (1977) interpreted certain pecularities of the wave fields on the profiles in Central and Western Europe in terms of an anisotropy with birefringence of S-waves.

Vinnik and Egorkin (1980) suggested azimuthal anisotropy as a possible reason for the differences between the velocity models obtained on the perpendicular profiles across the Siberian platform (see Fig. 5 and 7a). However, a similar effect could be produced by a lateral mantle heterogeneity.

High P-velocities (8.6 - 8.9 km/s) at depths of around 50 - 100 km were sometimes regarded as indirect evidence for velocity anisotropy (Green and Liebermann, 1976; Fuchs, 1977, 1979, 1980) . Although the high velocities in the continental lithosphere derived from later arrivals may sometimes be doubtful, the observations in Southern Germany (Ansorge et al., 1979) where high apparent velocities occur as first arrivals and are confirmed by reversed observations suggest that these high velocities actually occur in the subcrustal lithosphere of continents. The highest P-velocity derived in this experiment is 8.6 - 8.7 km/s at about a depth around 45 km. The vertical velocity gradient reaches .03 s^{-1} in Southern Germany which exceeds the gradient in a homogeneous geothermally heated medium under self compression by about two orders of magnitude. In the absence of phase transitions with sufficiently strong velocity increase the shallow high velocities together with the high vertical gradient in Southern Germany provide very strong indirect evidence for the presence of anisotropy in the subcrustal lithosphere. However, a conclusive 'true' anisotropy experiment probing the deeper subcrustal lithosphere with a common depth point techniques has so far not been performed on the continents.

Studies of the oceanic upper mantle by long-range seismic experiments are in their early stage and the fine structure of the oceanic lithosphere and the asthenosphere is not yet well understood. A successful experiment has been performed on the meridional profile to the East of Japan (Asada and Shimamura, 1976); (the record section is displayed in Fig. 6). The model (Fig. 7c) of the upper mantle has a low-velocity layer between the 8.6 km/s layers. The lower lithosphere has a two-layered structure while the low-velocity layer could belong to the asthenosphere. This velocity distribution of the mantle beneath the oldest oceanic crust, disregarding differences in the crust, does resemble to a certain degree the velocity distribution for old continental plates. On latitudinal profile in the same region lower velocities were observed but it is not clear whether or not this difference is caused by anisotropy. Lower velocities were also observed on the profile in the northeastern region of the Pacific (Orcutt and Dorman, 1977).

Strong lateral variations in the structure of the upper mantle depending on the crustal age have been observed near the Mid- atlantic ridge (Steinmetz et al., 1977). The layer with the velocity of 8.3 km/s at a depth of 9 km below the ocean bottom was detected on the profile parallel to the ridge axis (4 my isochron). On the profile corresponding to the 9 my isochron the same velocity was found at a depth of 30 km. Higher velocities on profiles normal to ridge axis were explained by a dip of the refractor. Different views are expressed on the nature of the layer with a velocity of 7.7 km/s; it can be interpreted either as the lower crust or as the anomalous upper mantle (e.g. Steinmetz et al., 1977; Orcutt et al., 1976).

Long-range refraction experiments have been performed also in marginal seas: in the provencal basin (Western Mediterranean Fig. 7c) by Hirn et al. (1977) and in the Gulf of Mexico by Hales et al. (1970, Fig. 7b). The preferred model for the Western Mediterranean contains a layer of strong positive gradient at shallow depths, a thin low velocity layer at a depth around 30 km and a strong decrease of velocity at a depth of 60 km. In the Gulf of Mexico according to Hales et al (1970) the lower lithosphere is two-layered, the high-velocity layer begins at a depth of about 60 km. The highest velocity (8.67 km/s) is found at a depth of 75 km. The model contains a low-velocity zone that is terminated by a velocity increase at a depth of about 200 km.

Observations on long-range profiles if performed with calibrated equipment provide a possibility to estimate the mechanicical quality factor Q in the upper mantle. Asada and Shimamura (1976) observed a remarkably small attenuation of high-frequency oscillations with distance in the Western Pacific basin. They inferred Q in the order of about 6000 in the lithosphere and much lower values in the low-velocity layer. At the same time Egorkin and Kun (1978) inferred a Q of about 200 - 300 in the continental lower lithosphere. The reason for this difference was not discussed but in our opinion the stronger attenuation in the continental lithosphere could be explained by a stronger scattering of waves at small-scale lateral heterogeneities. Pronounced variations in the dominant frequencies were often observed along the continental profiles (see the examples of record sections in this report) but these variations were not systematically inverted into Q distributions.

Discussions and conclusions

The data on low-velocity layers are very important from the point of view of geodynamics since a lower velocity could be associated with a lower viscosity and a higher mobility of the mantle material.

From the detailed observations one or more thin low-velocity layers have been detected below the Moho discontinuity (e.g. Hirn et al., 1973). Bottinga et al. (1973) explain the existence of the topmost low velocity zone by partial dehydration of the pyrolitic mantle. As yet little is known about the lateral extent of this layer.

The nature of the deeper zones of low velocities in the subcrustal lithosphere which may also be regarded as zones of normal material between high-velocity layers is still under discussion. If the high-velocity layers are produced by a preferred orientation of olivine the lower velocities could be produced either by the absence of a preferred orientation or by a decreased concentration of olivine. It is difficult to decide where the top of the asthenosphere is located. In the Australian Precambrian platform the asthenosphere starts below the '200km-discontinuity' while the low-velocity zones possibly belonging to the asthenosphere of North America and of the Siberian platform appear at shallower depth. It is difficult to decide from the presently available data what is lithosphere and what asthenosphere. It is well possible that the asthenosphere under old continental platforms is not imaged in the seismic velocities by a single, broad zone of low velocity but forms a rather extended region where high- and low-velocity layers are intermingled. Hales et al. (1980a) discuss the correspondence between the observed mantle velocities and those of a garnet pyrolite mantle (57 % olivine, 29 % pyroxene, 14 % garnet applying the Precambrian shield geotherm after Clark and Ringwood (1964)). Fitting the garnet pyrolite velocities to the model of the Northern Australian platform this geotherm was modified by Leven (1980) and Leven et al. (1981). It is shown in Table 1 together with the garnet-pyrolite velocities. This geotherm is intermediate between Jordan's (1979) S1 and S2 conductive geotherms.

As already indicated the velocity distribution of Table 1 is included for reference in Fig. 7 where alle distributions are displayed. The comparison shows for the models of the old platforms quite remarkable differences: the garnet-pyrolite model does neither predict high(8.4 - 8.6 km/s) velocities systematically observed at depths less than 90 km nor low-velocity zones at depths of 90 - 150 km. In a broad range the observed velocities exceed those of the garnet-pyrolite model significantly. If these high velocities are not produced by a preferred orientation of olivine this discrepancy is a challenge to compositional and/or temperature models of the upper mantle.

Various interpretations of velocity distributions obtained from long-range experiments were discussed by a number of authors. For instance, according to Fuchs (1979, 1980) observed vertical P-velocity gradients are often well above the gradient of 0.004 s^{-1} which would be expected in a homogeneous medium under the effect of pressure if not influenced by temperature. The temperature along a geotherm would decrease this gradient even more. This excess of the observed gradient may indicate the change in composition with depth. It is also notable that the gradient through the high velocity layers in the models for France and Britain is higher than through the 'normal' mantle. This observation could be due to the increase in concentration of high-velocity minerals and/or to the increase in the amount of preferred orientation of anisotropic minerals.

Hirn et al. (1977) in their discussion of the P-velocity-depth distribution for the Western Mediterranean consider phase transformations in a chemically homogeneous medium and a layering in a chemically differentiated mantle. Hales (1969) explained the P-velocity increase from 8.1 to 8.4 km/s in the lower lithosphere of North America by the spinel-to-garnet transition. This seismic boundary exists in all velocity-cross sections of the old platforms. Minor compositional variations could change the depth of the spinel-to-garnet transition by tens of kilometres (Ringwood, 1975). According to Hall (1977) significant variations of the depth of the 8.1 - 8.4 km/s seismic boundary were observed in North America; see also the models in Fig. 7.

Anisotropic distribution of seismic velocities in the oceanic lithosphere is reasonably well known (for discussion of this topic see e.g. Bottinga et al., 1976; Chesnokov, 1979). The observation of the similar phenomenon in Southern Germany (see the preceding section) points to the problem of the origin of the anisotropy in the continental subcrustal lithosphere. Fuchs (1977) has found a correspondence between the directions of the anisotropy and of the tectonic stress in Europe. According to him the most likely mechanism to generate the anisotropy is coupling of dynamic processes in the asthenosphere to the subcrustal lithosphere whereby a preferred orientation of the olivine crystals could be induced. From available laboratory data on the generation of preferred orientation of the olivine crystals it cannot be predicted whether geological strain rates are sufficient to produce this orientation under present temperature conditions of the continental lower lithosphere. It is quite possible that the presently observed seismic anisotropy represents one or several palaeo-dynamic stages of the lithosphere-asthenosphere system.

In any case seismic studies of the anisotropy could contribute to the understanding of the dynamics and of the petrology of the upper mantle. Unfortunately much seismic evidence (high velocities, high gradients) of anisotropy in the continental upper mantle is indirect. The explanation of the observations could also be in other terms (e.g. compositional changes). Even if anisotropy were the cause for the high velocities it would be impossible to derive the magnitude and direction of anisotropy from single reversed profiles.

Therefore, a 'true' seismic anisotropy experiment probing the subcrustal lithosphere by a system of seismic long-range profiles crossing each other in the same relatively homogeneous regions with uniform coverage of azimuths and a common-depth area for rays bottoming in the subcrustal lithosphere is highly recommended. Such an anisotropy experiment may provide a new key to dynamical processes in the upper mantle and in its composition.

Long-range profiles produced strong evidence for lateral heterogeneities wherever the appropriate experiments were performed. The examples of lateral inhomogeneity in the mantle were discussed throughout this report. The upper-mantle heterogeneities are often in close and sometimes unexpected relationship with the surface tectonics. The future studies of these relationships could result in better understanding of the dynamics of the earth's interior.

Heterogeneities existing in the earth have a broad spectrum of horizontal scale lengths. Following to Jordan (1979) one can recognize large-scale (>3000 km), mesoscale (300 - 3000 km), smallscale (30 - 300 km) and microscale (<30 km) lateral heterogeneities. The observations on long-range profiles are an efficient tool for the studies of large-scale and mesoscale heterogeneities.

For detailed studies of heterogeneities with scale lengths of less than 500 - 1000 km profile systems consisting of a large number of reversed and overlapping observations are necessary. There is an increasing demand for efficient methods to compute wave propagation in laterally heterogeneous media and for corresponding data-inversion schemes. The combined inversion of long-range profiling data and of teleseismic travel-time residuals looks extremely promising in this report.

Microscale heterogeneities are responsible for strong scattering of waves at least in the continental lithosphere. The scattered waves interfere with overcritical reflections and this barrier for the very detailed velocity modelling has to be overcome by new processing methods. For that reason the better understanding of the contribution of scattering to the wave fields observed on long-range profiles is necessary.

Acknowledgements. This compilation and review would not have been completed without the help and advise from numerous colleagues in many parts of the world. We are especially obliged to Drs. Egorkin, Prodehl and Ryaboy. Dr. I. Kosminskaya kindly helped to provide the locations of long-range profiles in the USSR. We have to thank Dr. Prodehl for his assistance in the compilation of long-range profiles. He, Mrs. Di Pillo and Dr. Mechie helped to improve the English of the manuscript and Mrs. Kunz and Miss Welzenbach typed it. All this help is gratefully acknowledged.

References

Aki, K., Scattering of P-waves under the Montana Lasa. Journ. Geophys. Res., 78, 1334-1346, 1973.

Alpine Explosion Seismology Group, A lithospheric seismic profile along the axis of the Alps, 1975. First results. Pageoph, 114,1109-1130,1976.

Alpine Explosion Seismology Group. A lithospheric seismic profile along the axis of the Alps, 1975. First results. In: Proc. 15th Gen. Assem. of ESC Cracow, 1976. Inst. Geophys. Pol. Acad. Sci., A-4 (115), 435-444, 1977.

Ansorge, J., Die Feinstruktur des obersten Erdmantels unter Europa und dem mittleren Nordamerika. PhD-Thesis, Univ. Karlsruhe, 111 p., 1975.

Ansorge, J., K.-P. Bonjer, and D. Emter, Structure of the uppermost mantle from long-range seismic observations in Southern Germany and the Rhinegraben area. Tectonophysics, 56, 31-48, 1979.

Asada, T., and H. Shimamura, Observations of Earthquakes and explosions at the bottom of the western Pacific: structure of oceanic lithosphere revealed by longshot experiment. In: G.H. Sutton, M.H. Manghnani and R. Moberly (Editors), The Geophysics of the Pacific Ocean Basin and its Margin. Am. Geophys. Union, Washington D.D., Geophys. Mon., 19, 135-154, 1976.

Asada, T., and H. Shimamura, Long-range refraction experiments in deep ocean. Tectonophysics, 56, 67-82, 1979.

Berry, M.J., and D.A. Forsyth, Structure of the Canadian Cordillera from seismic refraction and other data. Can. J. Earth Sci., 12, 182-208, 1975.

Bamford, D., Refraction data in Western Germany - a time-term interpretation. Z. Geophys., 39, 907-927, 1973.

Bamford, D., An up-dated time-term interpretation of Pn data from quarry blasts and explosions in Western Germany. In: P. Giese, C. Prodehl, and A. Stein (Editors). Explosion Seismology in Central Europe, Data and Results, Springer, Berlin, 215-220, 1976.

Bamford, D., Pn velocity anisotropy in a continental upper mantle. Geophys. Journ. R.A.S., 44, 145-160, 1977.

Bamford, D., S. Faber, B. Jacob, W. Kaminski, K. Nunn, C. Prodehl, K. Fuchs, R. King, and P. Willmore, A lithospheric seismic profile in Britain. I. Preliminary results. Geophys. Journ. R.A.S. 44, 145-160, 1976.

Bamford, D., M. Jentsch, and C. Prodehl, Pn anisotropy studies in northern Britain and the eastern and western United States. Geophys. Journ. R.A.S. 57, 397-430, 1979.

Bates, A., and D.H. Hall, Upper mantle structure in southern Sascatchewan and western Manitoba from Project Edzoe. Canadian Journal of Earth Sciences, 12, 2134, 1975.

Berckhemer, H., MARS 66; Eine Magnetbandapparatur für seismische Tiefensondierung. Z. Geophys., 36, 501-518, 1970.

Blank, H.R., J.H. Healy, J. Roller, R. Lamson, F. Fisher, R. Mc Clearn, and S. Allen, Seismic refraction profile, Kingdom of Saudi Arabia: Field operations, instrumentation, and initial results. U.S. Geol. Survey, Saudi Arabian Mission, Project report, 259, 49 p., 1979.

Borcherdt, C.A., R.P. Meyer, and J.R. Van Schaack, Shot times and locations: Project Early Rise. Journ. Geophys. Res., 74, 4442-4452, 1969.

Bottinga, Y., A. Hirn, and L. Steinmetz, Implications de l'existence d'un canal à moindre vitesse sous Moho. Bull. Soc. Geol. Fr., (7) XV (5-6), 500-505, 1973.

Bottinga, Y., L. Steinmetz, and C.J. Allegre, Acoustic wave velocity anisotropy in the oceanic lithosphere. Bull. Soc. Geol. Fr. XVIII, 941-947, 1976.

Burmakov, Y.A., A.V. Egorkin, E.A. Popov, and V.Z. Ryaboy, The upper mantle structure in the East-European platform from seismic data. Dokl. Akad. Sci., USSR, 224, 84-87 (in Russian), 1975.

Cassel, B.R., and K. Fuchs, Seismic investigations of the subcrustal lithosphere beneath Fennoscandia. Journ. Geophys., 46, 369-384, 1979.

Clark, Jr., S.P., and A.E. Ringwood, Density distribution and constitution of the mantle. Rev. Geophys., 2, 35-88, 1964.

Chernov, L.A., Wave propagation in a Random Medium. Mc Graw-Hill, N.Y., 1960.

Chesnokov, E.M., Seismic anisotropy of the Earth's mantle. Nauka, Moscow, 144 pp. (in Russian), 1979.

Drummond, B.J., A crustal profile across the Archean Pilbara and northern Yilgarn cratons, northwest Australia. BMR Journ. Austral. Geology and Geophys., 4, 171-180, 1979.

Egorkin, A.V., V.Z. Ryaboy, L.N. Starobinets, and V.S. Druzhinin, Velocity profiles of the upper mantle according to DSS observation on land. Bull.(Izv.) Acad. Sci. USSR, Earth Physics, 7, 27-41 (in Russian), 1977.

Egorkin, A.V., and V.V. Kun, Attenuation of P waves in the upper mantle. Bull. (Izv.) Acad. Sci. USSR, Earth Physics, 7, 27-44 (in Russian), 1978.

Egorkin, A.V., and N.I. Pavlenkova, Studies of mantle structure of the USSR territory on long-range seismic profiles. Phys. Earth Planet Inter., 25, 12-26, 1981

Faber, S., Refraktionsseismische Untersuchung der Lithosphäre unter den britischen Inseln. PhD-Thesis, Univ. Karlsruhe, 132 pp., 1978.

Faber, S., and D. Bamford, Lithospheric crustal across the Caledonides of Northern Britain, Tectonophysics, 56, 17-30, 1979.

Finlayson, D.M., J.P. Cull, and B.J. Drummond, Upper mantle structure from the trans-Australia seismic survey (TASS) and other seismic refraction data. Journ. Geol. Soc. Australia, 21, 447-458, 1974.

Forsyth, D.A., M.J. Berry, and R.M. Ellis, A refraction survey across the Canadian Cordillera at 54° N. Can. J. Earth Sci., 11, 533-548,1974.

Fuchs, K.,Seismic anisotropy of the subcrustal lithosphere as evidence for dynamical processes in the upper mantle. Geophys. Journ. R.A.S., 49, 167-179, 1977.

Fuchs, K.,Structure, physical properties and lateral heterogeneities of the subcrustal lithosphere from long-range deep seismic sounding observations on continents. Tectonophysics, 56, 1-15, 1979.

Fuchs, K.,The subcrustal lithosphere - mechanical properties, differentiation and dynamical properties. In: H. Closs, K. v. Gehlen, H. Illies, E. Kuntz, J. Neumann, and E. Seibold (eds.). Mobile Earth, Int. Geodyn. Project, Final Report of the F.R. Germany. Deutsche Forschungsgemeinschaft. H. Boldt Verlag, Boppard, 241-253, 1980.

Gebrande, H., H. Miller, and P. Einarsson, Seismic structure of Island along RRISP - profile I. Journ. Geophys., 47, 239-249, 1980.

Ginzburg, A., J. Makris, K. Fuchs, C. Prodehl, W. Kaminski, and V. Amitai, A seismic study of the crust and upper mantle of the Jordan-Dead Sea. Rift and their transition towards the Mediterranean Sea. Journ. Geophys. Res., 84, 1569-1582, 1979a.

Ginzburg, A., J. Makris, K. Fuchs, B. Perathoner, and C. Prodehl, Detailed structure of the crust and upper mantle along the Jordan-Dead Sea Rift. Journ. Geophys. Res., 84, 5602-5612, 1979b.

Green, D.H., and A.L. Hales, The travel times of p-waves to 30° in the Central United States and upper mantle structure. Bull. Seism. Soc. Amer., 58, 267-289, 1968.

Gurbuz, B.,A study of the Earth's crust and upper mantle using travel times and spectrum characteristics of body waves. Bull. Seism. Amer., 60, 1921-1935, 1970.

Hales, A.L.,A seismic discontinuity in the lithosphere. Earth Planet. Sci. Lett., 7, 44-46,1969.

Hales, A.L., C.E. Hesley, and J.B. Nation, P travel times for an oceanic path. Journ. Geophys. Res., 75, 7362-7331, 1970.

Hales, A.L., and J. B. Nation, A seismic refraction survey in the northern Rocky Mountains: more evidence for an intermediate crustal layer. Geophys. Journ. R.A.S., 35, 381-399, 1973.

Hales, A.L., and J.H.W. Rynn, A long-range controlled source seismic profile in Northern Australia. Geophys. Journ. R.A.S., 55, 633-644, 1978.

Hales, A.L., K.J. Muirhead, and J.M.W. Rynn, A compressional velocity distribution for the upper mantle. Tectonophysics, 63, 309-348, 1980a.

Hales, A.L., K.J. Muirhead, and J.M.W. Rynn, Crust and upper mantle shear wave velocities from controlled sources. Geophys. Journ. R.A.S., 63, (3), 659-670, 1980b.

Hall, D.H., Partial melting and mineral stability boundaries and their bearing on the seismic exploration of the lithosphere in Canada. Can. J. Earth Sci., 14, 2638-2650, 1977.

Hirn, A., Anisotropy in the continental upper mantle: possible evidence from explosion seismology. Geophys. Journ. R.A.S. 49, 49-58, 1977.

Hirn, A., L. Steinmetz, R. Kind, and K. Fuchs, Long-range profiles in Western Europe: II. Fine structure of the lower lithosphere in France (Southern Bretagne). Z. Geophys., 39, 363-384, 1973.

Hirn, A., C. Prodehl, and L. Steinmetz, An experimental test of models of the lower lithosphere in Bretagne (France). Ann. Geophys., 31, 517-530, 1975.

Hirn, A.,Sondages sismiques profonds en France. Bull. Soc. Geol. Fr. XVIII (5), 1065-1071,1976.

Hirn, A., L. Steinmetz, and M. Sapin, A long-range seismic profile in the Western Mediterranean Basin: Structure of the upper mantle. Ann. Geophys., 33, 373-384, 1977.

Kaminski, W., D. Bamford, S. Faber, B. Jacob, K. Nunn, and C. Prodehl, A lithospheric seismic profile in Britain. II. Preliminary report on the recording of a local earthquake. Journ. Geophys., 42, 103-110, 1976.

Hill, D. P.,Crustal and upper mantle structure of the Columbia Plateau from long-range seismic refraction measurements. Geol. Soc. Am. Bull., 83, 1639-1648, 1972.

Hill, D.P., and L.C. Pakiser, Crustal structure between the Nevada Test Site and Boise, Idaho from seismic refraction measurements. In: Steinhart, J.S., and T.J. Smith (eds.). The earth beneath the continents: Am. Geophys. Union Mon. 10, 391-419, 1966.

Hirschleber, H.B., C.E. Lund, R. Meissner, A. Vogel, and W. Weinrebe, Seismic investigations along the Scandinavian "Blue Road" traverse. Journ. Geophys., 41, 135-148, 1975.

Iyer, H.M., L.C. Pakiser, D.J. Stuart, and D.H. Warren, Project Early Rise: seismic probing of the upper mantle. Journ. Geophys. Res., 74, 4409-4441, 1969.

Jordan, T.H., Structural geology of the Earth's interior. Proc. of the Nat. Acad. Sci. USA, 76, 4192-4200, 1979.

Kennett, B.L.N., A comparison of travel time inversions. Geophys. J. R.A.S., 44, 517-536, 1976.

Kind, R., Long range propagation of seismic energy in the lower lithosphere. Z. Geophys., 40, 189-202, 1974.

Kosminskaya, I.P., N.N. Puzyrev, A.S. Alekseev, Explosion seismology: its past, present and future. In : A.R. Ritsema (Editor), The Upper Mantle. Tectonophysics, 13, 309-323, 1972.

Lamson, R., and L. Leone, Saudi Arabia seismic refraction profile: data set, volume I. Office of Earthquake Studies, U.S. Geological Survey, Menlo Park, California, 861 p., 1980a.

Lamson, R., and L. Leone, Saudi Arabia seismic refraction profile: data set, volume II. Office of Earthquake Studies, U.S. Geological Survey, Menlo Park, California, 109 p., 1980b.

Leven, J.H., The application of synthetic seismograms to the interpretation of crustal and upper mantle structure. PhD thesis, Austr. Nat. University, Canberra, 237 p., 1980.

Leven, J.H., I. Jackson, and A.E. Ringwood, Upper mantle seismic anisotropy and lithospheric decoupling. Nature, 289, 234-239, 1981.

Lewis, B.T.R., and R.P. Meyer, A seismic investigation of the upper mantle to the West of Lake Superior. Bull. Seism. Soc. Amer. 58, 565-596, 1968.

Makris, J., A. Demnati, V. Niemann, J. Thiessen, and A. Ben Sari, Crustal and lithospheric structure of the Atlas and Antiatlas ranges of Marocco. Journ. Geophys., in press, 1981.

Massé, R.P., Compressional velocity distribution beneath central and eastern North America. Bull. Seism. Soc. Amer., 63, 911-935, 1973.

Mechie, J., K. Fuchs, W. Kaminski, and C. Prodehl, Data and results from the Rhenish Massif seismic experiment. Deutsch. Geophys. Gesellsch. Ann. Meeting (Abstract), 30.3.-3.4.1981

Mereu, R.F., and J.A. Hunter, Crustal and upper mantle structure under the Canadian Shield from Project Early Rise data. Bull. Seism. Soc. Amer., 59, 147-165, 1969.

Miller, H., H. Gebrande, and E. Schmedes, Ein verbessertes Strukturmodell für die Ostalpen, abgeleitet aus refraktionsseismischen Daten unter Berücksichtigung des Alpen-Längsprofils. Geol. Rundsch., 66, 289-308, 1977.

Miller, H., J. Ansorge, K. Aric, and G. Perrier, Preliminary results of the Alpine longitudinal profile, 1975 from France to Hungary. In: H. Closs, D. Roeder, and K. Schmidt (Editors), Alps, Apennines, Hellenides: Geodynamic Investigations along Geotraverses, Schweizerbart, Stuttgart, 620 pp., 1978.

Orcutt, J.A., B.L.N. Kennett, and L.M. Dorman, Structure of the East Pacific Rise from an ocean bottom seismometer survey. Geophys. J. R.A.S., 45, 305-320, 1976.

Orcutt, J.A., and L.M. Dorman, An oceanic long range explosion experiment: a preliminary report. Journ. Geophys., 43, 257-263, 1977.

Perathoner, B., K. Fuchs, C. Prodehl, and A. Ginzburg, Seismic investigation of crust-mantle transition in continental Rift systems- Jordan-Dead-Sea-Rift and Rhinegraben. Tectonophysics, (in press), 1981.

Ringwood, A.E., Composition and petrology of the earth's mantle. McGraw Hill, New York, 604pp., 1975.

Roller, J.C., and W.H. Jackson, Seismic wave propagation in the upper mantle: Lake Superior, Wisconsin, to Central Arizona. Journ. Geophys. Res., 71, 5933-5941, 1966.

RRISP Working Group, Reykjanes ridge Iceland seismic experiment (RRISP 77), Journ. Geophys., 47, 228-238, 1980.

Ryaboy, V.Z., Kinematic and dynamic characteristics of deep waves associated with boundaries in the crust and upper mantle. Bull. (Izv.) Acad. Sci. USSR. Geophys. Ser., A.G.U. Transl., 3, 177-184, 1966.

Ryaboy, V.Z., Study of the structure of the lower lithosphere by explosion seismology in the USSR. Journ. Geophys., 43, 593-610, 1977.

Ryall, A., and D.J. Stuart, Traveltimes and amplitudes from unclear explosions, Nevada Test Site to Ordway, Colorado. Journ. Geophys. Res., 68, 5821-5835, 1963.

Steinmetz, L., Hirn, A., and G. Perrier, Réflexions sismiques a la base de l'asthénosphère. Ann. Geophys., 30, 173, 1974.

Steinmetz, L., R.B. Whitmarsh, and V.S. Moreira, Upper mantle structure beneath the Mid-Atlantic ridge north of the Azores based on observations of compressional waves. Geophys. J. R.A.S., 50, 353-380, 1977.

Vinnik, L.P., Scattering of compressional waves in the Earth. Bull. (Izv.) Acad. Sci. USSR. Earth Physics, 11, 37-56, 1974.

Vinnik, L.P., V.Z. Ryaboy, L.N. Starobinetz, A.V. Egorkin, And N.M. Chernyshev, Velocities of P-waves in the upper mantle of the East-European platform. Proc. (Dokl.) Acad. Sci. USSR, 244, 70-73 (in Russian), 1978.

Vinnik, L.P., and A.V. Egorkin, Wave fields and models of the lithosphere-asthenosphere according to seismic observations in Siberia. Proc. (Dokl.) Acad. Sci. USSR, 250, 319-323 (in Russian), 1980.

Vinnik, L.P., and V.Z. Ryaboy, Deep structure of the East-European platform according to seismic data. Phys. Earth Planet. Inter., 25, 27-37, 1981.

Warren, D.H., J.H. Healy, J.C. Hoffman, K. Reinis, R. Srinivasreddy, and D.J. Stuart, Project Early Rise travel times and amplitudes. U.S. Geol. Surv. Open-File Rep., Menlo Par, Calif., 150 pp., 1968.

Wielandt, E., Anregung seismischer Wellen durch Unterwasserexplosionen. PhD Thesis, Univ. Karlsruhe, 120 pp., 1972.

Wielandt, E., Generation of seismic waves by underwater explosions. Geophys. Journ. R.A.S., 40, 421-439, 1975.

Working Group for Deep Seismic Sounding in Spain 1974-1975, Deep seismic soundings in Southern Spain. Pageoph., 115, 721-735, 1977.

Working Group of the European Seismological Commission. Reporter: C.E. Lund, Fennoscandian Long-range Project. EGS-ESC meeting, Budapest 21.-29. August 1980, abstract A 10.6., p. 84, 1980.

GRAVITY ANOMALIES OVER OCEANIC RIFTS

A. B. Watts

Lamont-Doherty Geological Observatory and Department of Geological Sciences
of Columbia University, Palisades, New York 10964

The earliest measurements of the gravity field over mid-ocean ridges were obtained by Vening Meinesz (1948) using a pendulum apparatus. Continuous profiling using surface-ship gravimeters (Talwani et al., 1961; 1965) showed that free-air gravity anomalies over ridges were generally small and that ridge crests were approximately in isostatic equilibrium. Talwani et al. (1965) by considering ridge crests as a mass excess, suggested they were compensated by an underlying mass deficiency. The form of the compensation could not be deduced from gravity data alone, but was generally considered to extend to depths greater than 30 km below the ridge axis.

Mid-ocean ridges are now recognized as the site of creation of oceanic lithosphere which subsequently evolves by cooling and solidification of hot mantle material. Simple cooling models have been constructed (for example, Sclater and Francheteau, 1970) which generally explain the thickness of the plates, the shape of their upper surface, heat flow and other geophysical parameters such as the dispersion of surface waves. In these models the excess elevation of ridges is assumed to be compensated at the base of the lithosphere.

Lambeck (1972) showed that the simple cooling models are associated with a small amplitude free-air gravity anomaly high over the ridge crest and flanking lows. The amplitude and wavelength of the gravity anomaly depends on the thermal structure of the ridge crest and hence the half-spreading rate. Cochran and Talwani (1977) showed that a small amplitude gravity anomaly high is observed over most mid-ocean ridges, although they could not distinguish a strong dependence on the spreading rate.

These previous studies were mainly concerned with the manner mid-ocean ridges were supported and thus were limited to the interpretation of the relatively long-wavelength gravity field (wavelength $\lambda \gtrsim 800$ km). A number of prominent features also occur in the short-wavelength gravity field ($\lambda \lesssim 250$ km) over ridge crests (Talwani et al., 1965; Van Andel and Bowin, 1968; Woodside, 1972; Cochran, 1979; Collette et al., 1980). The most striking of these are large amplitude free-air gravity anomaly lows (up to -75 mgal) over oceanic rifts.

The purpose of this paper is to review the current knowledge of gravity anomalies over oceanic rifts. A brief description is given of gravity anomalies over active and extinct oceanic rifts in the Indian, Pacific and Atlantic oceans and some of the significant results of the interpretation of the gravity field over oceanic rifts summarized.

Gravity Anomalies

Free-air gravity anomaly and bathymetry profiles of active oceanic rifts in the Pacific, Indian, and Atlantic oceans are summarized in Figs. 1 and 2. The sources and locations of each profile are given in Table 1.

Figure 1 shows that active oceanic rifts are associated with large amplitude (up to -70 mgal) short-wavelength ($\lambda \sim 40$ to 60 km) free-air gravity anomaly lows which are flanked by gravity anomaly highs. The oceanic rift of the relatively slow-spreading Mid-Atlantic and Sheba ridges (half-spreading rate 1.2 to 1.5 cm/yr) correlate with a gravity low of amplitude 50 to 70 mgal and wavelength about 80 km. The oceanic rift of the faster spreading Gorda ridge (2.9 cm/yr) correlates with a gravity low of amplitude 30 mgal and wavelength about 40 km. These differences in the shape of the gravity anomaly cannot be simply explained by differences in topography. The rift valley on profiles 1 and 3 (Fig. 1), for example, has a similar width and depth but is associated with different shape gravity anomalies.

Fig. 2 shows that rift valley 2 in a segment of the Mid-Atlantic ridge investigated as part of project FAMOUS (ARCYANA, 1975) is associated with a gravity anomaly low of amplitude 50 to 60 mgal and wavelength of about 60 km. The gravity anomaly minimum occurs over the rift valley terrace between each outer wall. The minimum does not, therefore, occur over the inner floor, which Needham and Francheteau (1974) consider to be the locus for crust of zero age.

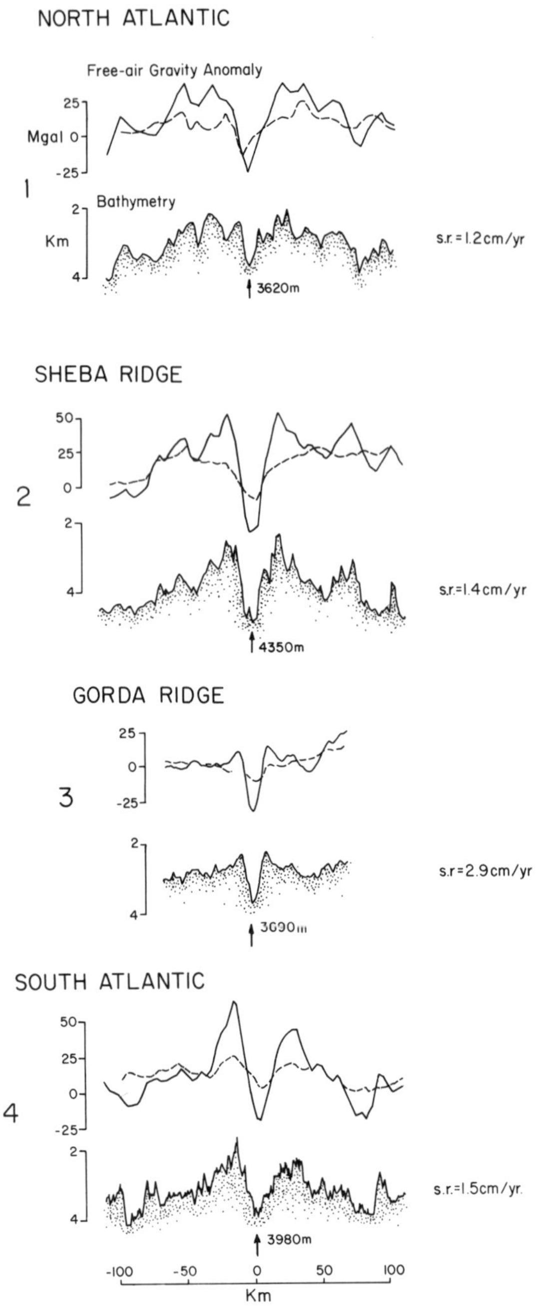

Figure 1. Selected free-air gravity anomaly, isostatic anomaly (based on Airy model of isostasy with T = 30 km), and bathymetry profiles of active oceanic rifts in the Atlantic (Profiles 1 and 4), Indian (Profile 2) and Pacific (Profile 3) oceans. Each profile has been projected normal to the local trend of the ridge crest. The half-spreading rates (s.r.), based on magnetic anomaly lineations over the ridge crest, are shown to the right of each bathymetry. The location of the profiles and sources of data are given in Table 1.

Isostasy

There is generally a strong correlation between short-wavelength free-air gravity anomalies and bathymetry over oceanic rifts and ridge flanks (for example, Figs. 1 and 2). This correlation provides information on the nature of the isostatic mechanism at ridge crests and the origin of oceanic rifts. In addition, since ridge flank topography is probably created at the ridge axis and then transported away from the ridge by sea-floor spreading (Needham and Francheteau, 1974), the isostatic mechanism at ridge flanks provides constraints on the physical processes occurring at the ridge axis.

Fig. 1 and 2b show isostatic gravity anomalies based on the Airy model (assumed depth of compensation T_c = 30 km) for each ridge crest profile. Although the isostatic anomalies over ridge crests are generally smaller than the free-air gravity anomalies, there are large departures from Airy isostasy particularly over the rifts. The oceanic rifts on each of the profiles are undercompensated, according to the Airy model.

The problem with using the Airy model at mid-ocean ridge crests, however, is that it predicts large variations in crustal thickness between the rim mountains and oceanic rifts. Seismic studies in the Atlantic (Keen and Tramontini, 1970) and Pacific oceans (Goslin et al., 1972; Stoffa et al., 1980) show that oceanic crust is of normal thickness at least to within about 10 km from the ridge axis. Thus the rim mountains do not appear to be supported by thickened crust, as predicted by the Airy model.

Cochran (1979) considered both the Airy model and plate or flexure model of isostasy at ridge crests. Isostatic anomalies based on the flexure model (assumed effective elastic thickness T_e = 7 to 13 km) showed only small departures from isostasy at oceanic rifts. However, this model, like the Airy model, also predicts significant crustal thickening beneath the rim mountains.

Although the nature of the isostatic mechanism at oceanic rifts is therefore not well understood, some insight into the physical processes at the ridge axis have come from studies of the isostatic mechanism of ridge flank topography. Talwani et al. (1972) showed that for wavelengths less than about 150 km the correlation between free-air gravity anomalies and topography could be explained by a simple model in which the topography of the ridge flank was uncompensated. McKenzie and Bowin (1976) showed, using linear transfer function techniques, that the correlation between gravity anomalies and topography could be best explained if ridge flank topography was compensated by the flexure model with T_e = 10 km. The analysis of McKenzie and Bowin (1976), included ridge flank topography as well as features which formed at a ridge crest, such as fracture zone topography and the aseismic Walvis ridge.

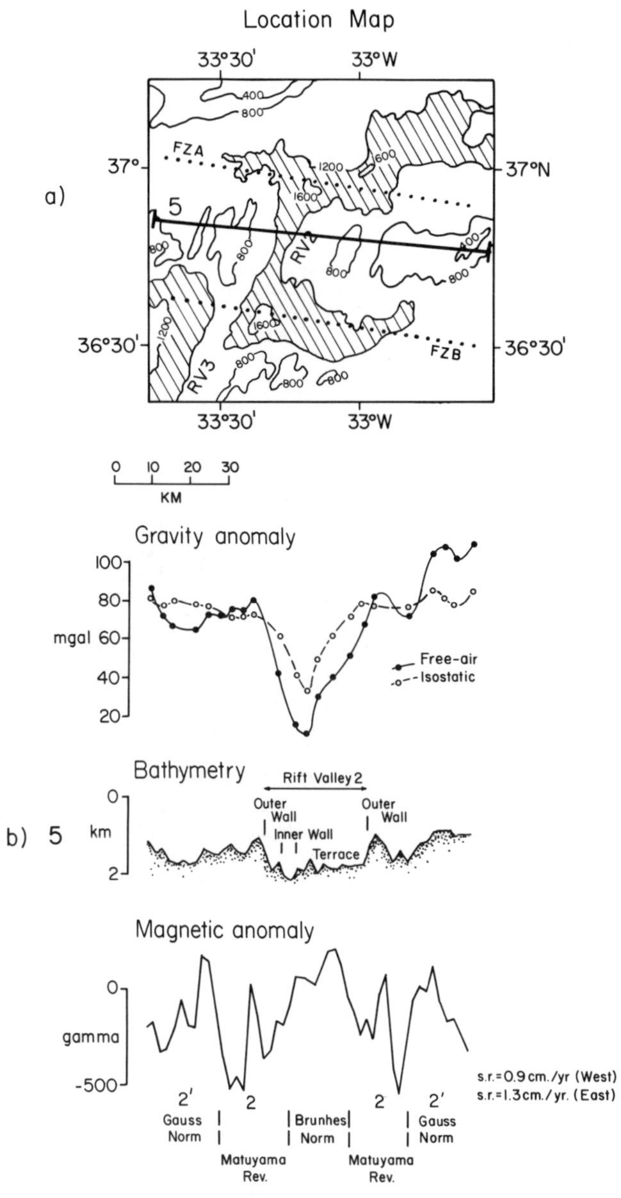

Figure 2. (a). Summary bathymetry map of the FAMOUS area (ARCYANA, 1975) of Mid-Atlantic ridge crest at 37° North (Ramberg et al., 1977). The contour interval is 400 fthm (732 meters). Heavy dots indicate fracture zones A and B. The continuous line shows the location of Profile 5 in Fig. 2b. RV2 = rift valley 2 and RV3 = rift valley 3 (Ramberg et al., 1977).
(b). Gravity anomaly, bathymetry and magnetic anomaly profile of rift valley 2 in the FAMOUS area of the Mid-Atlantic ridge crest at 37° North. The physiographic provinces on the bathymetry profile are based on terminology proposed by Needham and Francheteau (1974). The magnetic anomaly identifications and approximate boundaries between magnetic epochs are shown below the magnetic anomaly profile. Note the minimum in the free-air gravity anomaly and isostatic anomaly occurs not over the inner floor, which is considered the locus of crust of zero age, but over the terrace approximately equi-distant between the Outer walls. The half-spreading rates shown are based on Needham and Francheteau (1974) and assume zero age crust at the inner floor and their identification of the Brunhes-Matuyama boundary.

Recently, McNutt (1979) used similar transfer function techniques and analyzed free-air gravity anomaly and bathymetry data from the SURVEYOR survey of the Juan de Fuca and Gorda ridge crests. This study, which was based on gravity anomaly and bathymetry maps rather than individual profiles, showed that the gravity anomalies could be best explained if ridge flank topography was compensated by the flexure model with T_e = 2 to 6 km (Fig. 3).

Origin of the Rift Valley

Studies of the gravity field over ridge crests therefore provide substantial evidence for the existance of a continuous layer of strength at the ridge axis. The effective elastic thickness T_e of the ridge axis, as inferred from ridge flank gravity anomaly and bathymetry data, is in the range 2 to 13 km (McKenzie and Bowin, 1976; Cochran, 1979; McNutt, 1979). These estimates for the effective elastic thickness, which are significantly less than values associated with seamount loads in the plate interiors (20 to 30 km; Watts, 1978), can be generally interpreted as due to the origin of ridge flank topography in the axial region of the ridge crest, where the lithosphere is hot and weak.

The presence of a continuous layer of strength at the ridge axis has also been inferred from previous studies based on thermal and petrologic models (Turcotte, 1974; Tapponnier and Francheteau, 1978). Turcotte (1974), based on evidence for the depth to the 700°C isotherm beneath southwest Iceland, argued for a strong mechanical layer at the ridge axis. Tapponnier and Francheteau (1978) based on assumptions of the depth to the solidus temperature beneath the ridge axis, suggested the thickness of the mechanical layer was about 4 km.

Lonsdale and Speiss (1979) and Johnson and Vogt (1973) have mapped large topographic features near or at the ridge axis which require a strong mechanical layer to support them. Two small seamounts, which rise about 1.4 km above the ridge flank, occur either side of the East Pacific rise crest at lat. 8° 46'N on crust 0.6 m.y. old. Large volcanoes are generally absent in the axial region of the rise crest. High central volcanoes (up to a few hundred meters relief) sometimes occur, however, in the axial region of the Mid-Atlantic ridge (Johnson and

TABLE 1. Summary of Gravity Anomaly and Bathymetry Data

Profile Number	Ridge Crest	Ship	Cruise	Position of Ridge Crest Crossing		Azimuth	Reference for Location of Profile
1	North Atlantic	Snellius	SN07	28.0°N	44.0°W	90°	Cochran (1979), Fig. 6
2	Sheba	Vema	V3503	14.5°N	56.3°E	32°	Cochran (in press)
3	Gorda	Surveyor	IDOE	42.5°N	126.8°W	90°	IDOE 1971 Surveyor Seamap*, Line 36
4	South Atlantic	Vema	V2412	34.0°S	5.0°W	90°	Cochran (1979), Fig. 6
5	North Atlantic FAMOUS	Vema	V3203	36.8°N	33.3°W	113°	This paper, Fig. 2a
6	Labrador Sea	Vema	V2911	57.0°N	44.0°W	25°	Kristoffersen and Talwani (1977), Fig. 1
7	Coral Sea	Vema	V3313	13.8°S	152.1°E	26°	Weissel and Watts (1979), Fig. 1

* Data available from National Geophysical and Solar Terrestrial Data Center (NGSDC).

Vogt, 1973). The absence of large volcanoes in the axial region of the East Pacific rise may be due to the lesser strength of the lithosphere at fast-spreading ridges (Lonsdale, 1977). Thicker, stronger, lithosphere at slow-spreading ridges may explain the presence of high central volcanoes in the axial region of the Mid-Atlantic ridge.

A number of studies have now been carried out of the nature of the isostatic mechanism at seamounts and oceanic islands in the interiors of the plates (summarized in Watts et al., 1980). These studies suggest that the effective elastic thickness T_e of oceanic lithosphere is a strong function of the age of the lithosphere at the time it is loaded and that it does not change appreciably with time. Watts (1978) showed that T_e corresponds closely to the 300°C to 600°C oceanic isotherms, based on simple cooling models.

These studies of the plate interiors have implications for the gravity field over oceanic rifts. First, estimates of T_e for ridge crests (2 to 13 km) may generally constrain the average depth to the 300°C to 600°C oceanic isotherms at the ridge axis. Second, gravity anomalies associated with oceanic rifts like those associated with seamounts and oceanic islands should be supported for long periods of geological time.

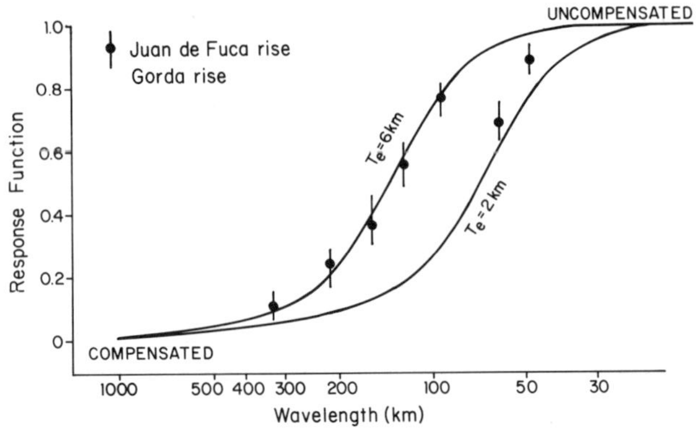

Figure 3. Isostatic response function obtained by McNutt (1980) from gravity anomaly and bathymetry data in the SURVEYOR area of the Juan de Fuca and Gorda rise crests. This function represents the variation in isostatic compensation of ridge crest topography as a function of wavelength. The critical wavelengths for which isostasy is important at ridge crest topography is in the range of about 40 to 300 km. For shorter wavelengths features are uncompensated but for longer wavelengths features are fully compensated. The solid line is a theoretical isostatic response function based on the flexure model for an assumed effective elastic thickness of T_e = 6 km and T_e = 2 km and a crustal structure based on a 2 km thick layer of density 2.6 g/cm^3 and a 3 km thick layer of density 2.9 g/cm^3 overlying a mantle of density 3.4 g/cm^3.

Figure 4 shows free-air gravity anomaly and bathymetry profiles over two oceanic rifts in the Labrador sea (Atlantic) and Coral Sea (Pacific) which ceased spreading 40 to 56 m.y.B.P. The amplitude and wavelength of the gravity anomaly associated with these extinct rifts are similar to the gravity anomalies associated with the active rifts (Fig. 1). Thus gravity anomalies over oceanic rifts do not appear to change appreciably with time, in agreement with results of flexure studies in the plate interiors.

A number of models have now been proposed to explain the existence of oceanic rifts. Deffeyes (1970) and Anderson and Noltimier (1973) consider that oceanic rifts occur because of an imbalance between the width of lava intrusion and the distance required to reach the mean spreading rate. Sleep (1969) and Sleep and Rosendahl (1979) suggest a viscodynamic explanation in which oceanic rifts form as a result of viscous head loss of the ascending hot mantle material. Collette et al. (1980), in contrast, consider that oceanic rifts originate due to a viscous lag in the mantle underlying the ridge axis.

These models for oceanic rifts therefore differ from that proposed by Vening Meinesz (1950) for

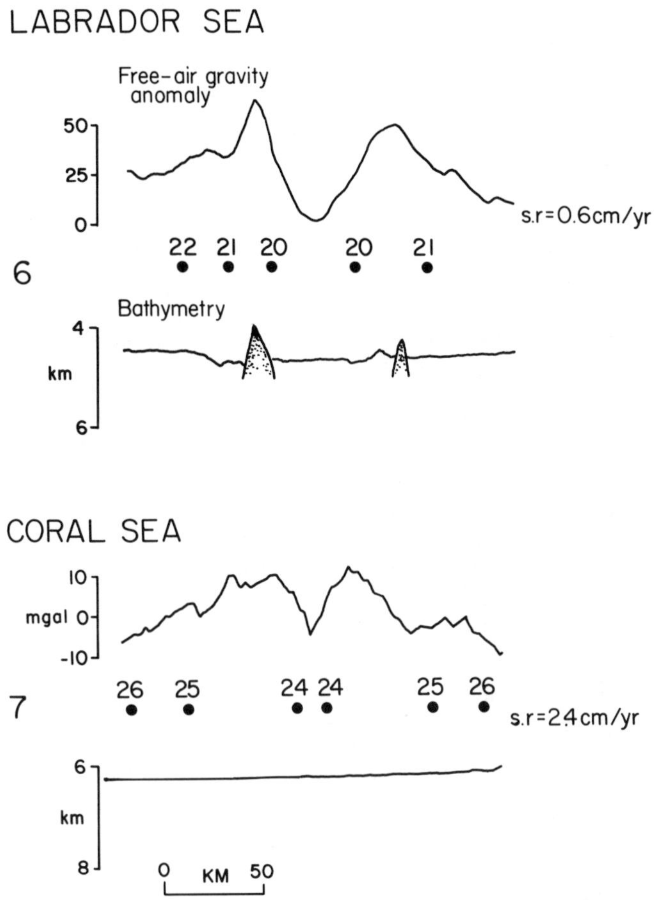

Figure 4. Selected free-air gravity anomaly and bathymetry profiles of the extinct ridge crests in the Labrador and Coral sea. The location of the profiles are given in Table 1. The magnetic anomaly identifications, based on Kristoffersen and Talwani (1977) for the Labrador sea and Weissel and Watts (1979) for the Coral sea, are shown above each bathymetry profile. The Labrador sea, which shows a major slowdown in half-spreading rate at anomaly 20 (about 46 m.y.B.P.), ceased spreading at anomaly 13 (about 40 m.y. B.P.). The Coral sea ceased spreading at anomaly 24 (about 56 m.y.B.P.). The topography of these extinct ridges is obscured by sediments but each spreading center is associated with a prominent free-air gravity anomaly low and a flanking gravity high.

continental rifts. In this model, steep normal faults formed by crustal extension produce a narrow wedge which subsides to form a rift valley. Vening Meinesz (1950) showed that the width of the rift would be expected to be less than about twice the elastic thickness of the crust. The elastic thickness at oceanic rifts, based on gravity and bathymetry studies as well as thermal and petrologic studies, is in the range 2 to 13 km suggesting, on the basis of Vening Meinesz's (1950) hypothesis, that the width of oceanic rifts should be less than about 4 to 26 km. This is in general agreement with the observed rifts in Fig. 1, which are typically less than 30 km in width. Le Pichon et al., (1973) rejected Vening Meinesz's (1950) hypothesis, however, since oceanic rifts are not a single down-dropped wedge between 60° fault scarps but are a series of low dipping (30°) blocks descending toward the axis of the rift.

Recently, Tapponnier and Francheteau (1978)

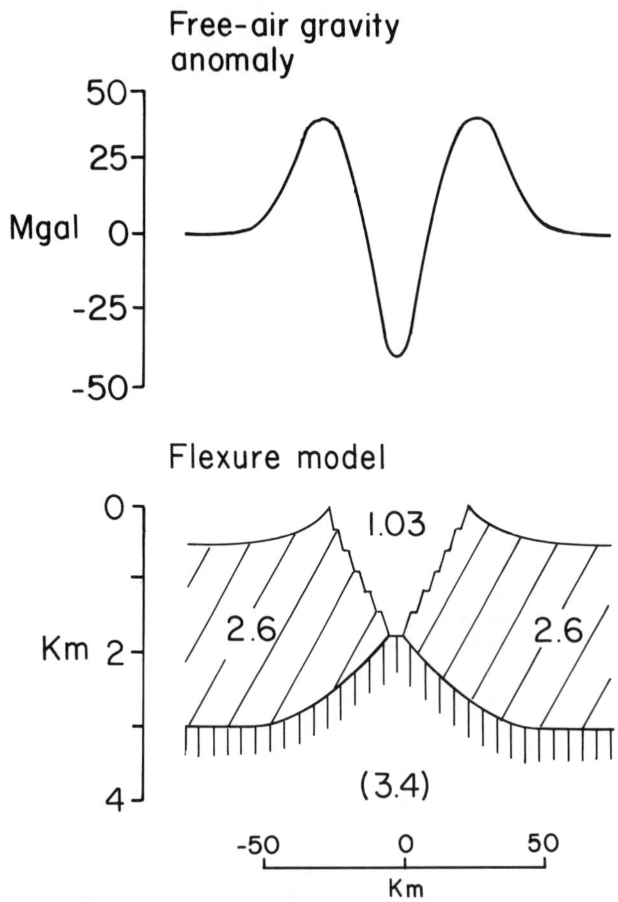

Figure 5. Computed free-air gravity anomaly based on the simple flexure model proposed by Tapponnier and Francheteau (1978) at oceanic rifts. In this model the mass deficit of the rift is balanced by a mass excess at depth. Since a layer of continuous strength is assumed at the ridge axis the balance is not achieved locally but regionally over a broad area. The flexure model shown has been computed assuming a net upward force, equal to the mass deficit, acts on a thin elastic plate with $T_e = 4$ km (Tapponnier and Francheteau, 1978). The gravity anomaly has been computed using the simple Bouguer slab approximation and the density distribution shown.

proposed a model for oceanic rifts which includes a thin elastic layer at the ridge axis. In this model the oceanic rift is a consequence of steady state necking or stretching of the lithosphere. The region of the rift is not stretched to zero thickness, however, since a uniform thickness of lithosphere is maintained at the rift. A result of the stretching is a restribution of mass from the peripheral regions into the axial region. Isostasy is not achieved locally by crustal thinning but regionally by a broad uparching of the rift. The wavelength of the uparching is wider than the oceanic rift itself and depends on the thickness of the elastic layer assumed.

Gravity data alone cannot, unfortunately, distinguish between these models for oceanic rifts. Sleep's (1969) model for slow-spreading ridges, in which the oceanic rift is caused by a loss of hydrostatic head, could explain the presence of large-amplitude negative gravity anomalies at rifts. This model cannot, however, explain the presence of small-amplitude gravity anomaly highs in flanking regions (Figs. 1 and 4). Tapponnier and Francheteau's (1978) model, which includes an elastic layer at the ridge axis, can explain the observed gravity anomaly patterns (Fig. 5). The main problem with this model, however, is in calculating the relative contribution of the gravity effect of the rift and the uparched crustal layers. Although the crustal structure at oceanic rifts is reasonably well known beneath the rim mountains, it is poorly known beneath the rift.

This review therefore suggests that there has been a significant increase in our understanding of the gravity field and tectonics of oceanic rifts during the past decade. Most information on the nature of the isostatic mechanism at oceanic rifts, however, has come not from studies of the rifts themselves but from geological features which form at or near rifts and are transported away from them by sea-floor spreading. We currently know very little about the nature of isostasy and the relative importance of viscous and elastic forces at rifts. There is accumulating evidence, however, for the existance of an elastic layer at both active and inactive rifts. Further studies using long seismic arrays, in combination with gravity data, appear to hold the most promise to address this problem during the next decade.

Acknowledgments. This work was supported by Office of Naval Research contract N00014-75-C-0210 Scope C and National Science Foundation grant OCE 79-18917. J. R. Cochran and B. J. Collette made some helpful comments on the manuscript. Lamont-Doherty Geological Observatory Contribution Number 3071.

References

Anderson, R.N. and H.C. Noltimier, A model for the horst and graben structure of mid-ocean ridge crests based upon spreading velocity and basalt delivery to the oceanic crust, Geophys. J. R. Astr. Soc., 34, 137-147, 1973.

ARCYANA, Transform fault and rift valley by bathyscaph and diving saucer, Science, 190, 108-116, 1975.

Cochran, J.R., An analysis of isostasy in the worlds oceans, Part 2, Mid-ocean ridge crests, J. Geophys. Res., 84, 4713-4729, 1979.

Cochran, J.R. and M. Talwani, Free-air gravity anomalies in the world's oceans and their relationship to residual elevation, Geophys. J. R. Astr. Soc., 50, 495-552, 1977.

Collette, B.J., J. Verhoef and A.F.J. de Mulder, Gravity and a model of the median valley, J. Geophys., 47, 91-98, 1980.

Deffeyes, K.S., The Axial Valley: A steady-state feature of the terrain, in: The Megatectonics of Continents and Oceans, edited by H. Johnson and B.L. Smith, Rutgers Univ. Press, N. J., 194-222, 1970.

Goslin, J., P. Beuzart, J. Francheteau, and X. Le Pichon, Thickening of the oceanic layer in the Pacific Ocean, Marine Geophys. Res., 1, 418-427, 1972.

Johnson, G.L. and P.R. Vogt, Mid-Atlantic Ridge from 47° to 51° North, Geol. Soc. Am. Bull., 84, 3443-3462, 1973.

Keen, C. and C. Tramontini, A seismic refraction survey on the Mid-Atlantic Ridge, Geophys. J. R. Astr. Soc., 20, 473-493, 1970.

Kristoffersen, Y. and M. Talwani, Extinct triple junction south of Greenland relative to North America, Geol. Soc. Amer. Bull., 88, 1037-1049, 1977.

Lambeck, K., Gravity anomalies over ocean ridges, Geophys. J. R. Astr. Soc., 30, 37-53, 1972.

Le Pichon, X., J. Bonin and J. Francheteau, Plate tectonics, Elsevier, N. Y., 300pp., 1973.

Lonsdale, P.F., Structural geomorphology of a fast-spreading rise crest: the East Pacific Rise near 3°25'S, Marine Geophys. Res., 3, 251-293, 1977.

Lonsdale P. and F.N. Spiess, A pair of young cratered volcanoes on the East Pacific Rise, J. of Geol., 87, 157-173, 1979.

McKenzie, D.P. and C. Bowin, The relationship between bathymetry and gravity in the Atlantic Ocean, J. Geophys. Res., 81, 1903-1915, 1976.

McNutt, M., Compensation of oceanic topography: An application of the response function technique to the Surveyor area, J. Geophys. Res., 84, 7589-7598, 1979.

Needham, H.D. and J. Francheteau, Some characteristics of the rift valley in the Atlantic Ocean near 36°48' north, Earth Planet. Sci. Lett., 22, 29-43, 1974.

Ramberg, I.B., D.F. Gray, and R.G.H. Raynolds, Tectonic evolution of the FAMOUS area of the Mid-Atlantic ridge, lat. 35°50' to 37°20' N., Geol. Soc. Amer. Bull., 88, 609-620, 1977.

Sclater, J.G. and J. Francheteau, The implications of terrestrial heat flow observations on current tectonic and geochemical models of the crust and upper mantle of the Earth,

Geophys. J. R. Astr. Soc., 20, 509-542, 1970.

Sleep, N.H., Sensitivity of heat flow and gravity to the mechanism of sea-floor spreading, J. Geophys. Res., 74, 542-549, 1969.

Sleep, N.H. and B. Rosendahl, Topography and tectonics of mid-ocean ridge axes, J. Geophys. Res., 84, 6831-6839, 1979.

Stoffa, P.L., P. Buhl, T.J. Herron, T.K. Kan, and W.J. Ludwig, Mantle reflections beneath the crestal zone of the East Pacific rise from multi-channel seismic data, Marine Geol., 35, 83-97, 1980.

Talwani, M., B.C. Heezen, and J.L. Worzel, Gravity anomalies, physiography and crustal structure of the Mid-Atlantic Ridge, Publ. 22, Bur. Central Seismol. Int. Ser. A. Trav. Sci. Facs., Int. Union of Geol. and Geophys., Paris, 81-111, 1961.

Talwani, M., X. Le Pichon, and M. Ewing, Crustal structure of the mid-ocean ridges, 2, Computed model from gravity and seismic refraction data, J. Geophys. Res., 70, 341-352, 1965.

Tapponnier, P. and J. Francheteau, Necking of the lithosphere and the mechanics of slowly accreting plate boundaries, J. Geophys. Res., 83, 3955-3970, 1978.

Turcotte, D.L., Are transform faults thermal contraction cracks?, J. Geophys. Res., 79, 2573-2577, 1974.

Van Andel, T.H. and C. Bowin, Mid-Atlantic ridge between 22° and 23°N latitude and the tectonics of mid-ocean rises, J. Geophys. Res., 73, 1279-1298.

Vening Meinesz, F.A., Gravity expeditions at sea, 1923-1938, vol. IV, Netherlands Geodetic Comm., Delft, 233 pp., 1948.

Vening Meinesz, F.A., Les "grabens' africains, résultat de compression ou de tension dans la croûte terrestre?, Inst. Roy. Colon. Belge. Bull., 21, 539-552, 1950.

Watts, A.B., An analysis of isostasy in the world's oceans, 1. Hawaiian-Emperor Seamount Chain, J. Geophys. Res., 83, 5989-6004, 1978.

Watts, A.B., J.H. Bodine, and N.M. Ribe, Observations of flexure and the geological evolution of the Pacific ocean basin, Nature, 283, 532-537, 1980.

Weissel, J.K., and A.B. Watts, Tectonic evolution of the Coral sea basin, J. Geophys. Res., 84, 4572-4582, 1979.

Woodside, J.M., The Mid-Atlantic ridge near 45°N, XX, The gravity field, Can. J. Earth Sci., 9, 942-959, 1972.

HEAT FLOW IN RIFT ZONES

Paul Morgan*

Department Earth Sciences and Physics, Box 3D, New Mexico State University, Las Cruces, New Mexico 88003

Abstract. Heat flow is fundamental to the understanding of the genesis of rift zones. A detailed compilation of heat flow data for Cenozoic continental rift zones gives mean heat flow values in the range 20 to 110 mW m^{-2} within the rifts, with large scatters in the data sets for the volcanic rifts, and significantly lower mean heat flow for non-volcanic rifts. For the volcanic rifts the transition to normal heat flow outside the rifts is sharp. Heat flow data from oceanic rift zones show a very large scatter, with means consistently lower than predicted from spreading models. The data indicate a large component of convective heat transport in the thinly sedimented crests of oceanic rifts. The continental rift thermal anomalies appear to be caused by igneous intrusion into the rifted crust, modified by groundwater heat transport. Regional and detailed local heat flow studies of oceanic rifts indicate that the total conductive and convective heat loss from the rifts is consistent with the thermal process of sea-floor spreading.

Introduction

Active thermal processes are dramatically manifested by volcanism in rift zones and are intrinsic to the very existence of rifts. Heat flow measurements provide information on which deductions can be made on the nature, magnitude, and history of these thermal processes. The great differences in the rates and effectiveness of heat transfer by conduction and convection, however, coupled with the temporal and spatial diffusion of conductive thermal anomalies, veil the power of surface heat flow data to resolve deep seated thermal processes. This should not be seen as a restriction. It is a challenge, because ultimately heat is the energy form responsible for rift genesis, and surface heat flow is the only directly measurable boundary condition for this important parameter.

*Present address: Lunar and Planetary Institute, 3303 NASA Road 1, Houston, Texas, 77058.

This chapter presents a summary of published heat flow data from Cenozoic continental and oceanic rift zones. Data are not presented from older rifts because the thermal processes associated with rifting leave a transient heat flow signature, which time has reduced to an insignificant level in pre-Cenozoic rifts. Models of continental and oceanic rifts incorporating the heat flow data are not presented in detail here.

Numerous discussions of rift models have been published during the past ten years, e.g. LePichon and Langseth (1969), Osmaston (1971), Bodvarsson and Lowell (1972), Pálmason (1973), Parker and Oldenburg (1973), Bottinga (1974), Lubimova and Nikitina (1975), Williams and Poehls (1975), Lachenbruch (1976), Lister (1977), Hurtig and Oelsner (1977), Parsons and Sclater (1977), Gass et al. (1978), Parsons and McKenzie (1978), Sleep and Rosendahl (1979), Withjack (1979), Bridwell and Anderson (1980), Bridwell and Potzick (1980), and the reader is referred to these references and to the conclusions and references of other authors in this volume. In discussing the data here the thermal processes are considered with respect to the local and regional thermal anomalies associated with rifts. The constraints placed on models by the heat flow data are emphasized.

The widely differing physical environments of rift zones, ranging from the deep ocean floor, to freshwater lakes, to desert climates, require different measurement techniques for heat flow, all of which have their advantages and disadvantages. A brief summary of measurement and data reduction and correction techniques is therefore given as it relates to the final analysis of the data. No compilation of heat flow data from continental rifts has previously been published, and so heat flow maps from the four major Cenozoic continental rifts are presented below. Several compilations of oceanic heat flow data have been published and it is not the purpose of this work to present a comprehensive review of these data. Only details and generalizations of oceanic heat flow relevant to oceanic rift zones are presented.

Heat Flow Determination Techniques

For measurement purposes, the parameter defined as terrestrial heat flow is the vertical component of the outward flow of heat from the earth, given by

$$q = K \, \partial T/\partial z \qquad (1)$$

where $\partial T/\partial z$ is the vertical temperature gradient, temperatures increasing with depth, K is the thermal conductivity of the earth material in which the gradient measurement is made, and q is the vertical heat flux, positive upward by convention. Two quantities, the gradient and the conductivity must therefore be measured for each heat flow determination.

Equation (1), however, is only a gross approximation to the complete heat transfer equation, it fails to account for lateral heat flow, heat generation within the measurement interval, non-conductive heat transport, and temporal variations in heat flow. The aim in heat flow data collection is to make measurements under conditions as close as possible to those assumed in equation (1), and to be able to correct for any departure from these conditions. Rift zones, unfortunately, are not ideal locations to meet these conditions, and part of the large scatter in heat flow values commonly observed in rift zones certainly reflects our inability to make adequate heat flow determinations. A summary of the application of the theory of heat conduction to geothermal measurements has been given by Jaeger (1965).

On land heat flow determinations are usually made by measuring the temperature gradient in a borehole greater than 100m in depth, and making laboratory thermal conductivity measurements of the rocks penetrated by the borehole. The basic techniques for these measurements have been described by Beck (1965). In the oceans and deep lakes the gradient is measured by a temperature probe which penetrates into the ocean or lake sediments, and conductivities are either measured _in situ_, or on sediment cores brought to the surface. The basic marine techniques have been described by Langseth (1965). Individual heat flow data derived from borehole measurements are generally regarded to be more reliable than a single marine determination. Marine determinations can be made in far greater numbers, however, which gives increased confidence in results if they show internal consistency.

Continental Rifts

Four major rift systems have been identified in continental settings which show evidence of Cenozoic thermal activity, the Baikal rift, the East African rift system, the Rhinegraben, and the Rio Grande rift. Heat flow data from these rift systems are presented in this section. Data from older continental rift systems are not presented because the heat flow anomalies associated with pre-Cenozoic rifting events have decayed to immeasurable levels. Minor thermal anomalies have been measured over older rifts, e.g. a 16 mW m^{-2} anomaly associated with the Oslo Graben (Haenel _et al._, 1974, 1979; Swanberg _et al._, 1974), but these anomalies appear to be caused by lateral changes in the thermal properties of the crustal rocks rather than residual heat associated with rift tectogenesis. Data are not presented from the broad extensional zone of the Basin and Range province of the western United States. This area has many geological and geophysical properties in common with the classic continental rifts, but is not directly analogous to the rift zones. For heat flow data and models for the Basin and Range province the reader is referred to Lachenbruch and Sass (1977, 1978), Lachenbruch (1979) and Blackwell (1978).

The Baikal Rift

The most complete compilation of heat flow data for the Baikal rift has been presented by Lubimova _et al._ (1972), and it is on this compilation that Figure 1 is based. Data are derived from marine type measurements in the sediments of Baikal Lake and from borehole determinations on land. Additional data have been presented by Moiseenko _et al._ (1973), Lysak (1976), and Duchkov _et al._ (1976, 1977), but because of inconsistencies between the data sets most of these data have not been integrated into Figure 1. The mean of 52 determinations reported by Lubimova _et al._ (1972) for the rift zone is 97±22 (standard deviation) mW m^{-2}, whereas Lysak (1976) presents 39 values with a mean of 83 ± 34 mW m^{-2}, and Moiseenko _et al._ (1973) report 17 determinations with a mean of 88 ± 28 mW m^{-2}. For the Siberian platform to the northwest of the rift zone the three groups of authors report means of 45 ± 6 (n = 13), 46 ± 10 (n = 34), and 45 ± 7 (n = 4), respectively, and means of 0 (no data), 54 ± 12 (n = 14), and 56 ± 10 (n = 12), respectively, for the Caledonian fold belt to the southeast of the rift. As Lubimova _et al._ (1972) do not present data from the fold belt, data from Lysak (1976) are shown in Figure 1 in parentheses. Detailed profiles of heat flow determinations by Duchkov _et al._ (1976, 1977) have identified a belt of low heat flow (42-54 mW m^{-2}) running along the lake in its central portion, but these workers tentatively conclude that these low values are partly or wholly due to surface factors, mainly temperature fluctuations in the bottom layer of water. In summary, all the published data are consistent with high heat flow in the rift zone, 80-100 mW m^{-2}, with an abrupt transition to low heat flow, 45-46 and 54-56 mW m^{-2} outside the rift zone. The standard deviations of the means of the data for the rift zone are greater than the standard deviations outside the rift by a factor of two or three.

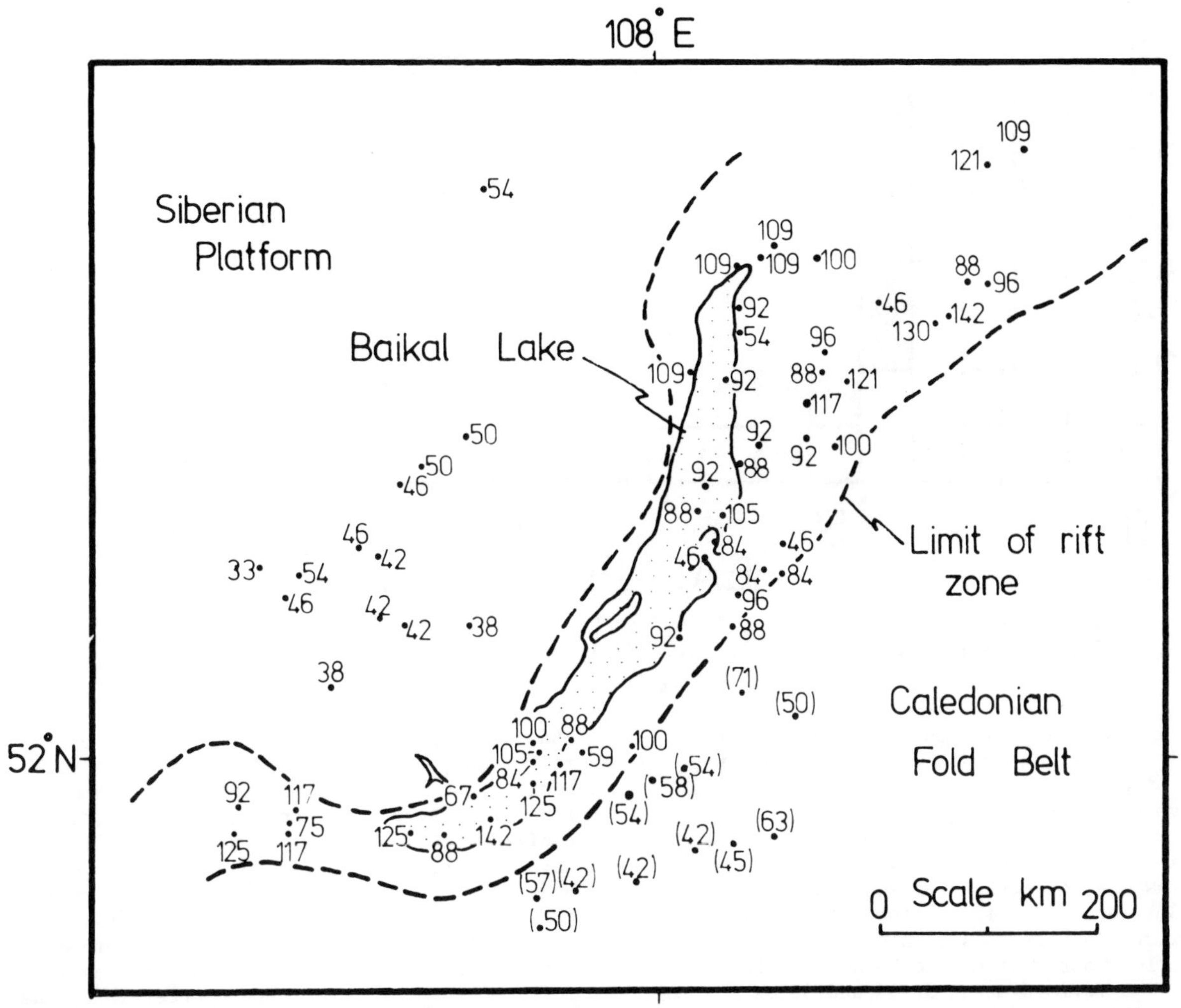

Fig. 1. Heat flow data in mW m^{-2} from the Baikal rift, adapted from Lubimova *et al*. (1972) with additional values in parentheses from Lysak (1976).

The East African Rift System

The East African rift system is divisible into the western and eastern rifts (e.g. see Baker *et al*., 1972), and the types of heat flow data available for East Africa are similarly divided. Lake determinations have been made in three deep lakes in the western rift, and borehole based determinations have been made in Kenya, across the eastern rift.

The data from the lake determinations are summarized in Figure 2. Twenty values from Lake Malawi (Von Herzen and Vacquier, 1967) range from 6 to 120 mW m^{-2} with a systematic areal distribution of the values with regions of low heat flow at the northern and southern ends of the lake (22 ± 13 mW m^{-2}, n = 12, and 31 ± 3 mW m^{-2}, n = 3, respectively) separated by a central region of high heat flow (96 ± 25 mW m^{-2}, n = 5). All but one of twelve heat flow values from Lake Tanganyika (Degens *et al*., 1971) are less than 60 mW m^{-2}, and excluding the high value the mean of the eleven remaining determinations is 38 ± 13 mW m^{-2}. The single value of 150 mW m^{-2} is located near a shift in the linear trend of the lake. Heat flow values from five localities in Lake Kivu have a wide range from 17 to 185 mW m^{-2}, although values reproduced well at stations where measurements were repeated (Degens *et al*., 1973). It is evident therefore that the heat flow is laterally highly variable in Lake Kivu.

Von Herzen and Vacquier (1967) and Degens *et al*. (1971) present similar comments on the data from Lakes Malawi and Tanganyika. The high values from both lakes are located near structural shifts in the lakes, and it is possible that heat transfer by groundwater convection, controlled by faulting transverse to the lakes causes local thermal anomalies. Von Herzen and Vacquier (1967) also suggest that the high values in Lake Malawi may result from shallow (<20 km) magmatic intrusions, but although there are hot springs along the margin of Lake Malawi,

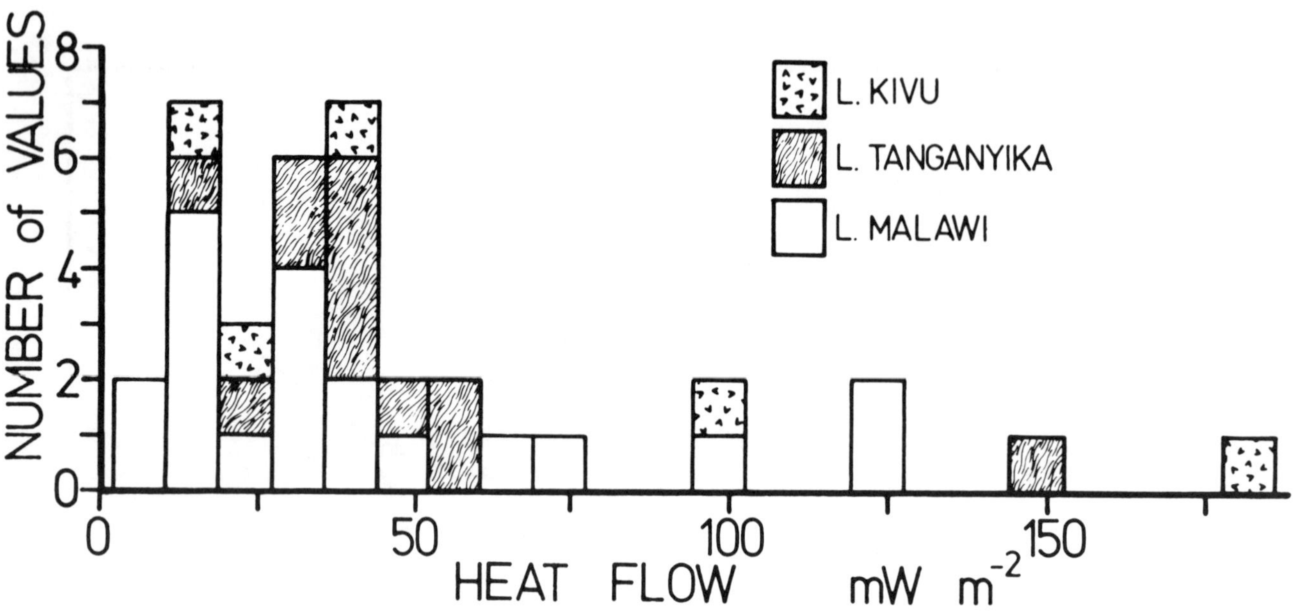

Fig. 2. Histogram of heat flow values from Lakes Malawi (after Von Herzen and Vacquier, 1967), Tanganyika (after Degens _et al_., 1971), and Kivu (after Degens _et al_., 1973).

there is no visible volcanism close to Lake Malawi or Lake Tanganyika to support this hypothesis. The low values from these lakes may be depressed by locally high sedimentation rates, or warming of the deep waters since the last glacial period. Degens _et al_. (1971) suggest a 13-17 mW m^{-2} correction for sedimentation and up to a 30% correction for bottom water warming. If maximum corrections are applied to the means of the low values from Lake Malawi and Lake Tanganyika revised means of 46 and 57 mW m^{-2} for Lake Malawi and 66 mW m^{-2} for Lake Tanganyika are obtained. The heat flow for these areas of the western rift is thought to lie between the limits set by the uncorrected and corrected data, i.e., in the range 22-57 mW m^{-2} for Lake Malawi and 38-66 mW m^{-2} for Lake Tanganyika.

Lake Kivu is in an area of the rift associated with extensive volcanism and Degens _et al_. (1973) report hydrothermal discharges in the lake. Hydrothermal discharge and circulation can explain the wide range of heat flow values in Lake Kivu by two mechanisms, convective heat transport and bottom water temperature fluctuations. Little can be concluded except that the heat flow is highly variable.

Preliminary heat flow measurements across the eastern rift in Kenya have been reported by Morgan (1973) and Morgan and Wheildon (1980), and these data are shown in Figure 3. The data are from borehole based measurements and although the temperature data indicate significant convective groundwater disturbances at many of the sites, the data reveal a definite heat flow pattern across the rift. The heat flow varies from 57 ± 17 mW m^{-2} (n = 4) on the western rift shoulder, up to 98 ± 48 mW m^{-2} (n = 7) on the rift floor, back down to 39 ± 21 mW m^{-2} (n = 10) east of the rift. Subsequent unpublished heat flow measurements in Kenya (Williamson, 1975) confirm this heat flow pattern, and if the new data are included in the above averages, a mean heat flow of 105 ± 51 mW m^{-2} (n = 15) is obtained for the rift floor, and 52 ± 17 mW m^{-2} (n = 13) for the rift flanks. There is therefore a significant heat flow anomaly within the eastern rift in Kenya, with a large scatter in the data. Heat flow values for the rift flanks are similar to the values determined for the coastal region of East Africa (Evans and Tammemagi, 1974; Evans, 1975, Williamson, 1975) which have a mean of 58 ± 16 mW m^{-2} (n = 55).

The distribution of heat flow values in the Kenya rift does not follow a simple trend and there is certainly redistribution of heat in the upper crust by groundwater convection. Another indication of the high heat flow is by the geothermal areas within the rift, which Searle (1970) has correlated with the axis of a positive Bouguer gravity anomaly running north-south along the axis of the rift. A clustering of Quaternary central volcanoes along this trend also indicates a region of very high heat flow along the rift axis (Searle, 1970). There was extensive late Tertiary and Quaternary volcanic activity in the Kenya rift (Baker _et al_., 1972), and source of the high heat flow is clearly residual magmatic heat.

Chapman and Pollack (1975, 1977) suggest that heat flow data from Zambia support the existance of a zone of incipient rifting in the central African plateau with a heat flow anomaly of at

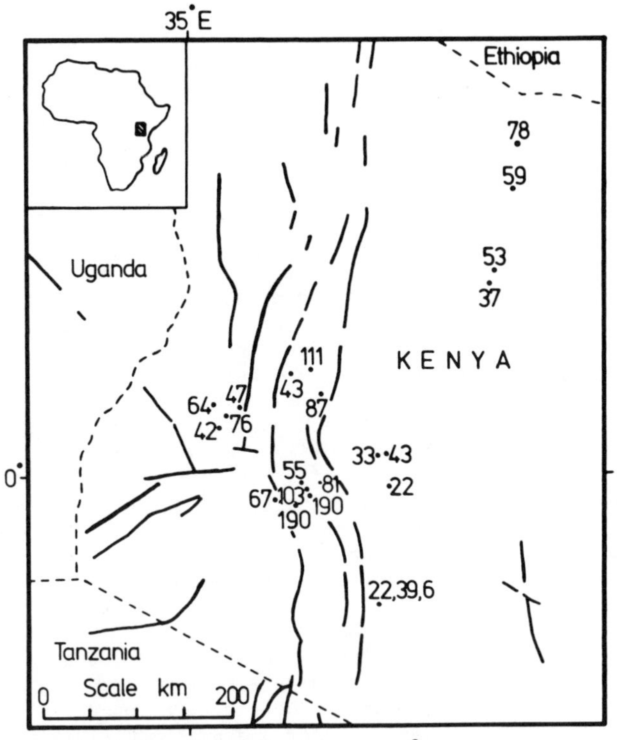

Fig. 3. Heat flow data in mW m^{-2} from the Kenya rift, adapted from Morgan and Weildon (1980).

least 15 mW m^{-2}. This anomaly is linked by other geophysical data to the East African rift system (Fairhead and Girdler, 1971), but surface geological evidence of rifting in this area is lacking.

The Rhinegraben

In presenting the results of a compilation of heat flow data from Europe, Hurtig and Oelsner (1977) report a mean heat flow of 107 ± 35 mW m^{-2} (n = 26) for the upper Rhinegraben. In contrast they report a heat flow of 73 ± 20 mW m^{-2} (n = 147) for the Hercynian foldbelt to the north and east of the rift. Published heat flow values for the Rhinegraben and surrounding areas are shown in Figure 4, based on a compilation by Bram (1979). From these data the mean heat flow in the rift zone is 112 ± 34 mW m^{-2} (n = 8) in contrast to 70 ± 10 mW m^{-2} (n = 22) for the surrounding areas, in good agreement with the mean values presented by Hurtig and Oelsner (op. cit.)

More detail of the geothermal field of the Rhinegraben is presented in studies of temperature measurements in oil exploration drill holes presented by Delattre et al. (1970) and Werner and Doebl (1974). These data indicate significant local thermal anomalies in the rift with widths as small as 1 kilometer, and whose intensity increases with depth. Werner and Doebl (op. cit.) conclude that these anomalies are probably caused by heat transfer by water circulation. The heat flow and temperature data therefore show high heat flow in the rift with a large scatter in the data, probably resulting from water circulation.

A third source of heat flow information for the Rhinegraben comes from studies of coal rank in the sedimentary rocks of the graben (Teichmüller, 1970; Doebl et al., 1974; Buntebarth, 1979). These studies give an indication of the thermal history of the sediments, and both Doebl et al. (op. cit.) and Buntebarth (op. cit.) report that the Rhinegraben was an area of high heat flow, 90-148 mW m^{-2}, during the peak in volcanic activity in the area and early rift genesis in the Lower Tertiary. Buntebarth (op. cit.) continues to report a decrease in heat flow to 64-94 mW m^{-2} at some locations in the Upper Tertiary, and that the recent high heat flow is associated with the upward migration of pore fluids related to Quaternary tectonic activity.

The Rio Grande Rift

Heat flow data from borehole measurements along the Rio Grande rift have recently been summarized by Reiter et al. (1979), and, excluding values greater than 167 mW m^{-2}, give a mean heat flow for the rift of 107 ± 27 mW m^{-2} (n = 25). Several attempts have been made to contour the Rio Grande rift heat flow anomaly (Reiter et al., 1975, 1979; Lachenbruch and Sass, 1977; Blackwell, 1978; Seager and Morgan, 1979; Swanberg 1979) but the scatter in the data allow different workers to contour the same data in a variety of patterns. Lachenbruch and Sass (1977) contour the Rio Grande rift as a series of discontinuous highs (>100 mW m^{-2}); all

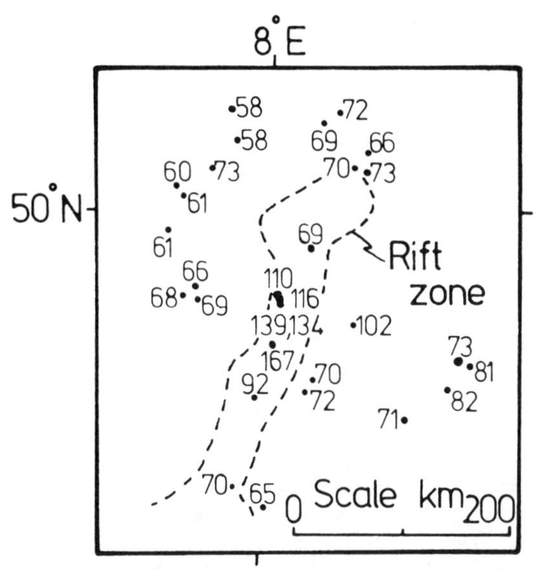

Fig. 4. Heat flow data in mW m^{-2} from the Rhinegraben, adapted from Bram (1979), with additional data from Haenel (1970).

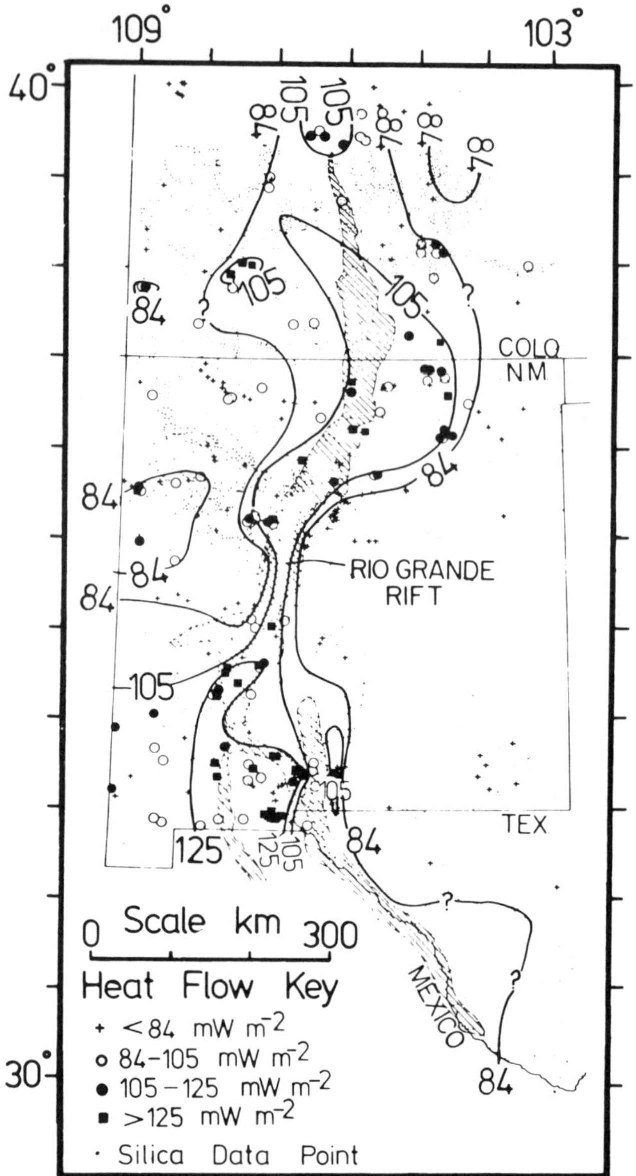

Fig. 5. Heat flow in mW m^{-2} in the Rio Grande rift, adapted from Seager and Morgan (1979) and Swanberg (1979).

other workers encompass most of the rift in a single envelope of high heat flow (>100 mW m^{-2}), although this interpretation inevitably includes several heat flow values less than 100 mW m^{-2} within the high heat flow contour. Seager and Morgan (1979) and Swanberg (1979) added heat flow values estimated from groundwater silica analyses by the technique of Swanberg and Morgan (1979, 1980) to the conventional heat flow data to produce the heat flow map from which Figure 5 is derived. This map, in common with the other heat flow maps, illustrates the basic pattern of high heat flow in the rift, but shows how little the high values are constrained to the actual rift. There is a more general correlation between areas of high heat flow and areas of Upper Tertiary and Quaternary volcanism in the area, not all of which is clearly related to the rift tectogenesis. Outside the rift anomaly, heat flow values decrease to less than 60 mW m^{-2} in the Great Plains to the east and in the central Colorado Plateau to the west and northwest, and to values of 80-100 mW m^{-2} in the Basin and Range and margin of the Colorado Plateau to the southwest (Lachenbruch and Sass, 1977).

Reiter _et al_. (1978) report heat flow values of 250 to 670 mW m^{-2} at four locations along the rift, and evaluations of geothermal resources in the area (e.g. Swanberg, 1979) indicate that numerous other locations along the rift may have very high heat flow. Unlike the Kenya rift, however, the high values do not cluster along the center of the rift, in fact many fall on the margins of the rift basins. Harder _et al_. (1980) argue that most of these high values are caused by forced groundwater convection, and do not indicate higher crustal heat flow values within the rift high heat flow zone.

Evidence for high heat flow from the lower crust beneath the Rio Grande rift comes from surface heat flow/radiogenic heat generation and petrologic studies. Decker and Smithson (1975) and Edwards _et al_. (1978) present heat flow/heat generation data which show that the heat flow anomaly is not caused by upper crustal radiogenic heat generation. Cook _et al_. (1978) report a 20 mW m^{-2} lower crustal heat flow anomaly for the southern Rio Grande rift, relative to the adja-

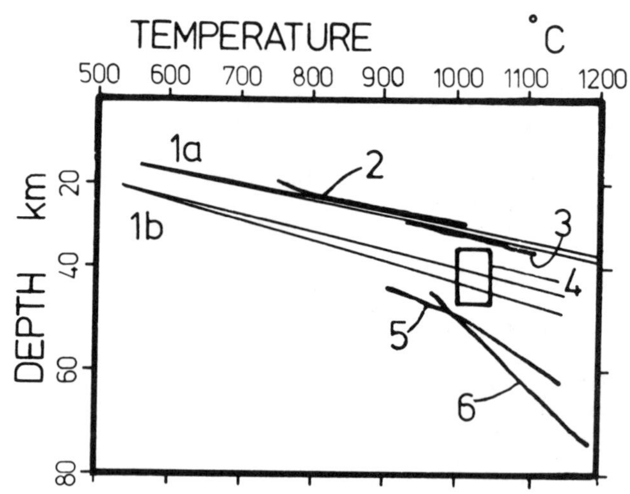

Fig. 6. Calculated geotherm segments for the southern Rio Grande rift, adapted from Seager and Morgan (1979). Curves 1a, 1b, extrapolated from heat flow; 1a:100 mW m^{-2}, 1b:84 mW m^{-2} (after Decker and Smithson, 1975). Curves 2-6, petrologic geotherms; 2: Padovani and Carter (1977), 3: Warren (1978), 4: Fodar (1978), 5: Reid (1976), 6: Mercier (1977).

cent Basin and Range province, which they model with multiple magmatic intrusions into the lower crust during the Upper Tertiary. Lower crustal and upper mantle geotherms beneath the rift inferred from xenolith petrologic studies are shown in Figure 6 from a summary by Seager and Morgan (1979). These data support the extension of high geothermal gradients down into the lower crust consistent with a surface heat flow of 100 mW m^{-2}. Lower gradients are indicated in the upper mantle, however, indicating significant convective heat transport below the Moho.

Data have been presented to suggest that the Rio Grande rift thermal anomaly extends north and south beyond the mapped geomorphological expression of the rift. Decker et al. (1980) suggest that heat flow values of 92-125 mW m^{-2} in the Rocky Mountains of northern Colorado extend the anomaly north to the Colorado-Wyoming border (latitude 41°N). Smith and Jones (1979) and Smith et al. (1979) report surface and reduced heat flow value of 96-121 and 50-88 mW m^{-2} respectively from northern Mexico which suggest a southeast trending, but discontinuous, southern extension of the rift thermal anomaly to approximately latitude 22°N. The direct association of these thermal anomalies with the morphological rift is tenuous, however.

Oceanic Rifts

Since the first heat flow determinations were made in the oceans in the early 1950's it has been recognized that high heat flow is associated with mid-ocean ridges (Bullard, 1954). The rapid increase in the quantity of oceanic heat flow data during the 1960's has defined a relatively consistent, if somewhat confused pattern to the heat flow across ridges. Oceanic heat flow data compilations by Lee and Uyeda (1965), von Herzen and Langseth (1965), Von Herzen and Lee (1969), Langseth and Von Herzen (1970) have shown that although the highest heat flow values are usually measured close to the rift zone, low heat flow values are also common in this region. Furthermore, there appears to be a zone of low heat flow in the 5-10 m.y. crust flanking many of the ridges, as shown in Figure 7, adapted from a compilation by Anderson (1972). Heat flow profiles over ridges based on theoretical models by McKenzie (1967), McKenzie and Sclater (1969), Sleep (1969), Sclater and Francheteau (1970), and Sclater et al. (1971) have not predicted the low heat flow on the ridge flanks, and apparently overestimate the heat flow at the ridge crests.

During the past decade or so there has been an important realization of the major role of hydrothermal circulation in heat transfer in young oceanic crust (Pálmason, 1967; Talwani et al., 1971; Bodvarsson and Lowell, 1972; Hyndman and Rankin, 1972; Lister, 1972; Sclater and Klitgord, 1973; Williams et al., 1974). Sclater et al. (1976) have shown that the scatter in the heat flow data is reduced if only values are used from areas where the crust is sealed from seawater by

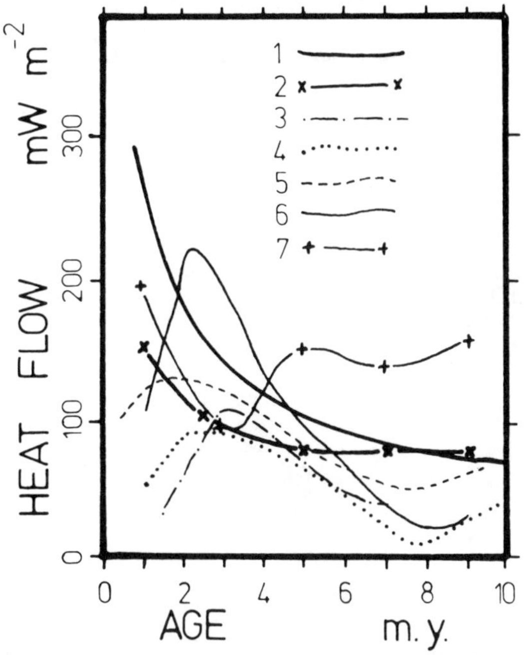

Fig. 7. Plot of mean heat flow against age of ocean floor, adapted from Anderson (1972). 1: Theoretical curve (McKenzie and Sclater, 1969); 2: East Pacific Rise (Anderson and Hobart, 1976); 3: Mid-Atlantic Ridge, 46°N (Hyndman and Rankin, 1972); 4: Reykjanes Ridge (Talwani et al., 1971); 5: Atlantic-Indian Ocean mean (LePichon and Langseth, 1969); 6: Juan de Fuca (Lister, 1970); 7: Galapagos rift zone (Anderson and Hobart, 1976).

an extensive and thick layer of impermeable sediment. Unfortunately oceanic rift zones rarely meet these conditions, and the closest to the ridge axis that Sclater et al. (op. cit.) are able to give a "reliable" heat flow is in the range 165-220 km (3-4 m.y. age crust) where the mean is 265 ± 120 mW m^{-2}. Lister (1977) has shown that the "reliable" data agree with the estimator: heat flow = $500(t)^{-\frac{1}{2}}$, where t is the sea-floor age in m.y., within the range t = 2 to 100 m.y. This estimator predicts infinite heat flow at the ridge axis, however, as do the early theoretical models. Models have been proposed which remove the heat flow singularity at the ridge crests (e.g. Parker and Oldenburg, 1973 ; Lubimova and Nikitina, 1975; Parsons and Sclater, 1977), but the parameters for these models are not well enough defined to accurately predict rift zone heat flow.

Heat flow studies have revealed a complex pattern of heat transfer in rift zones, high heat flow dramatically redistributed by hydrothermal circulation. A summary of these data are presented below, geographically subdivided for convenience.

The Atlantic

Early summaries of heat flow data from the Mid-Atlantic and Reykjanes Ridges all indicated a

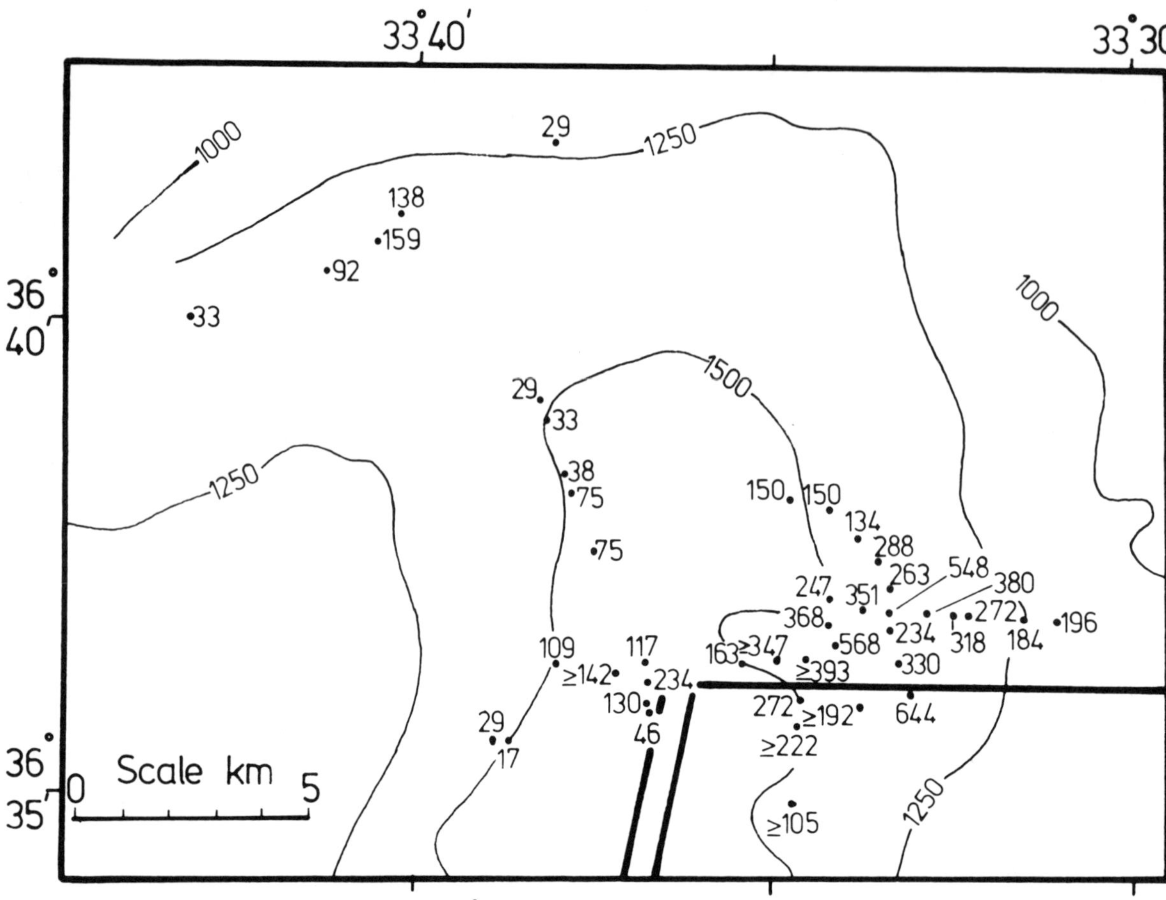

Fig. 8. Heat flow data in mW m^{-2} in topographic depression in Mid-Atlantic Ridge at 36°37'N, adapted from William et al., 1976. Uncorrected contour depth given in fathoms. Single heavy line and double heavy lines indicate approximate locations of transform fracture zone and spreading axis respectively.

decrease in heat flow towards the ridge crests in crust younger than 3 m.y. (LePichon and Langseth, 1969; Talwani et al., 1971; Hyndman and Rankin, 1972), as shown in Figure 7. Because of the general lack of sediments in the median rift of the ridges there are relatively few measurements from the actual rift zone. Hyndman and Rankin (1972) reported a mean of 30 ± 28 mW m^{-2} for 4 values within 20 km of the center of the median valley near 45°N, flanked by higher values at distances of 25 to 45 km. Hyndman et al. (1976) have shown that these low heat flow values are not restricted to the upper sediments, and measured a constant very low heat flow of 25 ± 4 mW m^{-2} in three deep boreholes penetrating up to 400m into basaltic basement in 3.5 m.y. sea-floor near 37°N. The data indicate very extensive lateral heat transfer in the basement by hydrothermal circulation. A compilation of data for the Reykjanes Ridge by Sclater and Crowe (1979, Figure 1) again shows low heat flow, 50 ± 16 (n = 7) mW m^{-2} close to the ridge axis, flanked by higher values. Data presented by Bram (1980) reveal an asymmetric thermal behaviour of the Reykjanes Ridge area, the average heat flow east of the ridge being nearly twice the heat flow to the west. Low to intermediate values, 17 - 175 mW m^{-2}, are reported by Lee and Von Herzen (1975) in the depressions of fracture zones and axial valleys near the South Atlantic triple junction (55°S, 0°E).

A detailed geothermal study of the Mid-Atlantic Ridge near 37°N has been made by Williams et al. (1976). Surveys were made in two topographic depressions marking the intersection of transform faults with spreading axes.

In the first depression, at approximately 36° 37'N, predominently high values were measured, up to 644 mW m^{-2}, as shown in Figure 8. The mean of 7 values within 2 km of the ridge axis is 168 ± 97 mW m^{-2}, although the highest values were measured approximately 5 km east of the axis. In the second depression, at approximately 36° 57'N, very low values, 19 ± 22 (n = 5) mW m^{-2}, were measured close to the end of the spreading axis. Higher values, up to 230 mW m^{-2}, were measured 8-9 km east along the fracture zone. The data do

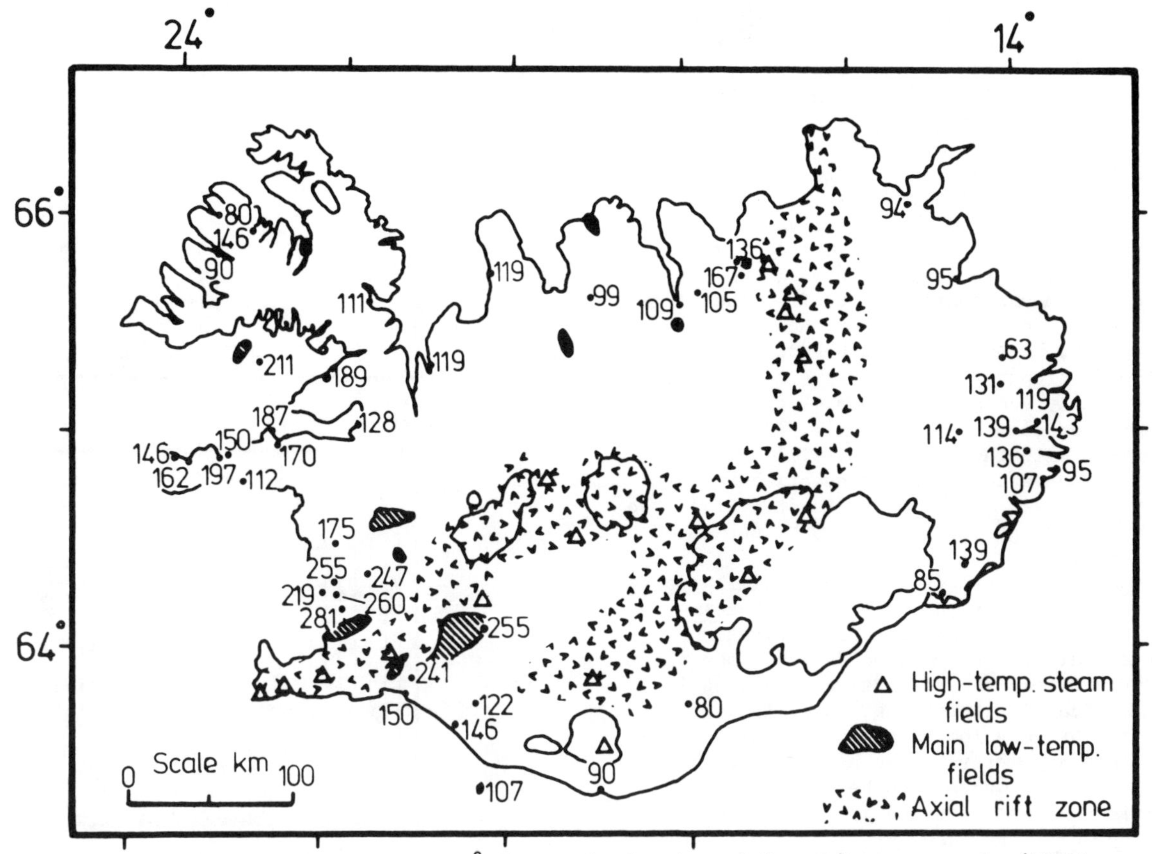

Fig. 9. Heat flow data in mW m^{-2} for Iceland, adapted from Pálmason et al. (1979).

not therefore indicate that high conducted heat flow is generally associated with the Atlantic central rift zones.

Iceland

Summaries of conductive heat flow in Iceland have recently been presented by Pálmason and Saemundsson (1979) and Pálmason et al. (1979), and these workers have shown that the measured conducted heat flow outside the rift zone is generally in good agreement with theoretical curves based on crustal accretion. The conductive heat flow data are shown in Figure 9. High values away from the rift in northwest Iceland are thought to be related to an extinct spreading axis. There are no conductive heat flow values within the rift zone, and drill hole temperature profiles from this zone indicate that hydrothermal circulation modifies the heat flow down to depths well over 3 km in some areas. High temperature steam fields in the rift zone indicate high heat flow, and Pálmason et al. (1979) report temperatures of about 340°C at a depth of 2 km in the Krafla field. Temperature data from wells in low temperature geothermal fields outside the rift zone indicate that the hydrothermal circulation is not restricted to the rift zone. Pálmason (1973) has calculated that the heat loss through the rift zone is 56 MW per kilometer of rift length by volcanic activity, heat conduction, and geothermal activity. If lateral heat transport in the lithosphere is included, the total average heat discharge is 134 MW per km of rift length.

The Pacific

Several summaries and detailed reports of heat flow data over the spreading centers of the eastern Pacific have been published during the past decade. Data from the East Pacific Rise and Galapagos spreading center have been reported and commented on by Sclater et al. (1971, 1974, 1976), Von Herzen and Anderson (1972), Sclater and Klitgord (1973), Williams et al. (1974, 1979a), Anderson and Hobart (1976), and Anderson et al. (1978). Detailed surveys of the Juan de Fuca ridge system have been reported by Davis and Lister (1977), and Davis et al. (1980), and Anderson and Von Herzen (1978) present data from the Pacific-Antarctic ridge. Profiles of heat flow averages as a function of age of the sea-floor for the East Pacific Rise, the Galapagos rift zone, and the Juan de Fuca ridge system are shown in Figure 7.

Measured heat flow values in the rift zone of the East Pacific Rise are low: Anderson et al.

(1978) report two values around 40 mW m^{-2} from the base of the central horst. Means of the measured heat flows in the young crust close to the rift are higher; Anderson and Hobart (1976) give means of 153 ± 93 mW m^{-2} (n = 28) for 0-1 m.y. sea-floor and 152 ± 84 mW m^{-2} (n = 58) for 0-2 m.y. sea-floor of the East Pacific Rise south of the equator, and 140 ± 74 mW m^{-2} (n = 21) for the 0-2 m.y. zone to the north. These values are well below the predictions of the theoretical curves, however, and this discrepancy between measured and predicted means continues out from the ridge in the area of sea-floor younger than approximately 10 m.y. Anderson and Hobart (1976) ascribe the discrepancy to convective heat loss from the thinly sedimented areas of the young sea-floor (see also Sclater et al., 1976), and in a comparison with the results of Williams et al. (1974) they suggest that the convective heat loss may vary by a factor of at least 5 along the ridge, with a range of 31.5 to 155 MW per km of ridge length.

Anderson and Hobart (1976) give a similar pattern of heat loss for the Galapagos spreading center, except that more rapid sedimentation in this area seals the convection in sea-floor older than approximately 5 m.y. Active spreading in the Costa Rica Rift spreading center is again accompanied by very low measured heat flow (Figure 10), and asymmetry in the heat flow either side of the crest appears to be controlled by an asymmetric sediment distribution (Anderson and Hobart, op. cit.). Anderson et al. (1976) report low heat flow, 58 ± 47 mW m^{-2} (n = 33), from a fossil ridge (4 m.y. old) on the north flank of the Galapagos spreading center with high heat flow 197 ± 32 mW m^{-2} (n = 13), on the flanks of the fossil ridge. In contrast, from an area of active hydrothermal mounds on the Galapagos spreading center, Williams et al. (1979a) report very high heat flow values. Total convective and conductive heat flux values of 630 to 31,700 mW m^{-2} were measured on the mounds. The mounds and immediately adjacent sediments are characterized by 76 measurements of 420 mW m^{-2} and greater. Fifty nine values ranging from 134 to 420 mW m^{-2} were measured in the area surrounding the mounds, and Williams et al. (op. cit.) estimate a mean heat flux of 545-630 mW m^{-2} for the mounds area, in good agreement with theoretical estimates of 500 to 670 mW m^{-2} for the 0.5 to 0.9 m.y. crust.

Regional averages heat flow values over the Juan de Fuca ridge system are generally high compared with the means of the young crust in the Galapagos and East Pacific Rise systems. This is shown by the profile in Figure 7, and is confirmed by detailed studies by Davis and Lister (1977) and Davis et al. (1980). Data presented by Davis and Lister (op. cit.) give means of 188 ± 135 mW m^{-2} (n = 15) and 334 ± 86 mW m^{-2} (n = 6) for the Middle and West Valleys of the northern Juan de Fuca ridge respectively. These values are still well below the heat flow predicted by spreading models, however, indicating a convective component to the heat flow. Detailed heat flow profiles over different sections of the ridge system show some periodicity in the lateral variations of heat flow, indicating cellular thermal convection (Davis et al., 1980). Data from two of the profiles close to the ridge axes are shown in Figure 11.

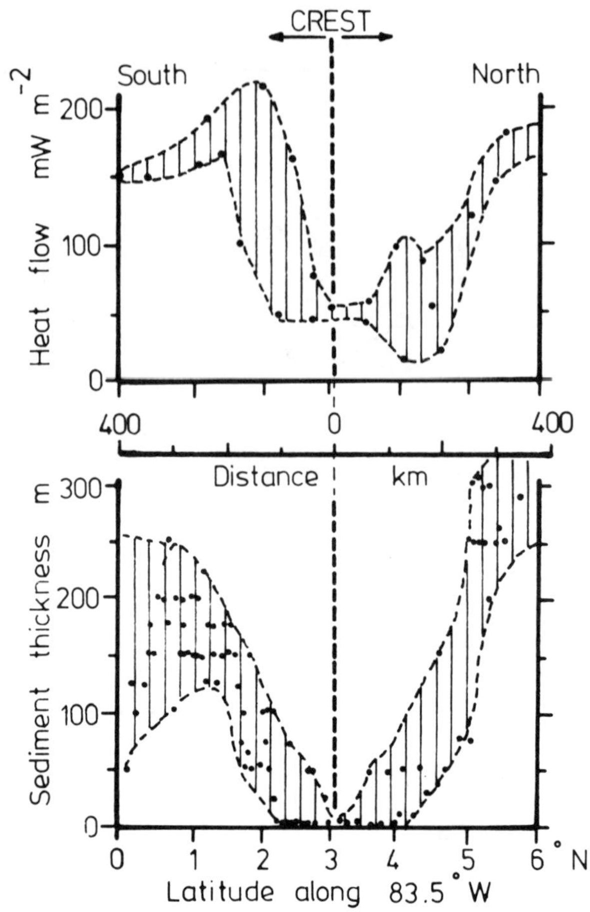

Fig. 10. Heat flow data across the Costa Rica rift, adapted from Anderson and Hobart (1976). Shaded area shows maximum scatter of measurements.

Data presented by Anderson and Von Herzen (1978) from the young (0-21 m.y.) sea-floor on the Pacific-Antarctic ridge crest give a low mean, 74 ± 31 mW m^{-2} (n = 10), approximately half the theoretical heat flow. Anderson and Von Herzen (op. cit.) suggest that approximately 50% of the heat flux is lost by convection, a similar proportion to that reported by Anderson and Hobart (1976) for the East Pacific Rise.

The Gulf of California

Detailed heat flow studies in the Gulf of California indicate that high heat flow is associated with crustal spreading in the Gulf (Lawver et al., 1973, 1975; Lawver and Williams, 1979;

Williams et al., 1979b). Lawver and Williams (op. cit.) give mean heat flow values of 222 and 242 mW m^{-2} for the Guaymas and Farallon basins respectively in which thick sediment blankets cover active spreading centers. Excluding data from local anomalies they recompute means of 180 and 192 mW m^{-2} for the two basins respectively. In the Farallon basin relatively low values were measured over the central depression, but in the Guaymas basin the highest values, greater than 1300 mW m^{-2}, were measured in the central trough, as shown in Figure 12. Detailed heat flow profiles across the spreading centers of the Guaymas basin by Williams et al. (1979b) support the hypothesis that they are the sites of active intrusion. The mean values from the basin are well below the theoretical values for the very young crust, indicating a large component of convective heat transfer.

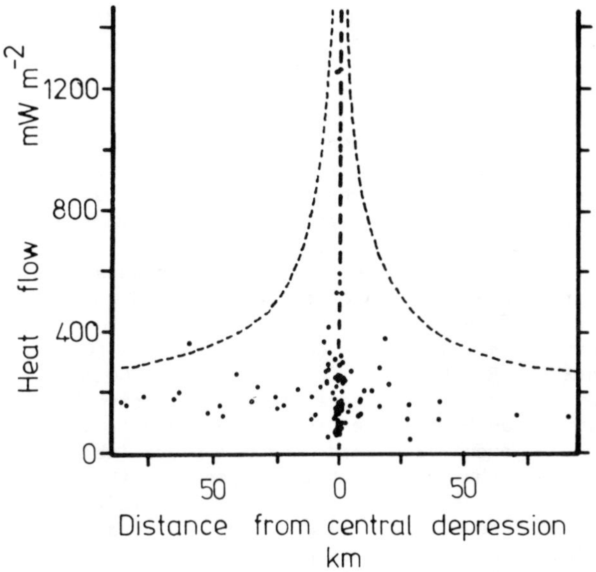

Fig. 12. Heat flow as a function of distance from the central trough in the Guaymas basin, Gulf of California, adapted from Williams et al. (1979b).

Indian Ocean

Anderson et al. (1977) have compiled the reliable heat flow data from the Indian Ocean, and concluded that the variation of mean heat flow with age is very similar to that in the eastern Pacific and Atlantic oceans. There is a large scatter in the data and a low mean in the crestal zone, but after a sedimentary blanket has accumulated (sea-floor older than 40 - 60 m.y.) blocking convective heat loss, the values are in agreement with theoretical predictions. The highest heat flow values (up to 335 mW m^{-2}) were measured close to the crests of the Indian Ocean and Carlsberg ridges, but, as with other rift zones, numerous low values in the young crust indicate a large convective heat loss in these areas.

The Gulf of Aden, Red Sea and Dead Sea Rifts

Heat flow data from the Red Sea and Gulf of Aden show a strong positive thermal anomaly in the axial rift zones (Haenel, 1972; Girdler and Evans, 1977). Girdler and Evans (op. cit.) give a mean of 467 ± 116 mW m^{-2} (n = 38) for heat flow within 5 km of the deepest water of the axial trough of the Red Sea. The heat flow drops dramatically to 89 ± 27 mW m^{-2} (n = 6) from 5 to 10 km from the axis increasing to 140 and 164 mW m^{-2} in the intervals 10-30 km and 30-50 km from the axis respectively. The heat flow remains relatively high, 111 ± 5 mW m^{-2} (n = 13), right to the Red Sea margin, and Morgan and Swanberg (1979) have recently reported a narrow zone of high heat flow in the continental crust of Egypt bordering the Red Sea. To the north in the Jordan-Dead Sea rift, Eckstein (1979) reports a small regional geothermal anomaly with values of 57 to 85 mW m^{-2} associated with the rift. Within this zone Eckstein (op. cit.) reports a range of values from 7 to 464 mW m^{-2} resulting from heat transport by groundwater.

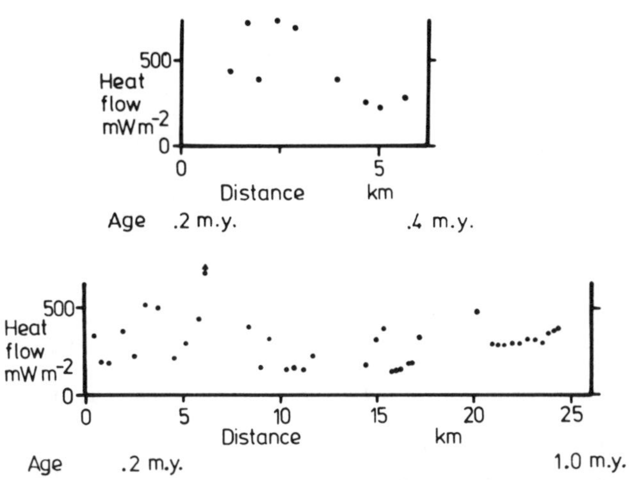

Fig. 11. Detailed heat flow profiles close to the Juan de Fuca ridge axes, adapted from Davis et al. (1980).

Discussion and Conclusions

From the data presented, it is clear that the present heat flow measurements in both continental and oceanic rift zones do not accurately define the heat flux from these environments. The mean heat flow values within the rift grabens from most of the continental rift zones are in the range 90-110 mW m^{-2}, but all data sets have a high standard deviation. Heat flow values from predominantly non-volcanic sections of the rifts, e.g. Lakes Malawi and Tanganyika, and possibly the Jordan-Dead Sea Rift, have relatively low mean heat flows, although uncertainties in the data do not place good constraints on the magnitudes of any anomalies. The anomalies are probably in the range 0 to 20 mW m^{-2} for the rift zones of Malawi and Tanganyika. The correlation of high mean heat flows for volcanic rifts and low mean heat flows for non-volcanic rifts, and the rapid lateral transition to normal heat flow values outside the rifts suggest that the thermal anomalies are derived primarily from igneous activity in the rifted crust. The large scatters in the data sets indicate that there is redistribution of the igneous heat by groundwater movements. Petrologically derived geotherms for the southern Rio Grande Rift suggest that 100 mW m^{-2} may be a reasonable estimate of conductive heat flow through the rift crust after an extensive igneous event or events. From the available data, only the Kenya rift may depart significantly from this estimate as geothermal data indicate a higher intensity of igneous activity along the axial zone of the rift, possibly related to a greater development of crustal separation than is seen in the other Cenozoic rifts.

There is a very large scatter in the heat flow data over oceanic rifts, and means of the data are consistently lower than the heat flow predicted by thermal models of sea-floor spreading. The scatter in the data decreases dramatically, and the heat flow values agree with theoretical curves as the age of the crust increases and a sedimentary blanket is developed over the crust. In the rift zones there is clearly a large component of convective heat loss which is not measured by the traditional conductive measurements. Two methods are being used to determine the heat flux through ocean rifts: i) carefully selected conductive data from well sedimented areas are used to constrain theoretical thermal models, which can in turn be used to predict rift heat flow where conductive measurements are inadequate; and ii) detailed studies are made in the rift zones to investigate the convective processes so that the magnitude of the convective heat component can be estimated. Both methods are making significant contributions to the understanding of the rift heat flow budget.

Acknowledgements. I am grateful to V. Cermák, L. Coulter and V. Harder for assistance in compiling data for this review, and to D. L. Williams and E. E. Davis for sending me preprints of their publications. G. Pálmason and an anonymous reviewer are thanked for critically reading the manuscript, and I am grateful to N. Denzler, M. K. Morgan, O. Trost and J. Vigil for typing, proofreading and photographic assistance. I received partial support during this study from National Science Foundation Division of Earth Sciences grant no. EAR77-23354, and Public Law Fund 480 grant no. INT78-16649 from the National Science Foundation Office of International Programs.

References

Anderson, R. N., Petrologic significance of low heat flow on the flanks of slow-spreading mid-ocean ridges, Geol. Soc., Am. Bull., 83, 2947-2956, 1972.

Anderson, R. N., and M. A. Hobart, The relation between heat flow, sediment thickness, and age in the eastern Pacific, J. Geophys. Res., 81 2968-2989, 1976.

Anderson, R. N., and R. P. Von Herzen, Heat flow on the Pacific-Antartic ridge, Earth Planet. Sci. Lett., 41, 451-460, 1978.

Anderson, R. N., C. F. Moore, S. S. Schilt, R. C. Cardwell, A. Tréhu, and V. Vacquier, Heat flow near a fossil ridge on the north flank of the Galapagos spreading center, J. Geophys. Res., 81, 1828-1838, 1976.

Anderson, R. N., M. G. Langseth, and J. G. Sclater, The mechanisms of heat transfer through the floor of the Indian Ocean, J. Geophys. Res., 82, 3391-3409, 1977.

Anderson, R. N., M. A. Hobart, R. P. Von Herzen, and D. J. Fornari, Geophysical surveys on the East Pacific Rise - Galapagos Rise system, Geophys. J. Roy. Astr. Soc., 54, 141-166, 1978.

Baker, B. H., P. A. Mohr, and L. A. J. Williams, Geology of the eastern rift system of Africa, Geol. Soc. Am., Spec. Pap. 136, 67 pp., 1972.

Beck, A. E., Techniques of measuring heat flow on land, in Terrestrial Heat Flow, Geophys. Monogr. Ser., v. 8, ed. by W. H. K. Lee, pp. 24-57, Am. Geophys. Un., Washington, D.C., 1965.

Blackwell, D. D., Heat flow and energy loss in the western United States, in Cenozoic Tectonic and Regional Geophysics of the Western Cordillera, Geol. Soc. Am., Mem. 152, ed. by R. B. Smith and G. P. Eaton, pp. 175-208, 1978.

Bridwell, R. J., and C. A. Anderson, Thermomechanical models of the Rio Grande rift, in NATO Advanced Study Institute The Mechanisms of Continental Drift and Plate Tectonics, ed. by S. K. Runcorn, Newcastle-upon-Tyne, England, in press, 1980.

Bridwell, R. J., and C. Potzick, Thermal regimes, mantle diapirs, and crustal stresses of continental rifts, in Mechanisms of Graben Formation, ed. by H. Illies, Tectonophysics, in press, 1980.

Bodvarsson, G., and R. P. Lowell, Ocean floor heat flow and the circulation of interstitial waters, J. Geophys. Res., 77, 4472-4475, 1972.

Bottinga, Y., Thermal aspects of sea-floor spreading, and the nature of the suboceanic lithosphere, Tectonophysics, 21, 15-38, 1974.

Bram, K., Heat flow measurements in the Federal Republic of Germany, in Terrestrial Heat Flow in Europe, ed. by V. Cermák and L. Rybach, pp. 191-196, Springer-Verlag, New York, 1979.

Bram, K., New heat flow observations on the Reykjanes Ridge, J. Geophys., 47, 86-90, 1980.

Bullard, E. C., The flow of heat through the floor of the Atlantic Ocean, Proc. Roy. Soc. London, A, 222, 408-429, 1954.

Buntebarth, G., The degree of metamorphism of organic matter in sedimentary rocks as a paleogeothermometer, applied to the upper Rhinegraben, Pure Appl. Geophys., 117, 83-91, 1979.

Chapman, D. S., and H. N. Pollack, Heat flow and incipient rifting in the central African plateau, Nature, 256, 28-30, 1975.

Chapman, D. S., and H. N. Pollack, Heat flow and heat production in Zambia: evidence for lithospheric thinning in central Africa, Tectonophysics, 41, 79-100, 1977.

Cook, F. A., E. R. Decker, and S. B. Smithson, Preliminary transient heat flow model of the Rio Grande rift in southern New Mexico, Earth Planet. Sci. Lett., 40, 316-326, 1978.

Davis, E. E., and C. R. B. Lister, Heat flow measured over the Juan de Fuca ridge: evidence for widespread hydrothermal circulation in a highly heat transportive crust, J. Geophys. Res., 82, 4845-4860, 1977.

Davis, E. E., C. R. B. Lister, U. S. Wade, and R. D. Hyndman, Detailed heat flow measurements over the Juan de Fuca ridge system, J. Geophys. Res., 85, 299-310, 1980.

Decker, E. R., and S. B. Smithson, Heat flow and gravity interpretation across the Rio Grande rift in southern New Mexico and west Texas, J. Geophys. Res., 80, 2542-2552, 1975.

Decker, E. R., K. R. Barker, G. J. Bucher, and H. P. Heasler, Preliminary heat flow and radioactivity studies in Wyoming, J. Geophys. Res., 85, 311-321, 1980.

Degens, E. T., R. P. Von Herzen, and H. K. Wong, Lake Tanganyika: water chemistry, sediments, geological structure, Naturwissenschaften, 58, 229-241, 1971.

Degens, E. T., R. P. Von Herzen, H. K. Wong, W. G. Deuser, and H. W. Jannasch, Lake Kivu: structure, chemistry and biology of an East African rift lake, Geol. Rundschau, 62, 245-277, 1973.

Delattre, J. N., R. Hentinger, and J. P. Lauer, A provisional geothermal map of the Rhinegraben (Alsatian part), in Graben Problems, ed. by J. H. Illies and S. Mueller, pp. 107-110, Schweizerbart, Stuttgart, 1970.

Doebl, F., D. Heling, W. Homann, J. Karweil, M. Teichmüller, and D. Welte, Digenesis of Tertiary clayey sediments and included dispersed organic matter in relationship to geothermics in the upper Rhinegraben, in Approaches to Taphrogenesis, ed. by J. H. Illies and K. Fuchs, pp. 192-207, Schweizerbart, Stuttgart, 1974.

Duchkov, A. D., S. A. Kazantsev, V. A. Golubev, S. V. Lysak, and E. S. Khaikovskii, Geothermal studies of Lake Baikal, Geologiya i Geofizika (Soviet Geology and Geophysics), 17, 104-107, 1976.

Duchkov, A. D., S. A. Kazantsev, V. A. Golubev, and S. V. Lysak, Geothermal studies of Lake Baikal, Geologiya i Geofizika (Soviet Geology and Geophysics), 18, 126-130, 1977.

Eckstein, Y., Heat flow and the hydrologic cycle: examples from Israel, in Terrestrial Heat Flow in Europe, ed. by V. Cermák and L. Rybach, pp. 88-97, Springer-Verlag, New York, 1979.

Edwards, C. L., M. Reiter, C. Shearer, and W. Young, Terrestrial heat flow and crustal radioactivity in northeastern New Mexico and southeastern Colorado, Geol. Soc. Am. Bull., 89, 1341-1350, 1978.

Evans, T. R., Terrestrial heat flow in eastern Africa and the North Sea, Ph.D. Dissertation, Univ. London, England, 1975.

Evans, T. R., and H. Y. Tammemagi, Heat flow and heat production in northeast Africa, Earth Planet. Sci. Lett., 23, 349-356, 1974.

Fairhead, J. D., and R. W. Girdler, The seismicity of Africa, Geophys. J. Roy. Astr. Soc., 24, 271-301, 1971.

Fodor, R. V., Ultramafic and mafic inclusions and megacrysts in Pliocene basalt, Black Range, New Mexico, Geol. Soc. Am. Bull., 89, 451-459, 1978.

Gass, I. G., D. S. Chapman, H. N. Pollack, and R. S. Thorpe, Geological and geophysical parameters of mid-plate volcanism, Phil. Trans. R. Soc. Lond. A., 288, 581-597, 1978.

Girdler, R. W., and T. R. Evans, Red Sea heat flow, Geophys. J. Roy. Astr. Soc., 51, 245-251, 1977.

Haenel, R., Interpretation of the terrestrial heat flow in the Rhinegraben, in Graben Problems, ed. by J. H. Illies and S. Mueller, pp. 116-120, Schweizerbart, Stuttgart, 1970.

Haenel, R., Heat flow measurements in the Red Sea and the Gulf of Aden, Z. Geophys., 38, 1035-1047, 1972.

Haenel, R., G. Grønlie, and K. S. Heier, Terrestrial heat flow determinations from lakes in southern Norway, Nor. Geol. Tidskr., 54, 423-428, 1974.

Haenel, R., G. Grønlie, and K. S. Heier, Terrestrial heat flow determinations in Norway and an attempted interpretation, in Terrestrial Heat Flow in Europe, ed. by V. Cermák and L. Ryback, pp. 232-239, Springer-Verlag, New York, 1979.

Harder, V., P. Morgan and C. A. Swanberg, Geothermal anomalies in the Rio Grande rift: origins and potential, Geothermal Resources Council, Transactions, 4, in press, 1980.

Hurtig, E., and C. Oelsner, Heat flow, temperature distribution and geothermal models in Europe: some tectonic implications, Tectonophysics, 41, 147-156, 1977.

Hyndman, R. D., and S. J. Rankin, The Mid-Atlantic Ridge near 45°N heat flow measurements, Can. J. Earth Sci., 9, 664-670, 1972.

Hyndman, R. D., R. P. Von Herzen, A. J. Erickson, and J. Jolivet, Heat flow measurements in deep crustal holes on the Mid-Atlantic Ridge, J. Geophys. Res., 81, 4053-4060, 1976.

Jaeger, J. C., Application of the theory of heat conduction to geothermal measurements, in Terrestrial Heat Flow, Geophys. Monogr. Ser., v. 8, ed. by W. H. K. Lee, pp. 7-23, Am. Geophys. Un., Washington, D.C., 1965.

Lachenbruch, A. H., Dynamics of a passive spreading center, J. Geophys. Res., 81, 1883-1902,1976.

Lachenbruch, A. H., Heat flow in the Basin and Range province and thermal effects of tectonic extension, Pure Appl. Geophys., 117, 34-50,1979.

Lachenbruch, A. H., and J. H. Sass, Heat flow in the United States and the thermal regime of the crust, in The Earth's Crust, Geophys. Monogr. Ser., v. 20, ed. by J. G. Heacock, pp. 626-675, Am. Geophys. Un., Washington, D.C., 1977.

Lachenbruch, A. H., and J. H. Sass, Models of an extending lithosphere and heat flow in the Basin and Range province, in Cenozoic Tectonics and Regional Geophysics of the Western Cordillera, Geol. Soc. Am., Mem. 152, ed. by R. B. Smith and G. P. Eaton, pp. 209-250, 1978.

Langseth, M. G., Techniques of measuring heat flow through the ocean floor, in Terrestrial Heat Flow, Geophys. Monogr. Ser., v. 8, ed. by W. H. K. Lee, pp. 58-77, Am. Geophys. Un., Washington, D.C., 1965.

Langseth, M. G., and R. P. Von Herzen, Heat flow through the floor of the world oceans, in The Sea, v. 4, pt. 1, ed. by A. E. Maxwell, pp. 299-352, Wiley-Interscience, New York, 1970.

Lawver, L. A., and D. L. Williams, Heat flow in the central Gulf of California, J. Geophys. Res., 84, 3465-3478, 1979.

Lawver, L. A., J. G. Sclater, T. L. Henyey, and J. Rogers, Heat flow measurements in the southern portion of the Gulf of California, Earth Planet. Sci. Lett., 12, 198-208, 1973.

Lawver, L. A., D. L. Williams, and R. P. Von Herzen, A major geothermal anomaly in the Gulf of California, Nature, 257, 23-28, 1975.

Lee, T. C., and R. P. Von Herzen, Heat flow near the South Atlantic triple junction, 55°S, 0°E, Geophys. Res. Lett., 2, 201-204, 1975.

Lee, W. H. K., and S. Uyeda, Review of heat flow data, in Terrestrial Heat Flow, Geophys. Monogr. Ser., v. 8, ed. by W. H. K. Lee, pp. 87-190, Am. Geophys. Un., Washington, D.C., 1965.

LePichon, X., and M. G. Langseth, Heat flow from the mid-ocean ridges and sea-floor spreading, Tectonophysics, 8, 319-344, 1969.

Lister, C. R. B., Heat flow west of the Juan de Fuca ridge, J. Geophys. Res., 75, 2648-2654, 1970.

Lister, C. R. B., On the thermal balance of a mid-ocean ridge, Geophys. J. Roy. Astr. Soc., 26, 515-535, 1972.

Lister, C. R. B., Estimators for heat flow and deep rock properties based on boundary layer theory, Tectonophysics, 41, 157-171, 1977.

Lubimova, E. A., and V. N. Nikitina, On heat flow singularities over mid-ocean ridges, J. Geophys. Res., 80, 232-243, 1975.

Lubimova, E. A., U. A. Zorin, and S. V. Lysak, Geothermal anomaly in the Baikal Lake region, Geothermics, 1, 31-34, 1972.

Lysak, S. V., Heat flow, geology and geophysics in the Baikal rift zone and adjacent regions, in Geoelectric and Geothermal Studies (East-Central Europe Soviet Asia), ed. by A. Adam, pp. 455-462, Arad. Kiado, Budapest, 1976.

McKenzie, D. P., Some remarks on heat flow and gravity anomalies, J. Geophys. Res., 72, 6261-6271, 1967.

McKenzie, D. P., and J. G. Sclater, Heat flow in the eastern Pacific and sea floor spreading, Bull. Volcanologique, 33-1, 101-118, 1969.

Mercier, J. C., Peridotite xenoliths: lateral and vertical heterogenities in the upper mantle, Ph.D. Dissertation, State Univ. New York at Stonybrook, USA, 1977.

Moiseenko, U. I., A. D. Duchkov, and L. S. Sokolova, Heat flow in some regions of Siberia and the far eastern USSR, Geothermics, 2, 17-23, 1973.

Morgan, P., Terrestrial heat flow studies in Cyprus and Kenya, Ph.D. Dissertation, Univ. London, England, 1973.

Morgan, P., and C. A. Swanberg, Heat flow and the geothermal potential of Egypt, Pure Appl. Geophys., 117, 213-226, 1979.

Morgan, P., and J. Wheildon, Heat flow measurements across the Kenya rift, submitted to Geophys. J. Roy. Astr. Soc., 1980.

Osmaston, M. F., Genesis of ocean ridge median valleys and continental rift valleys, Tectonophysics, 11, 387-405, 1971.

Padovani, E. R., and J. L. Carter, Aspects of the deep crustal evolution beneath south central New Mexico, in The Earth's Crust, Geophys. Monogr. Ser., v. 20, ed. by J. G. Heacock, 19-55, Am. Geophys. Un., Washington, D.C., 1977.

Pálmason, G., On heat flow in Iceland in relation to the mid-Atlantic ridge, in Iceland and Mid-Ocean Ridges, publ. 38, ed. by S. Bjornsson, pp. 111-127, Soc. Sci. Islandica, Reykjavik, 1967.

Pálmason, G., Kinematics and heat flow in a volcanic rift zone, with application to Iceland, Geophys. J. Roy. Astr. Soc., 33, 451-481, 1973.

Pálmason, G., and K. Saemundsson, Summary of conductive heat flow in Iceland, in Terrestrial Heat Flow in Europe, ed. by V. Cermák and L. Rybach, pp. 218-220, Springer-Verlag, New York, 1979.

Pálmason, G., S. Arnórsson, I. B. Fridleifsson, H. Kristmannsdóttir, K. Saemundsson, V. Stefansson, B. Steingrímsson, J. Tómasson, and L. Kristjánsson, The Iceland crust: evidence from drill hole data on structure and processes, in Deep Drilling Results in the Atlantic Ocean: Ocean Crust, ed. by M. Talwani, C. G. Harrison, and D. E. Hayes, pp. 43-65, Am. Geophys. Un., Washington, D.C., 1979.

Parker, R. L., and D. W. Oldenburg, Thermal model of ocean ridges, Nature Phys. Sci., 242, 137-139, 1973.

Parsons, B., and D. McKenzie, Mantle convection and the thermal structure of the plates, J. Geophys. Res., 83, 4485-4496, 1978.

Parsons, B., and J. G. Sclater, An analysis of the variation of ocean floor bathymetry and heat flow with age, J. Geophys. Res., 82, 803-827, 1977.

Reid, J. B., Upper and lower limits to the geothermal gradient beneath the southern Rio Grande rift, abs., Geol. Soc. Am., Abstracts with Program, v. 8, no. 5, pp. 621-622, 1976.

Reiter, M., C. L. Edwards, H. Harman, and C. Weidman, Terrestrial heat flow along the Rio Grande rift, New Mexico and southern Colorado, Geol. Soc. Am. Bull., 86, 811-818, 1975.

Reiter, M., C. Shearer, and C. L. Edwards, Geothermal anomalies along the Rio Grande rift in New Mexico, Geology, 6, 85-88, 1978.

Reiter, M., A. J. Mansure, and C. Shearer, Geothermal characteristics of the Rio Grande rift within the southern Rocky Mountain complex, in Rio Grande Rift: Tectonics and Magmatism, ed. by R. E. Riecker, pp. 253-267, Am. Geophys. Un., Washington, D.C., 1979.

Sclater, J. G., and J. Crowe, A heat flow survey at anomaly 13 on the Reykjanes Ridge: a critical test on the relation between heat flow and age, J. Geophys. Res., 84, 1593-1602, 1979.

Sclater, J. G., and J. Francheteau, The implications of terrestrial heat flow observations on current tectonic and geochemical models of the crust and upper mantle of the earth, Geophys. J. Roy. Astr. Soc., 20, 509-542, 1970.

Sclater, J. G., and K. D. Klitgord, A detailed heat flow, topographic and magnetic survey across the Galapagos spreading center at 86°W, J. Geophys. Res., 78, 6951-6975, 1973.

Sclater, J. G., R. N. Anderson, and M. L. Bell, Elevation of ridges and evolution of the central eastern Pacific, J. Geophys. Res., 76, 7888-7915, 1971.

Sclater, J. G., R. P. Von Herzen, D. L. Williams, R. N. Anderson, and K. Klitgord, The Galapagos spreading center: heat flow low on the north flank, Geophys. J. Roy. Astr. Soc., 38, 609-626, 1974.

Sclater, J. G., J. Crowe, and R. N. Anderson, On the reliability of oceanic heat flow averages, J. Geophys. Res., 81, 2997-3006, 1976.

Seager, W. R., and P. Morgan, Rio Grande rift in southern New Mexico, west Texas, and northern Chihuahua, in Rio Grande Rift: Tectonics and Magmatism, ed. by R. E. Riecker, pp. 87-106, Am. Geophys. Un., Washington, D.C., 1979.

Searle, R. C., Evidence from gravity anomalies for thinning of the lithosphere beneath the rift valley in Kenya, Geophys. J. Roy. Astr. Soc., 21, 13-31, 1970.

Sleep, N. H., Sensitivity of heat flow and gravity to the mechanism of sea floor spreading, J. Geophys. Res., 74, 542-549, 1969.

Sleep, N. H., and B. R. Rosendahl, Topography and tectonics of mid-oceanic ridge axes, J. Geophys. Res., 84, 6831-6839, 1979.

Smith, D. L., and R. L. Jones, Thermal anomaly in northern Mexico: an extension of the Rio Grande rift?, in Rio Grande Rift: Tectonics and Magmatism, ed. by R. E. Riecker, pp. 269-278, Am. Geophys. Un., Washington, D.C., 1979.

Smith, D. L., C. E. Nuckels, R. L. Jones, and G. A. Cook, Distribution of heat flow and radioactive heat generation in northern Mexico, J. Geophys. Res., 84, 2371-2379, 1979.

Swanberg, C. A., Chemistry of thermal and nonthermal groundwaters in the Rio Grande rift and adjacent tectonic provinces, in Rio Grande Rift: Tectonics and Magmatism, ed. by R. E. Riecker, pp. 279-288, Am. Geophys. Un., Washington, D.C., 1979.

Swanberg, C. A., and P. Morgan, The linear relation between temperatures based on the silica content of groundwaters and regional heat flow: a new heat flow map of the United States, Pure Appl. Geophys., 117, 227-241, 1979.

Swanberg, C. A., and P. Morgan, The silica heat flow interpretation technique: assumptions and applications, J. Geophys. Res., in press, 1980.

Swanberg, C. A., M. D. Chessman, G. Simmons, S. B. Smithson, G. Grønlie, and K. S. Heier, Heat flow-heat generation studies in Norway, Tectonophysics, 23, 31-48, 1974.

Talwani, M., C. C. Windisch, and M. G. Langseth, Reykjanes ridge crest: a detailed geophysical study, J. Geophys. Res., 76, 473-517, 1971.

Teichmüller, M., Bestimmung des inkohlungsgrades von kohligen einschlüssen in sedimenten des Oberrheingrabens-ein hilfsmittel bei der klärung geothermischer fragen, in Graben Problems, ed. by J. H. Illies and S. Mueller, pp. 124-142, Schweizerbart, Stuttgart, 1970.

Von Herzen, R. P., and R. N. Anderson, Implications of heat flow and bottom water temperature in the eastern equatorial Pacific, Geophys. J. Roy. Astr. Soc., 26, 427-458, 1972.

Von Herzen, R. P., and M. G. Langseth, Present status of oceanic heat flow measurements, Phys. Chem. Earth, 6, 367-407, 1965.

Von Herzen, R. P., and W. H. K. Lee, Heat flow in oceanic regions, in The Earth's Crust and Upper Mantle, Geophys. Monogr., ser., v. 13, ed. by P. J. Hart, pp. 88-95, Am. Geophys. Un., Washington, D.C., 1969.

Von Herzen, R. P., and V. Vacquier, Terrestrial heat flow in Lake Malawi, Africa, J. Geophys. Res., 72, 4221-4226, 1967.

Warren, R. G., Characterization of the lower crust-upper mantle of the Engle Basin, Rio Grande rift, from a petrochemical and field geologic study of basalts and their inclusions, Ph.D. Dissertation, Univ. of New Mexico, Albuquerque, U.S.A., 1978.

Werner, D., and F. Doebl, Eine geothermische karte des Rheingrabenuntergrundes, in Approaches to Taphrogenesis, ed. by J. H. Illies and K. Fuchs, pp. 182-191, Schweizerbart, Stuttgart, 1974.

Williams, D. L., and K. A. Poehls, On the thermal evolution of the oceanic lithosphere, Geophys. Res. Lett., 2, 321-323, 1975.

Williams, D. L., R. P. Von Herzen, J. G. Sclater, and R. N. Anderson, The Galapagos spreading center: lithospheric cooling and hydrothermal circulation, Geophys. J. Roy. Astr. Soc., 38, 587-608, 1974.

Williams, D. L., T. C. Lee, R. P. Von Herzen, K.

E. Green, and M. A. Hobart, A geothermal study of the Mid-Atlantic Ridge near 37°N, Geol. Soc. Am. Bull., 88, 531-540, 1976.
Williams, D. L., K. Green, T. H. Van Andel, R. P. Von Herzen, J. R. Dymond, and K. Crane, The hydrothermal mounds of the Galapagos rift: observations with DSRV ALVIN and detailed heat flow studies, J. Geophys. Res., 84, 7467-7484, 1979a.
Williams, D. L., K. Becker, L. A. Lawver, and R. P. Von Herzen, Heat flow at the spreading centers of the Guaymas Basin, Gulf of California, J. Geophys. Res., 84, 6757-6769, 1979b.
Williamson, K. H., Terrestrial heat flow studies in Kenya, Ph.D. Dissertation, Univ. London, England, 1975.
Withjack, M., A convective heat transfer model for lithospheric thinning and crustal uplift, J. Geophys. Res., 84, 3008-3022, 1979.

PHYSICAL ASPECTS OF HYDROTHERMAL SYSTEMS

Valgarður Stefánsson
Orkustofnun, Grensásvegur 9, 108 Reykjavík, Iceland

Sveinbjörn Björnsson
Science Institute, University of Iceland, 107 Reykjavík, Iceland

Abstract. The paper reviews recent advances in the understanding of geothermal systems with emphasis on physical aspects of hydrothermal processes. Topics that receive special attention are the discovery of submarine geothermal systems, interaction of magma and hydrothermal fluids, convection of thermal fluids, the nature of boiling and numerical modelling of hydrothermal systems.

The discovery of submarine geothermal areas has revealed that sea water is most likely as important a geothermal fluid as meteoric water. Although the bulk of hydrothermal fluids is of meteoric and oceanic origin, there is a growing evidence for magmatic influence and mantle contribution to geothermal fluids.

Intimate contact between water and convecting magma or fractured rock near solidus temperature is required to explain the intense heat output of major geothermal areas.

Density and other physical properties of water are strongly dependent on temperature. These properties along with phase transitions (boiling and condensation), and the presence of gases, greatly affect the convection of hydrothermal fluids. Phase changes induce convection in hydrothermal fluids prior to the ordinary buoyancy driven thermal convection. Two phase convection might thus be more common in geothermal systems than previously assumed.

The rate of convective hydrothermal cooling of plutons determines the lifetime of hydrothermal activity. Phase-change driven convection is more effective in cooling the heat source than single phase convection where boiling does not occur. Numerical simulations indicate that hydrothermal systems drawing their energy from a single pluton are not likely to have a lifetime in excess of $2 \cdot 10^5$ years in the case of single phase convection. A lifetime of an order of magnitude less is predicted when phase changes govern the cooling rate. Geothermal systems are generally of a complex nature and contain many zones representing different physical states. By using three descriptive parameters, pressure, temperature and fluid saturation, four main classes of physical states are identified: liquid saturated, boiling, vapor saturated, and supercritical.

During the lifetime of hydrothermal activity, physical conditions change and the boundaries of zones might gradually change or zones might flip from one stable state to another. Disturbances induced by exploitation accelerate this development. Rather than stressing the difference between types of geothermal systems (e.g. hot water or vapor-dominated), we emphasize that these types are merely physical states of a geothermal system. The geological and physical conditions determine the physical state at each time.

Introduction

The loss of heat from the Earth's interior is an essential feature in its evolution as a planet. Before the discovery of radioactivity, the heat loss was a strong argument in the dispute on the age of the Earth. (Thomson (Lord Kelvin) 1846).

The bulk of the heat is conducted to the Earth's surface. A minor contribution comes from advective flow of material. Heat advected by volcanic material is uncertain but estimates range from 0.1 to 3% and for advection by thermal waters about 0.3% of the total global heat loss (Bodvarsson 1981, Verhoogen 1980). Although advective processes are not significant in the global average, their contribution becomes essential in the global rift system where the heat transported by advective processes is of the same order as thermal conduction (Bodvarsson 1981).

In recent years evidence has been growing for hydrothermal circulation in young ocean crust. There is good evidence that seawater is the fluid involved in the metamorphism of oceanic crust to a depth of several kilometers. In ocean rift zones heavy metals are extracted by a hot, acidic solution of sea water. A substantial proportion of the total oceanic heat flow is vented to the sea by hydrothermal processes

(Williams and Von Herzen 1974, Lister 1980).

Growing economic importance of geothermal energy has led to intensified geothermal research and exploration on land. Most of the geothermal energy stored in crustal rocks is too diffuse to be recovered economically. Resources of potential economic significance are concentrated along the constructive and destructive boundaries of the major crustal plates (Fig. 1). These margins are zones where either new material from the mantle is being added to the crust or where crustal material is being absorbed in subduction zones. In both situations magmatic and hydrothermal processes play a major role in transferring heat to the surface.

The increased research effort is borne out in a prolific literature on geothermal aspects. A comprehensive review of all aspects would be a long and difficult exercise. We have therefore chosen to limit the scope of this article to physical aspects of hydrothermal processes and emphasize recent advance in the understanding of geothermal systems. Among topics that receive special attention are (1) the discovery of submarine geothermal systems (2), convection of thermal fluids and the nature of boiling geothermal systems (3), numerical modelling of hydrothermal systems, and (4) the interaction of magma and the hydrothermal fluid.

The discovery of submarine geothermal areas is probably the most spectacular development in geothermal exploration of the last decade. Four active hydrothermal fields have been located near ocean spreading centres (Red Sea, the TAG field of the mid-Atlantic Ridge near 26°N, the Galapagos field and the field of the East Pacific Rise near 21°N). It is now assumed that such systems are common on mid-ocean ridges and seawater is thus of no less importance than meteoric water as a source fluid for hydrothermal processes in the crust.

Geothermal systems have been classified into two major types, hot-water systems and vapor-dominated systems. Although these types include most geothermal systems discovered to date, there is growing evidence for systems of intermediate and more complex character where boiling of the thermal fluid results in two phase conditions that have profound influence on the physical state and the production characteristics of the geothermal reservoirs (e.g. Broadlands, (Grant 1977a), Olkaria (Björnsson 1978), Krafla (Stefánsson 1981), Tongonan (Whittome and Smith 1979)).

A considerable effort has been devoted to math-

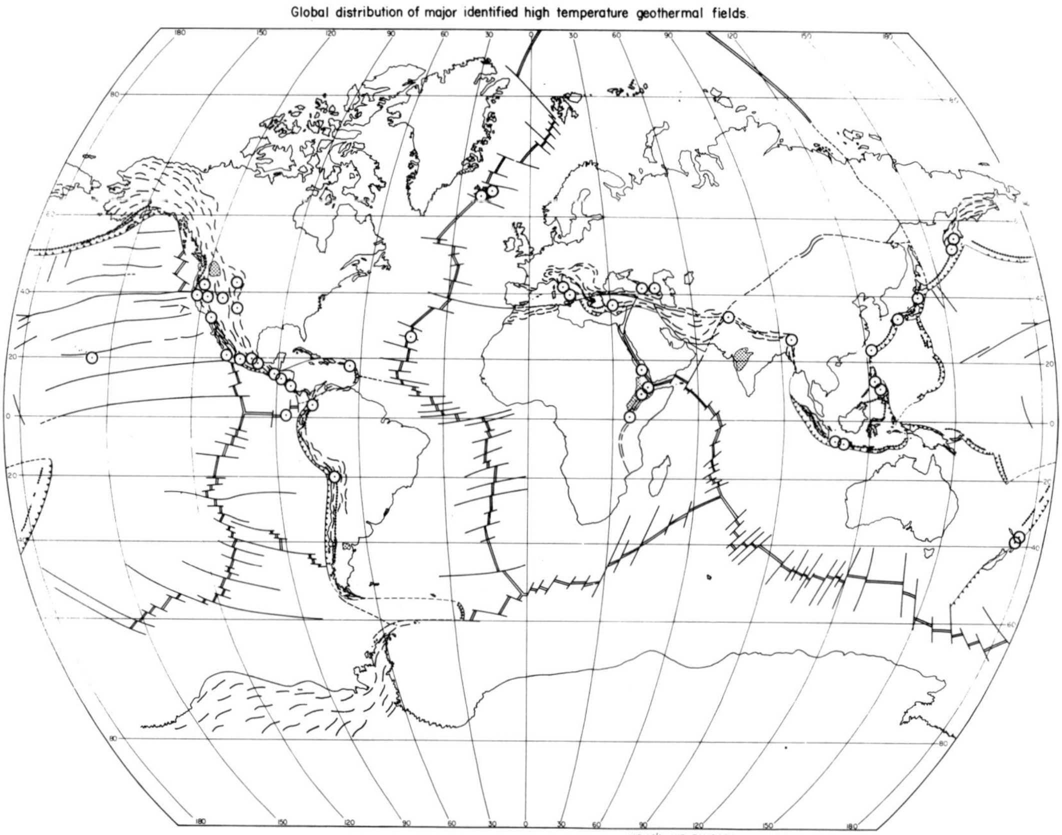

Fig. 1. Global distribution of major identified high temperature geothermal fields.

ematical modelling of geothermal systems. Numerical simulation has provided better understanding of the fundamental processes that govern two phase flow and enabled predictions of the response of geothermal systems to prolonged exploitation.

Close association between magmatic activity, rifting episodes and hydrothermal activity has recently been demonstrated in the Krafla area in Iceland. The relationship between volcanic and hydrothermal activity has been a common knowledge for a long time (e.g. Thoroddsen 1925) and intrusions or magma chambers have been described as the heat source for high temperature geothermal fields (Bodvarsson 1951, Banwell 1957, White 1957a). The direct observation of the interaction of magmatic and hydrothermal processes gives further insight into the nature of the heat source of geothermal systems. The improved understanding of the role of magma as a heat source is likely to focus attention on magmatic processes at mid-ocean ridges and in rift zones on land where magma chambers are inferred to exist at shallow depths.

This paper is mainly devoted to a review of the four topics mentioned above. In the light of the discovery of submarine geothermal systems and the experience gained during magmatic activity and rifting episodes in the Krafla area, a thorough treatment of the source of heat and fluid of hydrothermal systems is considered worthwhile. The progress in studies of two-phase geothermal systems is outlined by a treatment of different types of convection processes in hydrothermal systems. Rather than defining characteristic types of geothermal systems we stress an integrated point of view and examine different physical states of geothermal systems. Most systems are complex in nature and consist of a number of zones representing different physical states. The states of dominating liquid (hot-water) or dominating vapor are common stable states of geothermal systems. During the lifetime of the hydrothermal activity physical conditions gradually change. As a result zones of a geothermal system may gradually alter their physical state or flip from one stable state to another, if intermediate states are not stable. Disturbances induced by exploitation of geothermal fields may accelerate this development. For example, exploitation at Wairakei, New Zealand, has induced a vapor-dominated zone in the uppermost part of a liquid-dominated system, and due to the release of gas in the Bangore field, Italy, a vapor-dominated zone developed into a wet two-phase zone. These examples demonstrate that the "type" of a geothermal system is merely a physical state, determined by physical conditions and the governing hydrothermal processes.

Classification of Geothermal Systems

There have been many attempts to classify hydrothermal processes and geothermal activity. These classifications reflect naturally the background of their proponents and the status of knowledge at each time.

In the first attempts emphasis was on the surface manifestations of geothermal activity which were quite variable. A common classification was that of warm springs, boiling springs, fumaroles and solfataras. Water from fumaroles was often found to be acid whereas warm springs were usually alkaline. In this way one could distinguish surface activity by chemical characteristics. Another characteristic was the temperature of the ascending hydrothermal fluid. Bodvarsson (1961) distinguished between low-temperature areas where temperatures at the base of circulation were below 150°C and high-temperature areas with subsurface temperature above 200°C at a relatively shallow depth. Fumaroles, large areas of steaming ground and high degree of thermal alteration are surface characteristics of high-temperature areas, but warm and boiling springs associate with the low-temperature areas.

The geological environment of geothermal areas has been used to classify geothermal fields. McNitt (1970) reviewed various attempts at classification and proposed one based on geological processes and the positions of fields with respect to orogenic and volcanic belts.

By 1975 it was widely accepted that geothermal fields are localized in areas of young tectonism and volcanism, primarily along active or recently active plate boundaries (Muffler 1976b, Healy 1976).

The classification adopted by Muffler (1976b) is used as a basis for the assessment of geothermal resources of the U.S.A. (White and Williams 1975, Muffler 1979).

Ellis and Mahon (1977) used a classification based on cyclic and storage systems. In the cyclic systems meteoric water passes through a cycle of deep descent, heating and rising. In storage systems the water is stored in the rocks for geologically long periods and heated in situ either as a fluid within the formation, or as water of hydration in minerals.

Geothermal systems have often been classified with regard to their physical state. The discrimination between low- and high-temperature systems is of that kind. Distinction between conductive and convective systems refers to the dominating mode of heat transfer. Most hydrothermal convection systems contain liquid water as the dominant pressure-controlling fluid but in a few systems the pressure is evidently controlled by vapor rather than liquid. The former are classified as hot water systems and the latter are called vapor-dominated systems (Bodvarsson 1951, Banwell 1957, White 1964, James 1968, White et al. 1971, Truesdell and White 1973).

As the basis of knowledge of hydrothermal systems improves there is a trend towards classification based on processes and definition of basic parameters which describe the respective processes. The increasing emphasis on simulation of geothermal systems has urged this point of view. The main points of this approach are listed in Table 1. The hydrothermal system is defined as a volume of permeable rock saturated with a hydrothermal fluid consisting of water in liquid and vapor phase, and gases, dissolved in the liquid and free in the vapor. To describe the physical state of the system, a set of independent descriptive parameters must be defined. The parameters temperature, pressure, and liquid saturation are selected in Table 1, but other sets equally valid are density and enthalpy or density and internal energy. To describe the chemistry of the fluid, the concentration of a certain number of chemical components has to be specified. Important properties of the rock and fluid are thermal properties (heat capacity, thermal conductivity, latent heat), hydraulic properties (porosity, permeability, fluid viscosity), and chemical properties of the rock. Boundary conditions define heat and mass transfer, i.e. input and output of the system. Steady state conditions are generally assumed.

To define the physical state of a hydrothermal system the distribution of descriptive parameters and the properties of rock and fluid must be known. These internal conditions together with structural discontinuites constrain the processes taking place. The main physical processes are laminar flow, conductive flow of heat, convective transport of heat, boiling and condensation of the fluid. Chemical processes involve reactions between the rock and the fluid and distribution of chemical components. This parametric approach is used later on in this paper to define four physical states of hydrothermal systems.

TABLE 1

Parametric approach to the description of hydrothermal systems.

Descriptive parameters	
	Temperature
	Pressure
	Liquid saturation
	Concentration of chemical components
Properties of rock and fluid	
	Thermal
	Hydraulic
	Chemical
Boundary conditions	
	Transfer of mass and energy
Internal conditions	
	Distribution of descriptive parameters
	Distribution of properties of rock and fluid
	Structural discontinuities
Processes	Conduction
	Convection
	Boiling and condensation
	Laminar and/or turbulent flow
	Water-rock interaction

Source of fluid in continental and submarine hydrothermal systems.

In hydrothermal processes heat is mainly transported by the hydrothermal fluid. The fluid consists of liquid water with dissolved solids, water vapor, and gases dissolved in the liquid and free in the vapor.

The origin of thermal fluids was a subject of a controversy for a long time. An excellent review of the early discussion was presented by Allen and Day (1935). In general, two opinions have been defended. One, supporting volcanic or juvenile character of thermal waters, the other that the water is of meteoric origin.

This discussion can be traced at least to the middle of the last century. In 1846 Bunsen visited Iceland and made observations on hot springs there. He assumed (Bunsen 1847a, b) that atmospheric water descended to depths into the ground where it absorbed volcanic heat. In the same year Elie de Beaumont (1847) pointed out the similarities of magmatic gases and the gases escaping from thermal springs and concluded that both were of the same origin. For a century these two alternatives were discussed and disputed almost continuously. Some authors (Thoroddsen 1925, Allen and Day 1927, 1935) seem, however, to adopt both possibilities.

Generic types of water as a hydrothermal fluid were defined by White (1957a, b) and adapted by Ellis and Mahon (1977). The types are:

<u>Juvenile water</u>. "New" water from mantle derived magma and which has not previously been part of the hydrosphere.

<u>Magmatic water</u>. Water derived from magma, but not necessarily juvenile water, since magma may incorporate meteoric water of deep circulation or water from sedimentary material.

<u>Meteoric water</u>. Water recently involved in atmospheric circulation.

<u>Connate water</u>. "Fossil" water incorporated in sediments of their time of deposition.

<u>Metamorphic water</u>. Modified connate water, derived from hydrous minerals during their recrystallization to less hydrous minerals during metamorphic processes.

To the list White (1974) added <u>ocean water</u>, which is now obvious in view of the growing evidence for submarine geothermal systems.

Systematic studies of stable oxygen and hydrogen isotopes in geothermal water (Craig et al.

1956, Craig 1963, 1966, Årnason 1976) have now established meteoric water to be the dominant source of fluid in most active continental geothermal systems, with some exceptions, however, as noted by White et al. (1973). Evidence for this origin of thermal waters was further strengthened by Ellis and Mahon (1964, 1967) and Mahon (1967), who showed experimentally that the chemical composition of most waters could be attained by the solvent action of hot water on the local volcanic rocks.

Although the contribution of magmatic or juvenile water appears to be minor, there is growing evidence for magmatic influence on thermal fluids. Pronounced changes in the chemical composition of the fluid in the Krafla geothermal reservoir in Iceland are found to be associated with magmatic activity and a rifting episode. The concentration of CO_2 increased abruptly 100 times, followed by an increase in SO_4 which seemed to be caused by the release of magmatic SO_2 into the hydrothermal system (Gíslason et al. 1977, Årmannsson et al. 1981, Stefánsson 1981).

The excess of ^3He first verified in oceanic water (Clarke et al. 1969) and in volcanic gases (Mamyrin et al. 1969) is a further evidence for magmatic influence on thermal fluids. The $^3He/^4He$ ratio in the mantle is assumed to be approximately 10^{-5} compared with the ratio 10^{-6} for atmospheric helium and 10^{-7} - 10^{-8} for continental helium (Wasserburg et al. 1958). Measurements of helium isotope ratios in oceanic basaltic glasses (Gerling et al. 1971, Lupton and Craig 1975, Craig and Lupton 1976) have revealed values of 10^{-5}. Helium of mantle origin was also found in geothermal waters of the Kuril Islands, Kamchatka and Iceland (Kamensky et al. 1971, Mamyrin et al. 1972, Tolstikhin et al.1972, Baskov et al. 1973, Kononov et al. 1974, Gutsalo 1976, Kononov and Polak 1976). Subsequent helium isotope measurements in other geothermal areas (Craig et al. 1978a, Craig et al. 1978b, Welham et al 1979) as well as in submarine hydrothermal areas, (Craig et al. 1975, Lupton et al. 1976, Jenkins et al. 1978, Lupton 1979, Lupton et al. 1980, Jenkins et al. 1980) have revealed the same enrichment of mantle helium in hydrothermal fluids. Mantle helium is transported towards the surface by magma. The presence of mantle helium in hydrothermal fluids is either due to leak of gases from magma or interaction of water with young rock of mantle origin.

The role of ocean water, and the possibility of submarine geothermal systems on oceanic ridges was first pointed out by Elder (1965). Geothermal systems on land where ocean water is the main source of fluid have been described by Björnsson et al. (1970, 1972), Arnórsson et al. (1976), Kjaran et al. (1979). Submarine geothermal systems have recently been discovered at a number of sites near ocean spreading centers (Reviews by Rona and Lowell, 1978 and White and Guffianti 1979, Spiess et al. 1980). It is now well established that hydrothermal circulation plays a major role in the thermal balance of ocean ridges where ocean water is the dominant fluid source.

^3He has been found to be the most sensitive tracer to detect hydrothermal systems at the ocean floor. Lupton et al. (1980) show that hydrothermal water can be detected by ^3He measurements even though it is diluted 10^5 times with seawater, and Jenkins et al. (1980) point out that the detection limit corresponds to a thermal anomaly of the order 0.001°C. Further studies of helium in seawater have revealed that not only abundant ^3He relative to ^4He is found in the geothermal fluid, but the concentrations of both ^3He and ^4He are much higher in thermal water than in ambient seawater. Enrichment of the order of 10 times for ^4He and 50 times for ^3He have been reported (Jenkins et al. 1978, Lupton et al. 1980).

Distinction between ocean water and meteoric water might seem to be of little importance, but the chemical composition of oceanic water as an input parameter in geothermal systems has a large effect on the solvent action of the geothermal fluid. When seawater is heated within the rock matrix, the removal of Mg from seawater generates acidity which maintains heavy metals in solution at moderate temperature (about 300°C) (Seyfried and Bischoff 1981). This influences the chemical output of submarine geothermal systems towards much higher metal concentration as compared to geothermal systems fed by meteoric water (Seyfried and Bischoff 1977, Bischoff and Seyfried 1978).

In geothermal systems in the Imperial Valley, U.S.A., fluids of both high salinity and high temperature (350°C) are found. Metallic concentration is unusually high in these fluids (Muffler and White 1968). In the Krafla geothermal reservoir in Iceland, the fluid is of very low salinity but high metallic (mainly Fe) concentrations have been encountered in some wells. These high concentrations result from a very low pH value of the thermal fluid due to intermittent flow of volcanic gases (SO_2 , Cl_2) into the hydrothermal system (Arnórsson 1981, Årmannsson et al. 1981). Ore deposits in fossil hydrothermal brine systems are suggestive of a brine fluid at the time of deposition (White 1967, 1968a, White et al. 1971). One of the early signs of submarine hydrothermal systems was the observation of the metallic content of sediments near the East Pacific Rise (Bostrom and Peterson 1966, 1969). The significance of these observations was not generally recognized until the physical evidence for submarine hydrothermal systems became commonly known.

The new understanding of metallic deposition in the oceans has caused an explosive increase in research activities (Corliss 1971, Johnsen 1972, Dymond et al. 1973, Spooner and Fyfe 1973, Piper 1973, Hajash 1975, Bischoff and Dickson 1975, Wolery and Sleep 1976, Cronan 1976, Cann et al. 1977, Spooner 1977, Spooner and Bray 1977, Heath and Dymond 1977, Shanks and Bischoff 1977, Weiss et al. 1977, Paramentier and Spooner 1978, Mottl and Holland 1978, Anderson et al. 1978, Edmond et al. 1979a, 1979b, Corliss et al. 1978, 1979b, Williams et al. 1979, Lonsdale et al. 1980, Hekinian 1980; Review by White and Guffianti 1979).

Magma as heat source to the hydrothermal fluid.

The source of heat has been a subject of dispute no less than the origin of geothermal fluids. The close association of major geothermal activity with volcanic activity was noticed at an early stage (e.g. Thoroddsen 1925, Allen and Day 1927). Of the numerous speculations about the nature of thermal activity the work of Einarsson (1942) is the first quantitative treatment. He contended that the hot springs of Iceland were not physically different from ordinary cold springs, except for the greater depth of penetration of the water. The heat does not necessarily come from volcanism but comes simply from the conductive heat flux from the interior of the Earth. Bodvarsson (1951, 1961) elaborated this concept further. He agreed with Einarsson on the origin of low-temperature fields but concluded that this theory was not successful in explaining the heat output of the high temperature fumarolic fields. Neither was juvenile water likely to carry sufficient heat. The main source had to be the heat content of magma itself. Limited surface area and poor thermal conduction through solid rock require however intimate contact between the hot, recently solidified rock, and the percolating water. Banwell (Banwell et al. 1957) realized similar problems in explaining the heat output of large thermal areas in New Zealand. He suggested the presence of large chambers of convecting magma in order to explain the necessary flow of juvenile water. Isotopic studies of thermal waters are, however, not indicative of major juvenile components (Craig et al. 1956, Craig 1963, 1966). Banwell (1963) argued that the steam could be magmatic but still of meteoric origin. Water could penetrate to the magma through a few deep faults and there diffuse into the magma but leave again when it had circulated to some higher level where the pressure is lower. Convective transfer of heat from the magma by water vapor in this manner could result in heating of only comparatively minor volumes of rock in permeable formations near the surface, since large quantities of steam under pressure could escape from the magma along a few very limited fissures, and a significant permeability of the deeper rocks is not necessary. Banwell also pointed out that a volume of nearly 10^4 km^3 of magma was necessary to maintain the natural heat output of the Wairakei field (670 MW) for the estimated lifetime of about $5 \cdot 10^5$ years (Grindley 1961) assuming the general level of activity to be of the same order over most of the period.

White (1957a, 1968b) concluded that a magma supply of at least 10^2-10^3 km^3 was required to support the Steamboat Springs systems through its life of 10^5-10^6 years. He found that a batholith intruded into the shallow crust and then remaining static as it cools and crystallises is not a satisfactory model, unless the fissure system controlling the circulating water can gradually extend deeper into the batholith as stored heat is removed at higher levels by circulating water. As an alternative for the heat-flow problem he suggested convection within the magma chamber to maintain magmatic temperatures near the base of the hydrothermal circulation (White 1968b).

Irvine (1970) has studied the relation between temperature in a magma body and a crystallization mechanism where crystal fractionation is a major process. He described a convective process in the magma body where crystals are accumulated in the lower part of the intrusion but the temperature near the top remains close to or above the liquidus temperature of the magma. This process allows higher rate of heat loss and solidification than would occur if the crystals were frozen to the roof of the magma chamber. Convective processes of this nature appear to be capable of providing sufficient heat transfer to the hydrothermal fluid to explain the heat output of most geothermal systems.

Oxygen isotope analyses on rocks and coexisting minerals indicate large scale exchange between igneous bodies and heated meteoric groundwaters that have percolated down through the country rocks and penetrated into the plutons along fractures and faults (Taylor 1968, Taylor and Forester 1971). In the Skaergaard intrusion the bulk of the reacting water circulated through the pluton at temperature above 500°C. The magma body did not absorb any large quantities of water. The ^{18}O exchange occurred at sub-solidus temperatures and is clearly associated with fractures in the rock (Taylor and Forester 1979).

There has been growing evidence that hydrothermal circulation in young oceanic crust is a major factor in determining the thermal and physical state of the rocks at ridge crests (Elder 1965, Pålmason 1967, Talwani et al 1971, Bödvarsson and Lowell 1972, Hyndman and Rankin 1972, Lister 1972, Lister 1974, Klitgord and Mudie 1974, Williams et al. 1974, Corliss et al. 1979 a,b, Davies et al. 1980).

Lister (1974, 1976, 1977) has presented a conceptual model of the downward penetration of water into hot rocks by a process of cooling and thermal cracking. The process appears to be extraordinarily efficient in heat extraction. The cracks advance rapidly, cool a large volume of rock in a short time and provide an intense localized source of geothermal output (Lister 1980 a). A highly permeable matrix is produced for later groundwater flow and the cracks are densely spaced enough to allow major chemical interaction between the percolating fluid and rock. If these concepts prove correct active geothermal systems should be short-lived transient phenomena where ample water gains access to the magmatic source of heat. After solidification of the heat source relatively gentle circulation of cool water should persist in the residual permeable matrix, driven by heat conducted up from below. Lister (1977) has suggested the term "passive" geothermal system for the latter type of circulation to distinguish it from the intense high temperature "active" phase of initial penetration.

Evidence in support of water penetration into hot rock boundaries of solidifying magma is reported by Björnsson et al. (1980). Water-cooling of the recent lava flow in Heimaey, Iceland demonstrated a heat extraction efficiency of 40 kW/m^2 and a penetration rate of 12 meters in 14 days. The authors concluded that a process of this nature is required to explain the sustained intense heat output of the subglacial Grímsvötn geothermal area in Iceland which releases 5000 MW according to calorimetric observations of ice melting in the area. This process of heat extraction is likely to be effective where the rocks surrounding magma bodies are percolated by sufficient water, e.g. in subglacial and submarine rocks.

Hydrothermal vents jetting out water at 380°C have been discovered on the axis of the East Pacific Rise at a water depth of 2500 to 2900 m (Spiess et al. 1980). These hydrothermal springs are the sites of actively forming massive sulfide mineral deposits. The temperature of the fluid is unexpectedly high. It has been commonly considered that convective seawater cools the upper crust forming at ocean ridges and heavy metals are mobilized from the rock initially and are reprecipitated in the alteration phases. Mathematical modelling of this process (Lister 1972, Williams et al. 1974, Wolery and Sleep 1976) suggests that large volumes of seawater must cycle through spreading centers, and the effective water-rock ratios (total amount of water passing through the system/total amount of rock altered) may be in the range of 50 or 100:1. Experimental investigations (Seyfried and Bischoff 1977, Bischoff and Seyfried 1978, Seyfried and Bischoff 1981) have shown that the effective water/rock ratio determines the concentration of heavy metals in the temperature range between 300 and 400°C. At large effective water/rock ratios heavy metals are significantly leached from the rock and are maintained in solution, whereas temperatures in excess of 400°C are required for effective metal transport at low effective ratios (Bischoff 1980). Bischoff has discussed physical limits on the geothermal fluid issued from the vents. Both the chemistry of the discharging fluid and thermal balance considerations indicate that the effective water/rock ratios in the geothermal system are between 7 and 16. Such low effective ratios preclude effective metal transport at temperatures below 350°C, but metal solubilization at 400°C and above is effective even at such low effective ratios. Bischoff (1980) suggests that in the submarine systems at 21°N on the East Pacific Rise the hydrothermal circulation corresponds to Listers (1980a) "active" phase of hydrothermal cooling of the oceanic crust. During this phase, seawater extends to a zone immediately above a magma chamber, penetrating downward along a cracking front. Herron et al. (1978) suggested, on the basis of seismic data, a magma chamber to be located at approximately 2 km depth below the sea floor in this region. Hydrostatic pressure at this depth is estimated to be 450-500 bars. The maximum temperature attained by the seawater is limited by buoyancy effects and self-sealing by quartz deposition (Truesdell and Fournier 1976). Judging from the increasing rate of water expansion above 420°C at 450 bars, the maximum possible temperature cannot be much above 420°C. At 500 bars quartz solubility reaches a maximum at 420°C and falls steeply at higher temperatures (Kennedy 1950, Walther and Helgeson 1977). At higher hydrostatic pressures the expansion of water due to raised temperature becomes less pronounced and the point of maximum quartz solubility vanishes. The depth to the top of the magma chamber is thus controlling the maximum temperature attained by the seawater but maximum temperatures increase as the cooling front moves downward into higher pressure regimes (Bischoff 1980).

Although magma bodies are considered to be a common heat source of geothermal systems, direct evidence on the existence of these bodies and their relationship to the geothermal systems is rather scarce. Sanford and Einarsson (1981) reviewed evidence for magma chambers in rift zones. In continental rifts, lower than normal velocities and resistivities at mid-crustal and sub-crustal depths suggest a low degree of partial melting, but geophysical detection of significant accumulations of magma beneath or within the crust of continental rifts is rare. Sanford and Einarsson (1981) describe evidence for the presence of relatively large quantities of magma in the Rio Grande Rift and beneath the Yellowstone caldera. Ocean rifts seem to be generally underlain by a low velocity, low density mantle, suggesting partial melting. This amomalous mantle is particularly well developed under Iceland, where it is also characterized by a high P- to S-velocity ratio, high S-wave attenuation and a layer of low electrical resistivity near the crust-mantle boundary. Thermal accretion models also indicate partial melting under the crust of Iceland (Pålmason 1973, 1980, 1981).

Evidence of extensive crustal magma chambers is only found under the crest of fast spreading ocean ridges, but slowly diverging plate boundaries are not underlain by large magma chambers at high levels.

S-wave shadows have indicated a small magma body at 3 to 7 km depth beneath the Krafla geothermal system in the axial rift zone of NE-Iceland (Einarsson 1978). Periodic inflations and deflations of the Krafla volcano have been monitored by tilt meters, geodetic levelling, gravity surveys and earthquake studies during the present active episode which started in 1975 (Björnsson et al. 1977, Björnsson et al. 1979, Brandsdóttir and Einarsson 1979, Johnsen et al. 1980, Tryggvason 1980, Einarsson and Brandsdóttir 1980, Stefánsson 1981, Ármannsson et al. 1981). Dykes injected from the magma body into the associated fissure swarm have caused tectonic displacements and cut the roots of the geothermal system at Námafjall, 7 km distant (Brandsdóttir and Einarsson 1979, Larsen et al. 1979, Björnsson and Sigurdsson 1978, K. Ragnars, pers. comm.). The heat source under Námafjall may be an accumulation of magma drained from the magma body at Krafla. Similar conditions are apparently found in the geothermal system of the Puna district on Hawaii (Furumoto 1978) and its relation to the summit magma chamber of Kilauea. Gravimetric, magnetic and seismic data have shown anomalous zone under the Avachinsky volcano on the Kamchatka Peninsula. This is suspected to be a peripheral magma chamber (Fedotov et al. 1976).

Teleseismic P-wave delays have been used extensively to infer velocity structure at several geothermal systems, including The Geysers, California (Steeples and Iyer 1976a, Iyer et al. 1979), Yellowstone, Wyoming (Iyer 1975, Iyer and Stewart, 1977, Iyer 1979), Long Valley, California (Steeples and Iyer, 1976b), Kilauea Volcano, Hawaii (Ellsworth and Koyanagi, 1977) and the Coso Area, California (Reasenberg et al. 1980). At the Geysers a molten chamber about 14 km in diameter is inferred with its top about 7 km beneath the volcanic field (White and Guffianti 1979). At the Coso Geothermal Area an intense low-velocity body, which coincides with the surface expressions of late Pleistocene rhyolitic volcanism, high heat flow, and hydrothermal activity, is resolved between 5 and 20 km depth.

Eroded central volcanoes are widely distributed within the Tertiary basalt formations in Iceland (Walker 1963, 1966). The volcanic centres are places of unusually vigorous volcanic activity. This is shown by the great concentration of dykes and intrusive sheets. Walker (1966) estimates that the intrusions amount to at least 50% of the rock in some of the complexes. Each centre has a down-sagged core region. The hydrothermal alteration which is typically found in the collapsed core bears witness of ancient geothermal activity, attributed to the basic sheet swarm in the core. The zones of intense hydrothermal alteration appear to have hosted large hydrothermal reservoirs, each with a volume of the order of 100 km^3. The rate of hydrothermal processes puts contraints on the nature of the heat source and the nature of heat transport. As described above, the intense loss of heat observed in some geothermal fields is difficult to explain unless the heat transfer from the magma to the hydrothermal fluid is caused either by downward penetration of water and cracking of the hot intrusion or that convection is taking place within the magma body in such a way that the boundary between the magma and the hydrothermal fluid remains relatively thin for a considerable time.

The duration of hydrothermal activity is also an indicator on the nature of the heat source. Cooling of plutons by hydrothermal convection has been studied numerically by Cathles (1977), Norton and Knight (1977), Norton and Taylor (1979), and by Ribando et al. (1976). Single phase convection was applied by Norton and Knight (1977) and was further extended to include calculations of isotope reactions by Norton and Taylor (1979). Convection is found to predominate over conductive heat flow when the host rock permeability exceeds 10^{-18} m^2 (0.001 md). When the permeability of the pluton itself exceeds the same permeability the cooling is significantly accelerated. The lifetime of such systems was found to be of the order of 10^5 years. For the Skaergaard intrusion the most active period was found to be the first $1.5 \cdot 10^5$ years whereas the pluton had cooled to ambient temperature in $4 \cdot 10^5$ years. Cathles (1977) included phase transitions (boiling and condensation) in the convective cooling of intrusions. He found that vapor dominated steam zones are commonly formed above intrusions, but their lifetime is relatively short. In this model the latent heat of condensation allows faster upward migration of heat than can be achieved by single phase convection. Consequently the cooling is more rapid and the lifetime of the hydrothermal system generally less than $2 \cdot 10^4$ years. The lifetime of the vapor dominated zones was found to be less than $5 \cdot 10^3$ - 10^4 years.

In a simulation made by Pruess and Truesdell (1980) a heat source with constant heat flux was suddenly emplaced beneath a liquid saturated porous medium. The time required for new thermal equilibrium was found to be $9 \cdot 10^4$ years. During this time a resulting hydrothermal system evolved through liquid convection into boiling and ended up as a vapor dominated system.

These considerations indicate that hydrothermal systems drawing their energy from a single pluton are not likely to be maintained more than $2 \cdot 10^4$ years, when phase transitions in the fluid govern the cooling rate. Single phase convection is less effective and would allow a lifetime up to some $2 \cdot 10^5$ years, but lifetimes of the order of 10^6 years seem to need the presence of a very large intrusive body or that intrusions are formed continuously in the crust.

Some limits on the lifetime of geothermal

systems have been inferred from geological evidence. A minimum lifetime of 10^4 - $2 \cdot 10^4$ years has been estimated for active (Thorarinsson 1980) as well as extinct systems in Iceland (Fridleifsson 1973). In New Zealand an upper limit of $5 \cdot 10^5$ years has been inferred for Wairakei (Grindley 1965) and for Broadlands (Browne 1971) but a minimum of $2 \cdot 10^5$ years for Kawerau (Browne 1979). White (1968b) concludes that the lifetime of Steamboat Springs is at least 10^5 years and probably more than 10^6 years. The complex history of activity in this field started at least $2.5 \cdot 10^6$ years ago, and activity was perhaps most intense $1.1 \cdot 10^6$ years ago (White and Guffianti 1979).

The distribution of hydrothermal products at the TAG field indicates that hydrothermal activity has persisted in the TAG field during seafloor spreading for at least $1.4 \cdot 10^6$ years (Rona 1980).

To summarize this discussion on the heat source of geothermal systems we have found that in order to sustain the intense activity a magmatic heat source is required. The source is either found as a large convective magma chamber or as a complex of repeated intrusives at shallow level. Evidence from both geology and numerical simulation of hydrothermal processes indicates a lifetime of these systems of 10^4 - 10^5 years depending on the efficiency of the hydrothermal processes cooling the heat source. These conditions are mostly found at the margins of the crustal plates, which explains the global distribution of geothermal systems.

Convection and boiling in hydrothermal systems

As outlined in Table 1 the physical processes in hydrothermal systems are subject to external and internal boundary conditions of the reservoir rock as well as the properties of rock and fluid. Norton and Knapp (1977) defined porosity and permeability in fractured rocks and discussed transport phenomena in hydrothermal systems. These concepts were applied by Norton and Knight (1977) in a numerical model study of the nature of heat and mass transport in pluton environments. A review of the state of the art of geothermal reservoir modelling in porous and fractured rocks is given by Pinder (1979). Multiphase flow in porous media is theoretically well understood, but mathematical treatment of multiphase flow in fractured rocks is still in its infancy.

The following discussion will be limited to thermal properties of rock and fluid and processes of convection and phase transition in the hydrothermal fluid.

Thermal properties of rocks and water

Kappelmeyer and Haenel (1974) have compiled thermal properties for dry rocks and also included the effects of temperature and pressure on these properties. Thermal conductivities for most dry rocks range from 1.6 - 4 W/m°C, specific heats are approximately 1 kJ/kg°C and thermal diffusivities range from (5 to 15) 10^{-4} m^2/s. Anand et al. (1973) and Somerton et al. (1974) have pointed out that the thermal conductivity of fluid saturated rocks is a function of temperature, porosity and water saturation. The thermal conductivity increases with brine concentration and becomes more sensitive to temperature change at high concentrations. Since the dominant fraction of the total heat energy in hydrothermal systems is in the solid rock matrix, it is obvious that the thermal properties of rocks are of great importance for hydrothermal processes. The variation of these properties with temperature and pressure is though far less than that of the fluid properties and is therefore generally neglected in discussions and simulations of hydrothermal processes.

Most physical properties of water are strongly dependent on temperature and undergo extreme variations in the range of temperature and pressure relevant to geothermal systems (Goguel 1953, Sorey 1976, Straus and Schubert 1977). Between 25°C and 350°C, the viscosity of water changes by more than one order of magnitude and the thermal expansion coefficient by nearly two orders of magnitude. The density of water is reduced by a factor of two in this temperature range, whereas variation in other properties of water is less significant but by no means negligible.

The influence of pressure on the thermal properties of water is quantitatively less important than that of temperature. However, pressure is a significant parameter as it together with temperature controls boiling of the hydrothermal fluid. The fluid properties depend on the physical state, whether the fluid occurs as liquid or vapor.

Single phase convection of water

Thermal convection is a fundamental process of heat transport in hydrothermal systems. A thorough review of the basic characteristics of free convection of a single phase fluid in porous media is included in Witherspoon et al. (1975).

A linear stability analysis shows that thermal convection in a liquid-saturated porous layer is initiated when a critical value of the Rayleigh number, Ra, is exceeded. In a horizontal layer of thickness H and a temperature difference ΔT across, the Rayleigh number can be expressed,

$$Ra = \frac{\alpha \cdot g \cdot \Delta T \cdot H \cdot \rho^2 \cdot c_p \cdot K}{\mu \, . \, k}$$

where:

α = coefficient of thermal expansion of the fluid

g = acceleration of gravity

ρ = fluid density

c_p = specific heat of the fluid at constant pressure

K = permeability of the rock

μ = dynamic viscosity of the fluid

k = thermal conductivity of saturated rock

The most common approximation in the extensive literature on the subject of thermal convection is to consider the viscosity, the permeability, the thermal conductivity, the thermal expansivity, and the specific heat as constant values in the convection process (Boussinesq 1903). Variations in density are included in the buoyancy term of the vertical balances of forces but otherwise density is assumed to be constant. (see e.g. Horton and Rogers 1945, Lapwood 1948, Elder 1967, Palm et al. 1972, and Straus 1974).

The critical value of Ra is dependant upon the boundary conditions (Nield 1968 ; Ribando and Torrance 1976) and on the properties of the fluid and the rock matrix.

Straus and Schubert (1977) studied systematically the conditions for the onset of thermal convection in a water-saturated porous layer. They used an accurate representation of the equation of state for liquid water and an empirical formula for the viscosity of water as a function of temperature and pressure. The properties of water considered were: density ρ, thermal expansion coefficient α, isothermal compressibility β, specific heat at constant pressure c_p, adiabatic temperature gradient $\frac{\alpha g T}{c_p}$ and dynamic viscosity μ.

By allowing for variations in the thermal properties of water, Straus and Schubert determined the critical Rayleigh number for the onset of convection for various thicknesses of the porous layer as well as for various thermal gradients in the layer. In contrast to the results obtained when the properties of water are assumed constant it was found that the critical Rayleigh value depends on the thermal gradient as well as on the thickness of the layer. For very thick layers the critical Rayleigh number was found to be reduced by as much as a factor 30 below the value $4\pi^2$ valid for a layer containing fluid with constant thermal properties.

In a pioneering paper on thermal convection Goguel (1953) had already envisaged most of these results and the importance of considering the variation in thermal properties of water in estimates of the criteria for initiation of thermal convection.

The criterion for the onset of convection $Ra \geq Ra_{crit}$ can be interpreted as providing a minimum temperature difference for convection to appear in a layer of given thickness and permeability or as providing a minimum permeability for convection in a layer with a given thickness and temperature difference. The study of Straus and Schubert (1977) has shown that the permeability necessary for convection is seriously overestimated when the thermal properties of water are assumed to be constant values. Due to the effects of variable water properties convection can occur for smaller vertical temperature differences in rock of a given permeability or for smaller permeability at given temperature difference.

Critical values of the ratio of permeability to thermal conductivity $(K/k)_{crit}$ for the onset of convection in a water saturated porous layer were calculated as functions of layer thickness for different values of constant thermal gradient. In a layer of 1 km thickness and for a gradient of 200°C/km the ratio $(K/k)_{crit}$ = $1.5 \cdot 10^{-15}$ m s^3 ·C/kg was obtained for water compared to $1.5 \cdot 10^{-14}$ m s^3 ·C/kg for a constant property fluid.

The primary reasons for the increased tendency to initiate convection are the substantial increase of thermal expansivity and the decrease of viscosity with increased temperature. Variations in the specific heat, the adiabatic temperature gradient and the compressibility were found of minor importance in the cases considered by Straus and Schubert (1977).

Two-Phase Convection

So far the discussion has been limited to the properties of water and thermal convection of liquid water in a porous medium. In many geothermal systems the flowing water reaches the saturation pressure due to release of pressure and boiling is initiated. The fluid becomes a two-phase mixture of steam and water with thermodynamic and transport properties different from those of liquid water. Where steam and water are in thermodynamic equilibrium the fluid temperature and pressure are uniquely related by the Clapeyron equilibrium equation which determines the boiling (Clapeyron) curve separating the steam and water phases on a p-T diagram. The thermodynamic properties of each phase are unique functions of temperature (or pressure) only. The two-phase flow is generally assumed to be laminar. This assumption, however, might not be valid where rapid boiling occurs. Darcy's law is generally applied separately to the steam and water phases, introducing relative permeability factors to account for the restricted flow of each phase in the presence of the other. The relative permeability factors are expressed as functions of the volume fraction of each phase, S denoting the volume saturation of water and 1-S that of the steam. These functions are still poorly defined, however, and critical values such as the water saturation at which the water becomes immobile are inadequately known. The Corey equations (Corey 1954) are frequently used in the literature and more detailed relationships have been developed by Sigmund and McCafferty (1979). Experimental as well as theoretical studies of relative permeability and its influence on the characteristics of geothermal systems have yielded conflicting results (Kruger and Ramey 1974, Chen et al. 1978, Council and Ramey 1979, Mercer and Faust 1979, Grant 1977b, Sorey et al. 1980, Horne and Ramey 1978). In view of the difficulties met in defining the relative permeability, the use of complex relations is hardly warranted. The results obtained by Grant (1977b) based on analyses of the discharge characteristics of two phase wells at Wairakei, New Zealand, indicate that the mobility of steam is considerably larger than what is predicted by the conventional Corey equations. Many authors simply assume that the relative permeability factor for each phase is equal to the saturation value of the respective phase.

The Darcy flow of either phase is driven by the pressure gradient in excess of the static gravity gradient of each phase. Donaldson (1968) considered boiling processes within a one-dimensional steady upflow of hot water. He found that a two-phase zone of steam and water formed for vertical massflow rates above a threshold value. The steam ascended more rapidly than the water and was condensed at the upper boundary of the two phase zone. Sheu et al. (1979) extended the model of Donaldson (1968) to include a more complete energy equation and more realistic thermodynamic properties. For flow rates below a critical value only liquid water existed at all depths. Above the critical value, three zones exist, consisting of a near--surface water layer, an underlying two-phase zone of water undergoing pressure release boiling, and a deeper zone of liquid water. Assuming linear relative permeability curves Sheu et al. (1979) worked out illustrative examples for permeability $K=10^{-14}$ m^2 and thermal conductivity k=3 W/m°C. The study indicated, that the critical massflow rate u for the onset of a two-phase zone in water ascending from base reference conditions of 270°C and 90 bars was $u=0.383\cdot10^{-5}$ kg/m^{2}s. As the flow rate increases the thickness of the upper water zone decreases and the maximum volume fraction of steam increases. The maximum Darcy velocities in the two phase zone are more than an order of magnitude larger for the steam than for the water. For $u=0.776\cdot10^{-5}$ and $1.190\cdot10^{-5}$ kg/m^{2}s the maximum volume fraction of steam was found to be 0.03 and 0.09 respectively.

Schubert and Straus (1977) investigated the basic physical processes involved in three-dimensional porous medium convection with phase transitions. The steam-water mixtures were described by a homogeneous model, in which a single Darcy-velocity specifies the mass flow of the mixture, and the thermodynamic and transport properties of the mixture depend only on the properties of the individual phases and their relative amounts. The thermodynamic properties of the mixture are thus well defined. Transport properties such as viscosity are, however, not well known and various models were tried for this quantity. Schubert and Straus (1977) used this model to investigate the conditions for the onset of convection and the nature of that convection in a porous medium containing a steam-water mixture or water at saturation temperature at all depths. The tendency of these fluids to convection is quite different from that governing the instability of an ordinary single phase fluid. The ordinary Rayleigh instability is driven by buoyancy forces which cause relatively hotter, lighter fluid elements to rise and relatively colder, heavier fluid parcels to sink. The two-phase convection proceeds by way of a phase change instability mechanism associated with the requirement that the fluid temperature and pressure always lie on the equilibrium Clapeyron curve. Temperature variations are directly responsible for the pressure gradients which drive convection. A perturbed hotter region of the fluid is also at somewhat higher pressure than its surroundings, and fluid will flow horizontally away from the hot spot. Conservation of mass then requires that the horizontal divergence of fluid out of the hotter region be balanced by a vertical influx of fluid. Condensation and boiling occur to achieve a balance of forces in the vertical. The most striking aspects of this type of convection are the small lateral dimensions of the cells and the concentration of the flow, phase changes, and temperature variations toward the bottom of the porous layer. The saturated liquid convection cells are only about half as wide as those of ordinary buoyancy-driven convection in water, two-phase cells are still narrower and the flow more concentrated toward the bottom. This phase change instability mechanism induces convection prior to the onset of ordinary buoyancy driven thermal convection (Schubert and Straus 1977). Although buoyancy driven convection has been assumed to dominate in many hydrothermal systems, the real geometry of convection cells has never been observed in nature. One reason for this might be that phase change driven convection dominates the convective pattern with narrower cells and concentration of the convective flow near the bottom of the system.

The presence of gases in geothermal fluids may greatly enhance convective instability. In geothermal systems where chemical equilibrium is attained for all major components incorporated in alteration minerals, the concentration of gases (CO_2, H_2S, H_2 and CH_4) is directly related to the temperature of the geothermal fluid. Data on CO_2 concentrations in geothermal waters in Iceland (Arnórsson and Svavarsson 1980, Arnórsson et al. (in preparation)) show that the concentration and the partial pressure of CO_2 rise sharply with temperature in the 100-300°C range until a limit for the stability of calcite is reached in the range of 300°C. When calcite is no longer stable, CO_2 is not incorporated in any alteration mineral and becomes mobile. The partial pressure of CO_2 increases from 10^{-1} bars at 200°C to tens of bars in the range of 300°C.

The concentrations of CO_2 in geothermal systems near 300°C are sufficiently large to profoundly influence the physical state of the system. Examples of that are the Broadlands system, New Zealand (Mahon and Finlayson 1972, Sutton and McNabb 1977, Grant 1977a) and the Krafla system, Iceland (Stefánsson 1981, Ármannsson et al. 1981). The large amounts of CO_2 present in the Broadlands system apparently produced two-phase conditions in the upper 1-2 km of the undisturbed state. Boiling in a water - CO_2 fluid occurs where the sum of the partial pressures of CO_2 and steam exceeds the ambient pressure in the fluid. Boiling refers thus to the phenomenon in which CO_2 and vapor create a gaseous phase in equilibrium with the liquid phase at the ambient pressure. The boiling temperature of the water - CO_2 fluid is well below the saturation temperature for pure water at the same ambient pressure. The ascending fluid initiates boiling at greater depths than pure water. The boiling point curve within a water - CO_2 geothermal system is displaced to progressively greater depths as the partial pressure of the CO_2 increases with temperature and depth. Straus and Schubert (1979) showed that the buoyancy of the geothermal fluid depends critically on the presence of CO_2 because of the large volume changes that occur when CO_2 enters or leaves solution and forces water to simultaneously change phase. While the presence of CO_2 can, in principle, enhance or inhibit convection in geothermal fluids, the effect is generally to strongly enhance convection for most temperatures and pressures of interest.

Steam Water Counterflow in Vapor-Dominated Systems

In the discussion of single-phase and two-phase convection above it has implicitly been assumed that water is the continuous phase throughout the system and thus provides pressure control. Pressure in such systems is near hydrostatic values. This appears to be the most common state of geothermal reservoirs. There are however important exceptions where the steam is the continuous, pressure-controlling phase although liquid water is also present. White et al. (1971) described the characteristics of these vapor-dominated hydrothermal systems and compared them with hot-water systems. Concepts of vapor dominated systems were further worked out by Truesdell and White (1973). White et al. (1971) concluded that the main vapor-dominated reservoir actually is a two phase heat-transfer system, where vapor boiled off a deep water table and most of the condensate forming in the upper part of the system is returned to the deep water domain. Pressures throughout the reservoir are controlled primarily by the total vapor pressure at the boiling water table, modified by frictional resistance to the upward flow of vapor and by the weight of vapor. Liquid water favors small pores and channels because of its high surface tension relative to that of steam. Steam is largely excluded from the smaller spaces but dominates the wider channels.

This conceptual model has become generally accepted although the existence of the deep water table has been disputed in some cases (see Truesdell and White 1973). Recent deep (4.5 km) drilling at Larderello, Italy, has proved the existence of liquid fluid beneath the vapor dominated zone there (Truesdell, pers. comm.).

Production characteristics of vapor-dominated systems demonstrate, that liquid water must exist in these systems at depth or distributed within the reservoir. Truesdell and White (1973) estimate that the volume of water in pores in the Lardarello field is about 20-50 per cent. Estimates for The Geysers indicate that production is mostly coming from zones where the rock is saturated with liquid water or where the porosity of the reservoir is greater than 0.1. The water content in the Kawah Kamojang reservoir, Indonesia, was estimated to be 35% by Grant (1979, a,b).

An essential feature of a two phase vapor-dominated system is the counterflow of rising steam and descending water which has been termed one-dimensional convection as opposed to single-phase convection which can only occur in two or three dimensions. Large difference in density between steam and water provides the driving force that tends to segregate the two phases. Martin et al. (1976) demonstrated that one-dimensional, vertical two-phase, steady state, convective and conductive heat flow is theoretically possible in geothermal reservoirs. In their model steam is assumed to be generated at depth by heat conducted from below. The steam flows upward and an equal mass of hot water flows downward within the geothermal reservoir. At the top of the geothermal reservoir the steam condenses into hot water which then flows downwards. Above the reservoir the heat flow is again only conductive. Martin et al. (1976) found that in many cases two water/steam volume ratios satisfy the same heat flow rate. One is a high ratio in which liquid water is the principal mobile phase. The pressure gradient is approximately that of hydrostatic water. The other ratio corresponds to a high vapor saturation in which steam is the principal mobile phase and the pressure gradient is approximately that of static steam. Examples of results were calculated for a total heat flow of 0.8 W/m^2 (19HFU), a reservoir permeability of 10^{-4} m^2 (10 md) and the Corey relative permeability.

The overall fluid and heat flow were found to be relatively unsensitive to a wide range of conditions. For permeabilities greater than 10^{-15} m^2 and for high steam saturations, the convective fraction of the heat flow was generally found to be many times the conductive fraction. For high water saturations the two fractions were often of comparable magnitudes.

Schubert and Straus (1979) used a one-dimensional model of a counterflowing steam-water mixture above quiescent water in a porous medium to study two-phase processes in vapor-dominated systems. For small permeability to thermal conductivity ratios K/k, the thickness of the two phase region was essentially independent of K/k whereas for large K/k the thickness of the two-phase region was found to increase dramatically. For fixed K/k the thickness of the steam-water region also increases substantially with decreasing temperature gradient Γ in the underlying water. Kilometers thick region of counterflowing steam and water are possible with realistic values of K/k and Γ . In the model used by Schubert and Straus (1979), the heat flow through the system is controlled by the thermal gradient in the underlying water zone. Condensation of rising steam results in a steam flow which decreases with proximity to the surface and a water flow which increases with depth.

Most vapor-dominated reservoirs seem to be capped by a liquid-dominated condensate layer in rocks of lower permeability than the main reservoir rocks. This layer impedes the escape of steam from the reservoir. Only where the near surface rocks are relatively more permeable, does the steam rise closer to the surface and reach it in some locations (Grant 1979b, Schubert et al. 1980).

Sondergeld and Turcotte (1977) concluded on the basis of experimental studies that a two-phase convecting system can be stable with a layer of saturated steam present below a water-saturated layer in a porous medium. The experimental results suggested that there was no need for a cap rock on geothermal reservoirs. On the other hand, Schubert et al. (1980) and Schubert and Straus (1980) concluded that gravitional stability of water over steam under typical conditions in vapor-dominated geothermal systems requires that the permeability of the rocks at the depth of the steam-water boundary does not exceed about $4 \cdot 10^{-17}$ m^2 (0.04 md). An impermeable cap rock is thus not necessary but the permeability of near surface rocks must be more than an order of magnitude smaller than usually encountered in the main reservoirs of geothermal systems. The permeability of near surface rocks determines the thickness of the condensate layer. The top of the steam reservoir will extend upwards until condensate can be stabilized by rocks of sufficiently low permeability.

One characteristic of both liquid or vapor dominated two phase geothermal systems is that some wells produce superheated steam (Sestini 1970, Truesdell and White 1973, Stefánsson 1981). As first pointed out by Bodvarsson (1951) in a consideration of the thermodynamic behaviour of the Lardarello system, the superheat can be explained by assuming that steam flowing to producing wells receives heat from the reservoir rock. In this early study Bodvarsson (1951) also concludes that the steam produced at Lardarello must originate from liquid water in the reservoir, and that the heat capacity of the reservoir rock contributes significantly to the energy withdrawn from the system. These features of the Lardarello system are found in many later models (see e.g. Truesdell and White 1973).

Physical States in Hydrothermal Systems

Viewed on a geological time scale, geothermal systems are transient phenomena in the Earth's crust. Through their lifetime physical conditions are changing. The original reservoirs gradually alter their physical state and develop into a different type. White et al.(1971) suggested that vapor-dominated systems developed initially from hot-water systems characterized by very high heat supply and very low rates of recharge. If and when the heat supply of a developing system becomes great enough to boil off more water than is being replaced by recharge, a vapor-dominated reservoir begins to form.

One interesting characteristic of vapor-dominated reservoirs is the frequent occurrence of temperatures close to 240°C. It has been noted (James 1968, White et al. 1971) that the temperature and pressure of these reservoirs approximately coincide with those at the maximum enthalpy of saturated steam (2804 KJ/kg at 235°C and 30.6 bars abs., see Fig. 2). McNitt (1977) discussed the origin of steam in geothermal reservoirs. He pointed out that the process of boiling in a rising column of water is not likely to be isenthalpic. A porous medium saturated with boiling water that loses heat to produce a steam bubble cannot entrain that water in equilibrium with the bubble. Instead, the unconfined steam bubbles rise through the surrounding water leaving behind the water from which their latent heat was derived. At depths where the boiling water temperature is above the 235°C maximum enthalpy temperature, steam ascends through regions with which it is progressively farther out of thermodynamic equilibrium. To approach equilibrium part of the steam condenses, releasing latent heat to increase the enthalpy of the remaining steam. At higher levels where water temperatures are below 235°C the conditions are reversed. The rising steam carries more heat than it requires to attain thermodynamic equilibrium at these lower temperatures and the excess heat becomes available for increased evaporation.

Considerable progress has been made in the understanding of the physical behaviour of geo-

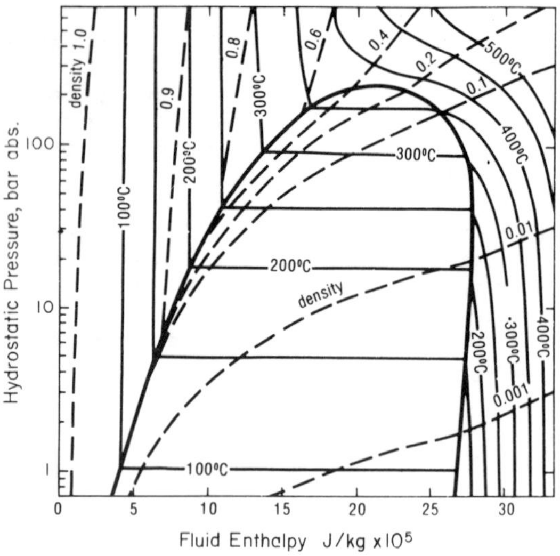

Fig. 2. Pressure-enthalpy diagram for pure water and vapor. Additional parameters are the temperature and the density of the thermal fluid. (Modified from White et al. 1971).

thermal systems from mathematical modelling and numerical simulation. For recent reviews of this subject the reader is referred to Witherspoon et al. (1975), Faust and Mercer (1978, 1979), Pinder (1979) and Mercer and Faust (1980). The approach is generally divided into: (1) Modelling the geothermal system under natural conditions in an effort to better understand how it forms and persists within the earth's crust, and (2) modelling the geothermal system during exploitation in order to predict its behavior and production characteristics. The model consists of a set of equations that describe the transport processes active within the system and the solution to these equations subject to conditions that prevail at a particular site. Existing models can simulate multiphase thermal fluids in a porous medium. The role of fractures in mass and energy transport is not well understood and the interaction between fractures and porous blocks is virtually unknown. It is believed, and in some reservoirs clearly demonstrated, that fractures are the primary conduits of mass and energy transport. The porous medium blocks delineated by these fractures, act as the long-term energy suppliers feeding the fracture system. According to Pinder (1979) there exist two schools of thought on how fractured reservoirs should be modelled: (1) the discrete fracture approach considering each fracture as a discrete entity defined by its size and orientation. Generally, a different set of governing equations will be assumed for the fracture than the adjacent porous medium (Coats 1977). The principal difficulty associated with this model is the inability to establish fracture geometry in the field. (2) the conceptual model for overlapping continua. This model assumes there exists a common region of dimensions where statistical properties of both the porous medium and the fracture medium can be represented in elementary volumes as though each were a continuum. Whether such a physical system is realizable in nature remains to be determined experimentally.

Geothermal systems are generally of a complex nature and contain many zones representing different physical states. There are various parameters that can adequately distinguish between these states. Fig. 2 shows a common presentation using the pressure p and the enthalpy h of the hydrothermal fluid as basic parameters. The critical point of water (T_{crit}, P_{crit}) is of great significance in this presentation. The Clapeyron curves for saturated liquid and vapor meet at that point and envelope the two-phase boiling region.

Another presentation using the pressure p and the density ρ of the fluid as basic parameters is given in Fig. 3. On this pressure-density diagram we recognize four regions of physical states in hydrothermal systems as shown in Fig.4. These are the vapor saturated region, the two phase boiling region, the liquid saturated region and the supercritical region. These regions are separated by the Clapeyron curves for saturated vapor and liquid, ρ ($P_{v\ sat}$) and ρ ($P_{l\ sat}$), the isothermal T_{crit} for $P > P_{crit}$ and the isobar P_{crit} for $T > T_{crit}$. In three of the regions the thermal fluid occurs as a single phase. The liquid saturated region contains the class of geothermal systems where temperature never reaches boiling. Most important of these are hydrothermal systems in the ocean crust. Hydrothermal circulation at hydrostatic pressures exceeding the critical pressure for water will not enter boiling unless it is induced by the release of volatiles from geothermal fluid. These pressures are found beneath oceans of 2.2 km depth. Hydrothermal circulation in the ocean crust is thus generally a single phase convection of seawater.

Geothermal systems belonging to the vapor-saturated region are found on active volcanoes and low pressure superheated steam is common at shallow depth in geothermal fields. Exploitation does also lead to dryout of water in vapordominated rocks near production wells.

Supercritical conditions are expected in geothermal systems that penetrate deep into the crust to supercritical pressures, where young igneous intrusions have generated supercritical temperatures. On land these conditions could be found below 3.5 km depth in the crust, assuming boiling conditions in the hydrostatic fluid above. On the sea floor the hydrostatic head of the ocean may exceed the critical pressure and supercritical temperatures can therefore exist at shallow depth beneath the floor. The presence of dissolved salts in geothermal fluids has an important effect on phase transitions. The

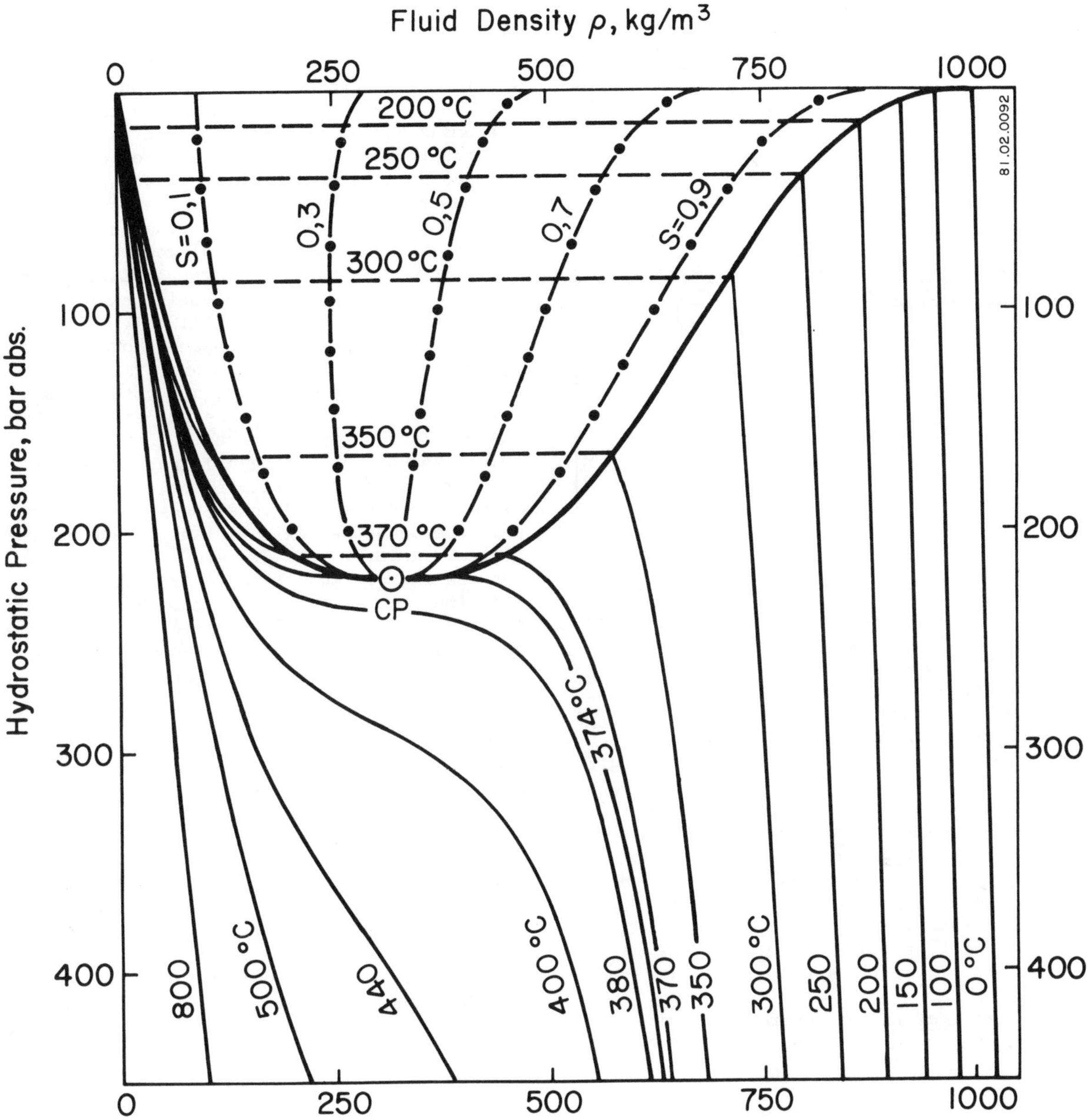

Fig. 3. Pressure-density diagram for pure water and vapor. Additional parameters are the temperature and the water saturation S. Based on tables presented by Schmidt (1979). CP: Critical point, S: Volume fraction of liquid in fluid.

temperature and pressure of the critical point increase with increased salinity (Sourirajan and Kennedy 1962) displacing the conditions for supercritical fluid to greater depth. Secondly, the boiling temperature of saline water at a given pressure is higher than that of pure water (Haas 1971). This effect delays the initiation of boiling in an ascending saline fluid.

The most important region for the discussion of geothermal systems is the two phase boiling region enveloped by the Clapeyron curves at sub-critical temperatures and pressures. To uniquely define physical conditions within this region an additional parameter such as the water saturation S, i.e. the volume fraction of water in the fluid, is needed. In the three single phase regions the vertical pressure gradient $dp/dz = \rho g$ is proportional to the density ρ of the fluid. In the two-phase region we could define the density as a weighted average of both phases, assuming that both are present in a homogeneous mixture. Whether this is valid for any geothermal system

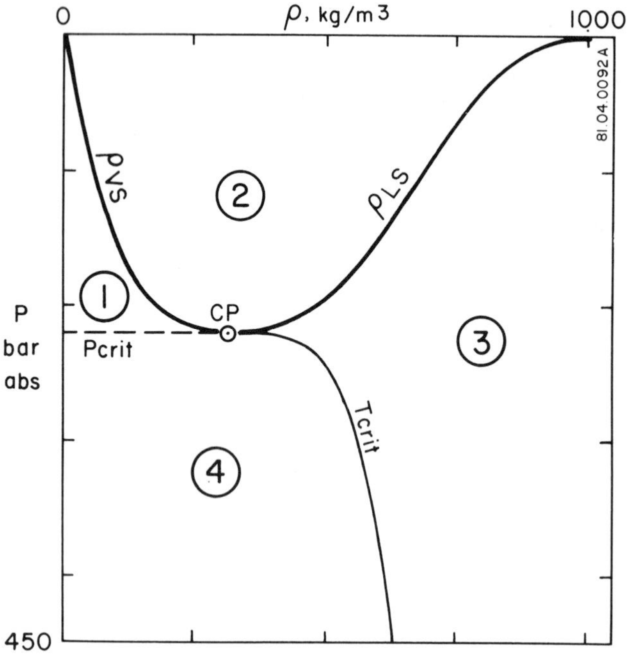

Fig. 4. Physical states in hydrothermal systems

1. Vapor saturated region, where the density is less than the density of saturated steam (ρ_{VS}) and the pressure is equal to or less than the critical pressure (P_{crit}).
2: Boiling region, where two phases are present. The region is enveloped by the Clapeyron curves for saturated vapor (ρ_{VS}) and saturated liquid (ρ_{ls}) at subcritical temperatures and pressures.
3. Liquid saturated region, where the temperature is less than the critical temperature (T_{crit}) and the density is greater than the density of saturated liquid (ρ_{ls}) for all $P \leq P_{crit}$ and greater than $\rho(T_{crit}, P)$ for all $P > P_{crit}$.
4: Supercritical region, where both temperatures and pressures exceed the critical point values (T_{crit}, P_{crit}).

under undisturbed natural conditions is not obvious. One would observe a vertical pressure gradient intermediate between that for static saturated steam and water, respectively. The intermediate gradient would enhance countercurrent flow of steam and water and aid gravity segregation of the phases. As mentioned above Martin et al.(1976)found that for a given rate of heat flow two stable conditions of countercurrent two-phase flow were theoretically possible, one with high water saturation and liquid dominating the pressure gradient, the other with low water saturation and vapor dominating. Reservoir performance and gravity segregation studies of oil and gas in oil reservoirs indicate that, in many cases, segregation occurs rapidly (Spivak 1974, Martin 1958 a and b). Similar segregation of steam and water should occur in geothermal reservoirs (Martin 1975). Intermediate saturations and pressure gradients are thus likely to occur mainly as transient phenomena during the transition of a zone in a reservoir from one stable state to another. Such conditions develop as a response of a geothermal reservoir to pressure drawdown during exploitation. Intermediate pressure gradients are predicted by mathematical models but direct observations in reservoirs are, so far, restricted to reservoir conditions during exploitation (Grant and Horne 1980, Grant 1980).

The physical states discussed above were defined by the aid of four parameters, pressure, temperature, fluid density, and water saturation, considering only the properties of pure water. The presence of chemical components (dissolved solids and gases) in the thermal fluid may shift the critical point and the Clapeyron curves, and thus change the boundaries between the regions of different physical states. However, the division into four states remains valid. This is also true during the lifetime of a geothermal system where certain physical states dominate depending on the physical conditions and the hydrothermal processes.

Acknowledgements. We thank Stefan Arnorsson, Gudmundur Bodvarsson, Malcolm Grant, Gudmundur Palmason, Kristjan Saemundsson, and Benedikt Steingrimsson for helpful suggestions and critical review of the manuscript.

References

Allen, E.T., and A.L. Day, Steam wells and other thermal activity at "The Geysers", California, Carnegie Inst. Washington Pub. 378, 106 pp, 1927.

Allen, E.T., and A.L. Day, Hot springs of the Yellowstone National Park, Carnegie Inst. Washington Pub. 466, 525 pp, 1935.

Anand, J., W.H. Somerton and E. Gomaa, Predicting thermal conductivities of formations from other known properties, Soc. Petrol. Engr. Jour. 13, 267-273, 1973.

Anderson, R.N., M.A. Hobart, and M.G. Langseth, Convective heat transfer in the oceanic crust and sediments on the flanks of midocean ridges in the Indian Ocean, EOS Trans. AGU, 59, 384, 1978.

Ármannsson, H., G. Gíslason, T. Hauksson, Magmatic gases in well fluids aid the mapping of the flow pattern in a geothermal system, Geochim. Cosmochim. Acta submitted 1981.

Árnason, B., Groundwater systems in Iceland traced by deuterium, "Rit" Soc. Sci. Islandica 42, 236 pp, 1976.

Arnórsson, S., Major element chemistry of the geothermal sea-water at Reykjanes and Svartsengi, Iceland. Min. Mag. 42, 209-220, 1978.

Arnórsson, S., Mineral Deposition From Icelandic Geothermal Waters: Environmental and Utilization Problems, J. Pet. Tech. 33, 181-187, 1981.

Arnórsson, S., and H. Svavarsson, CO_2 - gas geothermometer (in Icelandic), Abstract, Geothermal Conference, p. 35-37, Nov. 7, 1980, Reykjavík, Iceland.

Banwell, C.J., Thermal energy from the Earth's crust: Introduction and Part 1, Natural Hydrothermal systems, N.Z.J. Geol. Geophys. 6, 52-69, 1963.

Banwell, C.J., E.R. Cooper, G.E.K. Thompson, and K.J. McCree, Physics of the New Zealand thermal area, N.Z. Dept. Sci. and Indus. Research Bull. 123, 109 pp, 1957.

Baskov, E.A., V.E. Vetshtein, S.N. Surikov, I.N. Tolstikhin, G.A. Malyuk, and T.A. Mishnina, Isotope H , O , C , Ar , He composition of thermal waters and gases in the Kuril-Kamchatka volcanic region as an indicator of their formation, Geokhimya, 2, 180, 1973

Beaumont, E., Des emanations volcaniques et metallifères, Bull. Soc. Geol. France,(2), 4, 1249, 1847.

Bischoff, J.L., Geothermal System at 21°N, East Pacific Rise: Physical Limits on Geothermal Fluid and Role of Adiabatic Expansion, Science, 207, 1465-1469, 1980.

Bischoff, J.L., and F.W. Dickson, Seawater-basalt interaction at 200°C and 500 bars; Implications for origin of sea-floor heavy-metal deposits and regulation of seawater chemistry, Earth Planet. Sci. Lett., 25, 385-397, 1975.

Bischoff, J.L. and W. Seyfried, Hydrothermal chemistry of seawater from 25° to 350°C, Am. J. Sci. 278, 838-860, 1978.

Björnsson, A., K. Saemundsson, P. Einarsson, E. Tryggvason, K. Grönvold, Current rifting episode in north Iceland, Nature 266, 318-323, 1977.

Björnsson, A., and O. Sigurdsson, Volcanic eruption through a deep drillhole in the Krafla area, north Iceland (In Icelandic with English abstract), Náttúrufraedingurinn 48, 19-23, 1978.

Björnsson, A., G. Johnsen, S. Sigurdsson, G. Thorbergsson, and E. Tryggvason, Rifting of the plate boundary in north Iceland 1975-1978, J. Geophys. Res. 84, 3029-3038, 1979.

Björnsson, H., S. Björnsson and T. Sigurgeirsson, Geothermal effects of water penetrating into hot rock boundaries of magma bodies. Geothermal Resources Council, Trans. 4, 13-15, 1980.

Björnsson, S., Estimation of the reservoir potential of the Olkaria geothermal field in Kenya. Proc. Nordic Symposium on Geothermal Energy, May 29 - June 1, 1978, Chalmers Tekniska Högskolan, Göteborg, Supplement, p. 7-29, 1978.

Björnsson, S., S. Arnórsson, and J. Tómasson, Exploration of the Reykjanes brine area, Geothermics, Spec. iss. 2, 1640-1650, 1970.

Björnsson, S., S. Arnórsson, and J. Tómasson, Economic evaluation of Reykjanes thermal brine area. AAPG Bull. 56, 2380-2391, 1972.

Bodvarsson, G., Report on the Hengill Thermal Area (In Icelandic with a summary in English). J. Engin. Assoc. Iceland 36, 1-48, 1951.

Bodvarsson, G., Physical Characteristics of Natural Heat Resources in Iceland, Jökull 11, 29-38, 1961.

Bodvarsson, G., Terrestrial energy currents and transfer in Iceland, this volume, 1981.

Bodvarsson, G., and R.P. Lowell, Ocean floor heat flow and the circulation of interstitial waters, J. Geophys. Res. 77, 4472-4475, 1972.

Bostrom, K., and M.N.A. Peterson, Precipitates from hydrothermal exhalations of the East Pacific Rise, Econ. Geol. 61, 1258-1365, 1966.

Bostrom, K., and M.N.A. Peterson, The origin of aluminium ferromanganoan sediments in areas of high heat flow on the East Pacific Rise, Mar. Geol. 7, 427-447, 1969.

Boussinesq, J., Théorie Analytique de la Chaleur, 2, p. 172, Gauthier-Villars, Paris, 1903.

Brandsdottir, B., and P. Einarsson, Seismic activity associated with the September 1977 deflation of the Krafla central volcano in NE-Iceland, J. Volcanol. Geothermal Res. 6, 197-212, 1979.

Browne, P.R.L., Mineralization in the Broadlands Geothermal Field, Taupo Volcanic Zone, New Zealand, Soc. Mining Geol. Japan, Spec. Issue 2, 64-75, 1971.

Browne, P.R.L., Minimum age of the Kawerau geothermal field, New Zealand, J. Volcanol. Geothermal Res. 6, 213-215, 1979.

Bunsen, R., Über den inneren Zusammenhang der pseudo-vulkanischen Erscheinungen Islands, Annalen der Chemie und Pharmacie 62, 1-59, 1847.

Bunsen, R., Physikalische Beobachtungen über die hauptsächlichsten Geysir Islands, Poggendorffs Annalen 72, 159-170 , 1847.

Cann, J.R., C.K. Winter, and R.G. Pritchard, A hydrothermal deposit from the floor of the Gulf of Aden, Mineralog. Mag. London, 41, 193-199, 1977.

Cathles, L.M., An analysis of the cooling of intrusives by ground-water convection which includes boiling, Econ. Geology, 72, 804-826, 1977.

Chen, H.K., Counsil, J.R., and Ramey, H.J., Jr.: Experimental Steam-water Relative Permeability Curves. Geothermal Resources Council, Trans. 2, 103-104, 1978.

Clarke, W.B., M.A. Beg, and H. Craig, Excess ^3He in the sea; evidence for terrestrial primodial helium, Earth Planet. Sci. Lett. 6, 213-220 1969.

Coats, K.H., "Geothermal reservoir modelling", SPE 6892, presented at 52nd Annual Fall Technical Conference and Exhibition of the Soc. of Petrol. Engineers, Denver, Colorado, October 1977.

Corey, A.T., The interrelation between gas and oil relative permeabilities, Producers Monthly, 19, 38-41, 1954.

Corliss, J.B., The origin of metalbearing submarine hydrothermal solutions, J. Geophys. Res. 76, 8128-8138, 1971.

Corliss, J.B., M. Lyle, D. Diamond, and K. Crane, The chemistry of hydrothermal mounds near the Galapagos Rift, Earth Planet. Sci. Lett., 40, 12-24, 1978.

Corliss, J.B., J. Dymond, L.I. Gordon, J.M. Edmond, R.P. Von Herzen, R.D. Ballard, K. Green, D.L. Williams, A. Bainbridge, K. Crane, and Tj.H. Van Andel, Submarine thermal springs on the Galapagos Rift, Science 203, 1073-1083, 1979 a.

Corliss, J.B., L.I. Gordon, and J.M. Edmond. Some implications of heat/mass ratios in Galapagos rift hydrothermal fluids for models of seawater-rock interaction and the formation of oceanic crust, in Maurice Ewing Series 2, M. Talwani, Ch.G. Harrison and D.E. Hayes editors - A.G.U. Washington 1979b.

Counsil, J.R., and Ramey, H.J., Jr.,: Drainage Relative Permeabilities obtained from Steam-Water Boiling Flow and External Gas Drive Experiments. Geothermal Resources Council, Trans., 3, 141-143, 1979.

Craig, H., The isotopic geochemistry of water and carbon in geothermal areas, in Tongiorgi, E., (Ed.), Nuclear geology on geothermal areas, Spoleto, Pisa, Consiglio Nazionale Delle Ricerche, Lab. di Geol. Nucleare, 17-53, 1963.

Craig, H., The isotopic composition and origin of the Red Sea and Salton Sea geothermal brines. Science, 154, 1544-1548, 1966.

Craig, H., G. Boato, and D.E. White, Isotopic geochemistry of thermal waters, Chap. 5 of Nuclear processes in geologic settings, Nat. Research Council Comm. Nuclear Sci., Nuclear Sci. Ser. Rept. 19, 29-38, 1956.

Craig, H., W.B. Clarke, and M.A. Beg, Excess ^{3}He in deep water on the East Pacific Rise, Earth Planet. Sci. Lett. 26, 125-132, 1975.

Craig, H., and J.E. Lupton, Primodal neon, helium, and hydrogen in oceanic basalts, Earth Planet. Sci. Lett. 31, 369-385, 1976.

Craig, H., J.E. Lupton, and Y. Horibe, A mantle helium component in Circum - Pacific volcanic gases: Hakone, the Marianas, and Mt. Lassen, Adv. Earth Planet. Sci. 3, 3-16, 1978a.

Craig, H., J.E. Lupton, J.A. Welham, and R. Poreda, Helium isotope ratios in Yellowstone and Lassen Park volcanic gases. Geophys. Res. Lett. 5, 897-900, 1978b.

Cronan, D.S., Implications of metal dispersion from submarine hydrothermal systems for mineral exploitation on mid-ocean ridges and in island arcs, Nature, 262, 567-569, 1976.

Dasch, E.J., C.E. Hedge, and J. Dymond. Effect of seawater interaction on the strontium isotope composition of deepsea basalts, Earth Planet. Sci. Lett. 19, 177-183, 1973.

Davies, E.E., C.R.B. Lister, U.S. Wade and R.D. Hyndman, Detailed heat flow measurements over the Juan de Fuca Ridge system, J. Geophys. Res. 85, 299-310, 1980.

Donaldson, I.G. The flow of steam and water mixtures through permeable beds. A simple simulation of a natural undisturbed hydrothermal region, N.Z.J. Sci., 11, 3-23, 1968.

Dymond, J., J.B. Corliss, G.R. Heath, G.W. Field, E.J. Dasch, and H.H. Vech. Origin of metalliferous sediments from the Pacific Ocean, Geol. Soc. Am. Bull. 84, 3355-3372, 1973.

Edmond, J.M., C. Measures, R.E. McDuff, L.H. Chan, R. Collier, B. Grant, L.I. Gordon, and J.B. Corliss. Ridge crest hydrothermal activity and the balances of the major and minor elements in the Ocean; the Galapagos data, Earth Planet. Sci. Lett. 46, 1-18, 1979a.

Edmond, J.M., C. Measures, B. Mangum, B. Grant, F.R. Sclater, R. Collier, A. Hudson, L.I. Gordon, and J.B. Corliss. On the formation of metalrich deposits at ridge crests. Earth Planet. Sci. Lett. 46, 19-30, 1979b.

Einarsson, P., S-wave shadows in the Krafla caldera in NE-Iceland, evidence for a magma chamber in the crust. Bull. Volcanol. 41, 1-9, 1978.

Einarsson, P. and B. Brandsdóttir, Seismological evidence for lateral magma intrusion during the July 1978 deflation of the Krafla volcano in Northeast Iceland, J. Geophys. 47, 160-165, 1980.

Einarsson, T., Ueber das Wesen der Heissen Quellen Islands. "Rit" Soc. Sci. Islandica, 26, 91 pp, 1942.

Elder, J.W., Physical processes in geothermal areas, in Terrestrial Heat Flow, W.H.K. Lee (Ed.), Am. Geophys. Union Mon. 8, 211-239, 1965.

Elder, J.W. Steady free convection in a porous medium heated from below, J. Fluid Mech. 27, 29-48, 1967.

Ellis, A.J., and W.A.J. Mahon. Natural hydrothermal systems and experimental hot-water/rock interactions. Geochim. Cosmochim. Acta. 28, 1323-1357, 1964.

Ellis, A.J., and W.A.J. Mahon. Natural hydrothermal systems and experimental hot water rock interactions (Pt. II). Geochim. Cosmochim. Acta 31, 519-538, 1967.

Ellis, A.J., and W.A.J. Mahon, Chemistry and Geothermal Systems, 392 pp, Academic Press, New York, 1977.

Ellsworth, W.L., and R.Y. Koyanagi. Three-dimensional crust and mantle structure of Kilauea volcano, Hawaii, J. Geophys. Res., 82, 5379-5394, 1977.

Faust, C.R., and J.W. Mercer. Geothermal reservoir simulation 1. Mathematical models for liquid- and vapor-dominated hydrothermal systems. Water Resources Res. 15, 23-30, 1979.

Faust, C.R., and J.W. Mercer. Geothermal reservoir simulation 2; Numerical solutions techniques for liquid- and vapor-dominated hydrothermal systems. Water Resources Res., 15, 31-46, 1979.

Fedotov, S.A., S.T. Balesta, V.A. Droznin, Y.P. Masurenkov, and V.M. Sugrobov. On a possibility of heat utilization of the Avachinsky volcanic chamber, U.N. Symp. Development and Use Geothermal Resources, 2nd, 1975, Proc. 1, 363-369, 1976.

Fridleifsson, I.B., Petrology and structure of the Esja Quaternary volcanic region, southwest Iceland. (Ph.D. thesis) Oxford Univ. 208 pp, 1973.

Furumoto, A.S., The relationship of a geothermal reservoir to the geological structure of the East Rift of Kilauea volcano, Hawaii. Geothermal Res. Counc. Trans. 2, 199-201, 1978.

Gerling, E.K., B.A. Mamyrin, I.N. Tolstikhin, and S.S. Yakovleva. Isotopic helium composition in some rocks, Geokhimiya, 10, 1209, 1971.

Gíslason, G., K. Grönvold, N. Óskarsson, G. Sigvaldason, V. Stefánsson, Changes in fumaroles and drillhole discharge in the Krafla Caldera 1975-1977. Proceedings IASPEI-IAVCEI Durham 1977, 263, 1977.

Goguel, J., Le regime thermique de l'eau souterraine, Ann. Min. 10, 3-31, 1953.

Grant, M.A., Broadlands - A gas-dominated geothermal field, Geothermics, 6, 9-29, 1977a.

Grant, M.A.,: Permeability Reduction Factors at Wairakei, paper 77-HT-52, presented at AICHE-ASME Heat Transfer Conf., Salt Lake City, Utah, August, 15-17, 1977b.

Grant, M.A., Mapping Kamojang reservoir, Geothermal Resources Council, Trans., 3, 271-274, 1979a.

Grant, M.A., Water content of the Kawah Kamojang geothermal reservoir, Geothermics, 8, 21-30, 1979b.

Grant, M.A., Changes in fluid distribution at Ohaki, Proc. N.Z. Geotherm. Worksh. 1980, p. 69-74, Univ. Auckl. 1980.

Grant, M.A., and R.N. Horne, The initial state and response to exploitation of Wairakei geothermal field, G.R.C. Transactions, 4, 333-336, 1980.

Grindley, G.W., Geology of New Zealand Geothermal Steam Fields, U.N. Conf. on New Sources of Energy, Rome, Paper 35/G/34, 1961.

Grindley, G.W., The Geology, Structure and Exploitation of the Wairakei Geothermal Field, Taupo, New Zealand, New Zealand Geological Survey, Bulletin no. 75, 131 pp., 1965.

Gutsalo, L.K., Helium Isotopic Geochemistry in Thermal Waters of the Kuril Islands and Kamchatka. U.N. Symp. Development Use Geothermal Resources, 2nd, 1975, Proc. 1, 745-749, 1976.

Haas, J.L., Jr., The effect of salinity on the maximum thermal gradient of a hydrothermal system at hydrostatic pressure, Econ. Geol. 66, 940-946, 1971.

Hajash, A., Hydrothermal Processes along Mid-Ocean Ridges: An Experimental Investigation, Contrib. Mineral Petrol., 53, 205-226, 1975.

Healy, J., Geothermal Fields in Zones of Recent Volcanism, U.N. Symp. Development and Use Geothermal Resources, 2nd, 1975, Proc. 1, 415-422, 1976.

Heath, G.R., and J. Dymond, Genesis and transformations of metalliferous sediments from the East Pacific Rise, Bauer Deep and Central Basin, northweast Nasca plate, Geol. Soc. Am. Bull. 88, 723-733, 1977.

Hekinian, R., M. Fevrier, J.L. Bischoff, P. Picot, and W.C. Shanks, Sulfite deposits from the East Pacific Rise near 21°N, Science, 207, 1433-1444, 1980.

Herron, T.W.J. Ludwig, P. Stoffa, T. Kan, and P.Buhl, Structure of the East Pacific Rise Crest From Multichannel Seismic Reflection Data, J. Geophys. Res. 83, 798-804, 1978.

Horne, R.N., and Ramey, H.J., Steam/Water relative permeabilities from production data, Geothermal Resources Council, Trans. 2, 291-294, 1978.

Horton, C.W., and F.T. Rogers, Convection currents in a porous medium, J. Appl. Phys. 16, 367-370, 1945.

Hyndman, R.D., and D.S. Rankin, The Mid-Atlantic Ridge near 45°N XVIII. Heat flow measurements, Can. J. Earth Sci. 9, 664-670, 1972.

Iyer, H.M., D.H. Oppenheimer, and T. Hitchcock, Large teleseismic P-wave delays in the Geysers-Clear Lake geothermal area, California, Science 204, 495-497, 1979.

Iyer, H.M., Anomalous delays of teleseismic P-waves in Yellowstone National Park, Nature 253, 425-427, 1975.

Iyer, H.M., and R.M. Stewart, Teleseismic technique to locate magma in the crust and upper mantle, Bull. 96, Oreg. Dept. of Geol. and Miner. Ind., Portland 1977.

Iyer, H.M., Deep structure under Yellowstone National Park, U.S.A., A continental "hot spot", Tectonophysics 56, 165-197, 1979.

Irvine, T.N., Heat transfer during solidification of layered intrusions. I. Sheets and sills, Can. J. of Earth Sciences, 7, 1031-1061, 1970.

James, R., Wairakei and Lardarello; geothermal power systems compared, N.Z. Jour. Sci. and Tech. 11, 706-719, 1968.

Jenkins, W.J., J.M. Edmond, and J.B. Corliss, Excess ^{3}He and ^{4}He in Galapagos submarine hydrothermal waters. Nature 272, 156-158, 1978.

Jenkins, W.J., P.A. Rona, and J.M. Edmond, Excess ^{3}He in the deep water over the Mid-Atlantic Ridge at 26°N ; evidence of hydrothermal activity, Earth Planet. Sci. Lett. 49, 39-44, 1980.

Johnsen, G.V., A. Björnsson, and S. Sigurdsson, Gravity and elevation changes caused by magma movement beneath the Krafla Caldera, Northeast Iceland, J. Geophys. 47, 132-140, 1980.

Johnsen, A.E., Origin of Cyprus pyrite deposits, Proc. 24, Int. Geol. Congr. sect. 4, 291-198, 1972.

Kamensky, I.L., V.P. Yakutseni, B.A. Mamyrin,

S.G. Anufriev, and I.N. Tolstikhin, Helium isotopes in nature, Geokhimya, 8, 914, 1971.
Kappelmeyer, O., and R. Haenel, In: Geothermics with special reference to application, Geo-publication Associates, Berlin, Tables 6.3-6.6, 208-218, 1974.
Kennedy, G.C., A portion of the system silica-water, Econ. Geol. 45, 629-653, 1950.
Kjaran, S.P., G.K. Halldórsson, S. Thorhallsson and J. Elíasson, Reservoir engineering aspects of Svartsengi geothermal area, Geothermal Resources Council, Trans. 3, 337-339, 1979.
Klitgord, K.D., and J.D. Mudie, The Galapagos spreading centre: a near-bottom geophysical study, Geophys. J.R. astron. Soc. 38, 536-586, 1974.
Kononov, V.I., B.A. Mamyrin, B.G. Polak, and L.V. Khabarin, Helium isotopes in gases of Iceland thermal springs, Acad. Sci. USSR Reports, 217, no. 1, p. 172, 1974.
Kononov, V.I., and B.G. Polak, Indicators of abyssal heat recharge of recent hydrothermal phenomena, U.N. Symp. Development and Use Geothermal Resources, 2nd 1975, Proc. 1, 767-773, 1976.
Kruger, P., and Ramey, H.J., Simulation and reservoir engineering of geothermal resources, Stanford Geothermal Program Report SGP-TR-1, Stanford University, Stanford, California 1974.
Lapwood, E.R.: Convection of a fluid in a porous medium, Proc. Cambridge Phil. Soc. 44, 508-521, 1948.
Larsen, G., K. Grönvold and S. Thorarinsson, Volcanic eruption through a geothermal bore-hole at Námafjall, Iceland. Nature 278, 707-710, 1979
Lister, C.R.B., On the thermal balance of a mid-ocean ridge, Geophys. Journ. Royal Astron. Soc. 26, 515-535, 1972.
Lister, C.R.B., On the penetration of water into hot rock, Geophys. Jour. Royal Astron. Soc.,39, 465-509, 1974.
Lister, C.R.B., Qualitative theory on the deep end of geothermal systems. U.N. Symp. Development and Use Geothermal Resources, 2nd 1975, Proc. 1, 459-463, 1976.
Lister, C.R.B., Qualitative models of spreading-center processes, including hydrothermal penetration, Tectonophysics, 37, 203-218, 1977.
Lister, C.R.B., Heat Flow and Hydrothermal Circulation, Ann. Rev. Earth Planet. Sci. 8, 95-117, 1980.
Lonsdale, P.F., J.L. Bischoff, V.M. Burns, M. Kastner, and R.E. Sweeney, A high-temperature hydrothermal deposit on the seabed at a Gulf of California Spreading Center. Earth Planet. Sci. Lett., 49, 8-20, 1980.
Lupton, J.E., Helium - 3 in the Guaymas Basin ; Evidence for injection of mantle volatiles in the Gulf of California, J. Geophys. Res. 84, 7446-7452, 1979.
Lupton, J.E., and H. Craig, Excess ^{3}He in oceanic basalts; evidence for terrestrial primodial helium, Earth Planet. Sci. Lett. 26, 133-139, 1975.
Lupton, J.E., R.F. Weiss, and H. Craig, Mantle helium in the Red Sea brines, Nature 266, 244-246, 1977.
Lupton, J.E., G.P. Klinkhammer, W.R. Normark, R. Haymon, K.C. MacDonald, R.F. Weiss, and H. Craig, Helium - 3 and manganese at the 21°N East Pacific Rise hydrothermal site, Earth Planet. Sci. Lett. 50, 115-127, 1980.
Mahon, W.A.J., Natural hydrothermal systems and the reaction of hot water with sedimentary rocks, N.Z. Jour. Sci. 10, 206-221, 1967.
Mahon, W.A.J. and J.B. Finlayson: The chemistry of the Broadlands geothermal area, New Zealand, Am. J. Sci., 272, 48-68, 1972.
Mamyrin, B.A., I.N. Tolstikhin, G.S. Anufriev, and I.L. Kamensky, Anomalic isotopic composition of helium in volcanic gases, Acad. Sci. USSR Reports, 184, no. 5, p. 1197, 1969.
Mamyrin, B.A., I.N. Tolstikhin, G.S. Anufriev and I.L. Kamensky, Isotopic composition of helium in thermal springs of Iceland, Geokhimiya no. 11, p. 1396, 1972.
Martin, J.C., Reservoir Analysis for Pressure Maintenance Operations Based on Complete Segregation of Mobile Fluids, Trans., AIME 213, 220-227, 1958a.
Martin, J.C., Some Mathematical Aspects of Two-Phase Flows With Applications to Flooding and Gravity Segregation Problems, Prod. Monthly (April 1958) 22-35, 1958b.
Martin, J.C., Analysis of Internal Steam Drive in Geothermal Reservoirs, J. Pet. Tech. (Dec. 1975), 1493-1499, 1975.
Martin, J.C., R.E. Wegner, and F.J. Kelsey, One-Dimensional Convective and Conductive Geothermal Heat Flow, Summaries Second Workshop on Geothermal Reservoir Engineering (Dec. 1976), SGP-TR-20, Stanford University, p. 251-261, 1976.
McNitt, J.R., The Geologic Environment of Geothermal Fields as a Guide to Exploration, U.N. Symp. Development and Use Geothermal Resources, 1st, Proc. 1, Geothermics (Spec. Issue 2),24-31, 1970.
McNitt, J.R., Origin of steam in geothermal reservoirs, SPE 6764, presented at 52nd Annual Fall Technical Conference and Exhibition of the Soc. Petrol. Engr., Denver, Colorado, October 1977.
Mercer, J.W., and Faust, C.R.,: Geothermal Reservoir Simulation 3. Application of liquid - and vapor - dominated hydrothermal modelling techniques to Wairakei, New Zealand, Water Resources Res. 15, 653-671, 1979.
Mercer, J.W., and C.R. Faust, The physics of fluid flow and heat transport in geothermal systems, In: Sourcebook on the Production of Electricity from Geothermal Steam, edited by J. Kestin, U.S. Dept. of Energy, 121-135, 1980.
Mottl, M.J., and H.D. Holland, Chemical exchange

during hydrothermal alteration of basalt by seawater - I. Experimental result for major and minor components of seawater, Geochim. Cosmochim Acta, 42, 1103-1115, 1978.

Muffler, J.L.P., Tectonic and Hydrologic Control of the Nature and Distribution of Geothermal Resources, U.N. Symp. Development and Use Geothermal Resources, 2nd, 1975, Proc. 1, 499-507, 1976.

Muffler, L.J.P., (Ed.), Assessment of Geothermal Resources of the United States - 1978. U.S. Geol. Survey Circ. 790, 163 pp, 1979.

Muffler, L.J.P., and D.E. White, Active metamorphism of upper Cenozoic sediments in the Salton Sea Geothermal field and the Salton Trough, Southeastern California, Geol. Soc. Am. Bull. 80, 157-182, 1968.

Nield, D.A., Onset of Thermohaline convection in a porous medium, Water Resour. Res. 4, 553-560, 1968.

Norton, D., and R. Knapp, Transport phenomena in hydrothermal systems: The nature of porosity, Amer. J. Sci. 277, 913-936, 1977.

Norton, D., and J. Knight, Transport phenomena in hydrothermal systems: Cooling plutons, Amer.J.Sci.277, 937-981, 1977.

Norton, D., and H.P. Taylor, jr., Quantitative Simulation of the Hydrothermal Systems of Crystallizing Magmas on the Basis of Transport Theory and Oxygen Isotope Data: An analysis of the Skaergaard Intrusion, J. Petrology 20, 421-486, 1979.

Palm, E., J.E. Weber, and O. Kvernvold, On steady convection in a porous medium, J. Fluid Mech., 54, 153-161, 1972.

Pålmason, G., On heat flow in Iceland in relation to the Mid-Atlantic Ridge, in Iceland and Mid-Ocean Ridges, S. Björnsson (Ed.), "Rit" Soc. Sci. Islandica, 38, 111-127, 1967.

Pålmason, G., Kinematics and heat flow in a volcanic rift zone, with application to Iceland, Geophys. J.R. astr. Soc., 33, 451-481, 1973.

Pålmason, G., A continuum model of crustal generation in Iceland; kinematic aspects, J. Geophys. 47, 7-18, 1980.

Pålmason, G. Crustal rifting, and related thermo-mechanical processes in the lithosphere beneath Iceland, Geologische Rundschau, 70, 244-260, 1981.

Parmentier, E.M., and E.T.C. Spooner, A theoretical study of hydrothermal convection and the origin of the ophiolitic sulphite ore deposits of Cyprus., Earth Planet. Sci. Lett., 40, 33-44, 1978.

Pinder, G.F., State-of-the-Art Review of Geothermal Reservoir Modelling, Geothermal Subsidence Research Management Program 5, March 1979, 144 pp. Earth Sciences Division, Lawrence Berkeley Laboratory, University of California (LBL-9093), 1979.

Piper, D.Z., Origin of metalliferous sediments from the East Pacific Rise, Earth Planet. Sci. Lett. 19, 75-82, 1973.

Pruess, K., and A.H. Truesdell, A numerical simulation of the natural evolution of vapor--dominated hydrothermal systems, Summaries, Sixth Workshop on Geothermal Reservoir Engineering, Stanford, Dec. 1980.

Reasenberg, P., W.L. Ellsworth, and A. Walter, Teleseismic Evidence for a Low-Velocity Body Under the Coso Geothermal Area, J. Geophys. Res, 85, 2471-2483, 1980.

Ribando, R.J., and K.E. Torrance, Natural convection in a porous medium; Effects of confinement, variable permeability, and thermal boundary conditions, J. Heat Transfer, 95, 42-48, 1976.

Ribando, R.J., K.E. Torrance, and D.L. Turcotte, Numerical models for hydrothermal circulation in the oceanic crust, J. Geophys. Res. 81, 3007-3012, 1976.

Rona, P.A., TAG Hydrothermal Field: Mid-Atlantic Ridge crest at latitude 26°C , J. Geol. Soc. London, 137, 385-402, 1980.

Rona, P.A., and R.P. Lowell, Hydrothermal systems at ocean spreading centers, Geology 6, 299-300, 1978.

Sanford, A.R., and P. Einarsson, Magma Chambers in Rifts(this volume).

Schmidt, E., Properties of Water and Steam in SI-Units. Second revised edition (U.Grigull, Ed.) Springer-Verlag Berlin, Heidelberg, New York, R. Oldenbourg, München, 190 pp., 1979.

Schubert, G., and J.M. Straus , Two-phase convection in a porous medium. J. Geophys. Res. 82, 3411-3421, 1977.

Schubert, G., and J.M. Straus, Steam-water counterflow in porous media, J. Geophys. Res., 84, 1621-1628, 1979.

Schubert, G., J.M. Straus, and M.A. Grant,: A problem posed by vapor-dominated geothermal systems, Nature, 287, 423-425, 1980.

Schubert, G., and J.M. Straus, Gravitional stability of water over steam in vapor-dominated geothermal systems, J. Geophys. Res. 85, 6505-6512, 1980.

Sestini, G., Superheating of geothermal steam, Geothermics, spec. issue 2, v. 2, pt. 1, 622-648, 1970.

Seyfried, W. and J.L. Bischoff, Hydrothermal transport of heavy metals by seawater: the role of seawater/basalt ratios, Earth Planet. Sci. Lett. 34, 71-77, 1977.

Seyfried, W.E., and J.L. Bischoff, Experimental seawater-basalt interaction at 300°C, 500 bars, Chemical exchange, secondary mineral formation and implications for the transport of heavy metals, Geochim. Cosmochim. Acta, 45, 135-147, 1981.

Shanks, W.C., II, and J.L. Bischoff, Ore transport and deposition in the Red Sea geothermal system: a geochemical model, Geochim. et Cosmochim. Acta, 41, 1507-1519, 1977.

Sheu, J.P., K.E. Torrance, and D.L. Turcotte, On the structure of two-phase hydrothermal flows in porous media, J. Geophys. Res. 84, 7524-7532, 1979.

Sigmund, P.M., and F.G. McCufferty, An improved unsteady-state procedure for determining the relative-permeability characteristics of heterogeneous porous media, Soc. of Petroleum Engr. Journal, 19, 15-25, 1979.

Somerton, Witl., J.A. Keese and S.L. Chu, Thermal behaviour of unconsolidatet oil sands, Soc. Petrol. Engr. Jour., 14, 513-521, 1974.

Sondergeld, C.H., and D.L. Turcotte, An experimental study of two-phase convection in a porous medium with applications to geological problems, J. Geophys. Res. 82, 2045-2053, 1977.

Sorey, M.L. Numerical modelling of liquid geothermal systems, USGS Open File Rep. 75-613, US Geol. Surv., Menlo Park, Calif. 1976.

Sorey, M.L., M.A. Grant, and E. Bradford, Nonlinear effects in two-phase flow to wells in geothermal reservoirs. Water Resources Research, 16, 767-777, 1980.

Sourirajan, S., and G.C. Kennedy, The system H_2O - NaCl at elevated temperatures and pressures, Amer. J. Sci., 260, 115-141, 1962.

Spiess, F.N., K.C. MacDonald, T. Atwater, R. Ballard, A. Carranza, D. Cordoba, C. Cox, V.M. Diaz Garcia, J. Francheteau, J. Guerrero, J. Hawkins, R. Haymon, R. Hessler, T. Juteau, M. Kastner, R. Larson, B. Luyendyk, J.D. MacDougall, S. Miller, W. Normark, J. Orcutt, C. Rangin, East Pacific Rise; Hot springs and Geophysical Experiments, Science, 207, 1421--1433, 1980.

Spivak, A., Gravity Segregation in Two-Phase Displacement Processes, Soc. Pet. Eng. J. (Dec. 1974) 619-622; 1974.

Spooner, E.T.C., Hydrodynamic model for the origin of the ophiolitic cupriferous pyrite ore deposits of Cyprus, Geol. Soc. Lond. Spec. Publ., 7, 58-71, 1977.

Spooner, E.T.C., and W.S. Fyfe, Sub-sea floor metamorphism, heat and mass transfer, Contrib. Miner. Petrol. 42, 287-304, 1973.

Spooner, E.T.C., and C.J. Bray, Hydrothermal fluids of the sea water salinity in ophiolitic sulphite ore deposits in Cyprus, Nature, 266, 808-812, 1977.

Steeples, D.W., and H.M. Iyer, Teleseismic P-wave delays in geothermal exploration, U.N. Symp. Development and Use Geothermal Resources, 2nd, 1975, Proc. 2, 1199-1206,1976a.

Steeples, D.W., and H.M. Iyer, Low velocity zone under Long Valley as determined from teleseismic events, J. Geophys. Res. 81, 849-860, 1976b.

Stefánsson, V., The Krafla Geothermal Field, Northeast Iceland. In L. Rybach and L.P.L. Muffler (editors), Geothermal Systems and Case Histories, pp. 273-294, John Wiley and Sons 1981.

Straus, J.M., Large amplitude convection in porous media, J. Fluid Mech., 64, 51-63, 1974.

Straus, J.M., and G. Schubert; Thermal Convection of Water in a Porous Medium: Effects of Temperature - and Pressure - Dependent Thermodynamic and Transport Properties, J. Geophys. Res. 82, 325-333, 1977.

Straus, J.M., and G. Schubert, Effect of CO_2 on the buoyancy of geothermal fluids. Geophys. Res. Lett. 6, 5-8, 1979.

Sutton, F.M., and A. McNabb, Boiling curves at Broadlands geothermal field New Zealand, New Zealand, J. Sci., 20, 333-337, 1977.

Talwani, M.,C.C. Windish and M.G. Langseth, Jr., Reykjanes Ridge crest: a detailed geophysical study, J. Geophys. Res. 76, 473-517, 1971.

Taylor, H.P., Jr., The oxygen isotope geochemistry of igneous rocks. Contr. Miner. Petrol. 19, 1-71, 1968.

Taylor, H.P., Jr., and R.W. Forester, Low-O^{18} igneous rocks from the intrusive complexes of Skye, Mull, and Ardnamurchan, Western Scotland, J. of Petrology 12, 465-497, 1971.

Taylor, H.P., Jr., and R.W. Forester, An oxygen and hydrogen isotope study of the Skaergaard intrusion and its country rocks: a description of a 55 - M.y. old fossil hydrothermal system, J. of Petrology 20, 355-419, 1979.

Thomson, W. (Lord Kelvin), Age of the earth and its limitations as determined from the distribution and movement of heat within it. Ph.D. thesis. Univ. Glasgow, Scotland 1846.

Thorarinsson, S., The postglacial history of the Mývatn area, In, Ecology of eutrophic, subarctic Lake Mývatn and the River Laxá, ed. P.M. Jonasson p. 17-28, Copenhagen 1979.

Thoroddsen, Th., Die Geschichte der isländischen Vulkane, 458 pp , D. Kgl. Danske Vidensk. Selsk. Skrifter, 8. Raekke. Naturvidenskap og Matematik Afd., IX, Copenhagen, 1925.

Tolstikhin, I.N., B.A. Mamyrin, E.A. Baskov, I.L. Kamensky, G.S. Anufriev, and S.N. Surikov, Helium isotopes in gases of thermal springs of the Kurily - Kamchatka volcanic region ; Contributions to Recent Geochemistry and Analytical Chemistry, p. 405, Moscow, Nauka 1972.

Truesdell, A.H., and D.E. White, Production of superheated steam from vapor-dominated geothermal reservoirs, Geothermics 2,154-175,1973.

Truesdell, A.H. and R.O. Fournier, Conditions in the deepest parts of the hot spring systems in Yellowstone National Park, Wyoming, U.S. Geol. Surv. Open File Rep. 76-248, 22 pp, 1976.

Tryggvason, E., Subsidence events in the Krafla Area, North Iceland, 1975-1979, J. Geophys. 47, 141-153, 1980.

Verhoogen, J., Energetics of the Earth. National Academy Press, Washington D.C., 1980.

Walker, G.P.L., The Breiddalur central volcano, eastern Iceland, Quart. J. Geol. Soc. London 119, 29-63, 1963.

Walker, G.P.L., Acid Volcanic Rocks in Iceland Bull. Volc. 29, 375-406, 1966.

Walther, J.V., and H.C. Helgeson, Calculation of the thermodynamic properties of aqueous silica and the solubility of quartz and its polymorphs at high pressures and temperatures, Am. J. Sci. 277, 1315-1351, 1977.

Wasserburg, G.J., G. Gramanske, H. Faul, and R.

Hadgen, Cosmological and geological implications of isotope ratio variations. Natl. Acad. Sci. Publ. 572, no. 23, 1958.

Weiss, R.F., P. Lonsdale, J.E. Lupton, A.E. Bainbridge, H. Craig, Hydrothermal plumes in the Galapagos Rift, Nature, 267, 600-603,1977.

Welham, J.A., R. Poreda, J.E. Lupton, and H. Craig, Gas chemistry and helium isotopes at Cerro Prieto, Geothermics, 8, 241-244, 1979.

White, D.E., Thermal waters of volcanic origin, Geol. Soc. America Bull. 68, 1637-1658, 1957a.

White, D.E., Magmatic, connate, and metamorphic waters, Geol. Soc. America Bull., 68, 1659-1682, 1957b.

White, D.E., Preliminary evaluation of geothermal areas by geochemistry, geology, and shallow drilling, United Nations Conf. New Sources Energy, Rome 1961, Proc. 3, 403-409, 1964.

White, D.E., Mercury and base-metal deposits with associated thermal and mineral waters, in Barnes, H.L., Editor, Geochemistry of Hydrothermal Ore Deposits: New York, Holt, Rinehart, and Winston, Inc., 575-631, 1967.

White, D.E., Environments of generation of base--metal ore depth, Econ. Geol. 63, 301-335, 1968a.

White, D.E., Hydrology, Activity, and Heat Flow of the Steamboat Springs Thermal System, Washoe County Nevada, Geological Survey. Professional Paper, 458-C, 109 pp. 1968b.

White, D.E., Diverse origin of hydrothermal ore fluids, Econ. Geol. 6, 954-973, 1974.

White, D.E., L.J.P. Muffler, and A.H. Truesdell, Vapor-dominated hydrothermal systems compared with hot-water systems, Econ. Geology 66, 75-97, 1971.

White, D.E., I. Barnes, and J.R. O'Neil, Thermal and mineral water of non-meteoric origin, California Coast Ranges, Geol. Soc. America Bull. 84, 547-560, 1973.

White, D.E., and D.L. Williams (Eds.), Assessment of Geothermal Resources of the United States - 1975. U.S. Geol. Survey Circ. 726, 155 pp, 1975.

White, D.E., and M. Guffianti, Geothermal systems and their energy resources, Reviews Geophysics and Space Physics 17, 887-902, 1979.

Williams, D.L., and Von Herzen, R.P., Heat loss from the earth ; new estimate, Geology, 2, 327-328, 1974.

Williams, D.L., R.P. Von Herzen, J.G. Sclater, and R.N. Anderson, The Galapagos spreading center: Lithospheric cooling and hydrothermal circulation, Geophys. J.R. astron. Soc. 38, 587-608, 1974.

Williams, D.L., K. Green, T.H. van Andel, R.P. Von Herzen, J.R. Dymond, and K. Crane, The hydrothermal mounds of the Galapagos Rift ; Observations with DSRV Alvin and detailed heat flow studies, J. Geophys. Res. 84, 7467-7484, 1979.

Witherspoon, P.A., S.P. Neuman, M.L. Sorey, and M.J. Lippmann, Modelling Geothermal Systems, Paper presented at the International Meeting on Geothermal Phenomena and its Applications, Academia Nazionale dei Lincei, Rome, Italy, March 3-5, 1975. LBL - 3263, May 1975, 68 pp.

Whittome, A.J., and E.W. Smith, A model of the Tongonan geothermal field, Proc. N.Z. Geothermal Workshop 1979, 141-147, Univ. Auckland 1979.

Wolery, T.J., and N.H. Sleep, Hydrothermal circulation and geochemical flux at mid-ocean ridges, J. Geol., 84, 242-275, 1976.

MAGMA CHAMBERS IN RIFTS

Allan R. Sanford

Geoscience Department and Geophysical Research Center, New Mexico
Institute of Mining and Technology, Socorro, New Mexico 87801

Páll Einarsson

Science Institute, University of Iceland, Reykjavik, Iceland

Abstract. In continental rifts, lower than normal velocities and resistivities at mid-crustal and sub-crustal depths suggest a low degree of partial melting at these levels that could be a general feature of these structures. On the other hand, geophysical detection of significant accumulations of magma beneath or within the crust of continental rifts is rare. Two rift areas where a number of geophysical observations indicate the presence of relatively large quantities of magma are the Rio Grande rift in central New Mexico and Yellowstone National Park. In the former area, a thin nearly horizontal lens of magma with a minimum areal extent of 1700 km^2 has been seismically detected at mid-crustal levels. Historical surface uplift and the spatial and temporal distribution of seismic activity above the magma chamber indicate contemporaneous inflation. The shape of the magma body, as well as the composition of recent extrusives in the rift, suggests the chamber was formed by the injection of basalt. Significant accumulations of magma have been postulated beneath the Yellowstone caldera on the basis of (1) extraordinarily large heat-flow (900 to 1800 mW/m^2), (2) gravity and magnetic lows, (3) shallow seismicity, (4) attenuation of P and S waves, and (5) travel-time delays for teleseismic events.

Oceanic rifts seem to be generally underlain by a low velocity, low density mantle, suggesting partial melting. This anomalous mantle is particularly well developed under Iceland, where it is also characterized by a high P- to S- velocity ratio, high S-wave attenuation and a layer of low electrical resistivity near the crust-mantle boundary. This mantle condition prevails under the whole of Iceland.

Evidence of extensive crustal magma chambers is only found under the crest of fast spreading oceanic ridges, such as the East Pacific Rise. Slowly diverging plate boundaries such as the Mid-Atlantic Ridge and the axial rift zones in Iceland are not underlain by large magma chambers at high level. Evidence of small, localized magma bodies is found, however, e.g. at 3 km depth under the Krafla volcano in the axial rift zone of NE Iceland.

Introduction

An important facet of continental and oceanic rifting is the intrusion and extrusion of magma; massive quantities along oceanic ridges and lesser amounts along continental rifts. Important questions about rift magmatism are its point of origin and mode of intrusion and storage. The principal purpose of this paper is to present geophysical evidence for the existence of magma in and beneath rifts with emphasis on the detection of significant accumulations i.e. magma chambers. For technical reasons and the authors' backgrounds, the primary evidence cited is from seismic measurements. Detailed discussions of other geophysical observations, e.g. heat-flows and crustal resistivities, related to the presence of magma appear elsewhere in this volume.

Magma Chambers in Non-Rift Areas

Reports of detection of magma chambers by geophysical methods have been most prevalent from those regions of the earth where oceanic lithosphere is being or has recently been subducted beneath continental lithosphere. Considering the world-wide distribution of active volcanoes, detection in these areas should be statistically more probable than elsewhere. Of the 516 active volcanoes listed by McDonald (1972), about three-quarters occur above subduction zones, and the remainder are associated with oceanic and continental rifts and/or isolated mantle plumes beneath the interior of plates.

The physical characteristics of magma may also be a factor in making detection of chambers most probable in subduction areas. The viscous andesitic to rhyolitic magmas characteristic of these zones move slowly upward from their point of origin as large coherent masses. The low viscosity basaltic magmas found in rifts and above some mantle plumes move quickly from their point of origin in the mantle (Eaton and Murata, 1960). Unless special conditions exist for the concentration of basaltic magma, it can be disseminated in a complex zone of dikes and sills within the crust, a situation that might make detection of the magma by geophysical techniques very difficult.

Magma Chambers in Subduction Zones

Gorshkov (1956) appears to have been the first to report the direct detection of magma chambers anywhere in the deeper crust or upper mantle. On the basis of a sharp weakening of S waves passing beneath a group of volcanoes along the Kamchatka peninsula, he was able to map the position of a large magma chamber ($1 \cdot 10^4$ to $2 \cdot 10^4$ km^3) at a depth of 60 km. Ten years after Gorshkov's pioneering work, Fedotov and Farberov (1966) reported detection of another large (25 km diameter) magma chamber 20 to 80 km beneath a Kamchatka volcano using attenuation of S waves corrected for geologic conditions beneath the recording stations. The attenuation coefficient for S waves passing through the magma chamber, 0.039 ± 0.012/km, was found to be four times the value for a normal path in the region.

Following the Russian studies, two groups of American investigators reported detection of magma beneath the Mt. Katmai area of Alaska. Kubota and Berg (1967), using screening of mainly vertically polarized S waves, were able to map ten possible chambers, most at depths of less than 20 km. In a series of papers, Matumoto and his associates (Matumoto, 1971) reported an absence of S waves for some raypaths beneath Katmai peninsula volcanoes that cannot be attributed to source mechanism. Because of a close correlation of these paths with active volcanoes, S-wave shadowing from three shallow (<10 km) and one deep (about 20 to 30 km) magma chamber was proposed.

Within the last decade, geophysical studies in three areas of silicic volcanism in California have revealed gravity, magnetic, seismic and thermal anomalies which can reasonably be interpreted in terms of chambers of partial melt in the crust. For the volcanic center at Long Valley, relative travel-time residuals for teleseisms (Steeples and Iyer, 1976) can be explained by a spherical partial melt chamber of 7 km radius centered about 12 km beneath the Long Valley caldera. Although the interpretation is not unique, it is consistent with thermomechanical models (Lachenbruch and Sass, 1978) that require the existence of an upper crust magma chamber to explain the long-term geothermal characteristics of the area.

At the Coso volcanic center, 200 km south of Long Valley, relative travel-time residuals indicate a body of very low velocity at a depth of 5 to 20 km (Reasenberg et al., 1980). The body lies beneath a field of rhyolite domes of less than 0.3 m.y. age (Duffield et al., 1980); its shallowest part centered beneath the area of highest heat flow. Partial melt in the middle crust can explain the observed low velocities as well as the local seismicity and other geologic and geothermal data.

In the third area, The Geysers, a magma chamber in the crust appears to be the best explanation for the impressive geothermal phenomena, a substantial gravity low, shallow seismicity, and large relative travel-time residuals for teleseisms (Iyer et al., 1979). The latter measurements define an $\sim$25 km thick zone of crustal rock with severely reduced velocity (25%) starting $\sim$4 km beneath The Geysers. Although a number of factors can lead to a reduction in velocity (Iyer and Stewart, 1977), only a chamber containing a substantial fraction of melt appears capable of producing such a large decrease in velocity.

Intraplate Magma Bodies

The most comprehensive studies of intraplate volcanism have been made on the island of Hawaii. Using a combination of earthquake, harmonic tremor and tilt data, magma has been traced from its point of origin 50 to 60 km beneath the surface to complex holding chambers a few kilometers beneath the summit of Kilauea (Eaton and Murata, 1960). In another intraplate volcanic area, the Azores, Machado (1974) has used the distribution of earthquake isoseismals to infer the existence of magma chambers containing a mush of crystals at depths of ∿10 km beneath the islands.

Magma Chambers in Continental Rifts

Nearly all continental rifts have one or more geophysical observations that can be interpreted as indicating the presence of full or partial melts at different levels of the crust and mantle. Among these are observations of low-resistivity and low-velocity layers in the crust, lower than normal upper mantle velocity, and higher than normal heat-flows.

Juracek et al. (1979) summarized the results of magnetotelluric soundings in the Baikal, East African and Rio Grande rifts and the Rhinegraben. Detected within, but not outside of, each of these rifts was a low resistivity layer (< 50 Ω m) at depths of 10 to 30 km. Jūracek et al. believe the low resistivity layer is caused by a combination of four mechanisms; (1) enhanced pore fluid ionic mobility with increasing temperature, (2) mineral and electronic semiconduction, (3) partial melting, and (4) thermal alteration. However, Hermance and Pederson (1980) believe the best explanation for the low resistivity layer in the crust of Rio Grande rift is a zone of total to partial melt.

Mueller (1978) decribed the general structural features of the crust and mantle beneath continental rifts as derived primarily from seismic reflection and refraction observations. A prominent feature of his crustal model is an unstable low velocity zone ($V_p \simeq 5.5$ km/sec) at a depth of 8 to 10 km which Mueller believes consists of now solid mushroom-shaped granitic intrusions. Less certain is the existence of a second low velocity zone ($V_p \simeq 6.4$ km/sec) in the crust at depths of 16 to 22 km which Mueller believes could possibly be a magma chamber of differentiated monzodioritic material. The velocity of mantle rock immediately beneath the Moho is abnormally low for most continental rifts e.g. 7.3 km/sec in the Afar region (Berckhemer et al., 1975) and ≤ 7.9 km/sec in the Rio Grande rift (Toppozada and Sanford, 1976; and Olsen et al., 1979); this has been interpreted as possibly indicating the presence of a partial melt.

A well-known characteristic of continental rifts is that measured heat-flows average well above the normal established for stable continental areas (Morgan, this volume). Although the higher than average heat-flow values have been attributed by some investigators to deep convective motion of groundwater, Lachenbruch and Sass (1977) present a number of strong arguments against this mechanism and instead propose that the excess heat observed in rifts is supplied by magma rising from the asthenosphere.

In a later paper, Lachenbruch and Sass (1978) present steady-state thermo-mechanical models from which the amount of magma injected into or underplating the crust can be estimated from heat-flow data. When the average reduced heat flow is high (∿2.4 HFU), and the asthenospheric flux moderate (∿0.8 HFU), the intrusive model indicates injection of magma at the rate of about 1 km/m.y./unit cross-section of crust. Impressive as this amount appears, the volume of magma residing in the crust at any given time could be quite small. Only if the injection is assumed to occur at widely separated time intervals, e.g. 100,000 years, is the volume likely to reach detectable amounts and then only if it is concentrated in a specific area of the crust.

The small number of active volcanoes along continental rifts (MacDonald, 1972) suggests movement of detectable amounts of magma from the asthenosphere into the crust may be a relatively rare geologic occurrence. Geologic studies in one rift, the Rio Grande, indicate that large scale volcanic activity has indeed taken place at widely separated locations at widely spaced times (Lipman and Mehnert, 1979). Inasmuch as the low-resistivity and low-velocity layers observed in rifts appear to be stable long-term features, it seems unlikely that they are evidence for large concentrations of injected magma.

Injection of large quantities of magma, although rare, should produce a number of geophysical anomalies that should make detection possible and reasonably certain. One area of a continental rift where a suite of geophysical observations indicates the presence of significant quantities of magma in the crust is the Rio Grande rift near Socorro, New Mexico.

Rio Grande Rift Magma Body

The magma chamber mapped beneath the Rio Grande rift near Socorro, New Mexico is an extensive (>1700 km^2) sill-shaped body at mid-crustal depths (19.2 ± 0.6 km). Figure 1 shows the position of the magma body relative to the major structural features of the area. The general geologic and geophysical characteristics of the rift near Socorro are reviewed by Chapin et al. (1978) and Sanford et al. (1977) and will not be discussed here. Described below are the direct and indirect geophysical observations supporting the existence of the magma body. Except where noted, the material presented comes principally from Sanford, Alptekin, and Toppozada (1973), Sanford et al. (1977), Rinehart, Sanford, and Ward (1979), and Rinehart (1979).

Reflectivity of Upper Surface. A characteristic of a large percentage of the numerous microearthquakes recorded in the Socorro area is two strong sharply defined phases following the direct S phase by about 2.5 and 5.0 seconds (see examples in Figure 2). Analyses of arrival times indicate the late arrivals are S to P and S to S reflections off a crustal discontinuity at a depth 19.2 ± 0.6 km.

A remarkable feature of the reflections is their strength. Theoretical calulations demonstrate that ordinary mid-crustal discontinuities cannot account for the amplitude of the reflections even when it is assumed that SV energy radiated from the microearthquake foci is 10 times greater along the reflection raypaths than along the direct raypaths to the recording stations. On the other hand a singular discontinuity separating rigid and non-rigid crustal material can explain the absolute amplitude as well as the ratio of the S_zP to S_zS amplitudes as a function of the angle of incidence. The amplitude data are not considered adequate to estimate the exact nature of the non-rigid material, i.e. the degree of partial melt.

An alternate model for the reflecting interface is a mid-crustal sequence of thin solid layers which has velocities

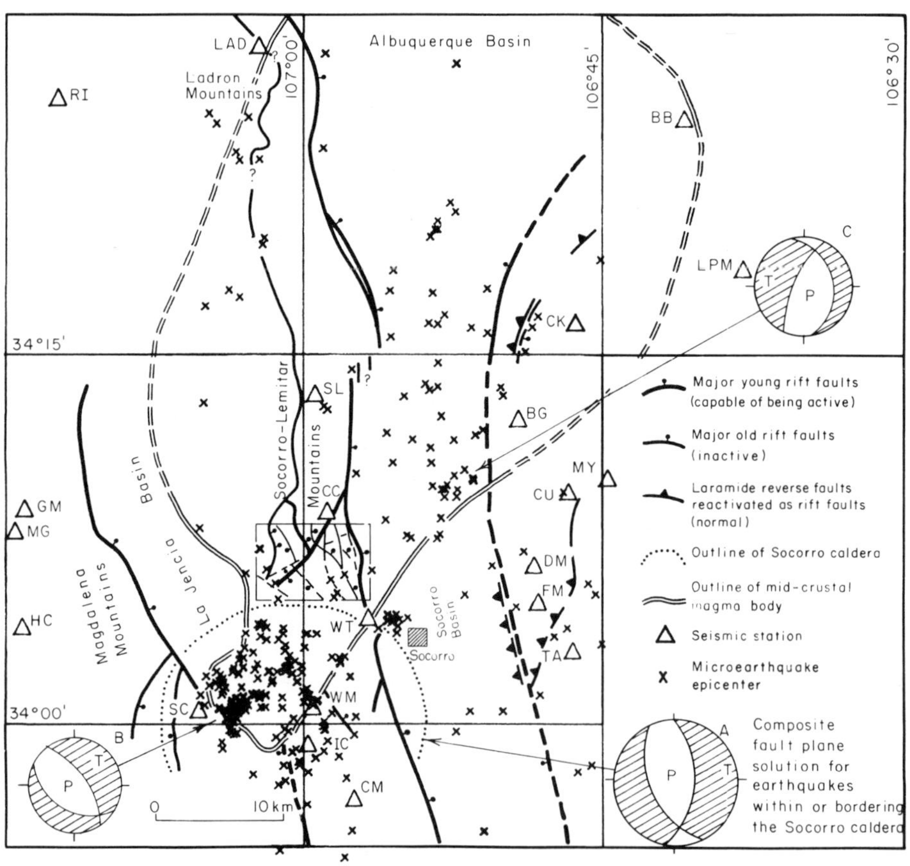

Fig. 1. Map of the Rio Grande rift in the vicinity of Socorro, New Mexico, showing epicenters of microearthquakes in relation to major faults, the margins of the mid-crustal layer of magma and the Socorro caldera (from Sanford et al., 1979).

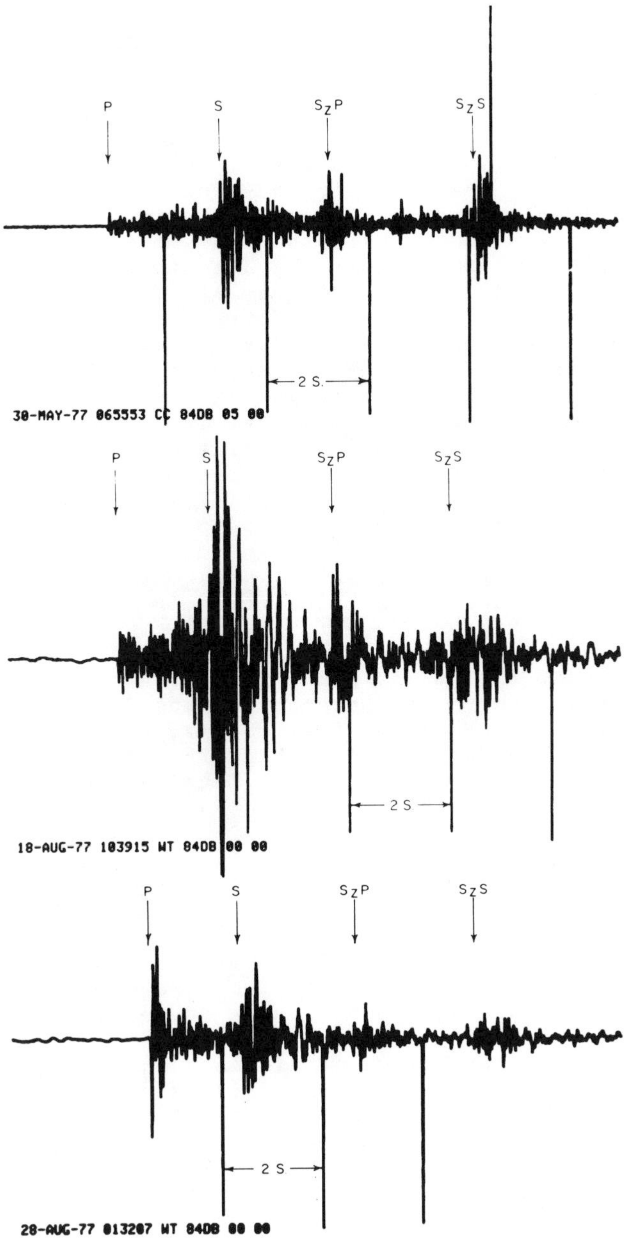

Fig. 2. Microearthquake seismograms with well-defined S to P and S to S reflections from the top of the mid-crustal magma layer.

and thicknesses which enhance the reflection of some of the high-frequency energy contained in the incident SV phase. This model has been rejected because the thicknesses and velocities of the thin layer sequence would have to be maintained over the entire lateral extent of the reflecting mid-crustal discontinuity (>1700 km^2), which appears improbable.

In addition, the spectra for the S_zS reflections match closely the fairly broad spectra (∿2 to 20 Hz) for the direct S phase; a clear indication that certain frequencies are not being preferentially reflected by a thin-layered sequence of crustal rock. Finally, the frequency content of the S_zP and S_zS reflections appear quite similar, an unexpected observation if the same layered rock sequence is to explain both strong reflections.

In 1975 and 1976, deep crustal reflection profiling was conducted by a consortium of universities (COCORP) along 155 km of line in the vicinity of Socorro using techniques developed by the petroleum industry (Brown et al., 1979 and 1980). The most distinctive P-wave reflections observed in this crustal profiling are closely coincident in depth and geographical extent with the S-phase reflections from the microearthquake studies. Important features of the P-wave reflections are their general coherency over fairly large distances, and their strength which averages 6db above background on one of the seismic profiles. The reflected signals, which have durations up to 1/4 second, appear to be a complex summation of several separate events which can be explained by reflection from closely spaced layers varying in number, velocity and thickness. Although the presence of partial or full melts in the layering is not required to explain the strength and character of the reflections, other observations suggest its existence. The first and strongest oscillation of the reflected pulse appears to be reversed in polarity from the incident pulse; this can be an indication of a discontinuity across which the velocity decreases. At least in some cases, a comparison between spectra for reflections from the 19 km discontinuity and deeper interfaces shows a loss of high-frequencies in the reflected signals from the deeper interfaces. The implication here is that the seismic signal has suffered more than normal attenuation by passage through partial or full melts.

Strong complex reflections, an average of 4.5 db above background on one profile, are observed from depths appropriate for the Moho discontinuity. These less prevalent reflected signals also appear to be the result of reflection from a sequence of thin layers in the crust. S-phase reflections from "Moho" levels in the crust have never been identified on seismograms of microearthquakes in the Socorro area. The observation that P-wave reflections of similar strength and character are generated at mid-crustal and "Moho" levels in the crust, but that S-wave reflections come from mid-crustal levels only, can be interpreted in several ways:

1. Magma only occurs at mid-crustal levels.
2. Magma occurs at "Moho" as well as mid-crustal levels. (Mid-crustal magma accumulations prevent S waves from penetrating deeper in the crust).
3. The mid-crustal P-wave reflections are from a layered sequence of rock associated with, but distinct from, the S-wave reflector.

Surface Uplift. An analysis of first-order level-line surveys in the Socorro area by Reilinger and Oliver (1976) and Reilinger et al. (1980) revealed surface uplift nearly centered on the extensive mid-crustal magma body (as presently mapped). Data from three level-lines were used; a north-south line following closely the axis of the rift, surveyed in 1912, 1950, 1951, and 1952; an east-west line entering the rift from the east about 60 km north of Socorro, surveyed in 1911 and 1958; and an east-west line entering the rift from the west at Socorro, surveyed in 1934 and 1978. An interpretation based on the combined results from the three lines suggests uplift has occured in an elliptical area of at least 7000 km^2 oriented along the axis of the rift. Maximum uplift, which occurs 25 km north of Socorro, has been 20 cm relative to the periphery. Considering the seismicity between 1911 and 1978, the level line data appear to be most consistent with more or less continuous uplift at a maximum rate of about 5mm/year. The quantity of magma that must be injected into the mid-crustal magma body to explain the uplift is also 5mm/yr/unit cross section or a total volume of about 10^{-2} km^3/yr. These are modest rates when compared to inflation values for major active volcanoes (Decker and Kinoshita, 1971).

Seismicity. The highest level of seismicity anywhere along the Rio Grande rift occurs in the 75 km section of rift from Socorro northward (Sanford et al., 1977 and 1979). Since the first published report of an earthquake in 1849, the numbers and strengths of shocks in this region of the rift have far exceeded anything observed elswhere along the entire length of the structure. This in itself indicates that the Socorro region is anomalous relative to other sections of the rift. In addition, the temporal and spatial distribution of seismic activity suggests that the relatively intense seismic activity is related to the injection of magma into the crust.

A characteristic of historical as well as recent seismic activity in the Socorro area is the occurrence of a majority of earthquakes in swarms. Swarms have been noted in this region since 1849 and one in 1906-1907 reached fairly severe levels. This intense and prolonged sequence of shocks, comparable in many respects to the Matsushiro swarm from 1965 to 1967 (Stuart and Johnston, 1975), commenced in July 1906 and lasted well into 1907. During this period shocks were felt nearly every day and at times reached a frequency of one perceptible tremor an hour. Three shocks in the swarm reached a Rossi-Forel intensity of VIII at Socorro and were felt over areas of about 275,000 km^2.

Earthquake swarms are generally considered evidence for injection of magma into the crust, particularly in regions where geologic and geophysical observations do not exclude the possibility. In the case of the Rio Grande rift, basaltic material reached the surface as recently as 0.140 m.y.a. (Kelly and Kudo, 1978). Also according to the Lachenbruch and Sass (1978) thermomechanical models, a probable consequence of rifting is injection of magma into the crust. Thus, an association between the Socorro earthquake swarms and magmatic intrusion seems likely although the mechanism is not well understood. Inasmuch as the level of earthquake swarming in 1906-07 exceeded anything noted before or after, it is belived it represents a period of intrusive activity well above the more or less steady-state injection rate suggested by the level-line data.

The distribution of microearthquake activity in the Socorro area is shown in Figure 1. Although the epicenters plotted on this figure are for shocks from May, 1975 to January 1978, their distribution is representative of the seismic activity over a much longer time period. Nearly all of the microearthquake activity is distributed over a 2300 km^2 area which is centered roughly above the extensive mid-crustal magma body. The high concentration of seismic activity west and southwest of station WT is of interest because geologic conditions are favorable for small shallow magma chambers in this region and there are some geophysical data suggesting their existence (Chapin et al., 1978). Elsewhere the epicenters are scattered and show no obvious correlation with known faults. A striking example is the absence of earthquakes along the major fault bordering the eastern margin of the Magdalena Mountains, a fault known to have had major movement in recent geologic time (Sanford et al., 1972).

Three composite focal mechanisms, plotted on the lower focal sphere are shown on Figure 1. All three indicate normal faulting with a minor amount of strike-slip movement, and T-axes nearly perpendicular to the dominant strike of mapped normal faults in the area. Although there is considerable variation in the direction of the T-axes, the average position of this axis indicates east-west extension at the crustal level of the microearthquake foci (less than 13.5 km).

Geodetic measurements were made in 1972,1976, and 1979 over an 11 station network extending roughly from 30 km east to 45 km west of Socorro and from 15 km north to 25 km south of Socorro. An analysis of the data over the entire period indicates no extension or compression in any direction within the error of measurements (J. Savage, personal communication, 1979). Thus it appears a mechanism other than general crustal extension is needed to explain the seismicity in the Socorro area. Vertical uplift arising from inflation of the mid-crustal magma chamber can lead to crustal extension immediately above the chamber which could be accomodated in the brittle upper crust by motion along pre-existing faults. Crustal strain produced by inflation of the mid-crustal magma chamber has not been demonstrated by the geodetic measurements (Prescott, Savage, and Kinoshita, 1979) because uncertainties in the observations are large relative to expected strains at least for some models.

Thickness of Magma Body. Neither the microearthquake studies nor the COCORP crustal profiling reveal obvious S or P phase reflections from the bottom of the magma layer. The absence of identifiable reflections suggests that either the bottom and top of the magma layer are closely spaced or that the bottom of the chamber is not a sharp seismic discontinuity. The latter is a definite possibility inasmuch as the early stages of cooling will produce a crystalline mush at the base of the magma chamber. In this case, density and velocity will increase gradually rather than sharply with depth.

Whether the magma layer has a sharp or gradational lower boundary has not been resolved by the reflection studies. However, studies of time-delays of P-waves passing through the magma layer indicate the amount of melt at the mid-crustal level cannot be large. Assuming that the observed time-delays, 0.1 to 0.2 secs, are totally the result of passage through a full-melt, the thickness of the chamber can range from 0.5 to 1.0 km. These are probably maximum estimates because abnormal conditions below the chamber could account for part of the observed time-delays.

Origin of Magma Chamber. Two resonable explanations for an accumulation of magma at mid-crustal depths in the Rio Grande rift are (1) intrusion of low viscosity basalt, and (2) in-situ melting of water-saturated granite. In the latter case, crystal settling and upward movement of the melt fraction can lead to lenses of full melt which would be seismically indistinguishable from an intruded basalt. A probable argument against in-situ melting is that the intersection of the geotherm and the wet-solidus for granite cannot explain the flat, nearly horizontal, upper surface of the extensive magma chamber. Approaching the edges of the melt, the intersection of the wet-solidus and geotherm should become progressively deeper which in turn should lead to a progressive increase in the depth of the upper surface of the chamber. Since this is not observed on the COCORP profiles or in the microearthquake reflection data, intrusion of low viscosity basalt is most likely to have occurred.

The nature of the nearest recent extrusives in the rift also supports intrusion of basalt as the best explanation for the accumulation of magma at mid-crustal levels. A series of volcanoes ranging in age from 2.78 to 0.140 m.y.a. are located near the axis of the rift 80 to 120 km north of Socorro. The extruded rocks are basalts whose composition (dominantly olivine tholeiites) indicates origin at shallow depth (35 to 40 km) in the upper mantle according to a model proposed by Baldridge (1979). Many of these magmas moved quickly through the crust to the surface, but others show evidence, e.g. plagioclase phenocrysts, of collecting in chambers at depths less than 35 km before extrusion.

Position of the Magma. Magma of olivine tholeiite composition has a density on the order of 2.6 gms/cm^3 (Murase and McBirney, 1973). The mean density of the Rio Grande crust calculated from the crustal velocity model of Toppozada and Sanford (1976), using the velocity-density relation obtained by Nafe and Drake (1963), is 2.7 gms/cm^3. Thus olivine-tholeiite magma generated anywhere in the mantle beneath the base of crust, which is at a depth of about 37 km in the vicinity of the magma body, can rise to the surface under simple lithostatic pressure if avenues are open. In a rift zone, the most likely avenues for upward movement of magma are faults. Inasmuch as an accumulation of magma is observed at mid-crustal levels, the upward rise of the magma must be blocked at that point by a crustal layer which accommodates tectonic stress by plastic deformation rather than brittle fracture.

Several observations support the existence of a plastic zone immediately above the magma layer. All earthquakes located above the magma body have depths of focus shallower than 13.5 km and 90 percent are less than 10 km. The magma layer, which is essentially flat, lies beneath intragraben structures which have cumulative vertical movements along their margins of up to 3.8 km (Sanford 1968). This observation suggests that these large movements have been accommodated by plastic flow before they reach mid-crustal levels. Even if magma was introduced into the crust after most of the vertical movement had occurred, the flatness of the upper surface of the magma body implies obliteration of any fault offsets at mid-crustal levels by plastic flow or creep. Magnetotelluric soundings by Juracek et al. (1979) above the magma body indicate a layer of low resistivity ($<10\Omega m$) starting at a depth of about 10 km. This observation suggests the beginning of a plastic zone because much higher than normal temperatures appear necessary for the processes leading to low rock resistivity.

The location of the magma, at a depth of 19.2 ± 0.6 km, coincides closely with the position of a mid-crustal discontinuity detected on seismic refraction profiles (Toppozada and Sanford, 1976; Olsen et al., 1979). At some places just beyond the margins of the magma body, weak microearthquake and COCORP reflections are observed at the same depth level as the magma. These two observations strongly suggest that magma is accumulating just below a major crustal discontinuity, the Conrad. Crustal models developed for the Rio Grande rift from seismic data do not have a low-velocity zone immediately above the Conrad as one would expect if that zone were plastic. However, seismic detection of a low velocity zone could be very difficult if its upper boundary is a broad transitional zone. It is worth noting that low-velocity layers appear in the upper crust of seismically derived crustal models for (1) the upper Rhinegraben (Mueller, 1978) and (2) the Basin and Range Privince and the Middle Southern Rocky Mountains (Prodehl, 1977).

Heat-Flow. Three heat-flow values have been obtained above the extensive mid-crustal magma body (Reiter et al., 1978a; Reiter and Smith, 1977, and Sanford, 1977). The measurements, which range from 400 to 490 mW/m^2 (9.6 to 11.7 HFU), were all made within a few hundred meters of station WT in Figure 1. In addition high-gradients were found in boreholes about 3 km south of WT, which in conjunction with the rock-type penetrated, suggest heat flows on the order of 250 to 290 mW/m^2 (6 to 7 HFU) in that area.

As Reiter et al. (1978b) point out, heat flows greater than 6 HFU cannot be attributed to simple heat conduction from magma residing at mid-crustal depths. These high heat flows require either water circulation at deep levels in fault zones or magma at high levels in the crust. The amount of magma at the 5 to 10 km level required to explain the heat flows is not great, thus detection by geophysical means could be difficult. However, seismic observation, e.g. S-wave screening, abnormal attenuation of P and S waves, and anomalously large values of Poisson's ratio, do suggest small pockets of magma may exist at 5 to 10 km levels in the crust near Socorro (Sanford, 1978; and Chapin et al., 1978).

Yellowstone Magma Body

Yellowstone National Park lies on the northeastern end of the eastern Snake River Plains which many investigators believe is a pull-apart structure (Eaton, 1979; Smith, 1978), but one that differs in important respects from other continental rifts (Kuntz, 1978). On the basis of the recent volcanic activity (1.9 to 0.07 m.y.a.) and the extraordinary hydrothermal activity, geologists have for many years postulated the existence of a shallow magma chamber beneath the Yellowstone rhyolite plateau. Within the last decade, a number of different geophysical observations have been made which strongly support the existence of silicic magma at shallow depths beneath the large (45 by 70 km) Yellowstone caldera. A brief description of the geophysical observations, taken from detailed papers by Eaton et al. (1975), Smith et al. (1977), Morgan et al. (1977) and Iyer and Stewart (1977), is given below.

Gravity mapping of the Yellowstone area has revealed a major gravity low (with a NE-SW elongation) that is roughly coincident with the caldera. The Bouguer anomaly with 50 mg closure and steep bounding gradients just beyond the margins of the caldera, can be interpreted, although not uniquely, as a relatively shallow continuous chamber of magma beneath the Yellowstone plateau. Like the gravity data, a major feature of the magnetic map of the region is a broad low over the caldera. The low can resonably be attributed to heating of shallow crustal material above the Curie point. Analyses of the magnetic data place the Curie isotherm at depths of 10 to 20 km beyond the caldera but much shallower within its margins.

Studies of the natural seismicity of the area indicate substantial earthquake activity in and around the caldera. Depths of focus range to 25 km beyond the caldera but none have been found deeper than 5 km within the caldera (A.M. Pitt, personal communication, 1979). The lack of deep activity suggests elevated temperatures that inhibit stick-slip movements. Like the Rio Grande rift near Socorro, an important characteristic of the natural seismic activity is its occurrence in swarms; up to several hundred events within a period of a few hours to a few days.

Studies of the amplitude and frequency of P and S waves generated by local earthquakes have revealed paths through the central part of the Yellowstone plateau for which (1) the S phase is absent or hard to distinguish, and (2) the initial P motion is emergent and its subsequent motion depleted in high frequencies. Not all paths are affected which suggests that the magma is not uniformly distributed in the upper crust.

Measured relative travel-time residuals for teleseismic events show substantial delays over a 100 km wide area centered on the caldera. A three-dimensional inversion of the residuals, which reach a maximum delay of $\sim$1.8 secs, defines a low velocity body that reaches to a depth of at least 100 km beneath the caldera. The reduction in velocity, which can be attributed to partial melt in the crust and mantle, is 15 percent in the upper 20 km and 7 percent between 20 and 100 km,

Geochemical and heat-flow measurements indicate a very large thermal flux, on the order of 900 to 1800 mW/m^2, through a 2500 km^2 region centered on the caldera. The same measurements show that the transfer of heat must be by hydrothermal convection. These observations can only be explained by an unusually shallow and/or young magma chamber beneath the convective system. In support of very recent intrusion of magma into the crust are relevelings of benchmarks in the caldera which show extensive uplift at rates as great as 14 mm/year (Pelton and Smith 1979).

Magma Chambers in Oceanic Rifts

Rift structures in the oceans are generally associated with the diverging plate boundaries where new oceanic crust is generated by intrusion and extrusion of magma. The questions of how and where magma is produced and stored are therefore intimately related to the questions of how the oceanic lithosphere evolves. Models of crustal generation by sea-floor spreading assume a magma chamber at small, but usually unspecified depth beneath the spreading center (e.g. Cann, 1974; Sleep, 1975, 1978, Rosendahl, 1976; Hall and Robinson, 1979). The chemical diversity of oceanic basalts seems to indicate a complex history of formation where processes such as partial melting, fractional crystallization, crystal accumulation and magma mixing play a role. For some of these processes the existence of large or small bodies of magma at different depths is required. Considerable effort has been spent in recent years searching for such magma reservoirs or zones of partial melting. Mostly for technical reasons, seismological methods provide the most important information on the location and extent of magma zones in the submarine environment. Evidence for magma includes abnormally low seismic velocities, high attenuation particularly of S-waves, and absence of seismic activity where, for other reasons, one would expect deformation to take place. Conversely, efficient propagation of S-waves and the occurrence of earthquakes throughout a crustal volume can be taken to indicate that the volume does not contain a substantial body of magma.

Teleseismic methods have been used in the search for zones of low Q and low velocities. Molnar and Oliver (1969) investigated the propagation of Sn along numerous paths across continents and oceans. The phase Sn is a shear wave that propagates over long distances in the lithosphere. They analysed paths crossing the Mid-Atlantic Ridge in various places, Juan de Fuca Ridge, Gorda Ridge and the East Pacific Rise in the Gulf of California, and found a narrow zone of inefficient propagation centered on the ridge crest. Similar zones appeared to be associated with the concave side of most island arcs. These zones where interpreted as gaps in the lithosphere, possibly associated with partial melting in the mantle.

Solomon (1973) studied the S-wave radiation field from an earthquake on the Charlie-Gibbs fracture zone, and found that S-waves leaving the source within the azimuthal range 273°-292° were attenuated. Duschenes and Solomon (1977) found an approximate correlation between attenuation and shear wave delays for the same event. The attenuated waves passed under the southern end of the Reykjanes Ridge where it joins with the Charlie-Gibbs fracture zone. The attenuation was inferred to take place within a localized zone of very low Q beneath the crest of the ridge. The top of the low Q body was then shallower than 50 km and its width was less than 100 km. Assuming that the body is two-dimensional, i.e. is continuous along the ridge, a depth of 150 km could be determined for its bottom. This depth determination depends strongly on the assumption, and the observations can also be explained by a low Q body of limited length.

The experiments of Solomon (1973) and Duschenes and Solomon (1977) where repeated by Muller et al. (1978) for events along other fracture zones of the Mid-Atlantic Ridge such as the Oceanographer, Romanche and the St. Pauls fracture zones. The negative results indicate that the effects seen on the Charlie-Gibbs fracture zone are either caused by localized structure or by an anomalous source of the particular earthquake.

Several authors have found that the nodal planes of the P-wave radiation field of normal faulting earthquakes along the mid-ocean ridges are not orthogonal when the rays are projected back to the focal sphere in the standard way (Sykes, 1967, 1970; Thatcher and Brune, 1971; Conant, 1972; Solomon and Julian, 1974; Klein et al., 1977; Einarsson, 1979). This effect was attributed by Solomon and Julian (1974) to lateral changes in the velocity structure in the oceanic mantle. In their model the downgoing rays are deflected towards the vertical when they pass through a lens-shaped low velocity body underlying the ridge crest. Their P-wave velocity model was derived from temperature models for the ocean-ridge mantle, the phase diagram for wet peridotite, and the expected dependence of velocity on temperature and phase. The low velocity region in the mantle in their model results from extensive partial melting, and must be several tens of kilometers wide. This model explains the observations but is by no means required by them. Other explanations of the apparent non-orthogonality of P-wave nodal planes include departure from a double couple source mechanism (Einarsson, 1979) and interference between the direct P, pP and sP on the seismogram (Hart, 1978; Nábělek and Tréhu, 1980). Tréhu et al. (1980) inverted the Rayleigh wave radiation pattern of two events on the Reykjanes Ridge to obtain the source moment tensor. The solutions indicated a pure double couple mechanism. They then compared the observed P waveforms with theoretical seismograms and showed that the apparent nonorthogonality was caused by interference of the direct P and surface reflected phases from a shallow source.

The teleseismic studies usually suffer from the lack of resolution, either because of the scarcity of data or low frequency of the waves used. More detailed investigations have been conducted in several small areas along the mid-ocean ridge system using explosions and small earthquakes as wave sources.

The Mid-Atlantic Ridge

Whitmarsh (1973, 1975) detected a 2 to 3 km wide low velocity zone associated with the uppermost 2.5 km of the crust beneath the median valley in the FAMOUS area of the Mid-Atlantic Ridge (near 37°N). In his study four, 18-50 km long refraction lines were shot in the crestal region. The longest one of the lines was along the median valley, another line was parallel to it at the distance of about 15 km. Two lines were shot across the ridge crest, where delays of up to 0.4 s. were found to be associated with the median valley. Since S-waves were found to propagate through at least the upper part of the low velocity zone, it was concluded that the low velocity did not represent molten magma. The lowering of the velocity was attributed to the relatively large proportion of voids and fractures in the tectonically disturbed zone above a deeper seated magma chamber. The low velocity zone was not clearly expressed in the travel times along the median valley line. This was interpreted as the result of lateral refraction or the low velocity zone being discontinuous. No evidence was found on the median valley line for a crustal magma chamber, and a mantle velocity of 8.09 ± 0.36 km s^{-1} was found at the depth of 9.5 km. This high mantle velocity has since been refuted by Fowler (1976) and Orcutt et al. (1976) who point out that the end of the profile crosses a fracture zone, and therefore probably represents the velocity in a considerably older crust. The absence of an extensive, shallow magma chamber under the median valley is supported by the data of Fowler and Matthews (1974) and Fowler (1976), who recorded S-waves on ocean bottom seismographs located in the western part of the median valley and about 7 km east of the valley from shots on the opposite side of the ridge axis out to a distance of 36 km. These S-waves propagated under the median valley at the depth of 2-7 km and show that only small pockets of magma can exist in this depth range. Some of the S-waves from earthquakes recorded by Francis et al. (1977) in the same region are likely to have propagated at similar depths across the median valley.

A shallow magma chamber under the median valley of the Mid-Atlantic Ridge near 45°N was suggested by Francis and Porter (1973), in order to explain the limited range to which earthquakes could be recorded by an ocean bottom seismograph located in the median valley. No earthquakes were recorded beyond the distance of about 12 km from the seismograph during the 4 days of operation. Assuming a uniform level of seismicity along the valley, waves from more distant events were interpreted to have been bent down by a low velocity zone at 3 km depth. Another explanation of the observations may be, that the seismicity is variable in space and time. For example, the level of seismicity along the mid-Atlantic plate boundary on the Reykjanes Peninsula in Iceland is known to vary over three orders of magnitude on the time scale of days (Klein et al., 1973). On the Reykjanes Ridge the seismicity is highly variable along the plate boundary with a pattern of quiet and active zones that seems to have been stable for decades (Francis, 1973; Tryggvason, 1973; Einarsson, 1979).

In a later study with ocean bottom seismographs, Lilwall et al. (1977, 1978) found that earthquakes in the region near 45°N were confined to the median valley. The hypocenters were distributed between depths of 2 to lo km, with maximum activity at 8 km depth. These earthquake locations can be taken as evidence against an extensive magma body under the median valley at crustal levels.

Steinmetz et al. (1977) conducted an extensive refraction survey in the crestal region of the Mid-Atlantic Ridge north of the Azores. Profiles were shot parallel to the axis at different distances, and two profiles were shot across the axis. The signals were recorded on ocean bottom seismographs and seismic stations on the Azores out to a range of 460 km. One of their conclusions is that the data did not indicate the presence of an intracrustal magma chamber in the crestal region. They found, however, that the ridge axis appeared to be a barrier to P-wave propagation in the upper mantle. A velocity model they constructed from the petrological model of Bottinga and Allégre (1976) was found to be qualitatively consistent with their travel time observations. The model contains a low velocity zone in the mantle under the ridge axis, caused by partial melting. Similar results were obtained for the ridge near 45°N by Fowler (1978).

Rowlett and Forsyth (1979) had the rare opportunity to study P-wave residuals from a South-American, intermediate depth earthquake recorded by four ocean-bottom seismographs in the Vema fracture zone close to its intersection with the Mid-Atlantic Ridge. A difference in residual of 0.5 s. was observed between stations only 12 to 25 km apart, the station closest to the intersection being late. Large lateral changes in the velocity structure are implied by these data. A model consistent with the observations has a 60 km thick zone of velocity lower than 7.1 km/s in the mantle under the station with the largest delay. These low mantle velocities suggest that a magma chamber or a narrow zone of pronounced partial melting extends into the fracture zone from beneath the axis of the ridge.

The seismological data for the Mid-Atlantic Ridge were reviewed in conjunction with other geophysical and petrological data by Nisbet and Fowler (1978), who constructed a model to satisfy all available constraints. In this model magma is segregated from the depth of 15-25 km, and rises rapidly to the base of the crust. Above this level injection takes place by crack propagation or by the development of a narrow, vertical magma chamber, but no large crustal chamber is present.

The East Pacific Rise

Reid et al. (1977) used seismic refraction and earthquake data to show that a narrow zone of low seismic velocities and high attenuation of S-waves exists at the depth of 2.5 km beneath the East Pacific Rise at 20°-21°N. Three ocean-bottom seismographs were operated in an array near the rise crest at 21°N and shots were fired on a split profile parallel to the crest out to a range of 40-65 km. Only a few shots were fired, but synthetic seismograms were calculated and compared to the recorded seismograms in order to increase the resolution of the data. The presence of the low velocity zone was primarily indicated by a conspicuous delay in the arrivals beyond 20 km if compared to a profile off the crest. The low velocity zone appears to be continuous along the rise crest. Furthermore, a microearthquake survey was conducted near 21°N, at the junction of the East Pacific Rise with the Rivera fracture zone. S-waves from an earthquake source slightly off the ridge axis were found to be strongly attenuated in the crestal region with respect to S-waves that travelled farther out on the flank. An apparent Q_β of no more than 50 was indicated within a narrow zone along the rise crest. Taken together, these data suggest that partial melting occurs in a continuous zone at a shallow level under the crest of the East Pacific Rise.

Slightly farther north, near 23°N, a crustal low velocity zone under the rise axis was indicated by travel time delays (McClain and Lewis, 1980). Several profiles were shot radially away from ocean bottom seismographs located east of the axis. The data indicated a very inhomogeneous velocity structure in the axial region, but a two-dimensional model could be constructed that gave a reasonable fit to the data. Some variability in structure along the axis was indicated. The delays attributed to the crustal low velocity zone were, for example, not observed on all the profiles that cross the axis. A low velocity zone in the mantle is included in the model to account for low amplitudes of mantle arrivals at large ranges. Interestingly, the rise crest in this region was found to have features characteristic of slow spreading ridges, such as a rift valley and significant seismicity.

The velocity structure of the East Pacific Rise crest at 9°N has been studied in some detail, both by ocean bottom seismographs (Orcutt et al., 1975 and 1976) and by sonobuoy refraction (Rosendahl et al., 1976). In the OBS study one profile was shot along the rise crest northwards from its intersection with the Siqueiros fracture zone. Another profile was shot parallel to the rise axis on a crust of 2.9 My age. This profile continued across the Siqueiros fracture zone onto a 5 My old crust. Evidence of a pronounced low velocity zone was only found along the axial profile. Here complex seismograms were recorded, and the arrival times beyond 30 km were significantly delayed with respect to arrivals at shorter range. The travel times and comparison with synthetic seismograms required a crustal model with 1.4 km thick low velocity zone at 2 km depth. The underlying mantle has a velocity of 7.7 km/s. The fit to the observations could be improved by reducing the ratio of S- to P-waves velocities in the low velocity zone and increasing the S-wave attenuation, which supports the interpretation of this zone as a zone of partial melting or a magma chamber. The sonobuoy study of Rosendahl et al. (1976) was conducted along the same line as the OBS study on the rise axis, but the many overlapping, split and reversed profile segments obtained allowed a more detailed and regionalized analysis. As in the OBS study, the presence of a low velocity zone under the summit province was indicated by an offset in the travel time curve. The crustal arrivals were abruptly terminated at the range of 22-30 km and the mantle arrivals beyond this distance were delayed. The mantle delay time was found to decrease with increasing water depth or decreasing elevation of the ridge. Many crustal models can be constructed to explain these observations. The model preferred by the authors contains a wedge shaped low velocity body in the crust, thickest under the rise axis and tapering out to a zero thickness at the margin of the axial block.

The top of the low velocity zone appears to be sharp enough to show up as a prominent reflection event in the multichannel seismic reflection data of Herron at al. (1978). At the ridge axis the two-way travel time of this event was 0.8 s. below the sea floor, consistent with the depth to the low velocity layer of 2 km as determined by Orcutt et al. (1975, 1976). The travel time increased with distance from the axis, which was interpreted to be caused by a dip of the reflecting interface away from the axis. This confirms the wedge shape of the low velocity body as derived by Rosendahl et al. (1976).

Iceland

The extensive volcanism of Iceland has for a long time stimulated speculations on the origin and evolution of magma. The mid-Atlantic plate boundary, as defined by the epicentral belt, crosses Iceland from the tip of the Reykjanes Peninsula in the southwest to Axarfjördur Bay in the north. This location of the country across a major diverging plate boundary makes it an ideal platform for studying processes associated with such boundaries. On the other hand, Iceland and its surroundings seem to be anomalous if compared to the rest of the Mid-Atlantic Ridge (see e.g. Vogt, 1974 for summary). The Iceland anomaly is seen in the topography, gravity field, distribution of tectonic activity, crustal and mantle structure, seismicity and petrochemistry. Because of this anomaly any generalisations of Icelandic processes to the rest of the ridge system should be done with caution.

The plate boundary is expressed on land by a series of tectonically active zones. Even though striking similarities can be found between these zones and the axial zones and the transform faults of the Mid-Atlantic Ridge, there are definite differences and the Icelandic zones seem to be more complex. The transform motion between the rift zones of Iceland and the submarine ridges, for example, is taken up by two transform zones, one near the north coast (Tjörnes fracture zone), the other in South Iceland. Both zones lack a clear topographic expression and are primarily defined by their seismicity, occurrence of large earthquakes, focal mechanisms and their geometric relationship with respect to the spreading axes (Sykes, 1967; Ward, 1971; Tryggvason, 1973; Björnsson and Einarsson, 1974; Einarsson, 1979). In neither of them is the tectonic movement concentrated on a single fault.

The volcanic rift zones of Iceland have been classified into two types, the axial zones and lateral zones (Saemundsson, 1978). This classification is based on the tectonic structure and the chemistry of the volcanic products (Jakobsson, 1972). In the axial zones rifting structures such as normal faults, open fissures and eruptive fissures are abundant, and tholeiitic basalts are erupted. The rifting structures are less conspicuous or even absent in the lateral zones, and here the eruptive products are transitional alkali and alkali olivine basalts.

The axial rift zones are generally thought to represent the diverging plate boundary, where crustal accretion takes place. These zones are 30-60 km wide, and in South Iceland two such zones run parallel to each other for a distance of about 100 km. The diverging plate boundary therefore appears to be considerably more complex than its counterpart on the ocean floor.

The rifting structures in the axial rift zones tend to occur in swarms (Saemundsson, 1974, 1978). Most of these fissure swarms run through a central volcanic complex, which is an area where the volcanic activity is higher than normal and silicic rocks occur. High temperature geothermal fields are usually found within the central volcanic complexes. The fissure swarm and its associated central volcano seem to act as a tectonic unit as demonstrated by the present activity of the Krafla volcano and fissure swarm in NE-Iceland (see e.g. Björnsson et al., 1977).

The fissure swarms are 2-20 km wide and generally have the structure of a graben, bordered by normal faults with throw of a few meters to a few tens of meters. Most faults also have a component of opening at the surface, and fissures without vertical displacement are common. A fissure swarm of the axial zone resembles the median valley of the Mid-Atlantic Ridge, only the scale of the vertical movements appears to be smaller by an order of magnitude. Furthermore, in Iceland a few fissure swarms may be arranged subparallel to each other within a broad axial zone (Fig. 3).

The crustal and upper mantle structure of Iceland has been studied by seismological, gravimetric and magnetotelluric methods. The data available prior to 1974 were reviewed by Pálmason and Saemundsson (1974), but since then considerable additional evidence has accumulated.

Early seismic studies of the Icelandic structure include the refraction work of Båth (1960), Tryggvason and Båth (1961) and the surface wave study of Tryggvason (1962). The most extensive surveys of crustal structure are by Pálmason (1963, 1971). In his interpretation, the Icelandic crust consists of four layers of basaltic material overlying an anomalous mantle with P-velocity of 7.2 km/s. The crustal thickness is variable between 8 and 14 km. The seismic refraction data of Pálmason have been reinterpreted by Flóvenz (1980) with the help of synthetic seismograms. A crust with velocity increasing continuously with depth was found to agree better with the data.

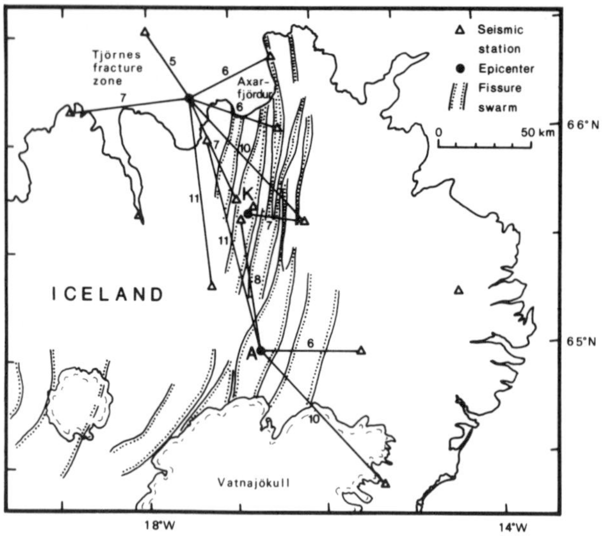

Fig. 3. S-wave propagation across the axial rift zone in NE-Iceland. The fissure swarms of the rift zone are shown as mapped by Saemundsson (1978). The Krafla and Askja central volcanoes are marked with K and A respectively. S-wave paths are shown with lines connecting epicenters and seismograph stations. Numbers indicate the depth of penetration of the respective ray. A surface focus is assumed for the Askja and Tjörnes events, the Krafla event occurred at the depth of 7 km. Part of the rift zone is covered by the Vatnajökull glacier.

Combining the seismic data with borehole temperature data Pálmason concluded that the temperature near the crust-mantle boundary is close to the melting range of basalts in SW- and W-Iceland, but somewhat below it in N- and SE-Iceland. Another important conclusion from the seismic data is that the source of the bowl-shaped Bouguer anomaly over Iceland discovered by Tr. Einarsson (1954) is not to be found in the crust. The low-velocity mantle therefore also seems to have low density. Bott (1965, 1974) suggested that this mass deficiency was due to the presence of low density basic magma produced by partial melting within a rising mantle convection current beneath Iceland. This interpretation is supported by Francis (1969) who found that the P- to S-velocity ratio in the mantle under Iceland is anomalously high. The thickness of the anomalous mantle layer was estimated by Tryggvason (1964) and Long and Mitchell (1970) to be 200-250 km, assuming a velocity of 7.4 km/s. This estimate is based on the delay of teleseismic P-waves recorded in Iceland, if compared with stations in Greenland and N-Europe.

Francis (1969) concluded from his study of travel times of P-waves in the Iceland region, that the E-W extent of the low-velocity mantle under Iceland and the Mid-Atlantic Ridge is 300 km. His conclusion depends critically on the assumption that the low velocity zone is centered on the Mid-Atlantic Ridge. His results have to be reinterpreted if a different geometry of the low velocity is assumed. Perhaps the best estimate of the lateral extent of the anomalous mantle can be made from the shape of the Bouguer anomaly over Iceland. This would imply that the anomalous mass is centered on central Iceland and extends to the coast or perhaps as far as the insular shelf margin.

Beblo and Björnsson (1978, 1980) used the magnetotelluric method to derive the resistivity structure of the lower crust and upper mantle under NE-Iceland along two intersecting profiles. A layer of low resistivity was found under the whole area imbedded between layers of higher resistivity. The resistivity in the layer is about 15 Ωm, and its thickness is about 5 km. The depth to the top of the layer is 10 km within the axial rift zone, and increases to 15-20 km outside the zone. These results are in a general way consistent with earlier magnetotelluric results of Hermance and Grillot (1970, 1974) and Hermance (1973) in other parts of Iceland. Comparing the results with laboratory data for electrical resistivity of basalts and peridotite, Beblo and Björnsson concluded that the low-resistivity layer most likely represents a layer of high melt fraction at the base of the crust, produced by partial melting of the mantle and subsequent ascent of the melt. They also calculated mean temperature gradients in the crust, assuming temperature of 1000-1100°C in the central part of the low resistivity layer. The calculated gradients are in good agreement with observed gradients in shallow drill holes in the same area.

A multinational seismic refraction experiment (RRISP) was conducted in 1977 to study the seismic properties of the anomalous upper mantle under Iceland, and the transition between the Icelandic structure and the more normal oceanic structure south of Iceland (Angenheister et al., 1979, 1980; Gebrande et al., 1980; Goldflam et al., 1980). Large and small explosions and recording stations were lined up along an 800 km long profile across Iceland and along the SE-flank of the Reykjanes Ridge. Shorter profiles were also measured parallel and perpendicular to the main profile both in Iceland and in the ocean. Some of the most important results will be summarized below.

A more or less normal oceanic crust was found south of Iceland, underlain by a subcrustal lithosphere, where P-wave velocities as high as 8.6 km/s were found at depths greater than 20 km. A lithospheric thickness of 50 km was inferred from the disappearance of the seismic signals at long range, but this is a rough estimate. The crustal layers continue into Iceland with increased thickness, whereas the subcrustal oceanic lithosphere seems to be abruptly terminated near the insular shelf margin. The subcrustal velocities under Iceland range between 7.0 and 7.6 km/s down to about 50 km, and no evidence for sharp velocity discontinuities could be found. The crustal thickness in Iceland varies between 10 and 15 km and no clear relationship with the axial rift zones was found. S-waves were transmitted through the crust out to a distance of 140 km with velocity corresponding to a P- to S-velocity ratio of 1.76. Beyond this distance the S-waves became progressively late with respect to the P-waves, indicating a velocity ratio approaching 2. At the same range the S-waves became attenuated with respect to the P-waves and no S-waves were recorded beyond 250 km. This was interpreted as the result of a fundamental change of physical properties at the base of the crust, most probably partial melting. The disappearance of the S-waves is not associated with the axial rift zones in particular, but seems to be a general feature of the anomalous mantle under Iceland. The whole of Iceland thus appears to be underlain by material in the state of partial fusion, which may explain the high mobility of the Icelandic crust and the widespread occurrence of volcanism, even outside the axial rift zones. Two regions of relatively low P-velocity were identified within the lower crust, possibly indicating zones of high melt concentration. One region is in Central Iceland, the other is under the Vestmannaeyjar volcanic center, i.e. in the lateral rift zone of S-Iceland.

A few other seismological studies also provide evidence for magma bodies in the Icelandic crust. Results of two reflection surveys across the axial rift zones have been interpreted to indicate zones of magma (Zverev et al., 1980a,b). Zones of high S-wave attenuation have been identified under the Krafla volcano (Einarsson, 1978), and gaps in the seismic zone on the Reykjanes Peninsula were interpreted as possible indicators of small magma bodies (Klein et al., 1977). However, the propagation of S-waves in the Icelandic crust, notably across the axial rift zones, seems to preclude the existence of a large scale magma body at high levels. Several examples of S-wave rays are shown in Fig. 3. These are high frequency S-waves from earthquakes, observed at vertical, short-period seismograph stations of the Icelandic network. In most of these cases the S-wave amplitude is larger than that of the P-wave. The four earthquakes used here as examples occurred northwest of Tjörnes on Nov. 14, 1979, in the Askja caldera on Nov. 8 and 10, 1979, and slightly south of the Krafla caldera on Dec. 7, 1979. The numbers shown indicate the depth of penetration of the rays assuming a surface focus of the earthquake. It is therefore a minimum depth. The depths are found from the ray tracings of Gebrande et al. (1980) through their crustal and upper mantle model of NE-Iceland. We see that high frequency S-waves travel through the axial rift zone down to the depth of at least 10 km. If a general, continuous magma chamber underlies the axial rift zones of Iceland, we must conclude that the depth to its top is at least 10 km.

Klein et al. (1973, 1977) made a detailed study of the seismicity of the Reykjanes Peninsula, the direct landward continuation of the Reykjanes Ridge, using a dense array of over 20 seismographs. They identified the Mid-Atlantic plate boundary as a narrow zone of seismicity that runs along the peninsula. Most hypocenters were in the depth range 2-5 km, but earthquakes occurred from the surface down to 9 km depth, which is an indication of the minimum thickness of the brittle crust near the plate boundary. Two prominent seismic gaps with a diameter of 1 km were found within the seismic zone near the tip of the peninsula. These gaps could represent either rigid and fault-bounded blocks or low-strength bodies such as magma chambers. No other constraints on this interpretation are available. Magma bodies of this size would probably not be detectable by S-waves from earthquakes. S-waves recorded on the peninsula are usually quite complex, probably because of the heterogeneous crust, but no zones of high S-wave attenuation have been found so far.

Combined reflection-refraction seismic measurements were carried out on profiles crossing the axial rift zones in SW- and NE-Iceland by Zverev et al. (1980a,b). They identified reflecting horizons down to depths of more than 15 km, generally dipping towards the axis of the zone. In both regions volumes of limited lateral extent were identified where no reflectors occurred. These bodies of no reflectors are located under the rift zones at the depth of 10-15 km, i.e. within the anomalous mantle or the lowest part of the crust, and were interpreted as possible magma chambers.

One of the best documented cases of magma accumulation and magmatic activity in the crust is the present rifting episode of the Krafla Volcano and the Krafla fissure swarm in NE-Iceland (for location see Fig. 3). Joint interpretation of structural, petrological, gravity, seismological and crustal deformation data has led to a fairly consistent model of these events (see e.g. Björnsson et al., 1977, 1979; Grönvold and Mäkipää, 1978; Einarsson, 1978; Brandsdóttir and Einarsson, 1979; Einarsson and Brandsdóttir, 1980; Tryggvason, 1980). Magma accumulates at a fairly constant rate at about 3 km depth under the volcano and causes a relatively slow inflation of the volcano. This magma flow probably began in 1975. The inflation is abruptly terminated by rapid deflation when a critical level is exceeded. Small batches of magma are erupted to the surface within the Krafla caldera during the initial phase of a few of the deflation events, but most of the deflation is associated with lateral flow of magma away from the reservoir that never reaches the surface. The direction and rate of propagation of the magma can be traced by the associated seismicity. The magma injects into the fissure swarm to the north or to the south of the volcano and forms a dyke. The rate of propagation is highest in the beginning, typically 0.5 m/s, but slows down as the dyke becomes longer and the deflation rate lower. The dyke may reach as far as 60 km laterally, and subsidence of up to 2.5 m is measured in the center of the Krafla caldera during a deflation event. Most events are considerably smaller than this, however. Extensive surface deformation, both elastic and discontinuous, occurs in the fissure swarm where the dyke is emplaced. Where measurements are available a consistent pattern of deformation is found. The central part of the fissure swarm subsides and is dilated by normal faulting and fissuring. The flanks of the fissure swarm are uplifted and contracted in elastic rebound. The observed movements are of the order of several meters, and about 80 km long segment of the plate boundary has been affected so far. Movement along the diverging plate boundaries in Iceland seems to be episodic, with events of the Krafla type occurring about twice a centry in the whole country. Between these rifting episodes the axial rift zones are remarkably quiet, even at the microearthquake level.

The propagation of S-waves under the Krafla volcano has been studied in some detail (Einarsson, 1978). The inflation of the Krafla volcano is associated with considerable seismic activity. A relatively large proportion of the seismograms of these earthquakes lack the S-waves completely. This absence of the S-wave cannot be explained by source effects. By locating the hypocenters and tracing the seismic rays between the source and the different receivers, it was possible to delineate two zones within the Krafla caldera, where the S-wave is apparently attenuated. The attenuative bodies have horizontal dimensions of 1-2 km and are located near the center of inflation (Fig. 4). The depth to the top of the bodies is 3 km or less. The bottom is not well constrained, but is shallower than 7 km. The Q_β of these bodies is less than 10, supporting the interpretation that they represent magma reservoirs or zones of high melt concentrations. Further support comes from the location in the roots of a central volcano and the depth distribution of hypocenters within the caldera. Earthquakes occur down to 3-4 km depth, below which no earthquakes occur until at 7 km depth.

The geographic association of the attenuative magma bodies with the center of inflation suggests that the present inflation is caused by steady inflow of magma into these magma bodies or some subdivisions of them. The seismic data give little information on the internal structure of the magma reservoir. It could be a massive lump of molten material, but it might also be a complex system of small, more or less interconnected chambers, pockets, sills and dikes.

Even though eruptions within the Krafla caldera and possibly other central volcanoes of the axial rift zones appear to be fed by crustal magma reservoirs, this is not the case for all Icelandic eruptions. The 1973 eruption on Heimaey

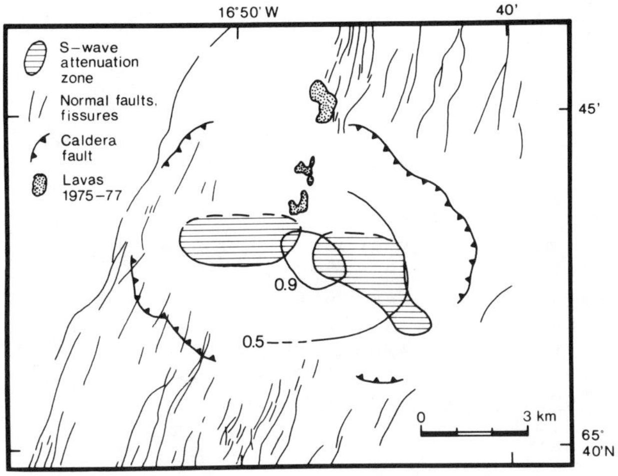

Fig. 4. Geographic relation between the Krafla caldera, the center of inflation and deflation, and the zones of high S-wave attenuation. The inflation/deflation area is shown by isolines of vertical ground movement, normalized to the maximum value and averaged over several cycles between March 1976 and July 1977 (from Björnsson et al., 1979). The S-wave attenuation zones are taken from Einarsson (1978), and the faults are from a map by K. Saemundsson in Björnsson et al. (1979). The fissure swarm continues through the caldera, but the faults in the central part of the caldera are omitted on the map to avoid crowding.

within the lateral rift zone in S-Iceland is an example. The eruption was accompanied by seismic activity that seemed to correlate well with the production rate of lava. The earthquakes were located under the volcano at depths ranging between 15 and 25 km, i.e. within the mantle (Björnsson and Einarsson, 1974: Einarsson 1979). It appears likely that the hypocenters delineate the source volume or a temporary storage volume of the magma. Whatever the causal relationship is between the earthquakes and the magma production, the hypocenters are not likely to be deeper than the magma source. The magma erupted during the Heimaey eruption therefore apparently came directly from the mantle without a temporary storage in the crust.

Discussion

A general characteristic of most active continental and oceanic rifts is that they are underlain by mantle of abnormally low velocity, resistivity, and density. These observations appear to require some degree of partial melting in the upper mantle. Thermo-mechanical models suggest that the higher than normal heat-flows found in rifts arise from movement of magma from points of generation in the mantle to the base of or into the crust. To date, few accumulations of melt in or immediately beneath the crust have been detected in a relatively unambiguous fashion. Specific examples discussed above are the magma bodies (1) at mid-crustal levels in the Rio Grande rift, (2) in the Yellowstone and Krafla calderas, and (3) along the East Pacific Rise. Only in the latter case does the geophysical evidence suggest extensive crustal magma chambers along the axis of a rift. Inasmuch as the East Pacific Rise is a fast spreading ridge, a relation between the amount of magma in the crust and rates of divergence appears probable, as suggested by the thermo-mechanical models of Lachenbruch and Sass (1978).

Along slowly diverging rifts, geologic observations indicate that magmatic activity is episodic, ranging from hundreds of years in a highly volcanic region like Iceland, to hundreds of thousands of years in the Rio Grande rift. In these rifts, the number of geophysically detectable accumulations of magma, at any given time, is likely to be small.

Acknowledgements. Some of the material presented in this paper is based on research supported by the Division of Earth Sciences - National Science Foundation (Grant Number EAR77-23166) and the Energy and Minerals Department of the State of New Mexico through the Energy Institute - New Mexico State University (Grant Number EMD 77-2312).

References

Angenheister, G., H. Gebrande, H. Miller, W. Weigel, P. Goldflam, W. Jacoby, G. Pálmason, S. Björnsson, P. Einarsson, S. Zverev, B. Loncarevic, and S. Solomon, First results from the Reykjanes Ridge Iceland seismic project 1977, Nature, 279, 56-60, 1979.

Angenheister, G., H. Gebrande, H. Miller, P. Goldflam, W. Weigel, W. Jacoby, G. Pálmason, S. Björnsson, P. Einarsson, N.I. Pavlenkova, S. M. Zverev, I.V. Litvinenko, B. Loncarevic, and S. Solomon, Reykjanes Ridge Iceland seismic experiment (RRISP 77), J. Geophys., 47, 228-238, 1980

Baldridge, W.S., Petrology and petrogenesis of Plio-Pleistocene basaltic rocks from central Rio Grande rift, New Mexico, and their relation to rift structure, in Rio Grande Rift: Tectonics and Magmatism, edited by R.E. Riecker, pp. 323-353, Amer. Geophys. Union, Washington, D.C., 1979.

Båth, M., Crustal structure of Iceland, J. Geophys. Res., 65, 1793-1807, 1960.

Beblo, M., and A. Björnsson, Magnetotelluric investigation of the lower crust and upper mantle beneath Iceland, J. Geophys., 45, 1-16, 1978.

Beblo, M., and A. Björnsson: A model of electrical resistivity beneath NE-Iceland, correlation with temperature. J. Geophys., 47, 184-190, 1980.

Berckhemer, H., B. Baier, H. Bartelson, A. Behle, H. Burkhardt, H. Gebrande, J. Makris, H. Menzel, H. Miller, and R. Vees, Deep seismic soundings in the Afar region and on the highland of Ethiopia, in Afar Depression of Ethiopia, edited by A. Pilger and A. Rösler, pp. 89-107, E. Schweizerbart'sche Verlagsbuchhandlung, Stuttgart, 1975.

Björnsson, A., K. Saemundsson, P. Einarsson, E. Tryggvason, and K. Grönvold, Current rifting episode in north Iceland, Nature, 266, 318-323, 1977.

Björnsson, A., G. Johnsen, S. Sigurdsson, G. Thorbergsson, and E. Tryggvason, Rifting of the plate boundary in north Iceland 1975-1978, J. Geophys. Res., 84, 3029-3038, 1979.

Björnsson, S., and P. Einarsson, Seismicity of Iceland, in Geodynamics of Iceland and the North Atlantic Area, edited by L. Kristjánsson, pp. 225-239, Reidel, Dordrecht, Holland 1974.

Bott, M.H.P., The upper mantle beneath Iceland, Geophys. J.R. Astr. Soc., 9, 275-277, 1965.

Bott. M.H.P., Deep structure, evolution and origin of the Icelandic transverse ridge, in Geodynamics of Iceland and the North Atlantic Area, edited by L. Kristjánsson, pp. 33-47, Reidel, Dordrecth, Holland 1974.

Bottinga, Y., C.J. Allégre. Geophysical, petrological and geochemical models of the oceanic lithosphere. Tectonophys., 32, 9-59, 1976

Brandsdóttir B., and P. Einarsson, Seismic activity associated with the September 1977 deflation of the Krafla central volcano in NE Iceland, J. Volcanol. Geothermal Res., 6, 197-212, 1979.

Brown. L.D., P.A. Krumhansl, G.E. Chapin, A.R. Sanford, F.A. Cook, S. Kaufman, J.E. Oliver, and F.S. Schilt, COCORP seismic reflection studies of the Rio Grande rift, in Rio Grande Rift: Tectonics and Magmatism, edited by R. E. Riecker, pp. 169-184, Amer. Geophys. Union, Washington, D.C., 1979.

Brown, L.D., C.E. Chapin, A.R. Stanford, S. Kaufman, and J. Oliver, Deep structure of the Rio Grande rift from seismic reflection profiling, J. Geophys. Res., 85, 4773-4800, 1980.

Cann, J.R., A model for oceanic crustal structure developed. Geophys. J.R. Astr. Soc., 39, 169-187, 1974.

Chapin, C.E., R.M. Chamberlain, G.R. Osburn, D.W. White, and A.R. Sanford, Exploration framework of the Socorro geothermal area, New Mexico, N. Mex. Geol. Soc. Spec. Publ. 7, 114-129, 1978.

Conant, D.A., Six new focal mechanism solutions for the Arctic and a center of rotation for plate movements, M.A. thesis, Columbia University, New York, 1972.

Decker, R.W., and W.T. Kinoshita, Geodetic measurements, in The Surveillance and Prediction of Volcanic Activity, pp. 47-74, Unesco, Paris, 1971.

Duffield, W.A., C.R. Bacon, and G.B. Dalrymple, Late Cenozoic volcanism, geochronology, and structure of the Coso range, Inyo County, California, J. Geophys. Res., 85, 2381-2404, 1980.

Duschenes, J.D. and S.C. Solomon, Shear wave travel time residuals from oceanic earthquakes and the evolution of oceanic lithosphere, J. Geophys. Res., 82, 1985-2000, 1977.

Eaton, G.P., A plate-tectonic model for late Cenozoic crustal spreading in the western United States, in Rio Grande Rift: Tectonics and Magmatism, edited by R.E. Riecker, pp. 7-32, Amer. Geophys. Union, Washington, D.C., 1979.

Eaton, G.P., R.L. Christiansen, H.M. Iyer, A.M. Pitt, D.R. Makey, H.R. Blank, Jr., I. Zietz, and M.E. Gettings, Magma beneath Yellowstone National Park, Science, 188, 787-796, 1975.

Eaton J.P., and K.J. Murata, How volcanoes grow, Science, 132, 925-938, 1960.

Einarsson, P., S-wave shadows in the Krafla caldera in NE-Iceland, evidence for a magma chamber in the crust, Bull. Volcanol., 41, 1-9, 1978.

Einarsson, P., Seismicity and earthquake focal mechanisms along the mid-Atlantic plate boundary between Iceland and the Azores, Tectonophysics, 55, 127-153, 1979.

Einarsson, P., and B. Brandsdóttir, Seismological evidence for lateral magma intrusion during the July 1978 deflation of the Krafla volcano in NE-Ice-

land, J. Geophys., 47, 160-165, 1980.
Einarsson, Tr., A survey of gravity in Iceland, Soc. Sci. Islandica, Rit 30, 22pp., 1954.
Fedotov, S.A., and A.I. Farberov, On screening of shear seismic waves and on magmatic chamber in the upper mantle under Avacha Volcano region, in Volcanism and Deep Structure of the Earth, pp. 43-48, Publ. Nauka. (in Russian), 1966.
Flóvenz, O., Seismic structure of the Icelandic crust above layer three and the relation between body wave velocity and the alteration of the basaltic crust, J. Geophys., 47, 211-220, 1980.
Fowler, C.M.R. The Mid-Atlantic Ridge: Structure at 45°N, Geophys. J.R. astr. Soc., 54, 167-183, 1978.
Fowler, C.M.R., Crustal structure of the Mid-Atlantic Ridge crest at 37°N, Geophys. J.R. Astr. Soc., 47, 459-491, 1976.
Fowler, C.M.R., and D.H. Matthews, A seismic refraction experiment, using sonobuoys and ocean bottom seismographs, on the Mid-Atlantic Ridge near 37°N, Nature, 249, p. 752-754, 1974.
Francis, T.J.G., Upper mantle structure along the axis of the mid-Atlantic ridge near Iceland, Geophys. J.R. Astr. Soc., 17, 507-520, 1969.
Francis, T.J.G., The seismicity of the Reykjanes Ridge, Earth Planet. Sci. Lett., 18, 119-124, 1973.
Francis, T.J.G., and I.T. Porter, Median valley seismology: The Mid-Atlantic Ridge near 45°N, Geophys. J.R. Astr. Soc., 34, 279-311, 1973.
Francis, T.J.G., I.T. Porter, and J.R. McGrath, Ocean-bottom seismograph observations on the Mid-Atlantic Ridge near 37°N, Geol. Soc. Amer. Bull., 88, 664-677, 1977.
Gebrande, H., H. Miller, and P. Einarsson, Seismic structure of Iceland along RRISP-Profile I, J. Geophys., 47, 239-249. 1980.
Goldflam, P., W. Weigel, and B. Loncarevic, Seismic structure along RRISP-Profile I on the south-east flank of the Reykjanes Ridge, J. Geophys., 47, 250-260, 1980.
Gorshkov, G.S., On the depth of magmatic chamber of Kliuchevskiy Volcano, Akad. Nauk. SSSR Doklady, 1o6, 703-705 (in Russian), 1956.
Grönvold, K., and H. Mäkipää, Chemical composition of Krafla lavas 1975-1977, Nordic Volcanological Institute, Report 78 16, 49 pp., 1978.
Hall, J.M., and P.T. Robinson, Deep crustal drilling in the North Atlantic Ocean, Science, 204, 573-586, 1979.
Hart, R.S., Body wave observations from the September, 1969, North Atlantic Ridge earthquake (abstract), EOS. Trans. Amer. Geophys. Union, 59, 326, 1978.
Hermance, J.F., An electrical model for the sub-Icelandic crust, Geophysics, 38, 3-13, 1973.
Hermance, J.F., and L.R. Grillot, Correlation of magnetotelluric, seismic, and temperature data from southwest Iceland, J. Geophys. Res., 75, 6582-6591, 1970.
Hermance, J.F., and L.R. Grillot, Constraints on temperatures beneath Iceland from magnetotelluric data, Phys. Earth Planet. Inter., 8, 1-12, 1974.
Hermance, J.F., and J. Pederson, Deep structure of the Rio Grande rift: A magnetotelluric interpretation, J. Geophys. Res., 85, 3899-3912, 1980.
Herron, T.J., W.J. Ludwig, P.L. Stoffa, T.K. Kan, and P. Buhl, Structure of the East Pacific Rise crest from multichannel seismic reflection data, J. Geophys. Res., 83, 798-804, 1978.
Iyer, H.M., D.H. Oppenheimer, and T. Hitchcock, Abnormal P-wave delays in the Geysers-Clear Lake Geothermal area, California, Science, 204, 495-497. 1979.
Iyer, H.M., and R.M. Stewart, Teleseismic technique to locate magma in the crust and upper mantle, in Magma Genesis, Proceedings of the Amer. Geophys. Union Chapman Conference on Partial Melting in the Earth's Upper Mantle, edited by H.J.B. Dick, pp. 281-299, Bull. 96, State of Oregon, Dept. of Geol. and Mineral Industries, 1977.
Jakobsson, S., Chemistry and distribution pattern of recent basaltic rocks in Iceland, Lithos, 5, 365-386, 1972.
Juracek, G.R., M.E. Ander, and H.T. Holcombe, Magnetotelluric soundings of crustal conductive zones in major continental rifts, in Rio Grande Rift: Tectonics and Magmatism, edited by R.E. Riecker, pp. 209-222, Amer. Geophys. Union, Washington, D.C., 1979.
Kelley, V.C., and A.M. Kudo, Volcanoes and related basalts of Albuquerque basin, Circ. 156, N. Mex. Bur. of Mines and Miner. Resour., Socorro, 1978.
Klein, F.W., Páll Einarsson, and M. Wyss, Microearthquakes on the mid-Atlantic plate boundary on the Reykjanes Peninsula in Iceland, J. Geophys. Res., 78, 5084-5099, 1973.
Klein, F.W., P. Einarsson, and M. Wyss, The Reykjanes Peninsula, Iceland, earthquake swarm of September 1972 and its tectonic significance, J. Geophys. Res., 82, 865-888, 1977.

Kubota, A., and E. Berg. Evidence for magma in the Katmai Volcanic Range, Bull. Volcanol., 31, 175-214, 1967.

Kuntz, M.A., Some differences between the eastern Snake River Plain and other continental rift systems, Program and Abstracts for 1978 International Symposium on the Rio Grande Rift, pp. 49-50. Los Alamos Scientific Laboratory, Report LA-7487-C, 1978.

Lachenbruch, A.H., and J.H. Sass, Heat flow in the United States and thermal regime of the crust, in The Earth's Crust, edited by J.G. Heacock, pp. 626-675, Amer. Geophys. Union Geophys. Mono. 20, 1977.

Lachenbruch, A.H., and J.H. Sass, Models of an extending lithosphere and heat flow in the Basin and Range province, Geol. Soc. Am. Memoir 152, edited by R.B. Smith and G.P. Eaton, pp. 209-250. 1978.

Lilwall, R.C., T.J.G. Francis, and I.T. Porter, Ocean-bottom seismograph observations on the Mid-Atlantic Ridge near 45°N, Geophys. J.R. Astr. Soc., 51, 357-370, 1977.

Lilwall, R.C., T.J.G. Francis, and I.T. Porter, Ocean-bottom seismograph observations on the Mid-Atlantic Ridge near 45°N - further results, Geophys. J.R. Astr. Soc., 55, 255-262, 1978.

Lipman, P.W., and H.H. Mehnert, The Taos Plateau volcanic field, northern Rio Grande rift, New Mexico, in Rio Grande Rift: Tectonics and Magmatism, edited by R.E. Riecker, pp. 289-311, Amer. Geophys. Union, Washington, D.C., 1979.

Long, R.E., and M.G. Mitchell, Teleseismic P-wave delay time in Iceland. Geophys. J.R. Astr. Soc., 20, 41-48, 1970.

MacDonald, G.A., Volcanoes. Prentice-Hall, Englewood Cliffs, New Jersey, 51o pp, 1972

Machado, F., The search for magmatic reservoirs, in Physical Volcanology, edited by L. Civetta, P. Gasparini, G. Luongo, and A. Rapolla, pp. 255-273, Developments in Solid Earth Geophysics, 6, Elsevier Scientific Publ. Co., Amsterdam, 1974.

Matumoto, T., Seismic body waves observed in the vicinity of Mount Katmai, Alaska, and evidence for the existence of melted chambers, Geol. Soc. Amer. Bull., 82, 2905-2920, 1971.

McClain, J.S., B.T.R. Lewis. A seismic experiment at the axis of the East Pacific Rise. Marine Geology, 35, 147-169, 1980.

Molnar, P., and J. Oliver, Lateral variations of attenuation in the upper mantle and discontinuities in the lithosphere, J. Geophys. Res., 74, 2648-2682, 1969.

Morgan, P., D.D. Blackwell, and R.E. Spafford, Heat flow measurements in Yellowstone Lake and the thermal structure of the Yellowstone caldera, J. Geophys. Res., 82, 3719-3732, 1977.

Mueller, S., Evolution of the Earth's crust, in Tectonics and Geophysics of Continental Rifts, edited by I.B. Ramberg and E.-R. Neumann, pp. 11-28, D. Reidel Publish. Co., Dordrecht, 1978.

Muller, J.L., J.L. Hall, and S.C. Solomon, Attenuation and travel time delays of shear waves beneath the mid-Atlantic Ridge from transform earthquakes (abstract), EOS Trans. Amer. Geophys. Union, 59, 370, 1978.

Murase, T., and A.R. McBirney, Properties of some common igneous rocks and their melts at higher temperatures, Geol. Soc. Amer. Bull., 84, 3563-3592, 1973.

Nábělek, J.L., A.M. Tréhu. Shallow earthquakes and nonorthogonal nodal planes. J. Geophys. Res., in press 1980.

Nafe, J.E., and C.L. Drake, Physical properties of marine sediments, in The Sea, Vol.3, edited by M.N. Hill, pp. 794-815, Interscience Publishers, New York, 1963.

Nisbet, E.G., C.M.R. Fowler, The Mid-Atlantic Ridge at 37 and 45°N: some geophysical and petrological constraints, Geophys. J.R. Astr. Soc., 54, 631-660, 1978.

Olsen, K.H., G.R. Keller, and J.N. Stewart, Crustal structure along the Rio Grande rift from seismic refraction profiles, in Rio Grande Rift: Tectonics and Magmatism, edited by R.E. Riecker, pp. 115-143, Amer. Geophys. Union, Washington, D.C., 1979.

Orcutt, J.A., B.L.N. Kennett, L.M. Dorman, and W.A. Prothero, Evidence for a low velocity zone underlying a fast spreading ridge crest, Nature, 256, 475-476, 1975.

Orcutt, J.A. B.L.N. Kennett, and L.M. Dorman, Structure of East Pacific Rise from an ocean-bottom seismometer survey, Geophys, J.R. Astr. Soc., 45, 305-320, 1976.

Pálmason, G., Seismic refraction investigation of the basalt lavas in northern and eastern Iceland, Jökull, 13, 40-60, 1963.

Pálmason, G., Crustal structure of Iceland from explosion seismology, Soc. Sci. Islandica, Rit 40, 187 pp., 1971

Pálmason, G., and K. Saemundsson, Iceland in relation to the Mid-Atlantic Ridge. Annual Review of Earth and Planetary Sciences, 2, 25-50, 1974.

Pelton, J.R., and R.B. Smith, Recent crustal uplift in Yellowstone National Park, Science, 206, 1179-1182, 1979.

Prescott, W.H., J.C. Savage, and W.T. Kinoshita, Strain accumulation rates in the western United States between 1970 and 1978, J. Geophys. Res., 84, 5423-5435, 1979.

Prodehl, C., The structure of the crust-mantle boundary beneath North America and Europe as derived from explosion seismology, in The Earth's Crust, edited by J.G. Heacock, pp. 349-369, Amer. Geophys. Union Geophys. Mono. 20, 1977.

Reasenberg, P., W. Ellsworth, and A. Walter, Teleseismic evidence for a low-velocity body under the Coso geothermal area, J. Geophys. Res., 85, 2471-2483, 1980.

Reid, I., J.A. Orcutt, and W.A. Prothero, Seismic evidence for a narrow zone of partial melting underlying the East Pacific Rise at 21°N, Geol. Soc. Amer. Bull., 88, 678-682, 1977.

Reilinger, R.E., and J.E. Oliver, Modern uplift associated with a proposed magma body in the vicinity of Socorro, New Mexico, Geology, 4, 583-586, 1976.

Reilinger, R.E., J.E. Oliver, L.D. Brown, A.R. Sanford, and E.I. Balazs, New measurements of crustal doming over the Socorro magma body, submitted to Jour. Geol. 1979.

Reiter, M.A., and R. Smith, Subsurface temperature data in the Socorro Peak KGRA, New Mexico, Geothermal Energy Magazine, 5, 37-41, 1977.

Reiter, M., C. Shearer, and C.L. Edwards, Geothermal anomalies along the Rio Grande rift in New Mexico, Geology, 6, 85-88, 1978, a.

Reiter, M., C.L. Edwards, A.J. Mansure, and C. Shearer, Heat-flow data and major geologic features along the Rio Grande rift in New Mexico, in Guidebook to the Rio Grande Rift in New Mexico and Colorado, compiled by J.W. Hawley, p. 234, N. Mex. Bur. of Mines and Miner. Resour., Circ. 163, 1978, b.

Rinehart, E.J., The determination of an upper crustal model for the Rio Grande rift near Socorro, New Mexico, employing S-wave reflections produced by local microearthquakes, Ph.D. dissertation, New Mex. Inst. Mining Technol., Socorro, 1979.

Rinehart, E.J., A.R. Sanford, and R.M. Ward, Geographic extent and shape of an extensive magma body at midcrustal depths in the Rio Grande rift near Socorro, New Mexico, in Rio Grande Rift: Tectonics and Magmatism, edited by R.E. Riecker, pp. 237-251, Amer. Geophys. Union, Washington, D.C., 1979.

Rosendahl, B.R., R.W. Raitt, L.M. Dorman, L.D. Bibee, D.M. Hussong, and G.H. Sutton, Evolution of oceanic crust, 1. A physical model of the East Pacific Rise crest derived from seismic refraction data, J. Geophys. Res., 81, 5294-5304. 1976.

Rosendahl, B.R., Evolution of oceanic crust, 2. Constraints, implications, and inferences, J. Geophys. Res., 81, 5305-5314, 1976.

Rowlett, H., and D. Forsyth, Teleseismic P-wave delay times in a major oceanic fracture zone, Geophys. Res. Letters, 6, 273-276, 1979.

Saemundsson, K., Evolution of the axial rifting zone in Northern Iceland and the Tjörnes fracture zone, Geol. Soc. Amer. Bull., 85, 495-504, 1974

Saemundsson, K., Fissure swarms and central volcanoes of the neovolcanic zones of Iceland, Geol. J. Special Issue No. 10, 415-432, 1978.

Sanford, A.R., Gravity survey in central Socorro County, New Mexico, Circ. 91, N. Mex. Bur. of Mines and Miner. Resour., Socorro, 1968.

Sanford, A.R., Temperature gradient and heat-flow measurements in the Socorro, New Mexico, area, N. Mex. Inst. of Mining and Technol., Geosc. Dept., Geophysics Open-File Rpt. 15, Socorro, 1977.

Sanford, A.R., Characteristics of the Rio Grande rift in vicinity of Socorro, New Mexico, from geophysical studies, in Guidebook to the Rio Grande Rift in New Mexico and Colorado, compiled by J.W. Hawley, pp. 116-121, N. Mex. Bur. of Mines and Miner. Resour., Circ. 163, 1978.

Sanford, A.R., A.J. Budding, J.P. Hoffman, O.S. Alptekin, C.A. Rush, and T.R. Toppozada, Seismicity of the Rio Grande rift in New Mexico, Circ. 120, N. Mex. Bur. of Mines and Miner. Resour., Socorro, 1972.

Sanford, A.R., O.S. Alptekin, and T.R. Toppozada, Use of reflection phases on microearthquake seismograms to map an unusual discontinuity beneath the Rio Grande rift, Bull. Seismol. Soc. Amer., 63, 2021-2034. 1973.

Sanford, A.R., R.P. Mott, P.J. Shuleski, E.J. Rinehart, F.J. Caravella, R.M. Ward, and T.C. Wallace, Geophysical evidence for a magma body in the crust in the vicinity of Socorro, N.M., in The Earth's Crust, edited by J.G.

Heacock, pp. 385-403, Amer. Geophys. Union Geophys. Mono. 20, 1977.

Sanford, A.R., K.H. Olsen, and L.H. Jaksha, Seismicity of the Rio Grande rift, in Rio Grande Rift: Tectonics and Magmatism, edited by R.E. Riecker, pp. 145-168, Amer. Geophys. Union, Washington, D.C., 1979.

Sleep, N.H., Formation of oceanic crust: Some thermal constraints, J. Geophys. Res., 80, 4037-4042, 1975.

Sleep, N.H., Thermal structure and kinematics of mid-oceanic ridge axis, some implications to basaltic volcanism, Geophys. Res. Letters, 5, 425-428, 1978.

Smith, R.B., Seismicity, crustal structure, and intraplate tectonics of the interior of the western Cordillera, in Cenozoic Tectonics and Regional Geophysics of the Western Cordillera, edited by R.B. Smith and G.P. Eaton, pp. 111-144, Geol. Soc. Amer. Memoir 152, 1978.

Smith, R.B., R.T. Shuey, J.R. Pelton, and J.P. Bailey, Yellowstone hot spot: Contemporary tectonics and crustal properties from earthquake and aeromagnetic data, J. Geophys. Res., 82, 3665-3676, 1977.

Solomon, S.C., Shear-wave attenuation and melting beneath the mid-Atlantic ridge, J. Geophys. Res., 78. 6044-6059, 1973.

Solomon, S.C., and B.R. Julian, Seismic constraints on ocean-ridge mantle structure: Anomalous fault-plane solutions from first motions, Geophys. J.R. Astr. Soc., 38, 265-285, 1974.

Steeples, D.W., and H.M. Iyer, Low-velocity zone under Long Valley as determined from teleseismic events, J. Geophys. Res., 81, 849-860, 1976.

Steinmetz, L., R.B. Whitmarsh, V.S. Moreira. Upper mantle structure beneath the Mid-Atlantic Ridge north of the Azores based on observations of compressional waves, Geophys. J. Roy. Astr. Soc., 50, 353-380, 1977.

Stuart, W.D., and M.J.S. Johnston, Intrusive origin of the Matsushiro earthquake swarm, Geology, 3, 63-67, 1975.

Sykes, L.R., Mechanism of earthquakes and nature of faulting on the mid-ocean ridges, J. Geophys. Res., 72, 2131-2153, 1967.

Sykes, L.R., Focal mechanism solutions for earthquakes along the world rift system, Bull. Seismol. Soc. Amer., 60, 1749-1752, 1970.

Thatcher, W., and J.N. Brune, Seismic study of an ocean ridge earthquake swarm in the Gulf of California, Geophys. J.R. Astr. Soc., 22, 473, 1971.

Toppozada, T.R., and A.R. Sanford, Crustal structure in central New Mexico interpreted from the Gasbuggy explosion, Bull. Seismol. Soc. Amer., 66, 877-886, 1976.

Tréhu, A.M., J.L. Nábělek. S.C. Solomon. Source characterization of two Reykjanes Ridge earthquakes: Surface waves and moment tensors, P-waveforms and non-orthogonal nodal planes. J. Geophys. Res., in press 1980.

Tryggvason, E., Crustal structure of the Iceland region from dispersion of surface waves, Bull. Seismol. Soc. Amer., 52, 359-388, 1962.

Tryggvason, E., Arrival times of P-waves and upper mantle structure, Bull. Seismol. Soc. Amer., 54, 727-736, 1964.

Tryggvason, E., Subsidence events in the Krafla area, J. Geophys., 47, 141- 153, 1980.

Tryggvason, E., Seismicity, earthquake swarms and plate boundaries in the Iceland region, Bull. Seismol. Soc. Amer., 63, 1327-1348, 1973.

Tryggvason, E., and M. Båth, Upper crustal structure of Iceland, J. Geophys. Res., 66, 1913-1925, 1961.

Vogt, P.R., The Iceland phenomenon: Imprints of a hot spot on the ocean crust, and implications for flow below the plates, in Geodynamics of Iceland and the North Atlantic area, edited by L. Kristjánsson, pp. 105-126, Reidel, Dordrecht, Holland, 1974.

Ward, P.L., New interpretation of the geology of Iceland, Geol. Soc. Amer. Bull., 82, 2991-3012, 1971.

Whitmarsh, R.B., Median valley refraction line, Mid-Atlantic Ridge at 37°N, Nature, 246, 297-298, 1973.

Whitmarsh, R.B., Axial intrusion zone beneath the median valley of the Mid-Atlantic Ridge at 37°N detected by explosion seismology, Geophys. J.R. Astr. Soc., 42, 189-215, 1975.

Zverev, S.M., I.V. Litvinenko, G. Pálmason, G.A. Yaroshevskaya, and N.N. Osokin, A seismic crustal study of the axial rift zone in Southwest Iceland, J. Geophys., 47, 202-210, 1980 a.

Zverev, S.M., I.V. Litvinenko, G. Pálmason, G.A. Yaroshevskaya, N.N. Osokin, and M.A. Akhmetjev, A seismic study of the rift zone in Northern Iceland, J. Geophys., 47, 191-201, 1980 b.

MAGNETOTELLURIC AND GEOMAGNETIC DEEP-SOUNDING STUDIES IN RIFTS AND ADJACENT AREAS: CONSTRAINTS ON PHYSICAL PROCESSES IN THE CRUST AND UPPER MANTLE

John F. Hermance

Department of Geological Sciences, Brown University, Providence, R.I. 02912

Abstract. Deep electrical studies are reviewed along with other supportive geophysical/geological investigations of five of the major rift zones of the world: the Baikal rift, and Rhine graben, the East African rift, Iceland and the Rio Grande rift. All of these regions exhibit anomalously low values of electrical resistivity, density and seismic velocity, either within the crust itself or at high levels in the mantle. Deep electrical studies support a model where ascending masses of material from the mantle are intimately coupled to the fractionation of a basalt melt and its accumulation at higher levels within the earth. In Iceland, an interplate rift, the accumulation and chilling of melt at the base of the crust apparently leads to a significant component of crustal underplating, whereas in intraplate rifts such as the Rio Grande rift, the emplacement of basaltic magma at high levels may lead to extensive remelting of the crust, triggering eruptive episodes of silicic magmas.

Introduction

Knowledge of the distribution of electrical conductivity in the crust and upper mantle places important constraints on models for physical properties beneath the active rift zones of the earth. Clearly the coupling of regional doming, rifting, and magma genesis in these regions involves a variety of thermal phenomena which also serve to modulate in situ values of electrical conductivity. Therefore regional electromagnetic studies, such as magnetotelluric and geomagnetic deep-sounding experiments, can establish limits not only on the high temperature experienced by materials in the solid state, but on a variety of other thermally related phenomena as well: the presence of hydrothermal fluids in pores and cracks, hydrated and dehydrated phases in metamorphosed materials, and the presence of highly conducting melt phases distributed along grain boundaries. The large number of factors influencing the electrical conductivity precludes, of course, the unequivocal interpretation of any deep electrical studies on its own merits. However, when combined with other geophysical and geological information, electromagnetic studies contribute substantially to discriminating between various hypotheses offered for processes associated with rift tectonics.

The following discussion compares deep electrical studies in five of the major rift zones of the world: the Baikal rift, the Rhine graben, the East African rift, Iceland and the Rio Grande rift. We first summarize the principal geophysical/geoelectrical characteristics of each of these regions. Then a model is proposed which embraces many of the geophysical constraints established by the studies and which also seems compatible with many of the megatectonic/magmatic elements associated with rifting environments.

Electromagnetic induction studies in rift zones using long-period natural electromagnetic fields have been reviewed previously by Garland (1975), Hutton (1977) and Jiracek et al. (1979). Although present discussion is to some degree an extension of this earlier work, our present emphasis is on the implications these studies have for constraining models of actual physical phenomena in the earth, particularly the interaction between crust and mantle processes. In the interest of keeping the present discussion of manageable size, we will not review the interesting implications of a geomagnetic deep-sounding in the Imperial Valley of California (White, 1973a,b) or of a magnetotelluric investigation in the Snake River Plains of Idaho (Stanley et al., 1977).

Overview of Rift Studies

The Baikal Rift Zone

The Baikal rift zone (Figure 1) extends for a distance of 1500 km along the southeastern edge of the Precambrian Siberian platform and, with vertical displacements of up to 6-8 km, represents the world's deepest intracontinental rift valley (Logatchev and Florensov, 1978). The morphotectonic development of this graben feature accompanied broad-scale domal uplift, apparently due to local diapiric flow in the astheno-

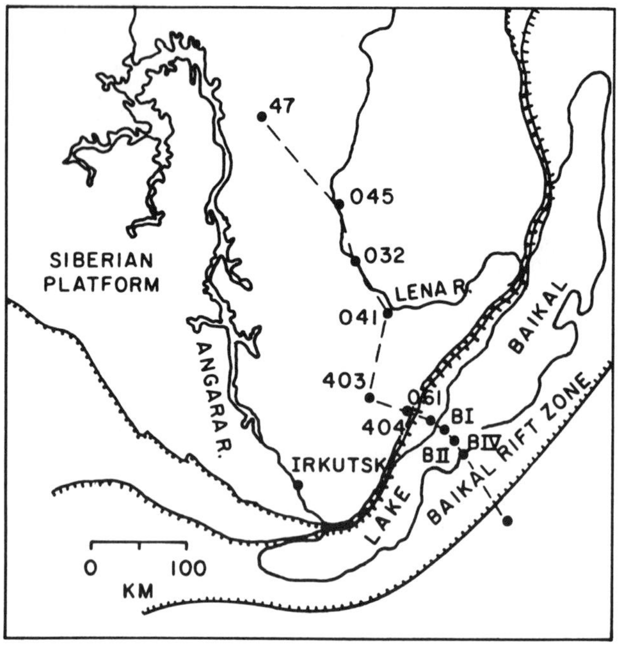

Fig. 1. Map of the Baikal region showing the location of sites for the magnetotelluric data reported by Gornostayev et al. (1970). Boundaries of the Siberian platform and the Baikal rift zone are generalized from Lubimova et al. (1972).

sphere. The Baikal rift system is isolated from the world rift system and appears to represent an autonomous complex of Cenozoic tectonism and volcanism (ibid.).

Although much of the research on the Baikal rift has not been reported in the English literature, this region is one of the most thoroughly studied of the continental rift systems (Puzirev et al., 1978). Soviet geophysicists have a long-standing tradition in the application of magnetotelluric and geomagnetic deep-sounding techniques to studying regional tectonic problems, and it is not surprising that magnetotelluric sounding experiments were among the first quantitative geophysical indicators of the anomalous nature of the crust and upper mantle beneath the Baikal rift and the adjacent regions (Gornostayev et al., 1970, Gornostayev, 1972; Pospeev and Mikhalevsky, 1976; Berdichevsky et al., 1980). Basically, the magnetotelluric interpretation indicates five general zones within the earth's interior (Figure 2): first, a conductive sedimentary layer several kilometers thick and having a resistivity on the order of 20-50 ohm-m; second, a resistive zone (> 500 ohm-m) representing crytalline crust and extending to a depth of 10-20 km; third, a conductive intracrustal zone at a depth of 12-25 km having a depth-integrated conductivity (section conductance) of S = 600 - 2500 Siemens (Mhos). The maximum thickness of this zone appears to be 20 km (if $\rho \simeq 30$ ohm-m); it could have a minimum thickness of less than 1 km if the resistivity is on the order of 1 ohm-m or less. The fourth zone appears at a depth of 20-40 km (depending on the thickness of the third zone) and seems to have a resistivity in excess of several hundred ohm-m. It is thought to be made up of peridotite containing emulsified basalt. Fifth and finally, data from several long-period sites indicate a more conductive layer (< 10 ohm-m) at depths in excess of 50-100 km beneath the rift axis and at depths exceeding 200 km to the northwest of the rift (Pospeev and Mikhalevsky, 1976). However, a recent statistical analysis of more than 300 magnetotelluric soundings in the Baikal region (Berdichevsky et al., 1980) shows that although a high-level conducting layer (depth ∿12 km; conductance ∿2000 s) is clearly required to satisfy the data, the presence and character of a possible deeper conducting feature (d > 50 km) is not very well resolved.

In other words, the most striking result of the magnetotelluric interpretation from the Baikal rift is the presence of the anomalous conducting zone (depth ∿12-15 km; conductance ∿2000 s) of limited thickness (< 30 km). The second noteworthy feature of the data is that the earth appears to be surprisingly resistive beneath the intracrustal conductor, perhaps more than several hundred ohm-m. It is likely that this deeper zone (from a depth as shallow as 20-40 km to a depth exceeding 100 km) represents an ultramafic material which, while having a relatively high temperature, may still be relatively resistive (Gornostayev, 1972).

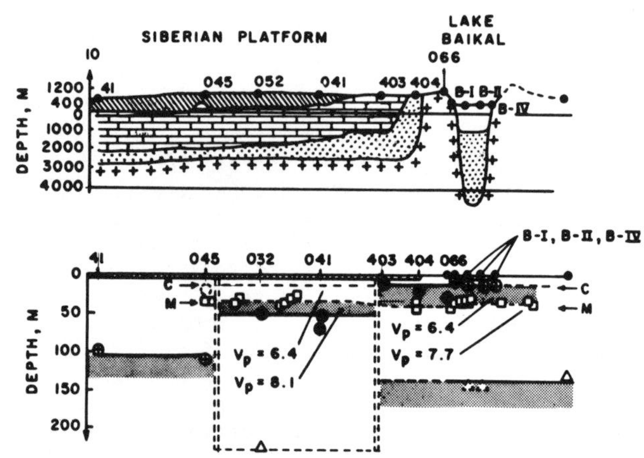

Fig. 2. Vertical section along Profile 1 (see map of Baikal region in Fig. 1) showing the detailed interpretation of magnetotelluric data by Gornostayev et al. (1970). The horizontal solid lines in the lower figure correspond to the depth to conductive interfaces. The velocities of compressional seismic waves are also shown, along with the approximate depths to the Mohorovicic discontinuity (M) and the Conrad discontinuity (C).

Heat flow work in the Baikal rift described by Lubimova et al. (1972) is compared with results from various rift areas of the world in Table 1. Clearly, the heat flow in the Baikal rift is anomalously high, although these values are restricted to a band at most 60 km wide, along the morphotectonic expression of the rift. Lubimova et al. (1972) have argued, on the basis of the narrow width of the Baikal thermal anomaly, for a relatively shallow depth of origin for this feature (< 24 km), and suggest the spatial coincidence of a long-term temperature perturbation with the high-level (10-20 km depth) electrical anomaly revealed through the geomagnetic variation and magnetotelluric deep-sounding studies described above. Moreover, Lubimova et al. point out that the static magnetic anomalies are restricted to a crustal depth less than 14 km in this region, suggesting a relatively shallow depth to the Curie isotherm (ibid.). From this evidence for high temperature at intermediate levels in the crust, they argue that the electrical anomaly may be associated with the accumulation of a basalt melt fraction derived either by partial melting in situ or by segregation and upward migration of melt from deeper zones.

Seismic investigations support many of the conclusions based on electromagnetic and deep thermal studies in the Baikal rift (Puzirev et al., 1970, 1974, 1978; Krylov et al., 1974, 1975). An intracrustal low-velocity layer (depth ∿ 15-17 km), with a velocity differential of -0.2 to -0.3 km/sec relative to values in normal material above and below this zone, has been detected under two sites along the margin of the rift approximately 150 km apart (Krylov et al., 1974, 1975). This low-velocity crustal layer is associated with the zone of high electrical conductivity (Puzirev et al., 1978; Berdichevsky et al., 1980). Earthquake foci beneath the rift are restricted to the volume above the anomalous intracrustal layer, and their absence at greater depth suggests a rapid transition to higher temperature, more plastic conditions (Puzirev et al., 1978). It is thought that the anomalous intracrustal layer is most likely to result from increased heating and partial melting (ibid.).

TABLE 1. Heat Flow in Major Rift Zones

Region	Heat Flow microcal/cm^2/sec	Reference
East African Rift	2.88	Von Herzen and Vacquier, 1967
Basin and Range	2.21-2.44	Warren et al., 1969
Rhine Graben	2.6	Meincke et al., 1967
Iceland NVZ	5-6	Palmason, 1973
Baikal Rift	2.0-3.4	Lubimova et al., 1972
Rio Grande Rift	2.56±0.65	Reiter et al., 1979

Large-scale seismic refraction experiments (line-lengths of up to 300 km) indicate the presence of anomalous low-velocity mantle material (7.6-7.8 km/sec) associated with the base of the crust beneath the entire region. This pillow-like zone (average thickness 17 km) has been mapped over an area of 200,000 km^2 and occupies a region 2-3 times wider than the actual geologic trace of the rift zone (Puzirev et al., 1978). Krylov (1976) has used estimates of the density, P-wave velocity and electrical resistivity to infer a partial melt fraction of 5-10%. Logatchev and Florensov (1978) reason that it is quite possible for mantle material to have penetrated into the lower crust, essentially in the form of semi-solid peridotite, with only an insignificant admixture of basalt magma, and feel that this interpretation is supported by the gravity, P-wave velocity, heat flow and deep electroconductivity soundings.

An alternative suggestion is that the anomalous conducting layer at intracrustal levels is caused by the release of crystalline-bound water through dehydration processes and its accumulation as an electrolytic fluid along pores and joints (Gornostayev, 1972; Pospeev and Mikhalevsky, 1976; Berdichevsky et al., 1980). Even if this is the case at high levels in the crust, there remains the distinct possibility that the electromagnetic data are indicating the accumulation of basaltic melt at depth below this, perhaps as shallow as 20 km or less (Gornostayev, 1972).

The Rhine Graben

The Rhine Graben (Figures 3 and 4) is a coherent structural feature, sharply bounded by master faults, for a distance of over 300 km from Frankfurt to Basel in the Federal Republic of Germany (Illies, 1970). The graben itself is superposed on a complex system of fractures and faults extending from the Netherlands into southern France, and seems to be a fundamental component in the dynamics of global tectonics (Illies, 1974).

The geophysical framework for regional studies of the Rhine graben and adjacent areas is summarized by Fuchs (1974). Several prominent features in the crust and upper mantle appear to play a profound role in the tectonic development of the region. First, a low-velocity layer (the sialic low-velocity zone) is detected at depths of only 10 km beneath the graben (Figure 5; Ansorge et al., 1970; Mueller, 1970). Second, significant crustal thinning seems to character-

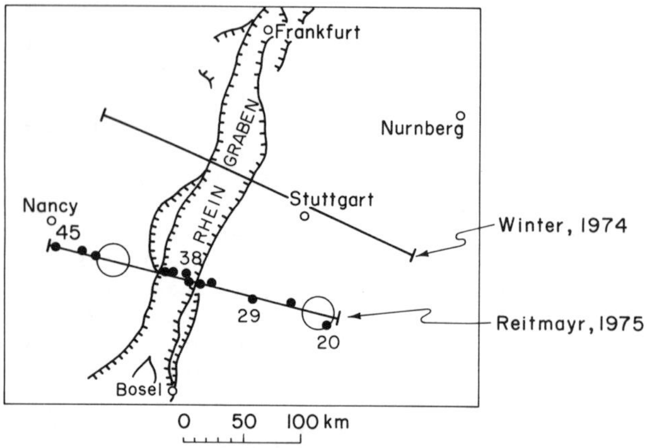

Fig. 3. Map of the Rhine graben (after Illies, 1970) showing the geomagnetic variation profile lines of Winter (1974) and Reitmayr (1975). The numbers on the southern profile refer to sites indicated in Fig. 7. The circles represent zones where, according to Reitmayr (1975), magnetotelluric measurements could be explained with one-dimensional models.

ize the graben; the crust-mantle boundary is elevated to a depth of only 25 km in the southern part of the graben (Ansorge et al., 1970), whereas in the northern graben, seismic reflection experiments indicate an uplift of the crust-mantle boundary along the flanks on either side of the graben (Meissner and Vetter, 1974). Third, there is some suggestion of a low-velocity pillow in the upper mantle (Meissner et al., 1970), and while reversed seismic refraction measurements indicate more normal mantle velocities at the base of the crust beneath the southern graben (Rhine Graben Research Group for Explosion Seismology, 1974; Edel et al., 1975), it is still necessary to invoke a low-density (hence low-velocity) pillow within the mantle itself to explain the regional gravity data (Fuchs, 1974). The pronounced thinning of the crust beneath the rift, seen seismically, would suggest a much larger positive Bouguer anomaly than actually measured; hence, an anomalous low-density (lower-velocity) compensating mass must be imbedded at some depth within more normal mantle material (Fuchs, 1974). Moreover, seismic surface wave studies using the fundamental Rayleigh wave mode (Reichenbach and Mueller, 1974) also support a low-velocity cushion in the upper mantle.

Mueller and Rybach (1974) argue that relatively high values of heat flow in the graben, along with an inferred decrease in the P-velocity of some 7 percent, are compatible with the view that the sialic low-velocity channel may be in a state of partial fusion (at a depth of 20-25 km). Local increases in the temperature gradient superposed on the regional field of increased heat flow may be explained by cooling bodies of acidic magma which have intruded into the crystalline basement from the sialic low-velocity zone (ibid.). Illies (1970) feels that abnormally high values of heat flow preclude the rupture of crustal material deeper than 10 km; rather, one would expect more plastic conditions to obtain. Fuchs (1974) proposes a model for the tectonic evolution of the rift in which the sialic low-velocity layer plays a major role in the rheological behavior of the crust.

Magnetotelluric, telluric and geomagnetic variation studies strongly support the possibility

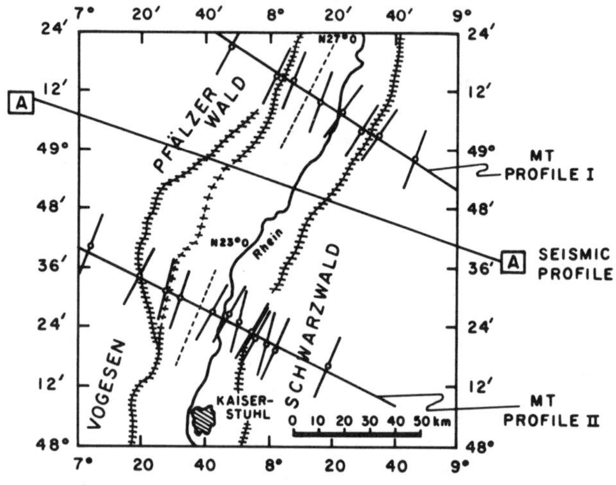

Fig. 4. The location of magnetotelluric profiles discussed by Scheelke (1972, 1974) and the seismic section described by Mueller and Rybach (1974). The seismic section shown in Fig. 5 is representative of the axial portion of the graben in this region.

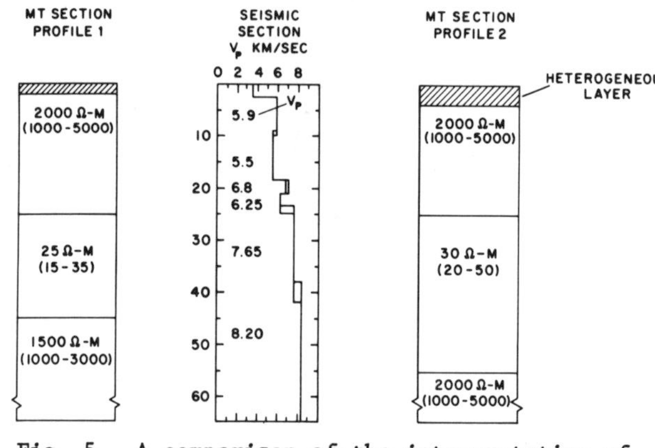

Fig. 5. A comparison of the interpretation of magnetotelluric data by Scheelke (1972, 1974) from the two profiles shown on the previous figure with the interpretation of seismic data by Mueller and Rybach (1974). The seismic velocities in the mantle are inferred from the surface wave studies of Reichenbach and Mueller (1974).

of very high temperature beneath the Rhine graben proper (Fuchs, 1974). Haak et al. (1970) used telluric measurements and Winter (1970) used geomagnetic variations to determine a depth of approximately 25 km to an anomalously conducting layer (< 30 ohm-m) beneath the axis of the rift. Because of the limited depth resolution due to the restricted period band for which data were used, there is some ambiguity intrinsic in these analyses. However, preliminary magnetotelluric measurements over a broader period range by Losecke (1970) suggest that this anomalous layer is at a depth of ∿28 km and may be associated with the pillow of low-velocity material in the upper mantle. The most reliable parameter that can be estimated for such a conducting layer, when sandwiched between resistive layers, is generally its total depth-integrated conductivity, or the so-called section-conductance. In this case, the layer conductance would be S = 1200 - 1700 Siemens (Mhos). Losecke (1970) favors an interpretation in which this is the conductance of a layer approximately 5 km thick and having a resistivity of 3-6 ohm-m. To interpret more refined magnetotelluric data with periods between 100 and 1000 seconds, Scheelke (1972, 1974) also found it necessary to assume an anomalously conducting layer (Figure 5)

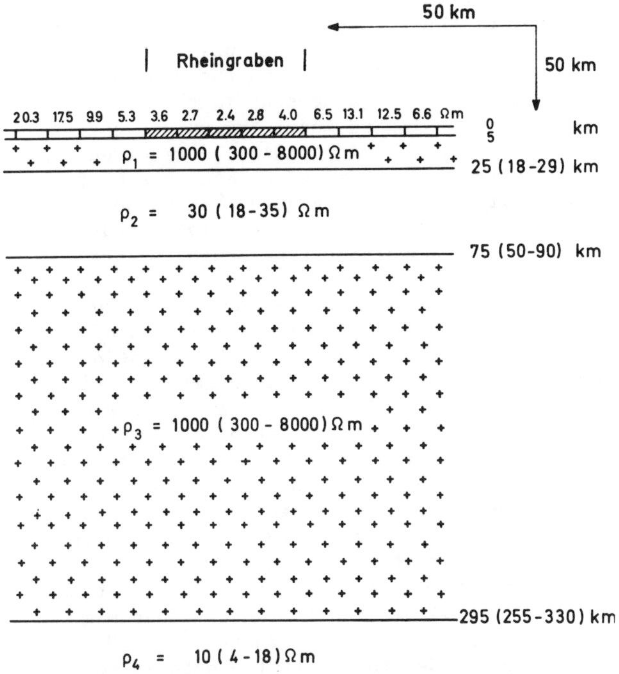

Fig. 6. The interpretation of geomagnetic variation data by Winter (1974) along the profile indicated in Fig. 3. Lateral changes in resistivity are restricted to a surface layer having a thickness of 5 km. The parameters and their estimated ranges of variability are shown for the best-fitting models.

beginning at a depth of about 25 km, although he assumed a resistivity of about 25 ohm-m and a thickness of 20 km (S = 800 Siemens).

Detailed interpretations of geomagnetic variation data along the two profiles shown in Figure 3 are presented by Winter (1974) and Reitmayr (1975). The magnetotelluric interpretation is compatible with the geomagnetic variation results, although, as shown in Figure 6, Winter (1974) favors a somewhat more resistive (∿30 ohm-m) and thicker (∿50 km) anomalous zone. The conducting layer proposed for the upper mantle by Reitmayr (Figure 7) is even more resistive (50 ohm-m), but is somewhat thinner (∿20 km) than the structure proposed by Winter. The section conductance of the layer proposed by Reitmayr (400 Siemens) is significantly smaller than that proposed by Winter (1700 Siemens). Some of this difference is explained by the fact that Reitmayr proposes the existence of a second conducting layer (∿25 ohm-m; 25 km thick) at a depth of 75 km. This layer (S = 1000 Siemens) is absent in Winter's interpretation, although he has an equivalent conductance above a depth of 75 km. These subtle differences are difficult to resolve from the geomagnetic variation data alone for the range of periods available. It seems clear, however, that the upper mantle has anomalously low resistivities (25-50 ohm-m) at depths as shallow as 25 km.

Magnetotelluric data have been analyzed by Haak and Reitmayr (1974) from two sites outside the morphotectonic expression of the rift (Figure 3): Bolstern, 120 km east of the Rhine graben, and Saint Stacl, 45 km west. The interpretation of data from both sides (Figure 7) indicates that the high-level (depth ∿25 km) conducting layer is either absent altogether or is significantly deeper (∿70-90 km) outside the rift proper. The upper mantle in these adjacent regions has a rather high resistivity (10^2 to 10^4 ohm-m) at depths on the order of 70 km (see Figure 7). It is interesting to notice the close similarity between the structure proposed for the Rhine graben by Reitmayr (Figure 7) and that proposed for the Baikal rift (Figure 2) by Gornostayev et al. (1974).

The East African Rift

The rift zone of East Africa is a system of normal faults bordering a trough-like feature 40 to 65 km wide which traverses two broad, contiguous domal uplifts in Ethiopia and Kenya (Baker et al., 1972). Domal uplift has occurred in three major pulses over the last 35-40 × 10^6 years - in late Eocene, in mid-Miocene and in Plio-Pleistocene times - and is associated with episodes of graben faulting and volcanism of intermediate to silicic type.

The Kenyan Rift. A large (350 km wide) negative-gravity anomaly (-50 milligal) is associated with the domal uplift and appears to reflect low-density, partially fused material in

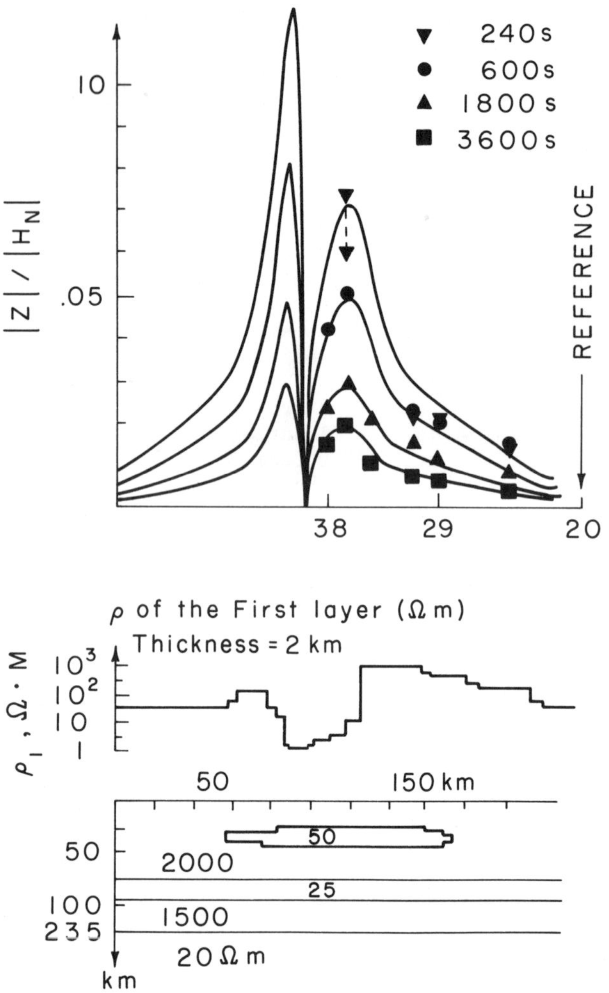

Fig. 7. The ratio of the amplitude of vertical to normal horizontal magnetic field variations from the sites occupied in the Rhine graben (Fig. 3) by Reitmayr (1975). His two-dimensional model and its theoretical $|Z|/|H_N|$ response at various periods is shown. Two low-resistivity zones are shown in the upper mantle: one laterally limited, at a depth of 25 to 45 km beneath the graben axis (ρ = 50 ohm-m); the other laterally continuous at a depth of 75 to 95 km (ρ = 25 ohm-m).

the upper mantle (Fairhead, 1976). The model proposed by Fairhead (1976) involves the upward movement of the lithosphere-asthenosphere boundary on a regional scale in which the cold, high-density lithosphere (3.34 gm/cc) is replaced by hot, lower-density asthenosphere (3.29 gm/cc) over a lateral distance of more than 250 km.

A seismic refraction experiment along the axis of the Kenyan rift south of Lake Rudolf (Griffiths, 1972) indicates an average crustal velocity of V_p = 6.4 km/sec and an upper mantle velocity of 7.5 km/sec. The crust appears to be only 20 km thick. The dispersion of seismic surface waves suggests a crust-mantle structure not unlike that of the Basin and Range province of western North America, with anomalous low-velocity mantle material (V_s = 4.25-4.45 km/sec) extending to a depth of 120-200 km (Knopoff and Schlue, 1972). Long et al. (1972) combined teleseismic observations with regional surface wave data in a study which suggested that the anomalous crust is of very limited lateral extent on either side of the morphotectonic rift feature. However, this study did confirm crustal, or lithospheric, thinning along the axis of the ridge, indicating a strong similarity between the mean compressional velocity of the mantle beneath East Africa and that beneath Iceland. The presence of such an ultra-low-velocity zone seems to indicate a mantle structure analogous to that beneath ocean ridges (ibid.).

Reconaissance geomagnetic variation studies using an array of six recording magnetic variometers (Figure 8) revealed an inductive anomaly associated with the Kenyan rift 100-150 km north of Nairobi (Banks and Ottey, 1974). These studies suggested the presence of a high-conductivity zone (< 0.1 S/m) elongated along the axis of the rift, at a depth of less than 20 km. Because of the restricted band of periods for which data were available (∿25 min), Banks and Ottey (1974) were unable to determine unequivocally the depth to the top of this conducting feature. According to these investigators, a conductor at the surface, such as might be associated with conducting sediments in the valley fill, could as readily explain the observed data as a conductor somewhat deeper in the crust. However, attempting to model the effect as due simply to the conducting sediments required a conductivity of 0.2 S/m and a thickness of 5 km for the rift fill; they argued that these values were unreasonably large, and concluded that the source of the anomaly must be sought at greater depths.

A magnetotelluric traverse (Rooney and Hutton, 1977) was undertaken in the same portion of the rift (Figure 8) as the geomagnetic variation studies described above. The magnetotelluric data (Figure 9) confirmed and extended the conclusions of the geomagnetic deep-sounding experiment. In fact, the magnetotelluric investigations, because they covered a much wider frequency band, were able to resolve a more detailed model for the electrical structure beneath the rift. It appears that the entire vertical section beneath the rift is dominated by lower than normal resistivities (approximately 15 ohm-m) from the surface to depths greater than 30 km. The simplest model that fits the data is essentially a homogeneous half-space having a resistivity of approximately 10 ohm-m from the surface to a depth exceeding 35 km. It is possible, however, that a resistive layer could be present beneath the conducting sediments and not be resolved in the analysis of the data available. According to Rooney and Hutton (1977), such a resistive layer could be at a depth somewhat greater than 5 km, and might be as thick as 5 to 10 km; it would reflect the presence of a resistive, crystalline basement. Whether or not such a resistive zone is present, however, the material beneath this zone would have to be strikingly conductive, and at a depth which seems to be of the order of (or less than) the depth to the crust-mantle boundary in this region.

Although data from only a single site were analyzed in detail by Rooney and Hutton (1977), there is remarkable agreement among long-period data from at least three sites along a profile over 100 km long, spanning the rift. This leaves little doubt that a thick section of material having a resistivity of 15 ohm-m or less must be associated with the sub-basement of the East African rift.

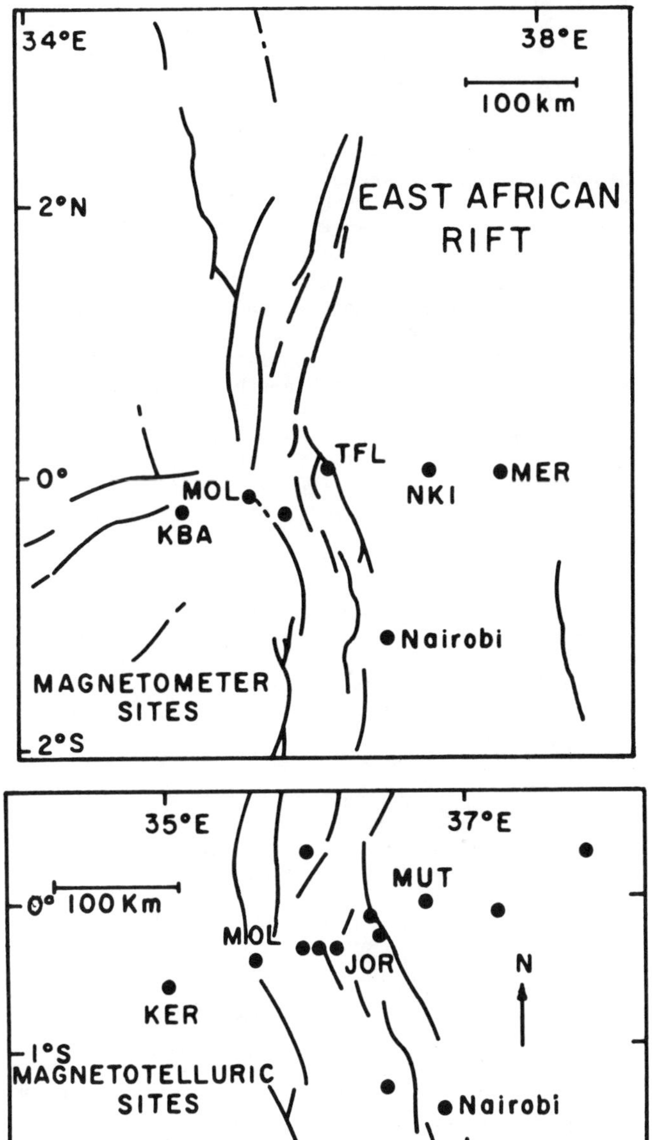

Fig. 8. The Kenyan rift in East Africa, showing the location of magnetic variometry sites (top figure, after Banks and Ottey, 1974) and magnetotelluric sites (bottom figure, after Rooney and Hutton, 1977).

Afar. The Afar triangle may be a supramarine manifestation of incipient oceanic development (Mohr, 1972). Basaltic volcanism of oceanic type is related to crustal separation, whereas large, essentially silicic central volcanoes are present along the margins of the depression and are thought to represent interactions between subcrustal magma and sialic crust (Barbieri et al., 1972). In this sense, the Afar region might be as representative of mid-ocean ridge processes as those observed in Iceland (Mohr, 1972).

Seismic sections show a strong similarity to the interpretation of Icelandic data (Lepine et al., 1972). An interpretation of gravity data suggests an attenuation of the crust from continental type in south Afar to largely oceanic type in north Afar (Makris et al., 1972); moreover, there is a strong indication of a low-density, low-velocity pod of mantle material beneath the entire region (ibid.).

Berktold et al. (1975), in discussing the interpretation of magnetotelluric measurements, note the gross similarity between the apparent resistivity data from Afar and those from Iceland described by Hermance (1973). Unusually high temperatures are inferred at depths of 15 km, and data from measuring sites outside the rift proper suggest that the conducting anomaly extends beyond the region of the rift itself, though at a somewhat greater depth. On the basis of the zone of very low resistivity which has been detected below the Afar depression and its west escarpment, a temperature of 800-1200°C is inferred at a depth of 15 km, corresponding to a mean gradient of 60°C/km. Such a temperature is compatible with partial fusion of material at this depth (Berktold et al., 1975).

The Rift in Southern Africa. It has been suggested, on the basis of recent faulting and seismic activity, that the region of southern Africa between Rhodesia and Botswana on one side and Angola and Zambia on the other may be a site for incipient rifting (Reeves, 1972; Scholz et al., 1976). Geomagnetic variation experiments by de Beer et al. (1975), using an array of 25 Gough-Reitzel recording variometers, indicate an electrical conductivity anomaly associated with this same regional trend. Even though a quantitative interpretive model of the data has not been advanced, the strength of using natural electromagnetic methods in a reconnaissance mode has been well demonstrated by these experiments.

Iceland

The position of Iceland on the Mid-Atlantic ridge has led to its extensive study as a site of

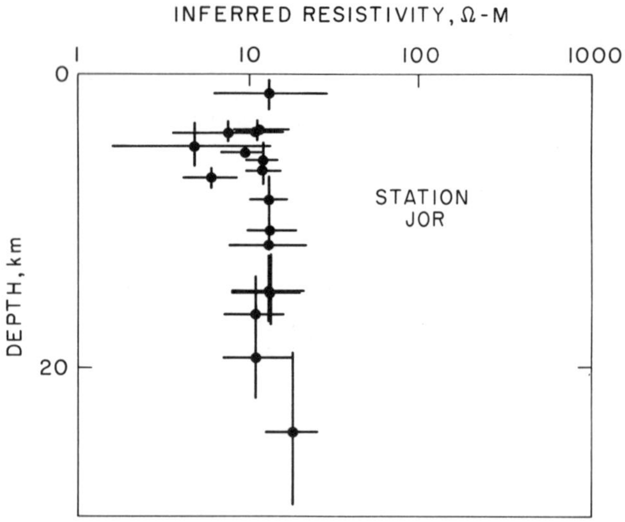

Fig. 9. Inversion of magnetotelluric parallel resistivity data from site JOR in the East African rift (see Fig. 8) (after Rooney and Hutton, 1977).

active accretion at plate margins (Bodvarsson and Walker, 1964; Ward, 1971; Palmason and Saemundsson, 1974; Beloussov and Milanovskiy, 1976). The neovolcanic zone (Figure 10), in particular, is characterized by the type of active linear volcanism associated with mid-ocean spreading centers (Palmason and Saemundsson, 1974). Figure 11 is a map of the area around Lake Myvatn. The geology between Myvatn and Jokulsa a Fjollum is characterized by a series of north-northeast striking fractures and fissures, dramatic evidence of the post-glacial tectonic activity which characterizes this region (Saemundsson, 1974). Although the neovolcanic zone has been volcanically active since the last glaciation and is apparently a center for present crustal spreading, the eastern and northwestern margins of the island, in contrast, are much more stable tectonically, having an age on the order of 10^7 years B.P. or greater.

A major episode of rifting along the boundary separating the North American and European plates began in northern Iceland on December 20, 1975 and is continuing at the present time (Bjornsson et al., 1977, 1979). During this time, crustal spreading of some 3 m has been observed in the Krafla caldera (Figure 11), and the caldera itself has been the focus of an 80-km-long fissure swarm. Needless to say, Iceland is one of the most active supramarine volcanic areas on earth (Palmason, 1976) and remains one of the best examples of the correlation of electrical conductivity anomalies and regional thermal processes in the deep crust and upper mantle (Garland, 1975).

A number of geophysical studies have been carried out in this area. Gravity measurements have been performed by Einarsson (1954), Schleusener et al. (1976) and Palmason (private communication, 1978); heat-flow work is summarized by Palmason (1973).

Seismic refraction studies in the neovolcanic zone (Palmason, 1971) suggest that velocities in the upper crust increase from 2-3 km/sec at the surface to approximately 6.5 km/sec at a depth of 4-5 km. The lower crust (V_p = 6.5 km/sec) is thought to be made up of basic intrusives and metabasalts (Palmason, 1971; Palmason and Saemundson, 1974). Material having a P-wave velocity of 7.2 to 7.4 km/sec is associated with an anomalous mantle and, although not seen on all seismic refraction lines in this region, is estimated from surface wave studies (Båth, 1960; Tryggvason, 1962) to be at an average depth of approximately 10 km. In comparison, seismic refraction studies by Palmason (1971) indicate that the upper mantle may be as little as 8 km beneath the southwestern rift zone and may deepen to as much as 15 km beneath the southeastern portion of the island. Francis (1969), using body-wave data from earthquakes along the Mid-Atlantic ridge, and Long and Mitchell (1970), using teleseismic signals recorded at four stations over Iceland, suggest that anomalously low-velocity mantle material may extend to a depth of 200-250 km beneath the entire island.

The first indication of anomalously high electrical conductivities beneath Iceland was provided by the geomagnetic variation studies of Hermance and Garland (1968), who used a primi-

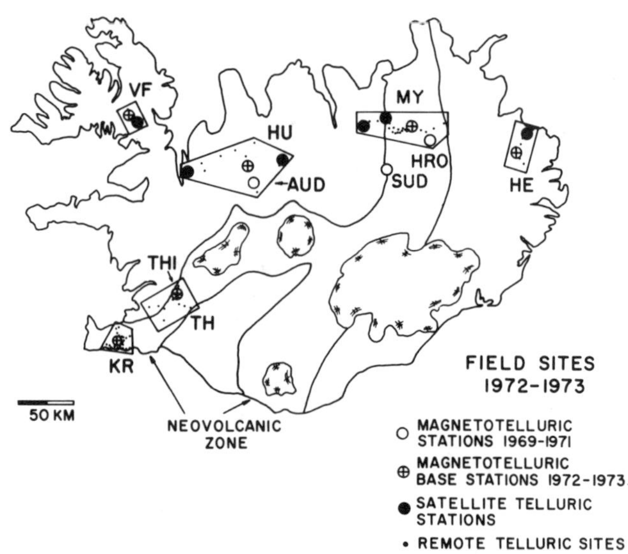

Fig. 10. Location of magnetotelluric survey areas in Iceland (after Hermance et al., 1976). Code letters refer to Krisuvik (KR), Thingvellir (TH), Westfjords (VF), Hunavatnssysla (HU), Myvatn (MY) and Herad (HE). The boundaries of the neovolcanic zone are generalized from Palmason and Saemundsson (1974).

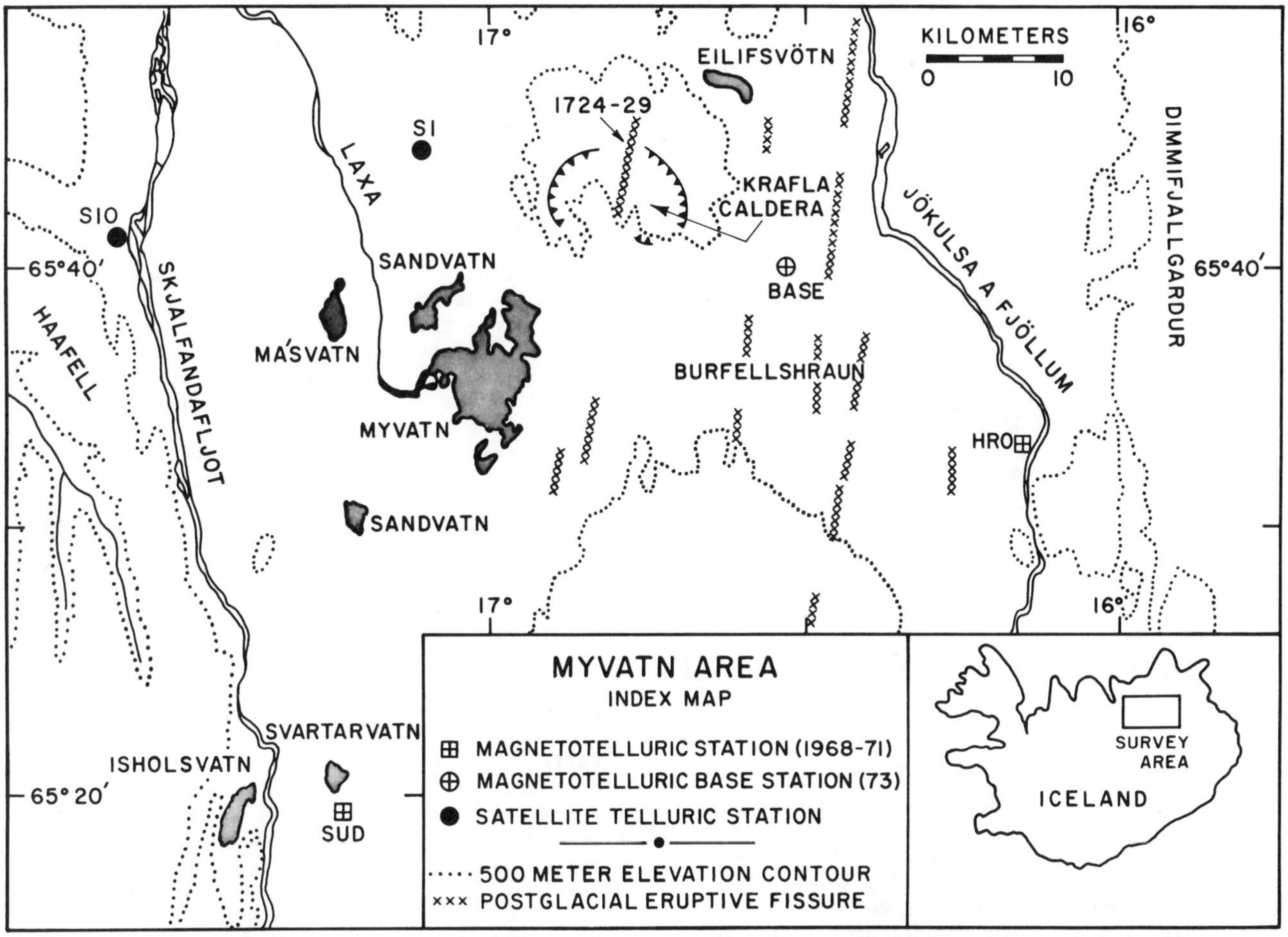

Fig. 11. The northern neovolcanic zone in Iceland showing Krafla Caldera (after Thayer, 1975; Thayer et al., 1980). Data from the site HRO are described in detail in Thayer et al. (1980), as is the interpretation of data from the base site, and the satellite sites S1 and S10. In general, features of the electrical structure beneath these sites are similar to those for the other neovolcanic zone sites shown in Figs. 13 and 14 below.

tive form of magnetic gradiometry in applying $\Delta Z/\Delta H$ ratios for magnetic bay-type events. More refined models of the deep electrical structure of the region were provided through subsequent broad-band magnetotelluric studies by Hermance and Grillot (1970, 1974), Grillot (1973), Thayer (1975), Hermance et al. (1976), Beblo and Bjornsson (1978, 1980) and Thayer et al. (1980). Electrical measurements of the upper crust ($d \leq 5$ km) using active techniques are summarized by Bjornsson (1976).

Hermance and Grillot (1974) concluded from magnetotelluric data from southwest Iceland that not only were temperatures in the upper mantle beneath this region higher than those predicted theoretically for the ocean basins and continental areas (Figure 12), but also the geothermal gradient in the mantle was less than a few °C/km, or nearly two orders of magnitude smaller than the geothermal gradient in the crust. Such low temperature gradients at shallow depth in the upper mantle, along with the implications from the teleseismic studies described above, suggest that the anomalous temperature regime beneath Iceland shown in Figure 12 extends over a minimum depth range of several hundred kilometers. Clearly, such a temperature profile calls for a significant revision of concepts regarding the petrological evolution of crustal material along the boundaries of accreting plates.

It is interesting to note that a number of sites in the neovolcanic zone exhibit a strikingly similar magnetotelluric response (Grillot, 1973; Thayer, 1975). Hermance et al. (1976) compared data from sites in the neovolcanic zone with data from one of the older (10^7 years B.P.) geologic provinces in Iceland. At the longest periods the data tend to appear quite similar, as shown in Figure 13; however, at shorter periods a significant difference can be seen.

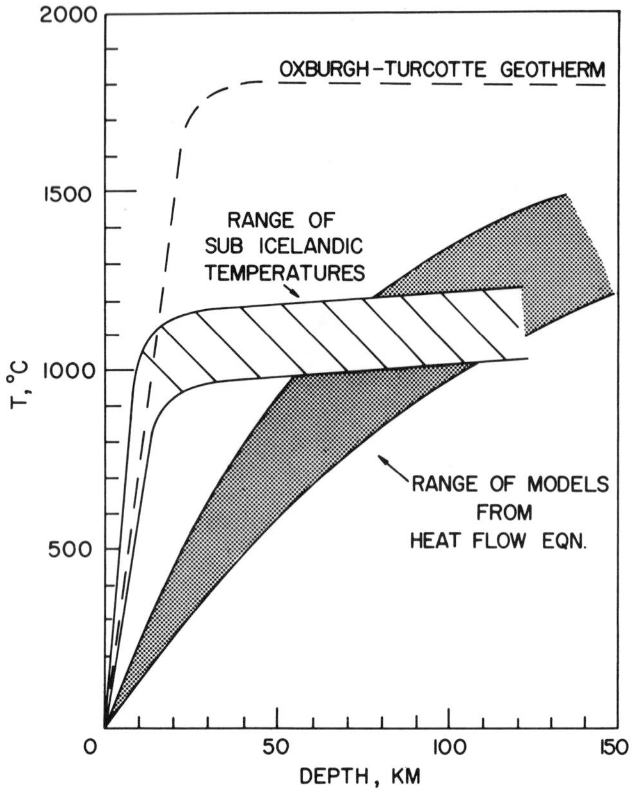

Fig. 12. The range of possible temperatures beneath Iceland as inferred from a synthesis of magnetotelluric, heat flow and seismic data (after Hermance and Grillot, 1970; 1974). Upper mantle temperatures beneath Iceland are significantly higher than those temperatures inferred from the conductive heat flow equation (e.g. Ringwood, 1975), whereas they are significantly lower than the temperatures implied by the thermal boundary layer model of Oxburgh and Turcotte (1968). The distribution of temperatures is much closer to that suggested by the kinematic modeling of Bottinga and Allegre (1976).

In order to explore possible reasons for this systematic difference, Hermance et al. (1976) applied a Monte Carlo inversion scheme to the data in Figure 13. For a given set of measurements a range of uncertainty was specified. Theoretical apparent resistivity values were then generated for a random set of layered models. Those models that generated values falling within the range of uncertainty of the observations were accepted and those that generated values outside the range of uncertainty were rejected. The results of two Monte Carlo runs are shown in Figure 14. Both runs were constrained to calculate values from models having the same range of possible resistivity. In one case, however, models were accepted that generated values falling within the envelope of data from the Tertiary basalt province; in the other case, values had to fall within the envelope of the data from the neovolcanic zone. These models were also constrained to have a constant value of 600 ohm-m for the uppermost 750 m (as required by surface active resistivity measurements). Similarly, below 100 km the value of resistivity was constrained to be 40 ohm-m. The resistivity in each of the three layers at intermediate depths was allowed to vary over the permissible ranges indicated. The only layer that shows a significant difference in the acceptable values of resistivity is that extending from 6 to 12 km in depth. Since this is approximately the base of the crust (Palmason and Saemundsson, 1974), we conclude from this study that the resistivity at the base of the crust is probably within the range of 5 to 10 ohm-m beneath the neovolcanic zone, whereas the resistivity beneath the Tertiary province is probably within the range of 15 to 30 ohm-m.

What causes a systematic lowering of the bulk resistivity of the deep crust beneath the neovolcanic zone? Three possibilities exist:

1. We could be seeing the effect of hydrothermal pore fluids. This seems improbable because the high P-wave velocity (6.5 km/sec) suggests a material having relatively low porosity. Moreover, the lowest resistivities that Hermance (1973) was able to synthesize using a highly idealized version of an electrolytic pore fluid model were above 12 ohm-m.

2. Temperatures may be high enough that electrical conduction in the solid rock lowers the resistivity to 5 ohm-m or so. Again, the results of Hermance (1973) suggest that with a geothermal gradient of 100°C/km one encounters resistivities higher than 100 ohm-m at depths less than

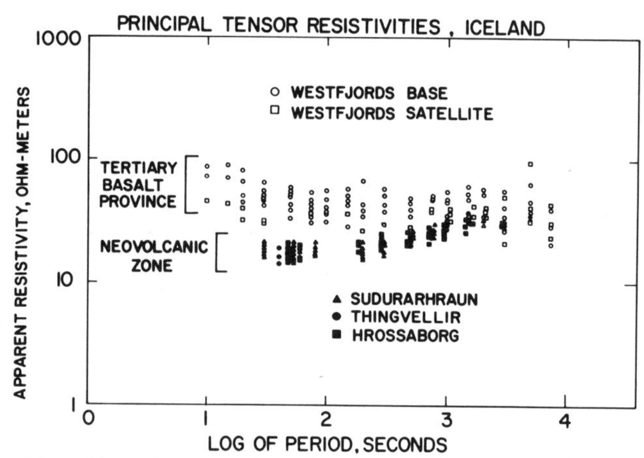

Fig. 13. A comparison between apparent resistivity data as a function of period from two sites in the Tertiary basalt province (Westfjords) and three sites in the currently active neovolcanic zone. These are principal resistivity values for the component which is felt to be least disturbed by lateral inhomogeneities (Hermance et al., 1976).

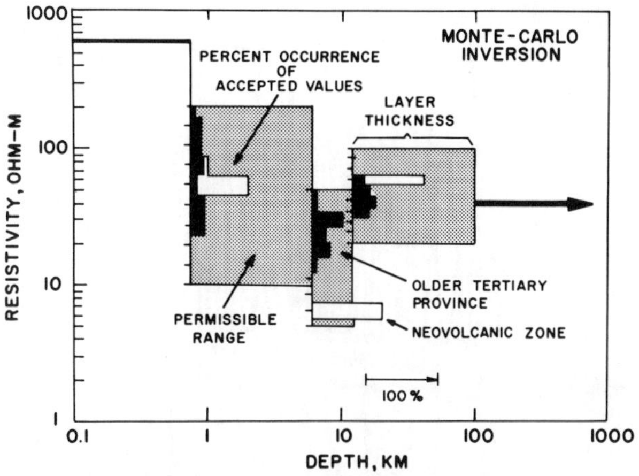

Fig. 14. Monte Carlo inversion of the two sets of data in the last figure. The surface layer and deepest layer were constrained to be 600 ohm-m and 40 ohm-m respectively. Theoretical models were calculated for which the resistivity in each layer could assume any of the values indicated by the tick-marks along the left-hand margin of the permissible range for each layer. The histograms show the percent occurrence of layer resistivities which generate theoretical values within the acceptable range of the observed data.

10 km. These values are too high to be compatible with the present results.

3. Perhaps there is a small fraction of basalt melt present in the crust on a regional scale beneath the neovolcanic zone. This would not be unlikely, considering the extensive current volcanism there. Palmason (1973), extending the earlier work of Bodvarsson and Walker (1964), concluded that a kinematic model with continuous replenishment of source material along a crustal spreading zone explained many of the geologic features in the area as well as the heat-flow observations across northern Iceland. Since crustal spreading is likely to occur over a relatively broad region, it is equally likely that the injection and accumulation of magma also occurs over a broad region. The present precision of seismic data seems able to accommodate such a possibility.

Figure 15, a schematic diagram from Thayer et al. (1980), shows the basic elements of the magnetotelluric interpretation within the framework of other investigations in Iceland. Seismic layers 1 and 2, to the left of the figure, represent the transitional zone from uncompacted pyroclastics at the surface to relatively competent basaltic material at a depth of 4 km (Palmason and Saemundsson, 1974). Layer 3 represents the lower crust (V_p = 6.5 km/sec), with the mantle ($V_p \sim$ 7.2 km/sec) at a depth of 10 km. Zones I through IV, on the right-hand side of the figure, are based on the regional electromagnetic studies described above. Zone I is a regional hydrothermal zone associated with seismic layers 1, 2 and perhaps the top of layer 3. The range in resistivity values for this layer (25-30 ohm-m) and the appropriate seismic velocities are compatible with electrolytic conduction in high-temperature hydrothermal fluids distributed in pores and cracks in relatively permeable crustal material (Hermance, 1973; Bjornsson, 1974). This leads one to the conclusion that the dominant proportion of hydrothermal activity in the Icelandic crust is regionally induced by the enormously high geothermal gradients associated with the neovolcanic zones (100-165°C; e.g. Palmason, 1973).

Zone II, beneath the zone of regional hydrothermal activity, is associated with seismic layer 3 (V_p = 6.5 km/sec). This zone seems to represent an intensely intruded region (perhaps consisting of up to 100% dikes), as proposed by Walker (1975) on geologic grounds and by Palmason (1973) on the basis of kinematic heat-flow models. Since the electrical resistivity ($\rho \geq$ 30 ohm-m) at this depth can be explained by less than a 2-5% concentration of melt (Waff, 1974; Shankland and Waff, 1977; Hermance, 1979), this implies that, although the zone could have been created completely by dike intrusives in response to crustal spreading, only a few percent of the intrustives are at any time molten.

Zone III is perhaps the most significant feature of the magnetotelluric interpretation: its manifestation is a conducting layer ($\rho <$ 10 ohm-m; $\Delta T <$ 4 km) which appears to be associated with the crust-mantle boundary in this area (depth $\sim$ 10-15 km). This layer is interpreted as a zone of magma accumulation (Hermance et al., 1976; Beblo and Bjornsson, 1978, 1980). The concentration of melt may be as low as 7.5-15% if distributed over a vertical thickness of 4 km, or it may be as high as 100% if distributed over a thickness of only 200 m (the resolution of the magnetotelluric method is not sufficient to discriminate between these alternatives). Moreover, we cannot determine from magnetotelluric measurements alone whether this zone consists of a single coherent lens or of a number of discrete tabular-like bodies of either microscale or macroscale proportions. Nevertheless, we can say with some confidence that over a depth interval of 4 km (at a depth of 10 km), there is the vertically integrated equivalent of at least 200 m of molten magma, and that, because this magma zone appears beneath a number of magnetotelluric sites over a large area, it must be a steady-state feature of the neovolcanic zone. The presence of this zone of magma has important implications for kinematic models depicting crustal spreading in Iceland (e.g. Palmason, 1973; Daignieres et al., 1975). Thayer et al. (1980) suggest that crustal underplating, as envisaged by Lachenbruch and Sass (1978) for the Basin and Range province in the western United States, may be playing an important role in crustal development in Iceland.

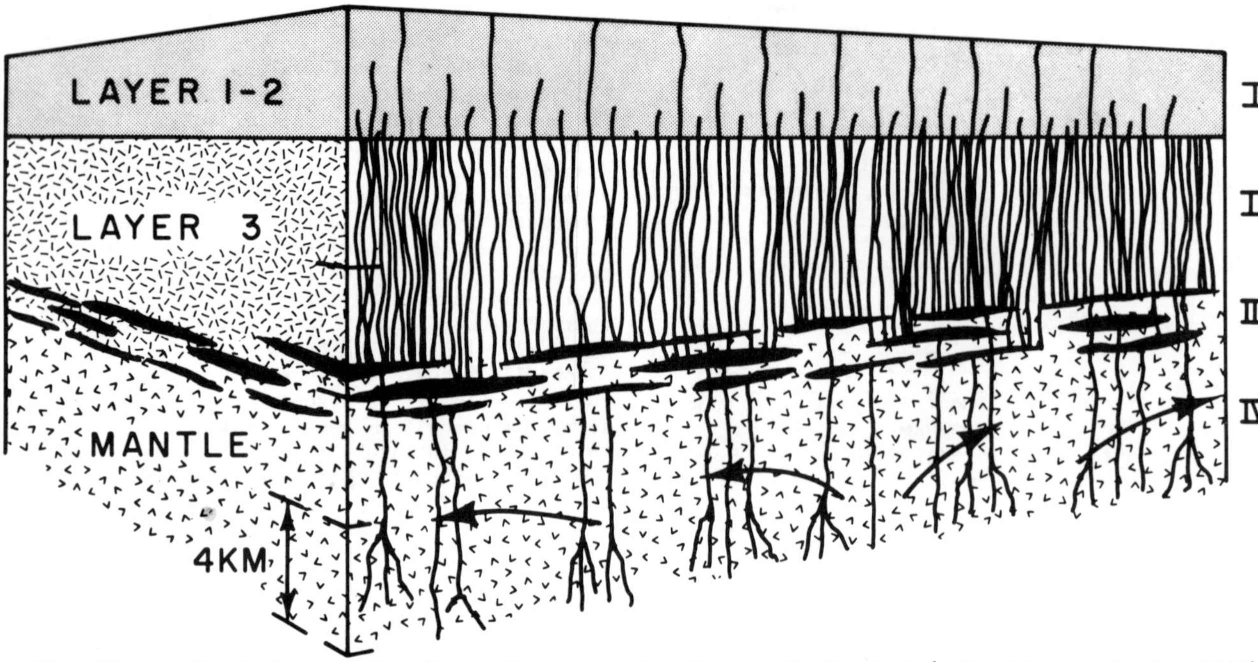

Fig. 15. A physical model for the northern neovolcanic zone in Iceland (after Thayer et al., 1980). The layers indicated on the left side of the figure are based on the seismic studies of Palmason (1971). On the right side of the figure, the Roman numerals denote zones which have distinctive electrical properties. Zone I is a permeable surface layer in which hydrothermal circulation is quite pronounced. Zone II has at most a few percent partial melt present. Zone III represents a conducting layer at the base of the crust which is thought to be a zone of magma accumulation and crustal underplating. Zone IV, the upper mantle, has only a few percent partial melt but seems to be in a state of mass or convective transport.

Underplating involves the accretion of a zone of molten magma to the base of the crust which, as it cools, leads to crustal thickening. Although deserving of a great deal more attention than the process has received up to the present time, one can nevertheless estimate the significance of such a process if it were operative. Based on the magnetotelluric interpretation described above, the crustal basal zone of magma accumulation may have a thickness on the order of 4 km. This zone is not totally molten at any one time, of course, because of transient cooling effects; in fact, the magnetotelluric interpretation suggests an effective thickness for the molten fraction of 200 m over a depth of approximately 5 km. Hermance et al. (1976) and Thayer et al. (1980) proposed, however, that melt is being more or less continuously supplied to this layer, so that an equivalent fraction of melt (.2 km/ 5 km ∿ 4%) is a steady-state feature of this zone. If one assumes that a total crustal spreading velocity of 1 cm/year is achieved over a lateral distance of 50 km in the northern neovolcanic zone (cf. Palmason, 1973; or Palmason and Saemundsson, 1974), one can invoke the mass-flow balance relation of Lachenbruch and Sass (1978) to argue that a two-dimensional steady-state flow of molten material representing a total flux of 4 km · 1 cm/year must be occurring over a lateral distance of 50 km. If this material simply solidifies at constant temperature, it releases a latent heat of fusion of approximately 100 cal/gm or 300 cal/cm^3. The average heat released from this underplating process, q_u, which is contributed to the regional surface heat flux, can be expressed by

$$q_u = \frac{L \cdot v_d \cdot T \cdot \ell}{W \cdot \ell} \tag{1}$$

where L is the latent heat of fusion (300 cal/ cm^3), v_d is the crustal drift velocity (1 cm/ year), T is the thickness of the underplating zone (4.10^5 cm), ℓ is a characteristic dimension along the strike of the spreading zone (a term which cancels), and W is the width of the active zone of extension ($\sim 5 \times 10^6$ cm). For the values of the parameters indicated above,

$$q_u \sim 0.8 \text{ hfu} \tag{2}$$

where 1 hfu = 10^{-6} cal/cm^2/sec. Clearly this component of the heat flux is strongly dependent on the thickness of the underplating zone, T, as well as on the total width of the region over which crustal extension is distributed, W.

If, for example, W is less than the figure assumed above, and has a value between 15 and 25 km, say, the heat flow contribution will be correspondingly greater: on the order of 2 to 3 hfu.

In short, it appears that crustal underplating, which is simply the accretion of material in the molten phase to the base of the crust along a zone of crustal extension, may play an important role in ridge processes. Although hitherto neglected as a significant contributor to the dynamics of accreting plate boundaries, the presence of such an underplating layer beneath Iceland suggests that an analogous feature may also be present beneath the submarine portions of the oceanic ridges.

In fact, Bottinga and Allegre (1976) have reflected on the relatively unsuccessful attempts by seismologists to detect and confirm the presence of high-level magma systems in the crust beneath the mid-ocean ridges. If such systems have the characteristics and dimensions typified by our studies in Iceland, they indeed may very likely represent difficult targets to resolve seismically. It is interesting in this regard to note that one of the features of the asthenospheric-upwelling model of Bottinga and Allegre (1976) is that far more liquid is produced by partial fusion in the upper mantle beneath the ridge crests than is erupted at the surface; perhaps 4 to 5 times as much liquid is formed as is actually erupted. As a consequence of these studies, along with the magnetotelluric interpretation in Iceland, one has independent lines of investigation converging on very similar models for magma genesis in the upper mantle.

The presence of the magma accumulation zone also has important implications for models explaining the emplacement of magma along fissure swarms up to 70 km away from central volcanoes (Sigurdsson and Sparks, 1978). These results suggest the possibility that, rather than flowing horizontally over these extreme distances (70 km), magma may flow vertically (< 10 km) from a pre-existing magma zone.

Zone IV, the upper mantle, is characterized by relatively small thermal gradients (Hermance and Grillot, 1974) indicative of the bulk transfer of material through diapiric-like motion. The present interpretation (Thayer et al., 1980) suggests a higher concentration of melt in the upper few tens of kilometers of the mantle than in the region directly below. Presumably the uppermost mantle is supplying magma to the accumulation zone which is underplating the crust as well as to the dike intrusives which compensate for the actual crustal spreading itself.

Within the resolution of present data, the entire area of Iceland (120,000 km^2) is underlain by mantle material at a depth greater than 10-15 km which is laterally uniform in composition and temperature - at least considerably more uniform than the crust appears to be. Such a result is similar to the results of magnetotelluric measurements in rift areas elsewhere, such as the Rhine graben (Scheelke, 1974).

The Rio Grande Rift

The Rio Grande rift (Figure 16) is characterized by a contiguous series of basin-type structures extending from southern Colorado through central New Mexico into northern Chihuahua, Mexico (Chapin, 1979). Throughout its development, the rift has undergone periods of extensive volcanic activity. Early volcanism was characterized by calc-alkalic andesites, but beginning approximately 5 million years ago and continuing to the present, basaltic lava flows have become more widespread from southern Colorado to northern Mexico (Chapin and Seager, 1975).

A number of geophysical studies suggest that major crustal and mantle processes are contributing to the evolution of the Rio Grande rift (Cordell, 1978). Some of the earliest geophysical studies in this region that dramatized the anomalous character of the rift feature itself, were the geomagnetic variation studies of Schmucker (1964, 1970) in the southern rift, and similar studies across other portions of the rift by Reitzel, Gough and Porath (see the review by Gough, 1974). This work clearly established the presence of an electrical conductivity anomaly associated with the axis of the extensional feature.

The heat flow studies of Reiter et al. (1975), Decker and Smithson (1975), Sass et al. (1976) and Reiter et al. (1978, 1979) reveal a belt of

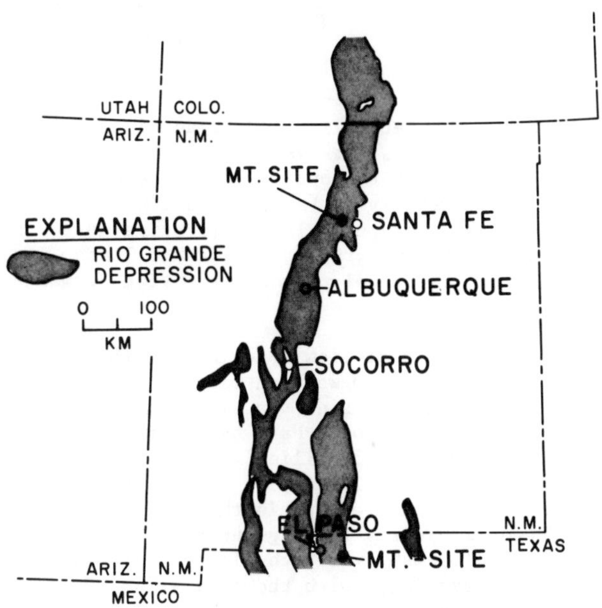

Fig. 16. The Rio Grande rift in New Mexico (after Chapin and Seager, 1975), showing the location of magnetotelluric deep-sounding experiments.

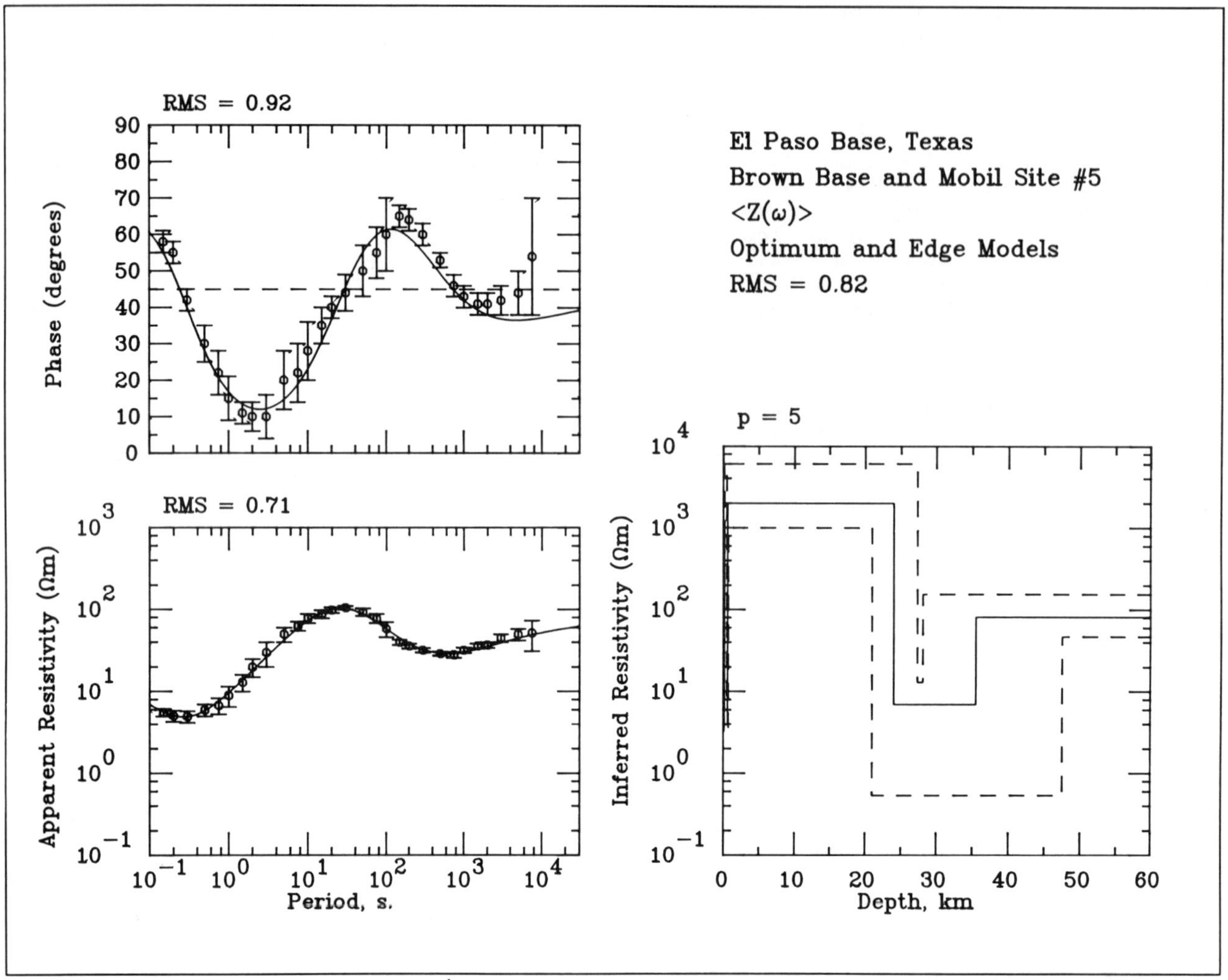

Fig. 17. Magnetotelluric data and interpretation from the southern Rio Grande rift near El Paso, Texas. The apparent resistivity, phase and associated error bars are shown as points along with the response of the optimum model determined by a nonlinear least-squares fit to both phase and apparent resistivity (continuous line). The optimum model (continuous line) and the extreme bounds on other possible models using up to the fourth eigenvalue (dashed lines) are shown in the lower right-hand graph. The dashed lines do not show allowable models, but define the allowable range for the model parameters (after Hermance and Pedersen, 1980).

anomalously high heat flow associated with the rift, particularly its western margin. Calculations by Cook (1975) using transient thermal models suggest that there have been three epochs of magma intrusion; 25 million years, 13.2 million years and 2.5 million years before present. The earliest thermal event, in this model, is associated with early rifting and calc-alkalic andesitic volcanism, while the latest intrusive episode is associated with the initiation of recent basaltic activity.

The regional seismic refraction interpretation of Toppozada and Sanford (1976) proposes a two-layer crust (V_p = 5.8 km/sec, 19 km thick; V_p = 6.5 km/sec, 21 km thick) under which lies a mantle having a lower-than-normal P-wave velocity of 7.9 km/sec. There is a suggestion that the crust thins to approximately 35 km beneath the north-central portion of the rift (Olsen et al., 1979), and that the upper mantle may have a somewhat lower velocity (7.6 km/sec); this model is in essential agreement with the surface-wave dispersion studies of Keller et al. (1979). On a smaller scale, Sanford and his colleagues (Sanford et al., 1973; Sanford et al., 1977; Shuleski et al., 1977; Rinehart et al., 1979) have used reflected P- and S-phases from local micro-earthquakes near Socorro to delineate a liquid-like layer, which they presume to be magma, having a thickness of less than several

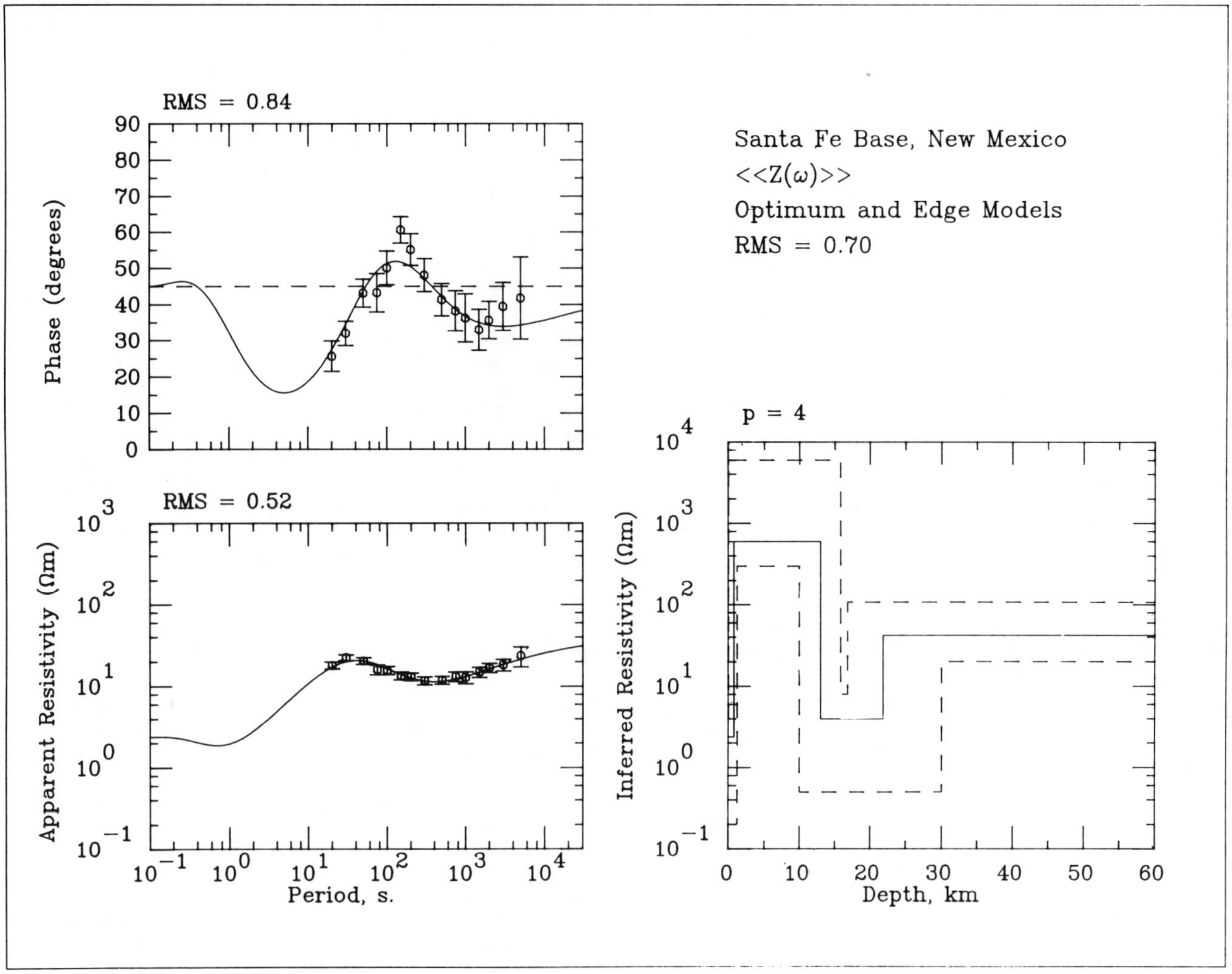

Fig. 18. Magnetotelluric data and interpretation from the north-central Rio Grande rift near Santa Fe, New Mexico (see Fig. 16). For an explanation of the presentation format, see the caption of Fig. 17.

kilometers, at a depth of 19 to 20 km beneath this portion of the rift. This zone has been mapped over a geographical extent of at least 1700 km^2. The presence of a coherent layer at this depth has recently been confirmed by the COCORP deep seismic reflection profiling in the Socorro area (Krumhansl et al., 1978; Brown et al., 1980).

Pedersen and Hermance (1976), using magnetotelluric measurements, suggest it is possible that this structure, or features related to it, may have a much wider extent. They report the presence of a conducting layer beneath El Paso at a depth of 21 to 28 km (Figure 17). Pedersen and Hermance (1978) report a similar layer beneath Santa Fe at a depth of 10 to 17 km (Figure 18). This work opened the possibility that the zone of enhanced electrical conductivity is contiguous with the tabular body previously delineated seismically by Sanford and his colleagues, and probably has a similar generic cause (Hermance and Pedersen, 1980).

Recent seismic refraction work by Olsen et al. (1979), near the site of the magnetotelluric experiment at Santa Fe described above, suggests the presence of a thin zone of anomalously low shear-wave velocity at mid-levels in the crust beneath the north-central rift. They feel this feature is a manifestation of a widespread intracrustal low-rigidity layer which is also generically related to the Socorro magma body.

Although the electrical conductivity itself can be affected by (1) electrolytic conduction in hydrothermal pore fluids, (2) conduction in the bulk silicate material itself, or (3) solid conduction through hydrated phases such as layered silicates, we find the suggestion offered by Sanford and his colleagues (e.g. Rinehart et al.,

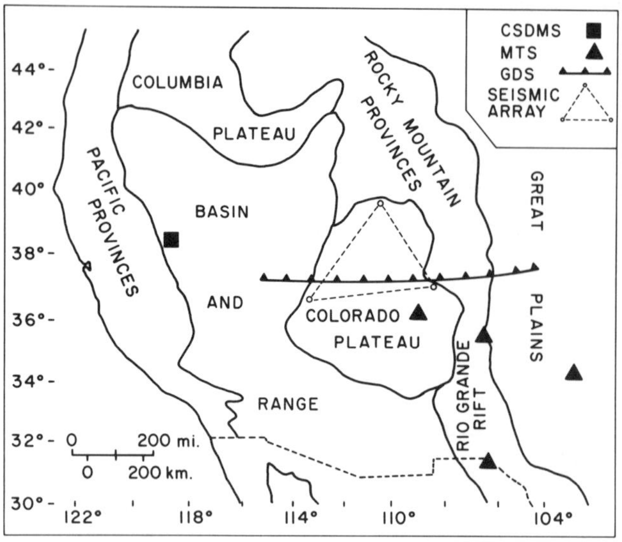

Fig. 19. A simplified tectonic map of the western United States. The sites for the following experiments are shown: the controlled-source deep magnetic sounding experiment (CSDMS) by Lienert and Bennett (1977); the geomagnetic deep-sounding (GDS) profile line of Porath (1971) and Porath and Gough (1971); the seismic array experiment of Bucher and Smith (1971) and the location of magnetotelluric sites (MTS) occupied by Brown University.

1978) for the Socorro structure the most attractive; (4) the intracrustal zone represents a generic class of phenomena associated with the accumulation of basaltic melt derived from sources at greater depth. If this is the case, we may be seeing the effects of a thin zone (1 km) comprised totally of melt, or a thicker zone containing as little as a 15-35% melt fraction.

Deep Structure of the Rio Grande Dome. As is true in many rift provinces, the Rio Grande rift is associated with regional uplift (doming) which is presently active (Cordell, 1978; Reilinger et al., 1979). The actual uplifted region bisected by the rift (Figure 19) extends laterally from the Colorado plateau to the High Plains provinces of western North America (ibid.). Long-path-length seismic surface wave studies (Biswas and Knopoff, 1974) indicate a general homogeneity of the mantle beneath this region (to a depth greater than 100 km). The seismic data presently available suggest that the mantle from the base of the crust to a depth greater than 100 km is grossly similar beneath the Colorado plateau, the Rio Grande rift and the High Plains province immediately adjacent. This region is characterized by low compressional and shear wave velocities, and apparently low densities as well.

It is instructive, therefore, to compare long-period magnetotelluric data from the Colorado plateau (Pedersen and Hermance, 1980) with similar measurements in the rift itself. The data from the Colorado plateau near Farmington, New Mexico (Figure 20) characterize the province as consisting of conductive surficial sediments a few kilometers thick, underlain by a more resistive crystalline crust down to about 28 kilometers. Beneath this zone the average resistivity is 13 ohm-m down to at least 130 km. On the other hand, seismic evidence indicates a crustal thickness of approximately 40 km for the interior of the plateau. The electrical data therefore require the transition from resistive crust to more conductive mantle material to occur in a lower crustal layer approximately 12 km thick.

Pedersen and Hermance (1980) point out that an upper limit on possible temperatures in the upper mantle is provided by comparing the range of possible upper mantle resistivities (10-15 ohm-m) to the laboratory measurements of Duba et al. (1974) on solid olivine; temperatures in excess of 1700°C would be required. Allowing for the uncertain effect due to impurities and more conductive grain boundaries, temperatures in excess of 1500°C are needed to yield resistivities less than 15 ohm-m (Shankland and Waff, 1977, Figure 2). Since reasonable temperatures in the upper mantle at the depths considered here are expected to be from 1000°C to 1400°C (Lachenbruch and Sass, 1977), it seems plausible that the bulk resistivity of the mantle may be influenced by the presence of some degree of partial melt (Waff, 1974). Shankland and Waff (1977) cite both field and experimental observations to support the hypothesis that such a melt fraction is thoroughly interconnected. Recently, petrologic and thermodynamic arguments have been advanced by Waff and Bulau (1979) to support a melt configuration which is thoroughly connected along grain edges, provided that the melt is in chemical equilibrium with the heat rock.

Using this model for the mantle beneath the Colorado plateau and the calculations of Shankland and Waff (1977), Pedersen and Hermance (1980) argue that a resistivity in the range 10 to 15 ohm-m would imply a mantle temperature of approximately 1200°C at 50 kilometers depth, assuming a water content of 0.1% for the upper mantle. At 100 kilometers depth the temperature would be about 1300°C. The partial melt fraction would vary from 8% to 3% over the same depth interval. From this analysis, a mantle geotherm for the Colorado plateau may be calculated yielding a gradient of 2.5°C/km (with an uncertainty of about the same magnitude).

An alternative estimate of the temperature gradient in the upper mantle was made by Pedersen and Hermance (1980) following a procedure originally outlined by Hermance and Grillot (1974), who argued that relative values of temperature are more precisely determined from deep electrical studies than are absolute values of temperature. Hermance and Grillot (1974) showed that the geothermal gradient, G_T, is related to the logarithmic gradient of resistivity through the expression

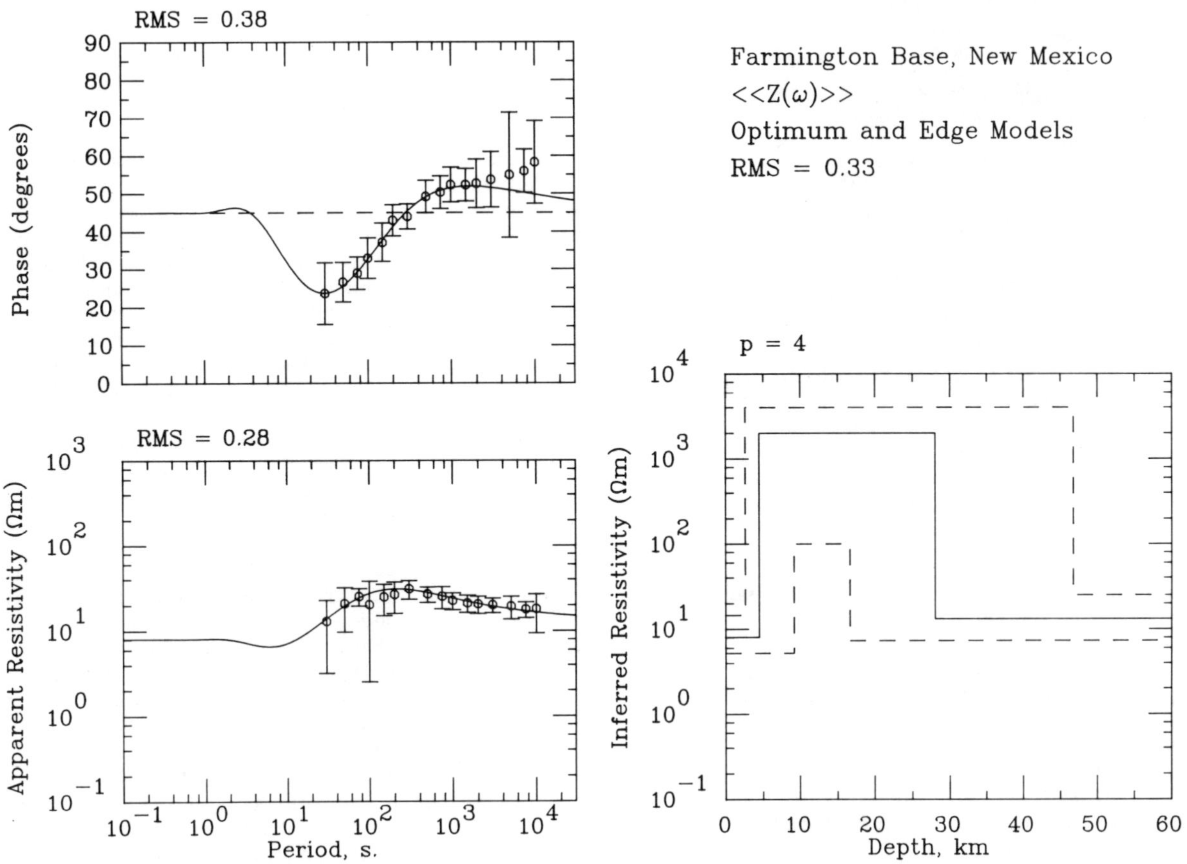

Fig. 20. Magnetotelluric data and interpretation from the Colorado plateau near Farmington, New Mexico (see Fig. 19). For an explanation of the presentation format see the caption of Fig. 17 (after Pedersen and Hermance, 1980).

$$G_T = (kT_1^2/E\Delta z)\ln(\rho_1/\rho_2) \qquad (3)$$

where k is the Maxwell-Boltzmann constant, T_1 is the temperature (°K) at the top of a vertical section, Δz is the thickness of the section, ρ_1 and ρ_2 are the resistivities at the top and bottom of the section, respectively, and E is the activation energy (on the order of 1-2 eV).

By considering the extreme range of possible resistivity gradients in the upper mantle, Pedersen and Hermance (1980) determined that the maximum variation in resistivity may be a factor of 3 over a depth range of 120 kilometers. Assuming a temperature of 1000°C at 30 kilometers depth and applying the above analysis, an estimate of 1.3°C/km was determined for the maximum geothermal gradient beneath the Colorado plateau. This value is consistent with the estimate obtained using the approach of Shankland and Waff (1977) described above. These low geothermal gradients suggest that vertical heat transfer beneath the Colorado plateau and the Rio Grande dome is dominated, as in Iceland, and presumably in many other rift areas, by a convective material transport mechanism.

Additional geophysical evidence in the form of uniform surface heat flow values of 1.5 to 1.8 hfu in the interior of the plateau (Reiter et al., 1978), the absence of long-wavelength aeromagnetic anomalies, and the need for isostatic compensation in the Colorado plateau to be largely in a low-density upper mantle (Thompson and Zoback, 1979) all supports the hypothesis of moderately high upper mantle temperature. The low (when compared to the stable continental interior) upper mantle compressional and shear velocities of 7.8 and 4.25 km/sec, respectively (Bucher and Smith, 1971; Keller et al., 1978),

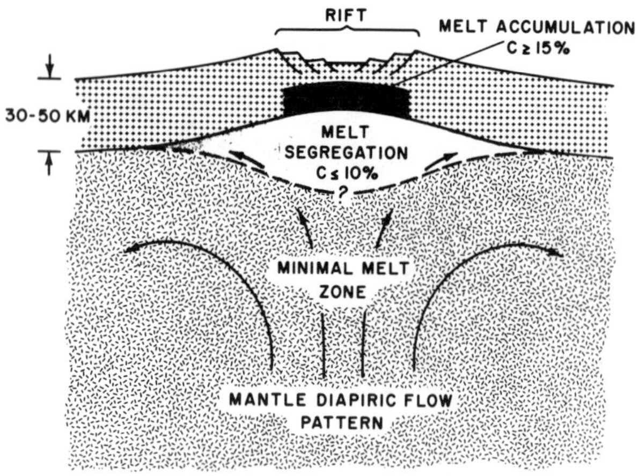

Fig. 21. A conceptual model for physical processes in the upper mantle and deep crust beneath active rift zones. The three zones indicated (the melt accumulation zone, the melt segregation zone, and the minimal melt zone exhibiting diapiric flow) each have distinctive geophysical/electromagnetic properties.

likewise indicate moderately high upper mantle temperature and are thus consistent with the magnetotelluric data. A comparison with electrical data available for the surrounding tectonic provinces indicates that the upper mantle beneath the Colorado plateau has resistivities at least as low as those for the Basin and Range, Rio Grande rift, and Great Plains, and may in fact be substantially more conducting than any of these surrounding tectonic provinces (Pedersen and Hermance, 1980).

Summary and Conclusions

A Conceptual Model for Deep Crust and Upper Mantle Processes

Many of the major rift zones of the world are very similar in morphotectonic appearance and development (Illies, 1970). Although significant differences exist in the character and degree of regional doming, rifting and magma genesis (Logatchev, 1978), in the present discussion we seek a basis for a hypothetical model that, while not satisfying all of the features of any specific rift zone, accounts for many of the features common to most of the rift zones studied.

Such a conceptual model is portrayed in Figure 21. Recent magnetotelluric investigations in Iceland and the Rio Grande rift (including the eastern Colorado plateau) support the idea of Illies (1970) that beneath rift systems (both mid-oceanic and intracontinental), ascending masses of material from the mantle are intimately coupled with regional doming and the morphotectonic development of rift features. This phenomenon is seen electromagnetically as essentially a constant value of electrical resistivity over a very large range of depths (from 30 to over 100 km), from which is inferred a small change in temperature over the same depth interval. Since crustal temperature gradients are on the order of 40 to 100°C/km in these regions, and we infer upper mantle gradients of less than a few °C/km, heat transfer mechanisms must be extremely efficient (i.e. by mass transport) in the upper mantle.

At the top of the mantle diapir, lowered values of the bulk resistivity indicate in some areas (e.g. Iceland) a slight enhancement of a partial melt fraction which may be associated with the low-velocity pillow beneath many rifts. We should keep in mind the fact that the seismic structure beneath some rifts appears to be systematically different from that beneath others. For example, the Baikal rift and the Rhine graben are associated with a low-velocity pillow at the base of the crust having a thickness of only 15-25 km. On the other hand, Iceland, the East African rift, and the Basin and Range province of western North America (of which the Rio Grande rift seems to be a part) all appear to have low-velocity material extending to very great depths (> 100 km).

The present interpretation of magnetotelluric and geomagnetic deep-sounding data from the Baikal rift and the Rhine graben seems also to favor a conductivity anomaly having a limited thickness in the crust and/or upper mantle. In contrast, magnetotelluric data from Iceland and from the vicinity of the Rio Grande rift indicate anomalous mantle conductivities over a very large depth interval (> 100 km). Electromagnetic studies in the East African rift are inconclusive on this point; although low resistivities are indicated for the full range of periods over which measurements were made, the effective depth of resolution is less than 35 km at the longest periods for which reliable data are available. Because of the close similarity between the electrical structure in the crust beneath Iceland and that beneath the most active portions of the East African rift, as well as because of the similarity in the seismic character of the upper mantle beneath these regions which extends to great depth (> 100 km), one might tentatively suggest that the character of the upper mantle beneath the East African rift is not unlike that beneath the Rio Grande rift or that beneath Iceland. Clearly this places important constraints on the possible distribution of partial melt in the upper mantle and, more importantly, on the physical character of possible diapiric motions responsible for regional tectonics.

On the other hand, areas such as the Baikal rift and the Rhine graben seem to have minimal melt in the mantle beneath the zone of melt segregation (Figure 2), whereas there seems to be clear evidence in the electrical data for a significant fraction of melt in the vicinity of the low-velocity pillow in these regions.

It does not seem plausible that melt, having segregated and formed a concentration of greater than a few percent of the total volume, can be dynamically stable (Walker et al., 1978). Therefore, melt in the pillow at the base of the crust probably migrates either laterally as part of the diapiric flow pattern, or vertically in response to a combination of buoyancy and lithostatic pressures, and accumulates at a preferred level in the crust which is presumably hydrostatically controlled.

Intracrustal low-velocity layers appear to be typical of continental rifts (Puzirev et al., 1978). Apart from the Baikal, the existence of such layers is noted under the Rhine graben (Mueller, 1970) and beneath the Rio Grande rift (Olsen et al., 1979). Although a low-velocity zone does not appear to have been detected at intracrustal depths beneath the East African rift, this may be due to a complexity of structure which masks the subtle effects of such a layer, rather than to a fundamental difference in the physical character of the rift itself (Griffiths, 1972; Long et al., 1973). Magnetotelluric interpretations place this zone of magma accumulation at the base of the crust in Iceland, whereas, from a combination of seismic and deep electrical studies, it seems to be at intracrustal levels beneath the Rio Grande rift, the Rhine graben and the Baikal rift.

The nature of the migratory zone (Figure 21) separating the region of melt segregation in the mantle from the zone of melt accumulation at the base of, or within, the crust is obscure at the present time, and its character undoubtedly varies from region to region. In some areas, such as portions of the Baikal rift and the north-central Rio Grande rift, the zone of melt accumulation seems to be a discrete layer at intermediate levels in the crust which can be clearly distinguished from its anomalous roots in the mantle. In other regions, such as the southern Rio Grande rift, Iceland and other portions of the Baikal rift, the zone of melt accumulation may be more intimately coupled to its mantle origins. The level at which magma accumulates must depend to a marked degree on its depth of origin as well as on the overall character of temperature and density gradients in the lithosphere and asthenosphere. Moreover, the frequency and mass flux with which the magma zone is replenished and sustained determines the extent to which the thermal regime of the lower crust is perturbed from a conductive environment to a convective environment (Lachenbruch and Sass, 1977).

Relationship to Extrusive Silicic Volcanism

It is well known that centers of silicic volcanism exist along some of the major rift zones such as the East Africa rift, the Rio Grande rift and Iceland. It has also been suggested by a number of workers that these silicic centers have developed through partial melting (anatexis) of crustal material through interaction with more primitive basaltic magma at depth. It is useful to speculate on the relationship of the conceptual model in Figure 21 to general aspects of silicic volcanism, as Hermance and Pedersen (1980) have done for the Rio Grande rift. Such speculation does not appear to be particularly germane to the Baikal rift (Logatchev and Florensov, 1978), but might be of interest for other of the world's rift areas.

To begin, we must recognize that many silicic centers on continents have lifetimes on the order of 10 million years (Doell et al., 1968; Smith and Bailey, 1968). Volcanism is of course not reversible, but proceeds in a specific volcanic center through a definite series of events, some of which seem to reflect similar stages among different volcanoes (Williams, 1941; Smith and Bailey, 1968). For example, the explosive eruption of a tuff phase occurs late in a volcano's history, and volcanism continues episodically (at intervals on the order of $4\text{-}6 \times 10^4$ years) within the caldera in the form of resurgent doming and rhyolite flows (Doell et al., 1968).

In our opinion, there may be a direct relationship between the high-level emplacement of basaltic magma at mid-crustal levels and surface volcanism. First, we suggest the impossibility of sustaining a single coherent silicic magma chamber for a period of 10 million years. Second, we suggest that the episodic volcanism so characteristic of the latter phases of the evolution of a volcanic center provides a clue to an important element in the evolution of these systems. In discussing Figure 21, we suggested that diapiric-type motion in the upper mantle leads

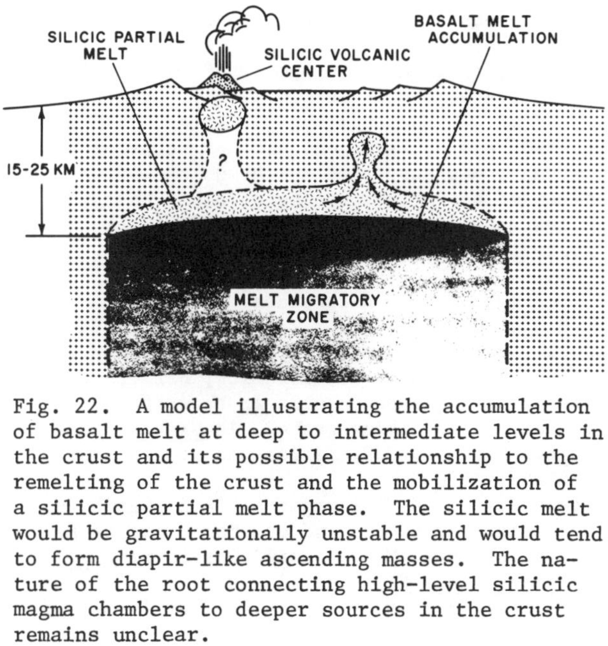

Fig. 22. A model illustrating the accumulation of basalt melt at deep to intermediate levels in the crust and its possible relationship to the remelting of the crust and the mobilization of a silicic partial melt phase. The silicic melt would be gravitationally unstable and would tend to form diapir-like ascending masses. The nature of the root connecting high-level silicic magma chambers to deeper sources in the crust remains unclear.

to partial fusion of mantle materials, resulting in the segregation and upward migration of a basaltic melt due to a combination of buoyancy and lithostatic pressures. The melt in turn reaches a preferred level in the crust (which is presumably hydraulically controlled) and a zone of magma rapidly accumulates, followed by a long cooling history. An appropriate time constant for a molten lens-like structure 2 km thick to chill below its solidus appears to be 10^4 to 5×10^4 years (Jaeger, 1964). In Figure 22, we illustrate that the emplacement of a molten (1200°C) basaltic lens at depth will perturb the surrounding country rock, which may undergo local secondary remelting (to temperatures apparently as low as 750-800°C (Bailey et al., 1976)). Assuming for the sake of illustration that the latent heat of fusion for basalt and granite are approximately equal (i.e. 100 cal/gm), then up to 400 cal/gm is available in the basalt intrusion to locally remelt the crust. In other words, in this grossly oversimplified model 1 unit of basalt can mobilize up to 4 units of silicic material.

The partial melt will segregate as a rhyolite magma (Figure 22) and, because of its lower density, will tend to rise as a silicic diapir, leading to extrusive volcanism at the surface. The episodic replenishment of the basaltic magma layer at depth is associated with the episodic silicic volcanism at the surface. Thus this model, although highly speculative, provides a basis for relating geophysically delineated structures in the deep earth, as conceptualized in Figure 21, to the genesis of major centers of silicic volcanic activity (Figure 22).

Acknowledgements. I'd like to acknowledge the contributions, while they were at Brown University, of my colleagues L.R. Grillot, R.E. Thayer, and J. Pedersen to our field programs in Iceland and the western United States. Of special importance during the development of ideas presented here were discussions with C. Chapin, A. Sanford, A. Lachenbruch, R. Bailey, A. Bjornsson and G. Palmason. The preparation of this paper was supported by the United States Department of Energy, Office of Basic Energy Studies under Contract DE-AC02-79ER10401 to Brown University.

References

Ansorge, J., D. Einter, K. Fuchs, J.P. Lauer, St. Mueller and E. Peterschmitt, Structure of the crust and upper mantle in the rift system around the Rhine graben, in Graben Problems, edited by J.H. Illies and St. Mueller, 184-197, 1970.

Bailey, R.A., G.B. Dalrymple and M.A. Lamphere, Volcanism, structure and geochronology of Long Valley caldera, Mono County, California, J. Geophys. Res., 81, 725-744, 1976.

Baker, B.H., P.A. Mohr, and L.A.J. Williams, Geology of the Eastern Rift system of Africa, Geol. Soc. Am., Spec. Paper 136, 1972.

Banks, R.J., and P. Ottey, Geomagnetic deep sounding in and around the Kenya Rift Valley, Geophys. J. Roy. Astron. Soc., 36, 321-335, 1974.

Barberi, F., H. Tazieff, and J. Varet, Volcanism in the Afar depression: its tectonic and magmatic significance, Tectonophys., 15, 19-29, 1972.

Båth, M., Crustal structure of Iceland, J. Geophys. Res., 65, 1793-1807, 1960.

Beblo, M., and A. Bjornsson, Magnetotelluric investigation of the lower crust and upper mantle beneath Iceland, J. Geophys., 45, 1-16, 1978.

Beblo, M., and A. Bjornsson, A model of electrical resistivity beneath NE-Iceland: correlation with temperature, J. Geophys., 47, 184-190, 1980.

Beloussov, V.V., and Ye. Ye. Milanovskiy, Tectonics and tectonic position of Iceland, Int. Geol. Rev., 18, 1373-1375, 1976.

Berdichevsky, M.N., L.L. Vanyan, V.A. Kuznetsov, V.T. Levadny, M.M. Mandelbaum, G.P. Nechaeva, B.A. Okulessky, P.P. Shilovsky and I.P. Shpak, Geoelectrical model of the Baikal region, Phys. Earth Planet. Int., 22, 1-11, 1980.

Berktold, A., V. Haak, and G. Pilger, eds., Magnetotelluric measurements in the Afar area, Inter-Union Commission on Geodynamics, 14, 66-79, 1975.

Biswas, N.N., and L. Knopoff, The structure of the upper mantle under the United States from the dispersion of Rayleigh waves, Geophys. J. Roy. Astron. Soc., 36, 515-539, 1974.

Bjornsson, S., and P. Einarsson, Seismicity of Iceland, in Geodynamics of Iceland and the North Atlantic area, edited by L. Kristjansson, 225-239, D. Reidel, Dodrecht-Holland, 1974.

Bjornsson, A., Electrical resistivity of layer 3 in the Icelandic crust, Greinar V, 1-22, Visindafelag Islendinga (Science Society of Iceland), 1976.

Bjornsson, A., K. Saemundsson, P. Einarsson, B. Tryggvason, K. Gronwold, Current rifting episode in north Iceland, Nature, 266, 318-323, 1977.

Bjornsson, A., G. Johnsen, S. Sigurdsson, G. Thorbergsson, and E. Tryggvason, Rifting of the plate boundary in north Iceland, 1975-1978, J. Geophys. Res., 84, 3029-3038, 1979.

Bodvarsson, G., and G.P.L. Walker, Crustal drift in Iceland, Geophys. J. Roy. Astron. Soc., 8, 285-306, 1964.

Bottinga, Y. and C. Allegre, Geophysical, petrological and geochemical models of the oceanic lithosphere, Tectonophys., 32, 9-59, 1976.

Brown, L.D., C.C. Chapin, A.R. Sanford, S. Kaufman and J. Oliver, Deep structure of the Rio Grande rift from seismic reflection profiling, J. Geophys. Res., 85, 4773-4800, 1980.

Bucher, R.L., and R.B. Smith, Crustal structure of the eastern Basin and Range province and northern Colorado plateau from phase velocities of Rayleigh waves, in The Structure and Properties of the Earth's Crust, Geophys. Mon. 14, edited by J.G. Heacock, American Geophysical Union, Washington, D.C., 59-70, 1971.

Chapin, C.E., and W.R. Seager, Evolution of the Rio Grande rift in the Socorro and Las Cruces areas, New Mexico Geol. Soc. Guidebook, 26th Field Conf., Las Cruces County, 297-322, 1975.

Chapin, C.E., Evolution of the Rio Grande rift: a summary, in Rio Grande Rift: Tectonics and Magmatism, edited by R.E. Riecker, American Geophysical Union, Washington, D.C., 1-6, 1979.

Cook, F.A., Transient heat flow models and gravity models in the Rio Grande rift of southern New Mexico, M.Sc. Thesis, Univ. of Wyoming, 1975.

Cordell, L., Regional geophysical setting of the Rio Grande rift, Geol. Soc. Amer. Bull., 89, 1073-1090, 1978.

Daignieres, M., V. Courtillot, R. Bayer, and P. Tapponnier, A model for the evolution of the axial zone of mid-ocean ridges as suggested by Icelandic tectonics, Earth Planet. Sci. Lett., 26, 222-232, 1975.

de Beer, J.H., D.I. Gough, and J.S.V. van Zijl, An electrical conductivity anomaly and rifting in southern Africa, Nature, 255, 678-680, 1975.

Decker , E.R., and S.B. Smithson, Heat flow and gravity interpretation across the Rio Grande rift in southern New Mexico and west Texas, J. Geophys. Res., 80, 2542-2552, 1975.

Doell, R.B., G.B. Dalrymple, R.L. Smith and R.A. Bailey, Paleomagnetism, potassium-argon ages, and geology of rhyolites and associated rocks of the Valles caldera, New Mexico, in Studies in Volcanology, Geol. Soc. Amer. Mem. 116, edited by R.R. Coats, R.L. Hay, and C.H. Anderson, 211-248, Geol. Soc. Amer., Boulder, Colorado, 1968.

Duba, A., H.C. Heard, and R.N. Schock, Electrical conductivity of olivine at high pressure and under controlled oxygen fugacity, J. Geophys. Res., 79, 1667-1673, 1974.

Duchkov, A.D., S.A. Kazantsev, V.A. Bolubev, S.V. Lysak, and E.S. Khaikovsky, Heat flow under the Baikal Lake, Geol. i Geofiz., 4, 112-121, 1976.

Edel, J.B., K. Fuchs, C. Gelbke, and C. Prodehl, Deep structure of the southern Rhinegraben area from seismic refraction investigations, J. Geophys., 41, 333-356, 1975.

Einarsson, P., S-wave shadows in the Krafla caldera in NE-Iceland: evidence for a magma chamber in the crust, Bull. Volcan., 41-3, 1-9, 1978.

Einarsson, T., A survey of gravity in Iceland, Soc. Sci. Islandica, 30, 1954.

Fairhead, J.D., The structure of the lithosphere beneath the Eastern Rift, East Africa, deduced from gravity studies, Tectonophys., 30, 269-298, 1976.

Francis, T.J.G., Upper mantle structure along the axis of the Mid-Atlantic Ridge near Iceland, Geophys. J. Roy. Astron. Soc., 17, 507-520, 1969.

Fuchs, K., Geophysical contributions to taphrogenesis, in Approaches to Taphrogenesis, edited by J.H. Illies and K. Fuchs, 420-432, Schweizenbart-Stuttgart, 1974.

Garland, G.D., Correlation between electrical conductivity and other geophysical parameters, Phys. Earth Planet. Int., 10, 220-230, 1975.

Gornostayev, V.P., V.I. Mikhalevsky, and V.I. Pospeev, Deep magnetotelluric surveys in the southern part of the Siberian platform and in the Baikal rift zone, Geol. i Geofiz., 4, 111-118, 1970.

Gornostayev, V.P., A deep geoelectric model of the Baikal region, Geol. i Geofiz., 6, 98-102, 1972.

Gough, D.I., Electrical conductivity under western North America in relation to heat flow, seismology, and structure, J. Geomag. Geoelec., 26, 105-124, 1974.

Griffiths, D.H., Some comments on the results of a seismic refraction experiment in the Kenya Rift, Tectonophys., 15, 151-156, 1972.

Grillot, L.R., Regional electrical structure beneath Iceland as determined from magnetotelluric data, Ph.D. Thesis, Dept. of Geological Sciences, Brown University, 1973.

Haak, V., K.P. Jengpiel, A. Berktold, and S. Grienwald, The telluric method: results of the measurements along a profile across the Rhine graben between Stuttgart and Kaiserslautern, in Graben Problems, edited by J.H. Illies and St. Mueller, 235-242, 1970.

Haak, V., and G. Reitmayr, The distribution of electrical resistivity in the Rhine graben area as determined by telluric and magnetotelluric methods, in Approaches to Taphrogenesis, edited by J.H. Illies and K. Fuchs, 366-369, Schweizenbart-Stuttgart, 1974.

Hermance, J.F. and Garland, G.D., Deep electrical structure under Iceland, J. Geophys. Res., 73, 3797-3800, 1968.

Hermance, J.F., An electrical model for the sub-Icelandic crust, Geophysics, 38, 3-13, 1973.

Hermance, J.F., The electrical conductivity of materials containing partial melt: a simple model from Archie's Law, Geophys. Res. Lett., 6, 613-616, 1979.

Hermance, J.F. and L.R. Grillot, Correlation of magnetotelluric, seismic and temperature data from southwest Iceland, J. Geophys. Res., 75, 6582-6591, 1970.

Hermance, J.F. and L.R. Grillot, Constraints on temperature beneath Iceland from magnetotelluric data, Phys. Earth Planet. Int., 8, 1-12, 1974.

Hermance, J.F., and J. Pedersen, The deep structure of the Rio Grande rift: a magnetotelluric study, J. Geophys. Res., 85, 3899-3912, 1980.

Hermance, J.F., R.E. Thayer, and A. Bjornsson, The telluric-magnetotelluric method in the regional assessment of geothermal potential, in Proc. Second U.N. Symposium on the Development and Use of Geothermal Resources, Vol. 2, 1037-1048, Stock No. UGO-000-0000S-1, Supt. Documents, U.S. Government Printing Office, Washington, D.C., 20402; 1976.

Hutton, R., Induction studies in rifts and other active regions, Acta Geod. Geoph. et Mont., in press, 1977.

Illies, J.H. Graben tectonics as related to crust-mantle interaction, in Graben Problems, edited by J.H. Illies and St. Mueller, 3-27, 1970.

Illies, J.H., Taphrogenesis and plate tectonics, in Approaches to Taphrogenesis, edited by J.H. Illies and K. Fuchs, 433-460, Schweizerbart-Stuttgart, 1974.

Jaeger, J.C., Thermal effects of intrusions, Rev. Geophys., 2, 443-456, 1964.

Jiracek, G.R., M.E. Ander, and H.T. Holcombe, Magnetotelluric soundings of crustal conductive zones in major continental rifts, in Rio Grande Rift: Tectonics and Magmatism, edited by R.E. Riecker, 209-222, Amer. Geophys. Union, Washington, D.C., 1979.

Keller, G.R., L.W. Braile, and P. Morgan, Regional crustal structure of the Colorado Plateau (abstract), in Conference on Plateau Uplift: Mode and Mechanism, Flagstaff, Arizona, compiled by the Lunar and Planetary Institute 3303 NASA Road One, Houston, Texas 77058, 1978.

Keller, G.R., L.W. Braile, and J.W. Schlue, Regional crustal structure of the Rio Grande rift from surface wave dispersion measurements, in Rio Grande Rift: Tectonics and Magmatism, edited by R.E. Riecker, 115-126, Amer. Geophys. Union, Washington, D.C., 1979.

Knopoff, L. and J.W. Schlue, Rayleigh wave phase velocities for the path Addis Ababa-Nairobi, Tectonophys., 15, 157-163, 1972.

Krumhansl, P., L.D. Brown, F.H. Cook, and S. Kaufman, Seismic reflection character of the Socorro magma body (abstract), EOS, 59, 390, 1978.

Krylov, S.V., S.I. Golenetsky, and G.I. Petrik, Agreement of seismology and DSS-data on the uppermost mantle structure of the Baikal rift zone, Geol. i Geofiz., 12, 61-65, 1974.

Krylov, S.V., B.P. Mishenkin, Z.R. Mishenkina, G.V. Petrick and V.S. Seleznev, Seismic cross-section of the lithosphere in the Baikal rift zone, Geol. i Geofiz., 3, 72-83, 1975.

Krylov, S.V., On the causes of anomalous properties of the upper mantle in rift zones, Geol. i Geofiz., 4, 3-17, 1976.

Lachenbruch, A.H., and J.H. Sass, Heat flow in the United States and the thermal regime of the crust, in The Earth's Crust, Geophys. Mono 20, edited by J.G. Heacock, 626-675, Amer. Geophys. Union, Washington, D.C., 1977.

Lachenbruch, A.H., and J.H. Sass, Models of an extending lithosphere and heat flow in the Basin and Range province, Geol. Soc. Amer. Memoir 152, R.B. Smith and G.P. Eaton (eds.), 209-250, 1978.

Lapine, J.C., J.C. Ruegg, and L. Steinmetz, Seismic profiles in the Djibouti area, Tectonophys., 15, 59-64, 1972.

Lienert, B.R., and D.J. Bennett, High electrical conductivities in the lower crust of the northwestern Basin and Range: an application of inverse theory to a controlled source deep magnetic sounding experiment, in The Earth's Crust, Geophys. Mono. 20, Edited by J.G. Heacock, 531-552, Amer. Geophys. Union, Washington, D.C., 1977.

Logatchev, N.A., and N.A. Florensov, The Baikal system of rift valleys, in Geodynamics of the Baikal Rift Zone, edited by N.A. Logatchev and P.A. Mohr, Tectonophys., 45, 1-13, 1978.

Long, R.E. and M.G. Mitchell, Teleseismic P-wave delay time in Iceland, Geophys. J. Roy. Astron. Soc., 20, 41-48, 1970.

Long, R.E., R.W. Backhouse, P.K.H. Maguire and K. Sundaralingham, The structure of East Africa using surface wave dispersion and Durham seismic array data, Tectonophys., 15, 165-178, 1972.

Long, R.E., K. Sundaralingham and P.K.H. Maguire, Crustal structure of the East African rift zone, Tectonophys., 20, 269-281, 1973.

Losecke, W., Ergebnisse magnetotellurischer Messungen bei Speyer, in Graben Problems, edited by J.H. Illies and St. Mueller, 242-248, 1970.

Lubimova, E.A., U.A. Zorin, and S.V. Lysak, Geothermal anomaly in the Baikal Lake region, Geothermics, 1, 31-34, 1972.

Makris, J., H. Menzel, and J. Zimmermann, A preliminary interpretation of the gravity field of Afar, northeast Ethiopia, Tectonophys., 15, 31-39, 1972.

Meincke, W., E. Hastig, and I. Weiner, Temperatur Verteilung Warmeleitfahigkeit und Warmefluss im Thuringer Becken, Geophys. und Geol., 11, 1967.

Meissner, R., H. Berckhemer, R. Wilde, and M. Poursadeg, Interpretation of seismic refraction measurements in the northern part of the Rhinegraben, in Graben Problems, edited by J. Illies and St. Mueller, 184-206, 1970.

Meissner, R., and U. Vetter, The northern end of the Rhine graben due to some geophysical measurements, in Approaches to Taphrogenesis, edited by J.H. Illies and K. Fuchs, 236-242, Schweizerbart-Stuttgart, 1974.

Mohr, P.A., Surface structure and plate tectonices of Afar, Tectonophys., 15, 3-18, 1972.

Mueller, St., Geophysical aspects of graben formation in continental rift systems, in Graben Problems, edited by J.H. Illies and St. Mueller, 27-37, 1970.

Mueller, St., and L. Ryback, Crustal dynamics in the central part of the Rhine Graben, in Approaches to Taphrogenesis, edited by J.H. Illies and K. Fuchs, 379-388, Schweizerbart-Stuttgart, 1974.

Olsen, K.H., G.R. Keller, and J.N. Stewart, Crustal structure along the Rio Grande rift from seismic refraction profiles, in Rio Grande Rift: Tectonics and Magmatism, edited by R.E. Riecker, 127-144, Amer. Geophys. Union, Washington, D.C., 1979.

Oxburgh, E.R. and D.L. Turcotte, Mid-ocean ridges and the geotherm distribution during mantle convection, J. Geophys. Res., 73, 2643-2661, 1968.

Palmason, G., Crustal structure of Iceland from explosion seismology, Soc. Sci. Islandica, 40, 1971.

Palmason, G., Kinematics and heat flow in a volcanic rift zone, with applications to Iceland, Geophys. J. Roy. Astron. Soc., 33, 451-471, 1973.

Palmason, G., and Saemundsson, K., Iceland in relation to the mid-Atlantic ridge, Annual Review of Earth and Planetary Sciences, 2, 25-50, 1974.

Palmason, G., Geophysical methods in geothermal exploration, Second U.N. Symposium on the Development and Use of Geothermal Resources, Vol. 2, 1175-1184, Stock No. UGO-000-00005-1, Supt. Documents, U.S. Government Printing Office, Washington, D.C., 20402, 1976.

Pedersen, J., and J.F. Hermance, Towards resolving the absence or presence of an active magma chamber under the southern Rio Grande rift zone (abstract), EOS, 57, 1014, 1976.

Pedersen, J., and J.F. Hermance, Evidence for molten material at shallow to intermediate crustal levels beneath the Rio Grande rift at Santa Fe (abstract), EOS, 59, 390, 1978.

Pedersen, J., and J.F. Hermance, The deep electrical structure of the Colorado plateau from magnetotelluric measurements, J. Geophys. Res., in press, 1980.

Porath, H., Magnetic variation anomalies and seismic low-velocity zones in the western United States, J. Geophys. Res., 76, 2643-2648, 1971.

Porath, H., and D.I. Gough, Mantle conductive structures in the western United States from magnetometer array studies, Geophys. J. Roy. Astron. Soc., 22, 261-275, 1971.

Pospeev, V.I., and V.I. Mikhalevsky, Deep magnetotelluric surveys of the south of the Siberian platform and in the Baikal Rift zone, in Geoelectric and Geothermal Studies (East-Central Europe, Soviet Asia), edited by A. Adam, 673-681, 1976.

Puzirev, N.N., M.M. Mandelbaum, S.V. Krylov, B.P. Mishenkin, and G.V. Krupskaya, Crustal seismic investigations in the Baikal Rift zone, J. Geomag. and Geoelec., 22, 169-175, 1974.

Puzirev, N.N., M.M. Mandelbaum, B.P. Mishenkin, G.V. Krupskaya and G.I. Petrik, The structure of the Baikalian rift according to the seismic shooting data, Geol. i Geofiz., 15, 155-167, 1974.

Puzirev, N.N., M.M. Mandelbaum, S.V. Krylov, B.P. Mishenkin, G.I. Petrik, and G.V. Krupskaya, Deep structure of the Baikal and other continental rift zones from seismic data, in Geodynamics of the Baikal Rift Zone, edited by N.A. Logatchev and P.A. Mohr, Tectonophysics, 45, 15-22, 1978.

Reeves, C.V., Rifting in the Kalahari?, Nature, 237, 95-96, 1972.

Reichenback, H., and St. Mueller, Ein Krusten-Mantel-Modell für das Riftsystem um den Rheingraben, abgeleitet aus der Dispersion von Rayleigh-Wellen, in Approaches to Taphrogenesis, edited by J.H. Illies and K. Fuchs, 348-365, Schweizerbart-Stuttgart, 1974.

Reilinger, R.E., O.D. Brown, and J.E. Oliver, Recent vertical crustal movements from leveling observations in the vicinity of the Rio Grande rift, in Rio Grande Rift: Tectonics and Magmatism, edited by R.E. Riecker, 223-236, Amer. Geophys. Union, Washington, D.C., 1979.

Reiter, M., C.L. Edwards, H. Hartman, and C. Weidman, Terrestrial heat flow along the Rio Grande rift, New Mexico and southern Colorado, Geol. Soc. Amer. Bull., 86, 811-819, 1975.

Reiter, M., C. Shearer, and C.L. Edwards, Geothermal anomalies along the Rio Grande rift of New Mexico, Geology, 6, 85-88, 1978.

Reiter, M., A.J. Mansure, and C. Shearer, Geothermal characteristics of the Rio Grande rift within the southern Rocky Mountain complex, in Rio Grande Rift: Tectonics and Magmatism, edited by R.E. Riecker, 253-263, Amer. Geophys. Union, Washington, D.C., 1979.

Reitmayr, G., An anomaly of the upper mantle below the Rhinegraben studied by the inductive response of natural electromagnetic fields, J. Geophys., 41, 651-658, 1975.

Rhine Graben Research Group for Explosion Seismology, The 1972 seismic refraction experiment on the Rhinegraben - first results, in Approaches to Taphrogenesis, edited by J.H. Illies and K. Fuchs, 122-137, Schweizerbart-Stuttgart, 1974.

Rinehart, E.J., A.K. Sanford, and R.M. Ward, Geographic extent and shape of an extensive magma body at midcrustal depths in the Rio Grande rift near Socorro, New Mexico, in Rio Grande Rift: Tectonics and Magmatism, edited by R.E. Riecker, 327-352, Amer. Geophys. Union, Washington, D.C., 1979.

Ringwood, A.E., Composition and Petrology of the Earth's mantle, McGraw-Hill, New York, 1975.

Rooney, D., and V.R.S. Hutton, A magnetotelluric and magnetovariational study of the Gregory Rift Valley, Kenya, Geophys. J. Roy. Astron. Soc., 51, 91-119, 1977.

Saemundsson, K., Evolution of the axial rifting zone in northern Iceland and the Tjörnes fracture zone, Geol. Soc. Am. Bull., 85, 495-504, 1974.

Sanford, A.R., O.S. Alptekin, and T.R. Toppozada, Use of reflection phases on microearthquake seismograms to map an unusual discontinuity beneath the Rio Grande rift, Seismol. Soc. Amer. Bull., 63, 2021-2134, 1973.

Sanford, A.R., R.P. Mott, Jr., P.J. Shuleski, E.J. Rinehart, F.J. Caravella, R.M. Ward, and T.C. Wallace, Geophysical evidence for a magma body in the vicinity of Socorro, New Mexico, in The Earth's Crust, Geophys. Mono. Ser., 20, edited by J.G. Heacock, 385-403, Amer. Geophys. Union, Washington, D.C., 1977.

Sass, J.H., W.B. Diment, A.H. Lachenbruch, B.V. Marshall, R.J. Munroe, T.H. Moses, Jr., and T.C. Urban, A new heat-flow contour map of the conterminuous United States, United States Geological Survey Open-File 76-756, 24 pp., 1976.

Scheelke, I., Magnetotellurische Messungen im Rheingraben und ihre Deutung mit zweidimensionalen Modellen, GAMMA, 20, 199 pp., 1972.

Scheelke, I., Models for the resistivity distribution from magnetotelluric soundings, in Approaches to Taphrogenesis, edited by J.H. Illies and K. Fuchs, 362-365, Schweizerbart-Stuttgart, 1974.

Schleusener, A., W. Torge and H. Drewes, The gravity field of northeastern Iceland, J. Geophys., 42, 27-45, 1976.

Schmucker, U., Anomalies of geomagnetic variations in the south-western United States, J. Geomag. Geoelec., 15, 193-221, 1964.

Schmucker, U., Anomalies of geomagnetic variations in the south-western United States, Bull. Scripps Inst. Oceanography, 13, 165 pp., Univ. of California Press, Richmond, California, 1970.

Scholz, C.H., T.A. Koczynski and D.G. Hutchins, Evidence for incipient rifting in southern Africa, Geophys. J. Roy. Astron. Soc., 44, 135-144, 1976.

Shankland, T.J. and H.S. Waff, Partial melting and electrical conductivity anomalies in the upper mantle, J. Geophys. Res., 82, 5409-5417, 1977.

Shuleski, D.J., F.J. Caravella, E.J. Rinehart, A.R. Sanford, T.C. Wallace, and R.M. Ward, Seismic studies of shallow magma bodies beneath the Rio Grande rift in the vicinity of Socorro, New Mexico, 16 pp., paper presented at South-Central Section, Geological Society of America, March 17, 1977.

Sigurdsson, H., and R.S.J. Sparks, Lateral magma flow within rifted Icelandic crust, Nature, 274, 126-130, 1978.

Smith, R.L., and R.H. Bailey, Resurgent cauldrons, in Studies in Volcanology, Geol. Soc. Amer. Mem., 116, edited by R.E. Coats, R.L. Hay, and C.A. Anderson, 613-632, Geol. Soc. Amer., Boulder, Colorado, 1968.

Stanley, W.D., J.E. Boehl, F.X. Bostick, and H.W. Smith, Geothermal significance of magnetotelluric sounding in the eastern Snake River Plain - Yellowstone region, J. Geophys. Res., 82, 2501-2514, 1977.

Thayer, R.E., Telluric-magnetotelluric investigations of regional geothermal processes in Iceland, Ph.D. Thesis, Dept. of Geological Sciences, Brown University, 276 pp., 1975.

Thayer, R.E., A. Bjornsson, L. Alvarez and J.F. Hermance, Magma genesis and crustal spreading in the northern neovolcanic zone of Iceland: telluric-magnetotelluric constraints, Geophys. J. Roy. Astron. Soc., in press, 1980.

Thompson, G.A. and M.L. Zoback, Regional geophysics of the Colorado Plateau, submitted to Tectonophys., Proc. of Symp. on Plateau Uplift, 1979.

Toppozada, T.B., and A.R. Sanford, Crustal structure in central New Mexico interpreted from the Gasbuggy explosion, Bull. Seismol. Soc. Amer., 66, 877-886, 1976.

Tryggvason, E., Crustal structure of the Iceland region from dispersion of surface waves, Bull. Seismol. Soc. Amer., 52, 359-388, 1962.

Von Herzen, R.P., and V. Vacquier, Terrestrial heat flow in Lake Malawi, Africa, J. Geophys. Res., 72, 16, 1967.

Waff, H.S., Theoretical considerations of electrical conductivity in a partially molten mantle and implications for geothermometry, J. Geophys. Res., 79, 4003-4010, 1974.

Waff, H.S. and J.R. Bulau, Equilibrium field distribution in an ultramafic partial melt under hydrostatic stress conditions, J. Geophys. Res., 84, 6109-6114, 1979.

Walker, D., E.M. Stolper, and J.F. Hays, A numerical treatment of melt/solid segregation: size of the eucrite parent body and stability of the terrestrial low-velocity zone, J. Geophys. Res., 83, 6005-6013, 1978.

Walker, G.P.L., Intrusive sheet swarms and the identity of crustal layer 3 in Iceland, Geol. Soc. London J., 131, Part 2, 143-161, 1975.

Ward, P.L., New interpretation of the geology of Iceland, Bull. Geol. Soc. Amer., 82, 2991-3012, 1971.

White, A., A geomagnetic variation anomaly across the northern Gulf of California, Geophys. J. Roy. Astron. Soc., 33, 1-25, 1973a.

White, A., Anomalies in geomagnetic variations across the central Gulf of California, Geophys. J. Roy. Astron. Soc., 33, 27-46, 1973b.

Williams, H., Calderas and their origin, Cal. Univ. Publs. Geol. Sci., 25, 239-346, 1941.

Winter, R., Erdmagnetische Tiefensondierung im Gebiet des Oberrheingrabens, in Graben Problems, edited by J.H. Illies and St. Mueller, 244-248, Schweizerbart-Stuttgart, 1970.

Winter, R., A model for the resistivity distribution from geomagnetic depth soundings, in Approaches to Taphrogenesis, edited by J.H. Illies and K. Fuchs, 369-375, Schweizerbart-Stuttgart, 1974.

PHYSICAL ASPECTS OF MAGMATISM IN CONTINENTAL RIFTS

L.A.J. Williams

Department of Environmental Sciences, University of Lancaster.
Lancaster LA1 4YQ, England

Abstract. A comparison of the tectonomagmatic evolution of the Upper Paleozoic Oslo graben of Norway, the Cenozoic Rio Grande rift of the south-western United States and the Cenozoic Kenya rift shows that the main episodes of igneous activity were spread over about 20-30 m.y. in each case. Contrary to widespread beliefs, no simple relationships can be demonstrated between volcanic events and major faults, or between igneous activity and pulses of uplift. Attention is drawn to fundamental differences in the suites of igneous rocks found in these rifts: occurrences of alkaline rocks are common but not characteristic features of continental rifts. Though estimates of volumes of igneous rocks are necessarily crude, the eastern rift system of Africa stands apart from all other examples in terms of the abundance of volcanics. The voluminous eruptions in Kenya led to complete infilling and overtopping of depressions at two stages of rift development.

Introduction

A comprehensive review of all the physical parameters of igneous rocks in continental rift environments cannot be attempted in a single chapter. This account concentrates, therefore, on a broad comparison of the tectonomagmatic evolution of three widely separated rifts.

Summaries of magmatic and tectonic events in the Upper Paleozoic Oslo graben of Norway and the Cenozoic Rio Grande rift of the south-western United States are taken largely from recent review articles which arose from symposia held during the term of the Geodynamics Project. In dealing with the Cenozoic East African rift system, emphasis is placed on the Kenya (Gregory) segment - the type example of a rift valley. The author uses his personal knowledge of this region to comment on some aspects of the voluminous volcanism.

The article sets out to examine issues such as the duration of magmatic activity; the broad compositions, proportions and distribution of exposed igneous rocks; the familiar problems encountered in attempts to estimate volumes of magmatic rocks; the role of volcanic rocks in determining the character, magnitude and timing of episodes of uplift and subsidence; and some of the demonstrated or claimed relationships between igneous and tectonic activity.

Magmatism in the Oslo Rift

The following account of the Oslo graben is based mainly on the excellent summaries which appeared in three recent volumes (Neumann and Ramberg, 1978; Ramberg and Neumann, 1978; Dons and Larsen, 1978). An earlier publication (Ramberg, 1976) describing the interpretation of a gravity study provides estimates of the areal and volume frequency distribution of the plutonic and volcanic rocks, and another review (Oftedahl, 1978a) deals specifically with one of the most striking features of the magmatism in this region - the development of numerous large cauldron structures.

Structure

The course of the NNE-trending Permian Oslo rift is well defined by plutonic and volcanic rocks exposed in a belt some 200 km long and about 40 km wide (Fig. 1). This igneous belt is only locally bounded by faults, so that it may represent merely the inner part of a deeply eroded trough which was originally much wider. Indeed, massive breccia zones in Precambrian rocks flanking the rift are generally accepted to be deep-seated expressions of 'master' faults in regions now stripped of volcanic and sedimentary cover. The precise age and magnitude of many of these faults understandably remain conjectural.

Oftedahl (1978b) questions the fact that the Oslo trough was a true graben because, although the western margin was defined by a prominent scarp, much of the eastern side may never have had an elevated rim. Ramberg (1976) also doubts the former existence of elevated rims and tilted graben shoulders, though clearly evidence bearing on the latter issue is likely to be tenuous.

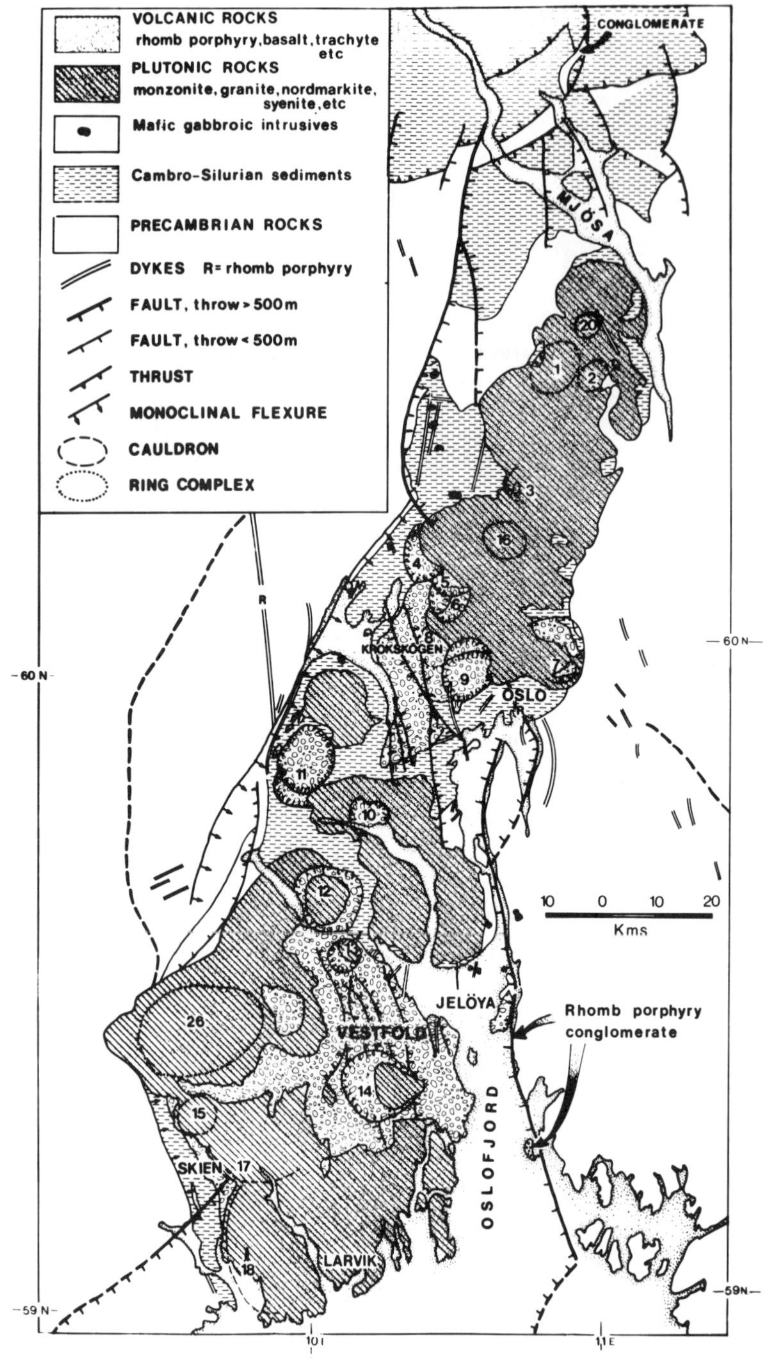

Fig. 1. Simplified geological map of the Oslo graben (after Ramberg, 1976).

The major fault along the eastern side of Oslofjord is one of the most rewarding examples for study because volcanics and thick conglomerates on the downthrow side are preserved on several islands. Vertical displacements amounting to 3000 m are inferred. In the northern part of the rift, downfaulted Cambro-Silurian sediments provide evidence of minimum vertical displacements of 500 m along the western side of the trough.

Farther south, much of the western margin of the igneous belt is characterized by monoclinal flexures affecting volcanics and sediments. Early basalts near Skien dip consistently to the northeast at about 35°, whereas the underlying Lower Paleozoic formations display dips of 18° in western outcrops and up to 36° near basalt contacts (Segalstad, 1978). Conflicting opinions have been expressed, however, about the age of monoclinal warping in this region. Segalstad contemplates downwarping before the deposition of Permian sediments and basalts in an attempt to show similar initial tectonic events in the Oslo and Kenya rifts. On the other hand, Ramberg and Larsen (1978) prefer development of flexures after the eruption of the Skien and other early basalts.

There is universal agreement that firm evidence of crustal doming is lacking in the Oslo region. Henningsmoen (1978) quotes Holtedahl (1953) in noting that sediments resting on a 'sub-Permian peneplain' show that it sloped gently southwards and westwards: but the 'peneplain' cannot be traced outside the graben area (Ramberg, 1976). In comparing the Oslo and Kenya rifts, Segalstad (1978) suggests that an episode of updoming possibly occurred in the former between undersaturated volcanism of the Skien area and later eruptions of alkali basalts. Ramberg and Larsen (1978) lean more towards pre-volcanic uplifts in speculating that the Skien basalts accumulated in an interdome basin.

Distribution and age of igneous rocks

The alkaline intrusive and extrusive rocks make up about 6500 km^2 of the total area (8500 km^2) of what is sometimes referred to as the graben 'floor' i.e. 75-80 per cent. These igneous rocks are spread out along nearly the full length of the graben (Fig. 1), but they are divisible into three major outcrops, each of which contains Permian volcanics and sediments as well as related sub-volcanic and deep-seated complexes. The outcrops are separated by relatively unmetamorphosed Lower Paleozoic sediments (marine Cambro-Silurian up to 1000 m thick, overlain by up to 1250 m of non-marine late Silurian red beds). Precambrian gneisses flank the igneous belt, and they are also exposed across the width of the graben near its poorly-defined northern limit.

Sparsely fossiliferous sub-volcanic deposits indicate an early Permian age for the onset of volcanism. Thin non-marine sediments are intercalated with some of the lavas, but fossils have been recovered only from sediments associated with the earliest basalts (Henningsmoen, 1978) so that there is no paleontological evidence to support the tentative Triassic age assigned to the youngest deposits.

Some of the results from isotopic dating projects suggest that volcanism commenced around 294 m.y. ago (Upper Carboniferous) in the central part of the graben; older dates can be expected from basalts farther south (Sundvoll, 1978b). The plutonics were evidently emplaced mainly over the period 275-284 m.y., but a more extended range of igneous activity than formerly envisaged is indicated by a date of 245 m.y. from a small granite pluton (Oftedahl, 1978b).

Plutonic rocks

Plutonic rocks are exposed over about 5000 km^2 of the rift (Table 1), whereas the preserved volcanics occupy not more than 1500 km^2. Two main groups of plutonic types are recognized (Neumann, 1978). One embraces the constituents of large batholiths which outcrop over 60 per cent of the graben area, ranging in composition from monzodiorite and granite to both nepheline-free and nepheline-bearing syenites; minor olivine gabbros are regarded as cumulates. Volumetrically far less significant are the rocks of a second group which occur as scattered mafic volcanic necks, linear patterns suggesting emplacement along major fissures. These 'Oslo essexites' are predominantly gabbroic (alkaline to quartz tholeiitic in affinities), but rock types range from pyroxenites to diorites.

TABLE 1. Areal distribution of rocks of the Oslo graben (Ramberg, 1976)

Rock type	Areal distribution (km^2)	(%)
Rhyolite, trachyte	31	0.4
Rhomb porphyry	1153	13.4
Basalt	220	2.6
Sum volcanics	1404	16.4
Basaltic volcanic necks ("Oslo essexite")	15	0.2
Monzonite		
larvikite	1670	19.5
kjelsaasite	201	2.3
akerite	52	0.6
Nepheline rocks (lardalite, foyaite)	65	0.8
Nordmarkite, syenite	1425	16.6
Granite		
biotite granite	840	9.8
ekerite	821	9.6
Sum plutonics	5089	59.4
Cambro-Silurian		19.1
shale and limestone	1545	
sandstone	95	
Precambrian within graben	440	5.1
Total	8573	100.0

Volcanics

Post-Permian erosion left only remnants of the volcanic rocks preserved within the rift, so that their original distribution and full compositional range remain speculative. There can be little doubt, however, that substantial thicknesses of lavas formerly occupied the entire graben because the now isolated outcrops contain flows with distinctive petrographic characteristics which enable convincing correlations to be made over wide areas.

The volcanics either overlie Lower Permian sediments which thicken southwards, or they rest with marked unconformity directly on strongly folded Cambro-Silurian deposits (northern areas) or on less-deformed Upper Silurian non-marine sediments (southern end of the graben).

Preserved lava sequences are locally 1.5-3 km thick, and the present total volume of volcanics is around 500 km^3 (Ramberg & Larsen, 1978). Basalts make up about 55 per cent by volume; rhomb porphyries (porphyritic trachytes/ trachyandesites with rhomb-shaped feldspar phenocrysts) 35 per cent; and trachytic to rhyolitic lavas and ignimbrites about 10 per cent. It has been suggested (e.g. Oftedahl, 1952) that the volcanics spread far beyond the graben, and estimates of the original volume exceed 10,000 km^3 (Ramberg & Larsen, 1978), but the evidence is inconclusive. It could well be argued, for instance, that occasional rhomb porphyry dykes and fragments of basalt in explosion breccias outside the graben denote localised eruptions rather than sheets continuous with the rift infill. Indeed, comparison with the East African rifts suggests that fissure sources outside the graben are unlikely to have been intimately associated with activity in the floor though, of course, coalescence of rift-derived and flanking volcanics cannot be ruled out.

Blocks of rhomb porphyry in conglomeratic deposits more than 1000 m thick on the eastern side of Oslofjord provide compelling evidence for the existence of a fault scarp formerly surmounted by lavas, but the occurrence neither sheds much light on the original extent of the rhomb porphyries beyond the graben nor does it illustrate the form of the sub-volcanic surface outside the rift.

The earliest basalts (designated B_1 but not necessarily contemporaneous) are widespread in central and southern parts of the rift. At the Krokskogen plateau, west of Oslo, a single flow of olivine tholeiite about 30 m thick is the only representative of B_1. Southwards from Krokskogen the basalts not only thicken considerably to some 2000 m by an increase in the number of flows, but they also become older and progressively more alkaline. At the southernmost outcrops around Skien, alkali olivine basalts are accompanied by basanitic, nephelinitic and melilitic rocks.

In the southwest, the B_1 Skien basalts and associated strongly alkaline types are considered to have been erupted just beyond early rift boundary faults which were later obliterated by intrusive activity. On the opposite side of the graben, the Jeløya volcanics consist of B_1 basalts, mugearites and intercalated pyroclastics, capped by early rhomb porphyry lavas and agglomerates. The basalts rest on coarse debris flows which testify to movements on the nearby Oslofjord fault earlier than the displacements indicated by rhomb porphyry conglomerates farther south. Minor intrusive complexes at Jeløya display basic and intermediate compositions which closely match the erupted products and confirm the proximity of source vents.

More centrally within the rift, the Vestfold and Krokskogen lava plateaus expose up to 1000 m of rhomb porphyries overlying B_1 basalts. Lengthy N-S dykes extending beyond the plateaus (Fig. 1) are taken to indicate fissure sources for the intermediate lavas. Units, commonly up to 100 m thick and consisting of surprisingly few individual flows, have been distinguished on the basis of the shape, size and abundance of anorthoclase-plagioclase phenocrysts. Such textural variations have proved invaluable for the correlation of these lavas which are otherwise chemically and mineralogically similar: they are undoubtedly the extrusive equivalents of larvikites, the commonest exposed intrusive rocks in the province.

Initial members of the rhomb porphyry flood lava sequence (RP_1 - RP_9) were surmounted by basaltic central volcanoes, remnants of which are now preserved mainly in cauldron successions. The B_2 lavas (alkali olivine basalts and basanites) and B_3 flows (olivine tholeiites and transitional basalts, together with subordinate felsic lavas and ignimbrites) are in turn overlain by trachytes, rhyolites, explosion breccias and lake sediments representing caldera infill. Common intercalations of flows high in the rhomb porphyry sequence show that flood eruptions were contemporaneous with and outlasted episodes of central vent activity and caldera collapse. The central volcanoes were periodically engulfed as a result of the massive scale of emission of the rhomb porphyries.

Cauldrons

Detailed studies of the Oslo graben have led to the identification of 15 undisputed cauldrons, three probable examples, and about half a dozen ring complexes exposing concentric patterns of plutonic rocks but no volcanics (Table 2).

The cauldrons range from 5 to 15 km in diameter, and subsidence exceeding 1 km can be demonstrated in many examples (Oftedahl, 1978a). The first stage in their evolution involved subsidence of caldera blocks of volcanic rocks into horizontally zoned syenitic/monzonitic magma chambers, intrusion of ring dykes 30-1000 m wide, and invasion of the blocks by central domes. This was followed by the emplacement of major composite batholiths by stoping, leaving undisturbed remnants of the older parts of each complex.

The subsided caldera blocks were 2-5 km thick, and the lava stratigraphy can frequently be matched with volcanic successions exposed outside the cauldrons. Locally, caldera infill (ignimbrites, breccias and lake sediments) is preserved.

Many workers (e.g. Ramberg and Larsen, 1978) accept a long-held view that the cauldrons developed on basaltic lava cones or shields. Oftedahl (1978a), however, maintains that caldera development was connected with the intermediate and felsic intrusions, and that it was not related to the basalt and rhomb porphyry flows.

The absence of resurgent domes in the Oslo cauldrons is attributed (Oftedahl, 1978a) to the composition and viscosity of the underlying intermediate magmas which moved upwards by stoping rather than forceful intrusion.

Tectonomagmatic evolution of the Oslo graben

Ramberg and Larsen (1978) describe six broad stages in the tectonic and magmatic evolution of the Oslo graben which developed in a region of late Precambrian to Lower Paleozoic troughs and basins. These were bounded locally by faults, some of which were rejuvenated during the Permian.

Permian non-marine sediments (Asker Group) accumulated in an early rift depression (Fig. 2). Similar broad downwarped troughs marked initial stages in the evolution of the Cenozoic Kenya, Rhine and Baikal rifts; these were all wider than the present rift structures (King, 1978).

Volcanism in the Oslo region commenced with the eruption of undersaturated basalts at the south-western margin of the trough. Subsequent migration of volcanic activity into central parts of the rift zone was accompanied by compositional changes in the basalts, undersaturated varieties ultimately giving way to quartz tholeiites.

The first stages of graben formation coincided with widespread rhomb porphyry fissure eruptions and the emplacement of early granites. The number of preserved rhomb porphyry flows increases southwards, but the extent to which these lavas overtopped the graben cannot be demonstrated.

The development of two en echelon rift segments (Ringerike-Vestfold graben and Akershus graben) was accompanied by a transition from fissure activity to build-up of central volcanoes. These cones and shields are predominantly basaltic, but some examples in the south are characterized by bimodal (basalt-felsic) or trachytic compositions. Four central volcanoes/cauldrons lie axially within the Ringerike-Vestfold graben; six others cluster where the two graben segments are offset (Fig. 2).

TABLE 2. Cauldrons and ring complexes in the Oslo graben (Oftedahl, 1978)

	Diameter (km)	Subsidence (km)
Cauldrons		
1. Lundbergkollen	10?	1
2. Hurdal	5	1?
3. Grua	5?	?
4. Øyangen	10 x 7.5	>2,3-4?
5. Svarten	11	1.5
6. Heggelia	7	1.0-1.5
7. Nittedal	11	0.8
8. Kampen	10?	ca. 1
9. Baerum	8.5 x 12	1-1.5
10. Drammen	7	1.0
11. Glitrevann	10 x 16	1.2-1.5
12. Sande	12	0.5
13. Hillestad	5	>0.5
14. Ramnes	12-16	ca. 2.0
15. Skrehelle	7	0.5-1.0
Probable cauldrons		
16. Stryken	ca. 10	∿1
17. Vealøs	10-20	?
18. Langesundsfjord	ca. 12	?
Ring complexes		
20. Nordlikampen	5	
21. Gjerdingen	4-5	
22. Katnosa	9	
23. Kikut	3-4	
24. Bjørnsjøen	7-9	
25. Tryvasshøgda	7	
26. Mykle	18 x 22	

Differences of opinion exist over the ages of the 'Oslo essexites'. Ramberg and Larsen (1978) adopt a minority view in suggesting that most of these gabbroic necks formed during the central volcano stage. The same authors attribute vent breccias to a late explosive phase in the evolution of the central volcanoes/cauldrons.

Intrusion of felsic plutons certainly commenced during the early stages of graben formation, but the emplacement of major composite batholiths marked an important late-stage event in the tectonomagmatic evolution of the rift.

Magmatism in the Rio Grande Rift

Over less than a decade, intensive studies of the Cenozoic Rio Grande rift of the south-

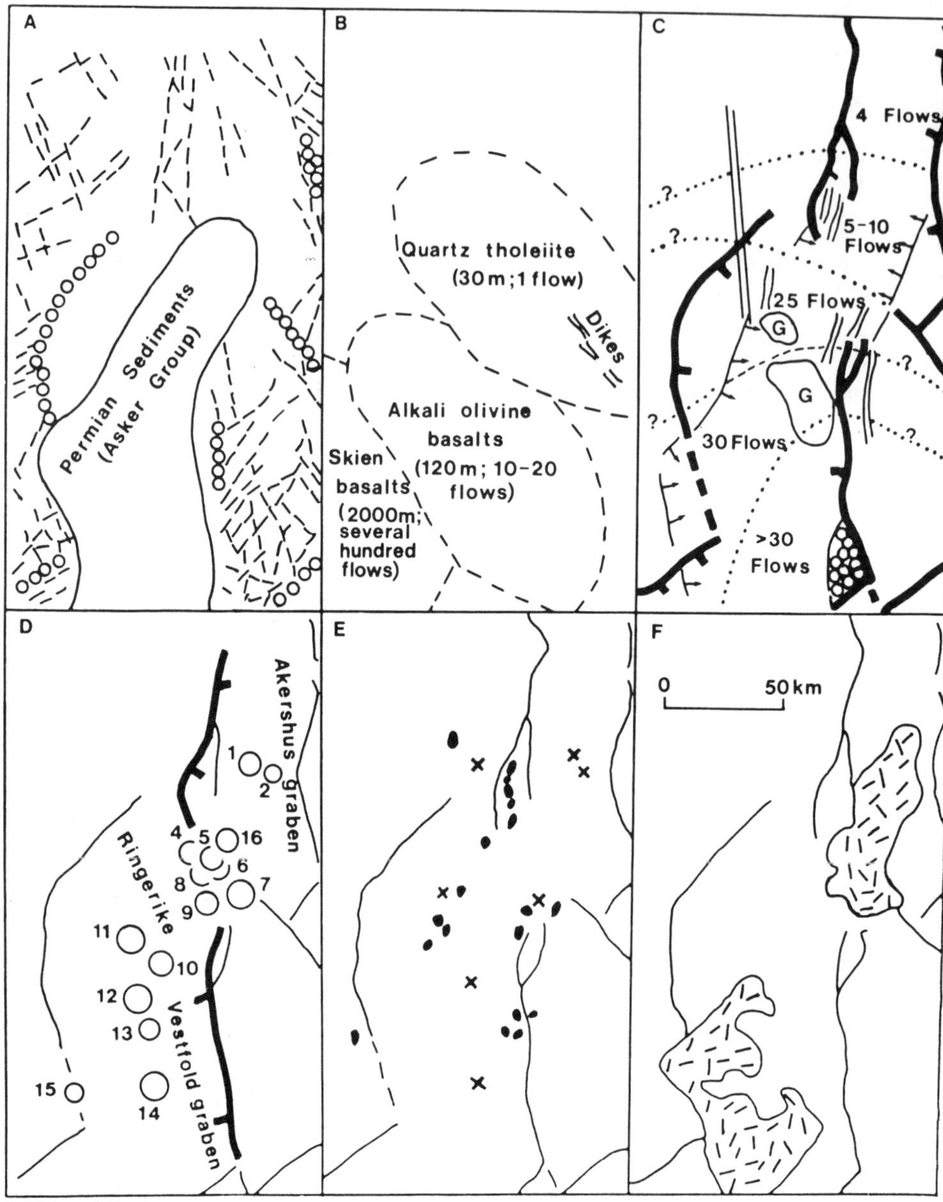

Fig. 2. Igneous and tectonic evolution of the Oslo graben (after Ramberg and Larsen, 1978).
(A) Accumulation of sediments in early trough. Dashed lines are pre-Permian faults, and circles indicate mylonite zones.
(B) Eruption of initial (B_1) basalts.
(C) Rhomb porphyry fissure eruptions (dikes shown with double lines); emplacement of early granites (G); graben formation (thick lines are major faults, and arrows denote flexures); accumulation of rhomb porphyry conglomerates (circles) in south.
(D) Development of Akershus and Ringerike-Vestfold graben with axial central volcanoes/cauldrons (numbered as in Table 2).
(E) Formation of gabbroic necks (black) and explosion breccias (crosses).
(F) Emplacement of major batholiths.

western United States have placed on record a wealth of information concerning the tectonic and magmatic evolution of this region. A recent volume (Riecker, 1979) contains a valuable selection of papers, and a guidebook (Hawley, 1978) covering the rift from central Colorado to El Paso in Texas provides a useful complementary source of information. The following summary of magmatic activity draws heavily on general reviews by Chapin (1979; entire rift system), Tweto (1979; Colorado segment) and Seager & Morgan (1979; southern New Mexico, west Texas and northern Mexico).

A review of post-40 m.y. volcanic and tectonic events in southwestern North America (Elston & Bornhost, 1979) and the presentation of a plate-tectonic model for late Cenozoic extension in the western United States (Eaton, 1979) both serve to emphasize problems which arise because the southern end of the Rio Grande rift system traverses or merges with part of the Basin and Range Province - a vast region of epeirogenic uplift, block faulting and basaltic magmatism which extends through Nevada, Utah, California, Arizona, southwestern New Mexico, and western Mexico.

Regional setting and structure

The Rio Grande depression slices through the southern end of the Rocky Mountains in a series of sags, graben and half-graben structures superimposed on a broad region of Laramide (Upper Cretaceous-Lower Tertiary) uplift between the Colorado Plateau on the west and the Great Plains to the east. Many of the Neogene faults defining the rift are rejuvenated structures with earlier displacements dating back to Precambrian, Paleozoic and Laramide orogenies but, on a regional scale, the depression cuts obliquely across older basins and uplifts. The northern half of the rift is topographically well defined, but to the south it loses its identity in a broad zone of parallel basins and ranges.

The rift depression trends SSE from Leadville in central Colorado to the New Mexico State line (250 km) in two major troughs, the upper Arkansas Valley and the wider San Luis Valley (Fig. 3). Through northern and central New Mexico, significant offsets of the San Luis, Espanola and Albuquerque basins give the rift an overall SSW trend over nearly 350 km to Socorro (Fig. 4). From the latitude of Socorro, three major troughs and intervening mountain ranges extend southwards for a further 250 km before swinging SE into Mexico and west Texas. Another trough (San Augustin Basin) branches WSW from Socorro towards the Arizona State line.

The depression is flanked largely by Precambrian rocks and Paleozoic to Lower Tertiary sedimentary formations, commonly standing about 1000 m above the level of deep Tertiary-Holocene basin-fill. The marginal ranges in Colorado and northern New Mexico have peaks rising to over 4000 m. Much of the present elevation was inherited from Laramide events, but further strong uplifts (locally more than 1000 m) occurred during the Miocene and Pliocene, mostly 4-7 m.y. ago. No regional pre-rift doming has been demonstrated, and it is clear that Cenozoic sediments and volcanics accumulated initially in broad depressions on a late Eocene, post-Laramide surface.

Despite the comprehensive nature of geological and geophysical investigations, there is still no general agreement about the full extent of the Rio Grande rift system. The question of a northward continuation beyond Leadville (Colorado) hinges largely on the definition adopted, for the upper Ankansas Valley graben passes into a belt of Neogene block faulting which extends to the Wyoming border. To the south, some workers maintain that the rift terminates at the end of the Albuquerque basin near Socorro. Others claim that it can be traced to Hatch in southern New Mexico; or to a line of basins along the Rio Grande in west Texas; or into the Los Muertos basin of Chihuahua in Mexico (about 900 km from Leadville). Yet others suggest that the southern continuation of the rift embraces a much wider belt of Basin and Range structures.

These differences of opinion over the limits of the rift make it impossible to define precisely the extent and character of associated magmatic activity. Additional problems arise because undisputed rift-related magmatism followed on or overlapped a widespread episode of earlier Tertiary (mainly Oligocene) igneous activity.

Older igneous activity

Pre-rift/early rift igneous rocks are preserved on the western side of the depression in two major fields (San Juan and Datil-Mogollon) about 450 km apart, but they were formerly much more extensive. Exposures, particularly within the Espanola basin, and scattered deep boreholes show that thick Oligocene-lower Miocene volcaniclastic sediments with interbedded intermediate to silicic lavas and tuffs are widespread in early basin-fill sequences.

Two distinct suites of volcanics are recognized in the San Juan Mountains of southern Colorado (Lipman et al., 1978). The earlier one embraces Oligocene andesitic lavas and breccias (35-30 m.y.) and silicic ash-flow tuffs (30-26.5 m.y.) which together cover much of the 40,000 km^2 field (Fig. 3). These volcanics define the western margin of the San Luis Valley, dipping beneath Miocene-Pliocene sediments of the Santa Fe Formation: what are probably equivalent volcanics appear in sporadic exposures on the upfaulted eastern side of the trough. Similarly, andesitic flows and breccias, rhyolitic ash-flow tuffs and silicic alkalic basalts (ranging in age from 35 to 5 m.y.) underlie younger volcanics of the Taos Plateau at the southern end of the San Luis Valley, passing beneath or inter-

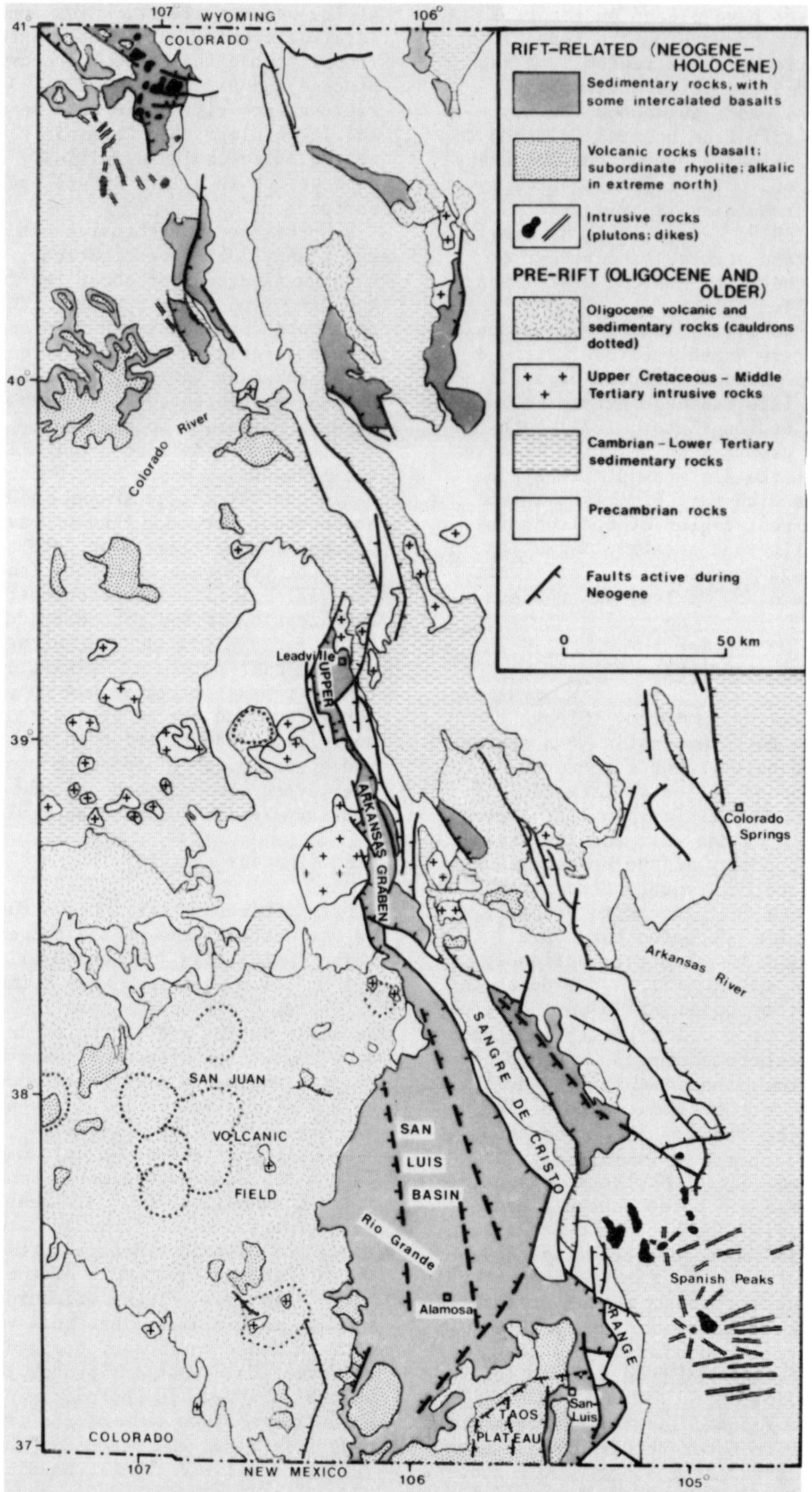

Fig. 3. The Rio Grande rift in Colorado (after geological map in Hawley, 1978).

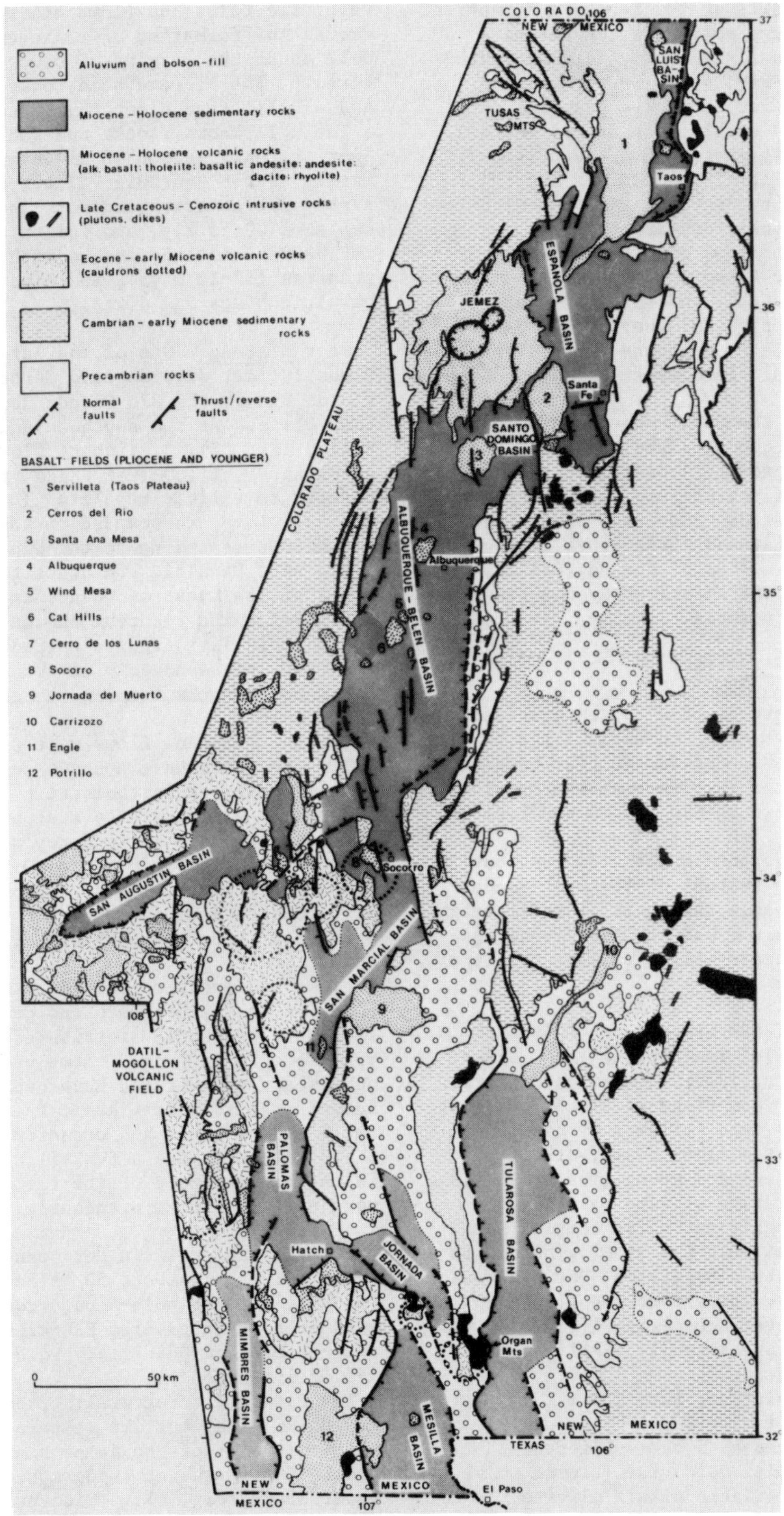

Fig. 4. The Rio Grande rift in New Mexico (after geological map in Hawley, 1978).

fingering with deposits of the Santa Fe Formation.

Eocene-lower Miocene volcanics constitute a major part of the Datil-Mogollon field of southwestern New Mexico (Fig. 4). Nearly all the flows were erupted from central volcanoes, and they belong to three overlapping suites (Elston, 1978). The earliest phases of activity (40-21 m.y.) produced calc-alkalic andesites and rhyolites, followed by ash-flow tuffs of high-silica rhyolite composition, and basaltic andesites. The ash flows were derived from large calderas up to 40 km in diameter. Separate plutons probably underlie the scattered cauldrons of an older group, whereas a cluster of high-silica rhyolite cauldrons is considered to be related to a single composite pluton about 125 km in diameter.

In many respects, the Datil-Mogollon and San Juan fields are analogous. They embrace considerable volumes of pre-rift volcanics, but show an extended range of activity which overlapped the formation of rift depressions. They form mountains or high plateaus rising locally to 4500 m on the western side of the depression, and both fields are characterized by numerous caldera/cauldron structures.

Rift-related igneous activity

Volcanics which were erupted contemporaneously with rifting are most conspicuous in large scattered fields of Miocene and younger basalts, basaltic andesites and rhyolites within or close to the depression through New Mexico and into southern Colorado. Substantial volumes of igneous rocks are present in these fields, but a much more striking feature of the Rio Grande rift is the considerable thickness of sedimentary infill (commonly 1-4 km, but locally exceeding 5 km) revealed by deep drilling or late uplift of basin-fill.

Colorado. Most of the basic lavas in the widely scattered fields in Colorado (Fig. 3) are alkali basalts (Tweto, 1979). Near the Colorado River, for instance, eruptions of alkali olivine basalts (24-20 m.y.) were followed by basaltic andesites and andesites (14-9 m.y.), alkali basalts (around 8 m.y.) and young (less than 1.5 m.y.) nepheline-normative olivine basalts. Farther north, a largely intrusive complex in the Elkhead Mountains near the Wyoming border is mainly of intermediate composition, but the rocks span a full range from basalt to rhyolite and they were emplaced over the period 11.5-7.5 m.y. Some of the basalts are strongly alkaline feldspathoidal varieties.

In southwestern Colorado, intermediate to silicic Oligocene activity in the San Juan Mountains was replaced by Miocene-Pliocene bimodal basalt-rhyolite volcanism (Lipman et al., 1978). Widespread silicic alkali olivine basalts and xenocrystic basaltic andesites (26-4 m.y.) are accompanied by more localized rhyolitic tuffs and flows which were erupted during the formation of a Miocene caldera as well as on the margins of older caldera structures. The Miocene acid rocks are more silicic and alkalic than older rhyolites in this region.

The well-known stocks and radial dikes at Spanish Peaks, about 30 km beyond the eastern margin of the San Luis Valley, are predominantly latite-trachyandesite-andesite porphyries emplaced 20-25 m.y. ago (Smith, 1979). Other intrusives contemporaneous with rifting are granites (17-12 m.y.), and minor rhyolitic and basaltic plugs and dikes.

Taos Plateau. One of the largest and compositionally most diverse late Cenozoic volcanic fields within the Rio Grande depression forms the Taos Plateau at the southern end of the San Luis Valley (Fig. 3). Pliocene flows (4.5-2 m.y.) covering about 1500 km^2 range from tholeiitic basalts to silicic rhyolites (Lipman & Mehnert, 1975). They are bounded on the west by rising ground corresponding to the dip slope of pre-rift/early rift basaltic and andesitic rocks (35-5 m.y.) of the Tusas and San Juan Mountains. These older volcanics reappear on the plateau in sporadic inliers along a buried intra-rift horst, and they are seen again in upfaulted remnants on Precambrian rocks forming the eastern side of the trough.

All the Pliocene flows were derived from central volcanoes which have a crude concentric arrangement with low-angle tholeiitic shields of an inner zone giving way to steeper olivine andesite shields which are in turn replaced by rhyodacite cones of an outer ring some 40-50 km across. Widespread flows of olivine tholeiite (3.5-4.5 m.y.) coalesce to form most of the plateau surface. Locally, they are at least 200 m thick.

Flows and spatter cones of xenocrystic basaltic andesite in the northern and central parts of the field are evidently distributed along a fissure zone some 20 km long. Some underlie tholeiites; others are younger and have been dated at 2 m.y. Subordinate silicic alkalic basalts occur in small cinder cones and occasional flows. All these volcanics and a few silicic rhyolite domes in the central part of the field were erupted more or less contemporaneously.

Jemez lineament. In northern New Mexico, major offsets of about 50 km between the San Luis-Espanola and Espanola-Albuquerque basins coincide with a broad transverse ENE-trending belt of intense Miocene-Quaternary volcanic activity (the 'Jemez lineament').

The Espanola structural depression is 40 km long and 65 km wide, but its western half is filled with volcanics of the Jemez Mountains so that the width of the present topographic basin is only about 35 km (Fig. 4). Widespread dissection of westward-tilted and faulted basin-fill has provided rewarding ground for detailed stratigraphic studies. Plio-Pleistocene alluvial deposits overlie more than 2 km of Miocene-Pliocene sediments of the Santa Fe Group (20-4 m.y.) which rest on and interdigitate with pre-rift/early rift tuffaceous sediments derived from source areas of older volcanics in the San Juan and Taos regions. These in turn are underlain by Eocene-(?)Oligocene sandstones, shales, conglomerates and localized volcaniclastic deposits. Mesozoic and Paleozoic formations undoubtedly floor western parts of the basin, but over large areas Tertiary deposits evidently rest directly on Precambrian basement rocks. Late Oligocene and early Miocene deposits extend beyond the confines of the present basin which did not become a fault-bounded feature until Miocene or later stages of development.

The explanation and sections attached to a geological map of the Jemez Mountains (Smith et al., 1970) provide an elegant summary of the structure and evolution of this classic complex which rises 3000 m above the Espanola basin. It is surmounted by two Quaternary calderas. The earlier Toledo structure is truncated by the larger and better known Valles caldera, nearly 17 km across east-west. The latter contains a resurgent dome which exposes, not only young rhyolites and ash flows, but also Tertiary andesitic lavas and breccias correlated with those widely distributed on the flanks of the volcano. Early Jemez flows are intercalated with sediments of the Santa Fe Group, but later volcanics overstep Tertiary, Mesozoic and late Paleozoic formations to rest on Carboniferous rocks close to contacts with the Precambrian.

The Jemez Mountains built up over largely concealed sites of Eocene-(?)Oligocene basaltic, andesitic and dacitic volcanism, but the voluminous Miocene-Pleistocene activity commenced with eruptions of thin multiple flows of olivine basalts (10.4 m.y.) in the southern part of the field, closely followed by rhyolite domes, flows and tuffs (about 10 m.y.). In a major episode of activity spanning the period 9-6.5 m.y., initial olivine-augite basalts were succeeded by domes and massive flows ranging in composition from rhyodacite to andesite, and these in turn were followed by rhyolitic flows, domes and pyroclastics. After a period of erosion, olivine-augite and augite basalts (7.4 m.y.) were erupted in the northern parts of the field, spreading out to rest directly on sediments of the Santa Fe Group. Activity continued with a great build-up of dacite, rhyodacite and quartz latite domes and flows (6.7-3.7 m.y.) at least 1000 m thick and having interbedded sediments, tuffs and lahars. This stage of volcanism culminated in construction of a few domes of rhyolitic obsidian and pumice around 2 m.y. ago. Two pulses of activity gave rise to the famed Bandelier Tuff, nonwelded to densely welded rhyolitic ash-flow deposits which form an extensive veneer over the flanks of the Jemez complex. A lower member, associated with formation of the Toledo caldera (1.4 m.y.), is overlain by ash flows which caused collapse of the Valles caldera (1.1 m.y.); in the summit region, the two members are separated by dome-forming rhyolites. The final phase of volcanism led to the construction of a ring of rhyolitic domes on the Valles caldera ring-fracture.

A large predominantly basaltic field of Plio-Pleistocene age occupies the Cerros del Rio area southeast of Jemez, separating the Espanola-Santa Fe and Santo Domingo basins at a 30 km offset in the course of the rift (Fig. 4). Activity here commenced around 2.6 m.y. with eruptions of alkali basalts (hawaiites) which formed numerous cinder cones and several maars (Aubele, 1978). The basal flows rest on sediments of the Santa Fe Group. A second stage of volcanism was characterized by domes and flows of andesites, dacites and basaltic andesites, but eruptions of hawaiites continued on a reduced scale. The final episodes of activity produced alkali basalts. Volcanism was in part contemporaneous with deposition of mid-Pliocene sediments and with eruption of the lower member of the Bandelier Tuff (1.4 m.y.), but activity had ceased before collapse of the Valles caldera (1.1 m.y.).

Albuquerque basin. Offset from the Espanola basin by the Jemez lineament, the Albuquerque trough extends the rift 150 km southwards to Socorro (Fig. 4). Deep boreholes, one of which penetrated Cretaceous strata, demonstrate up to 4 km of infill. Most of this is sedimentary, but Oligocene-lower Miocene volcaniclastic deposits and intermediate to silicic volcanics occur near the base of the succession.

Towards the top of the graben-fill, scattered Plio-Pleistocene basaltic fields include both small shield volcanoes and groups of cinder cones. These basalts display interesting petrological differences. Tholeiites of Santa Ana Mesa are chemically similar to flows in the Cerros del Rio and other fields on the Jemez lineament, whereas the Albuquerque volcanoes consist of olivine tholeiites with relatively lower alkalis (Baldridge, 1979). Farther south, olivine tholeiites are accompanied by basaltic andesites and alkali olivine basalts in the Isleta, Wind Mesa and Cat Hills fields.

Southern parts of the rift. Some investigators (e.g. Seager and Morgan, 1979) interpret a narrow zone of deep basins, young faulting and Pliocene-Quaternary volcanism as a continuation of the rift through southern New Mexico to Chihuahua in Mexico (Fig. 5).

South of Socorro, the volcanics in widely scattered fields are predominantly alkali olivine basalts and basanites; tholeiites are restricted to only one area.

The pre-rift/early rift volcanism in this region was dominated by eruptions of calc-alkalic

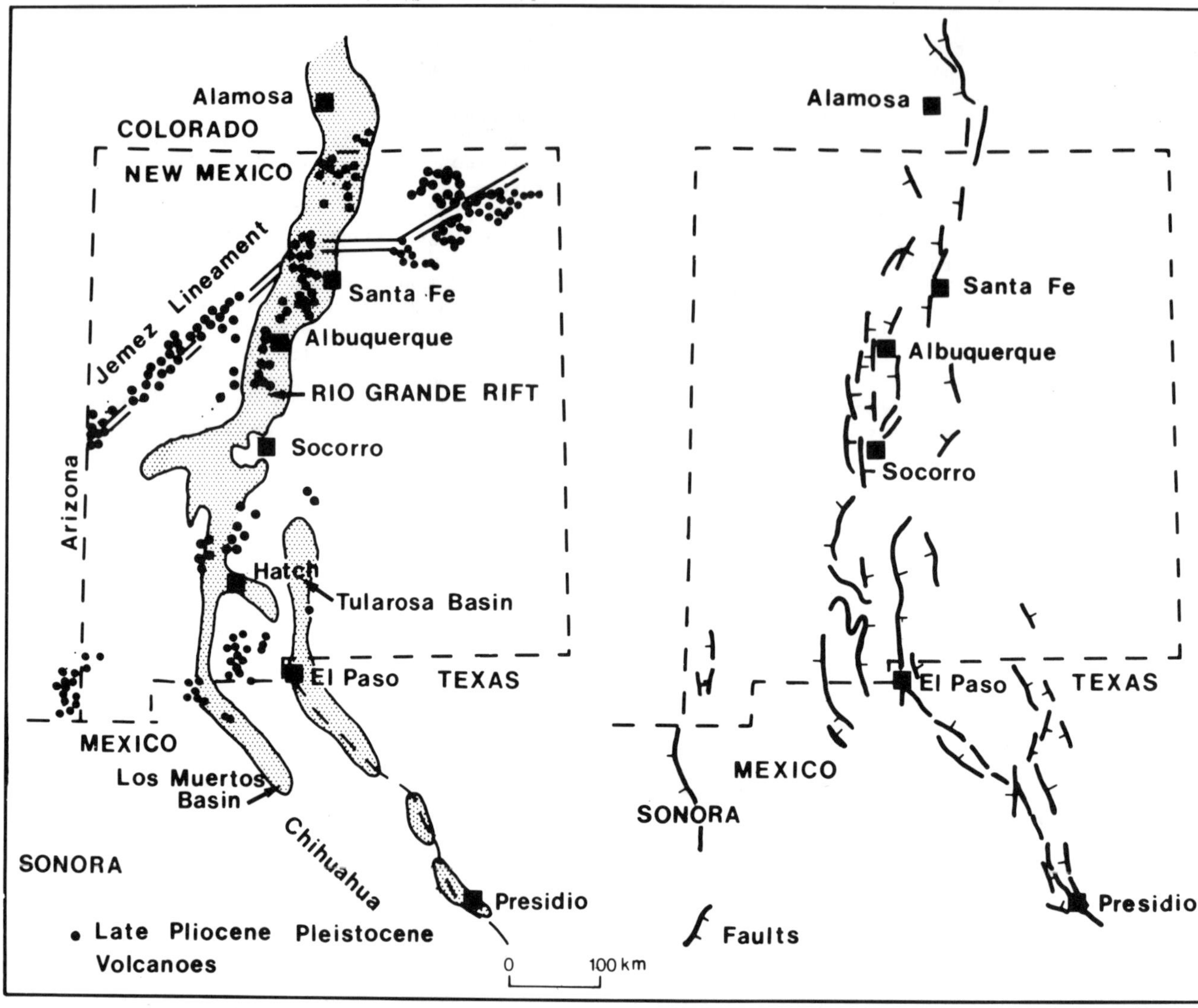

Fig. 5. Distribution of Plio-Pleistocene volcanics and late Quaternary faults with respect to basins of the Rio Grande rift (after Seager and Morgan, 1979).

suites, and the formation of overlapping and nested cauldrons. Andesites and rhyolitic ash-flow tuffs were followed by bimodal volcanism (interbedded calc-alkalic basaltic andesites and high-silica ash flows) some 26 m.y. ago. The first alkali olivine basalts did not appear until the Middle Miocene, after a rift-wide gap in magmatism (20-13 m.y.).

Basin development in this region had not commenced around 33 m.y. ago, but faulting was certainly in progress later in the Oligocene (by 26 m.y.), culminating in major movements between 9 and 2-3 m.y. (Hawley and Seager, 1978). Substantial uplifts occurred during the Tertiary, but at least half the present relief was inherited from Laramide events.

Tectonomagmatic evolution of the Rio Grande rift

The Rio Grande rift can be traced convincingly as a continuous feature from the Upper Arkansas Valley of Colorado to Socorro in New Mexico, but there is still lack of agreement over the extent of the rift system beyond these limits. Elston and Bornhost (1979), for instance, maintain that the case for a southern Rio Grande rift, distinct from the rest of the Basin and Range province, remains to be demonstrated. Inevitably, differences of opinion have arisen over the date of inception of rifting, and no universally accepted model of tectonomagmatic evolution can be offered at present for the Rio Grande graben system.

Elston and Bornhost (1979) relate major compositional changes in volcanics and sub-volcanic complexes to broad stages in a sequence of tectonomagmatic events ranging from those in an Andean arc situation (40-29 m.y.; calc-alkaline magmatism), through modified back-arc extension (30-18 m.y.; basaltic andesites and high-silica rhyolites) to intraplate block faulting (21 m.y. to present; basaltic and locally rhyolitic volcanism). They conclude that, viewed in this context, the formation of the Rio Grande graben system is a relatively recent and local event which commenced with eruption of true basalts after a hiatus in volcanism between 20 and 15 m.y. ago.

Other workers (e.g. Chapin, 1979; Seager and Morgan, 1979) claim that rifting in the north began about 27 m.y. ago and that the development of the southern segment can be traced back to 28 or even 32 m.y., preceding eruptions of the basaltic andesite lavas and interbedded high-silica ashflows which Elston and Bornhost relate to back-arc extension. The concept of a 'middle Miocene lull' in magmatism spanning the period 20-13 m.y. is certainly accepted though, as Chapin points out, this is not a characteristic of the Rio Grande region alone, for the sparsity of volcanics yielding dates in the range 20-17 m.y. has been reported over a wide area of the Basin and Range province.

Tweto (1979) follows Lipman and Mehnert (1975) in taking the change from intermediate compositions to bimodal basic-silicic volcanism as a convenient criterion for separating pre-rift igneous activity from that associated with the graben system in Colorado. This leads to a maximum age of 26-27 m.y. for rift-related complexes, though Tweto points out that some rhyolitic flows slightly younger than 26 m.y. may represent late differentiates of the Oligocene intermediate lavas, so that some overlap between the two episodes (together spanning 40-7 m.y.) must be anticipated. Recent work has shown that the Oligocene and later Tertiary suites in the San Juan volcanic field are genetically unrelated (Lipman et al., 1978), and it is concluded that a major change in the tectonomagmatic environment occurred 25 m.y. ago when magmatism associated with plate collision and subduction gave way to volcanism related to intraplate extension.

The late Cenozoic mafic lavas of the Rio Grande display an interesting overall compositional pattern. In the north, tholeiitic basalts of the Taos plateau within the rift are flanked by alkali olivine basalts beyond its margins (Lipman and Mehnert, 1975). Tholeiites are accompanied by alkali basalts in several fields at or near the top of basin fill in northern New Mexico (Baldridge, 1979). South of Socorro, the Plio-Pleistocene lavas are mainly alkali olivine basalts: but it remains a matter of opinion whether they are related to a southern extension of the rift, or whether they are simply local representatives of the Basin and Range volcanics. Elston and Bornhost (1979), for instance, maintain that alkali basalts are no more abundant along the southern Rio Grande than in other valleys to the west.

Penecontemporaneous eruptions of both tholeiitic and alkali olivine basalts occurred in Plio-Pleistocene fields where the rift intersects a major NE-SW structure - the Jemez lineament. Baldridge (1979) concludes that this transverse shear zone tapped magmas at various levels within a sub-crustal diapir. Though certainly supported by local geophysical data, the concept of a low-density mantle diapir is incorporated in a model partly by analogy with other continental rift zones such as the Oslo graben (Ramberg, 1976) and the Kenya rift (Baker et al., 1978). It is worth noting, however, that crustal doming is neither contemplated nor claimed in the model devised for the central Rio Grande rift.

It will be evident that many disputes over the date of inception of rifting along the Rio Grande hinge largely on matters such as the significance of compositional changes in igneous rocks. This is not to say that the earliest stages of basin development remain entirely unknown, although much of the evidence inevitably lies deeply buried in the undeformed basins. Even where successions dating back to the Eocene are exposed, as in the Espanola basin (Manley, 1979), there still remains the problem of how to identify events related to the commencement of rifting.

In an account of the Rio Grande depression in north-central New Mexico, Baltz (1978) deliberately avoids the issue of what features should be included in the 'rift'. He maintains, however, that there is scant evidence to support a traditional belief that the Miocene basins were downfaulted and so indicate the beginning of rifting. On the other hand, claims that faulting was not prominent until the late Miocene or early Pliocene cannot be used to disprove earlier commencement of rifting for it will be shown later that downwarping to form broad depressions, much wider than the present fault-bounded troughs, is recognized as an initial and integral part of rift evolution in East Africa.

Magmatism in the East African Rift System

The Afro-Arabian rift system extends some 6500 km from the Dead Sea rift in the north, along the neo-oceanic Red Sea (related to separation of the African and Arabian plates), and then through a complex pattern of intraplate structures in Ethiopia, Kenya, Uganda-Zaire border region, Tanzania, northern Zambia, Malawi and Mozambique. No single volume summarises the geology of the entire rift system, though individual segments have been taken as topics for discussion at numerous symposia over the last 15 years. These meetings reflect the emphasis which has been placed on the Kenya (Gregory) rift in research carried out during the Geodynamics Project, as well as the obvious importance of recent studies of the convergence zone in Afar where the East African system passes into structures of the Red Sea and the Gulf of Aden.

Completed and ongoing projects in the Kenya rift and in parts of Ethiopia are summarized in publications which arose from symposia in Irkutsk (Logatchev, 1977), London (Bishop, 1978), Oslo (Neumann & Ramberg, 1978; Ramberg & Neumann, 1978) and Rome (Accademia Nazionale dei Lincei, 1980); other accounts will be referred to in the pages that follow. Afar formed the theme for a symposium held in Bad Bergzabern in 1974, and for the ensuing ICG scientific report (Pilger & Rosler, 1975, 1976).

Structural trends in basement rocks

Over large areas in East Africa the various segments of the rift system closely follow the structural grain of the Precambrian basement which is commonly nearly north-south. McConnell (1972, 1977) stands virtually alone in maintaining that a genetic relationship exists between the rift valleys and ancient structural trends, and the more widely held view is that the rifts only follow the basement grain where this is favourably oriented. Bifurcation of rift structures produced two contrasting branches (the Eastern and Western rift systems) by deflection around the resistant Precambrian Tanganyika Shield in which structural trends are predominantly east-west.

Possible basement controls of fault trends and volcanic lineaments in parts of the Eastern Rift will be mentioned later. Closely related, but far less speculative, issues concern the influence of pre-volcanic basement topography in controlling the distribution of lavas and pyroclastics. In general, differential erosion of basement rocks accentuated their structural grain, but the effects are commonly underestimated.

Western Rift

Few recent studies have been carried out along the Western Rift, and only brief treatment is warranted here to draw attention to major structures and the unique character of associated magmatism.

In its broad structural pattern, the northern half of the Western Rift is reminiscent of the Rio Grande system in the United States, notably in having marked en echelon offsets of depressions framed by faults with total displacements locally exceeding 3 km. Moreover, the troughs in Western Uganda contain very thick accumulations of sediments. Most of the lakes in the Western Rift have been renamed, but earlier names are retained here to avoid confusion. Boreholes and gravity surveys demonstrated some 2.5 km of sedimentary infill in the Lake Albert region and nearly 1 km is reported in the Albert Nile sector (King, 1970).

Another point of similarity with the Rio Grande, albeit superficial, arises from the sporadic distribution of volcanics in the Western Rift: but the comparison breaks down completely when the compositions of the igneous rocks in the two rifts are considered. The mixed calc-alkalic, tholeiitic and alkalic basalt suites of the American rift have few counterparts in the Pliocene-Quaternary potash-rich, commonly highly undersaturated volcanics of western Uganda.

In a classic account, Holmes and Harwood (1937) showed that the lavas of the Birunga (Virunga) field of southwestern Uganda fall into six petrographic groups: an olivine-leucitite series, a group of melilite-feldspathoid rocks, a leucite-basanite series, and the closely related potash trachybasalt, potash trachyandesite (latite) and limburgite-trachybasalt series. Only in the last-named series and in rare melilite nephelinites does soda exceed potash. Detailed petrochemical comparisons are beyond the scope of this chapter, but it is worth noting that in early accounts attention was drawn to certain similarities between the restricted oversaturated volcanics (hypersthene latites) and xenolithic monzonites of Sabinyo volcano in southwest Uganda and quartz syenites of the Organ Mountains, New Mexico (Holmes & Harwood, 1937); but the latter are pre-rift, and the former are in no way representative. Much more important in showing the lack of similarity in magmatism in the Rio Grande and Western Rift environments are the carbonatite complexes of the latter (see Tuttle and Gittins, 1966).

Farther south, the volcanic field near Lake Kivu is of particular interest because, as Bailey (1974) has pointed out, it contains tholeiites which are rare elsewhere in the East African rifts. The tholeiites are accompanied by alkali basalts, but occurrences of trachytes and unusual potash-rich rhyolites preclude comparison with the volcanics of all but the Ethiopian rift where the association is reported from at least three localities on the plateau (P.A. Mohr, personal communication).

At the southern end of the Western Rift, an isolated area of late Tertiary-Quaternary volcanism in the Rungwe region of Tanzania displays nephelinites as well as more abundant representatives of basalt-trachyte-phonolite suites (Harkin, 1960). In the compositional range and problems which arise in attempts to explain the preponderance of felsic volcanics, the Rungwe field has more in common with the Eastern Rift than the petrologically unique province of Uganda and neighbouring regions.

Eastern Rift

A review of the geology of the Eastern Rift System of Africa (Baker et al., 1972) has yet to be superceded as the most informative single summary of the evolution of what is undoubtedly the leading example of a strongly magmatic continental rift, but much detailed work has been carried out in both Kenya and Ethiopia since that account was compiled and several sections require substantial modification in the light of recent research developments. In fact, it is becoming increasingly difficult to make valid generalizations about the stratigraphy and petrology of the soda-rich alkaline volcanics in this province without intimate knowledge of results which are not yet widely available for consultation and comparison.

The problems can best be illustrated by reference to results emerging from detailed studies in Kenya, and the remaining part of the article is devoted to that sector of the Eastern Rift system.

Summary of investigations in Kenya

The closing stages of the Upper Mantle Project were marked by outstanding progress in structural stratigraphic, petrological, geomorphological and paleontological studies in Kenya (King & Chapman, 1972; King et al., 1972; Williams, 1969, 1970, 1972; Baker at al., 1971; Macdonald et al., 1970; Bishop et al., 1969). The same period produced a spate of geophysical projects (see Girdler, 1972), stimulated by the impact of plate tectonics.

The most significant advances towards a clear understanding of the structural and volcanic evolution of the Kenya rift have come from work in the northern part of the country (e.g. King & Williams, 1976; King, 1978; Chapman et al., 1978; Williams & Truckle, 1980). Maps at a scale of 1:125,000 are in course of publication by the Geological Survey of Kenya. As and when available, the sheets will cover some 20,000 km^2, and the investigations which led to their compilation must be ranked among the major achievements in studies of continental rifts during the Geodynamics Project.

Detailed remapping in southern Kenya was carried out in parallel with the above project, producing revised and better-founded ideas about the history of volcanism along the western margin of the rift (Crossley, 1979). Independently, smaller areas have been examined in detail on the eastern side of the rift in southern Kenya (Baker & Mitchell, 1976), and in a region closer to the equator (Lamb, 1975).

Despite their obvious value alongside geochronological data from the volcanics, comments on paleontological studies must necessarily be excluded from this account; a recent symposium volume (Bishop, 1978) contains excellent summaries and references.

Topography and the 'Kenya dome'

The 750 km-long Kenya (Gregory) rift is an impressive topographic feature particularly where it transects the equatorial highland region in which ground flanking the depression commonly rises to altitudes of 3000-4000 m. The highest part of the rift floor stands at about 2000 m above sea level near the equator, descending to less than 300 m in the north at Lake Turkana (formerly Lake Rudolf), and to altitudes around 600 m at the Tanzanian border in the south.

Many review articles mention this topographic culmination along the rift floor, but few make any attempt to explain its existence in what is undoubtedly the deepest part of a largely infilled trough. To some, the high part of the rift floor is regarded as a feature associated with pulses of domal uplift because it coincides with high elevations of the rift shoulders; but no detailed evolutionary sequence has ever been presented to substantiate this view. Others accept or demand late uplift of the rift shoulders as an important structural event in the formation of the 'Kenya dome', but maintain that the present topographic features are more a reflection of volcanic accumulation than vertical uplift associated with rift evolution.

The culmination in the floor occurs in an area of maximum proved thickness of Quaternary rocks within the rift. It embraces one of the two major developments of rhyolitic volcanics in Kenya, but it cannot be explained simply as a single large central volcano. Finally, the possibility of a culmination arising by accumulation of volcanics together with contemporaneous or later tilting of the rift floor has yet to be tested in detail. This can only be done by precise correlation of volcanic and sedimentary formations, and studies of the evolution of internal drainage systems.

Because of these problems, the term 'Kenya dome' has come to be used in recent years to describe a topographic rather than a structural feature (King, 1978; Williams, 1978a). Studies of erosion surfaces in Ethiopia (Merla, 1963, Merla and Minucci, 1938) are still in their infancy, so that it is not yet possible to assess claims that a major topographic culmination bisected by the rift can be ascribed to three episodes of uplift. However, Mohr and Gouin (1976) make the point that the complicated pattern of block uplifts in the Horn of Africa, though involving slight tilting away from the rift, cannot be called 'domal'.

Major rift structures in Kenya

For descriptive purposes, the Kenya rift is conveniently divided into three sections (Fig. 6). What is frequently referred to as the central part is a 70 km-wide depression which trends SSE from the equator for about 150 km. The southern section has a south to south-southwest trend to the Kenya-Tanzania border, beyond which the rift soon loses its identity in a region of tilted blocks. In northern Kenya, a well, defined depression 60-70 km wide trends north then north-northeast for about 200 km from the equator before the rift structures fan out over a very wide area. Near the equator, a WSW-trending depression (the Nyanza trough; formerly termed the Kavirondo or Kisumu rift) some 20-30 km wide branches from the main rift towards Lake Victoria.

The sinuous course of the Kenya rift is attributable to intersecting fault patterns, rather than an en echelon arrangement of constituent basins as in the Rio Grande system. It is not unlikely that the SSE-trending central section in Kenya owes its orientation to strong basement control, but there are no outcrops of metamorphic rocks within this part of the trough and exposures close to the margins provide inadequate evidence to test the hypothesis.

In few areas in Kenya is there any approximation to a simple graben structure. Major faults with total displacements up to at least 3 km commonly define the western margin of the trough, and they expose Precambrian basement rocks beneath a cover of volcanics. These structures are interspersed with areas in which monoclinal downwarping played a significant role in the evolution of the depression. Downwarping is an even more characteristic feature of the eastern margin of the rift. Major faults are common, but basement rocks are not revealed in the scarps.

The evolution of the rift was marked by progressive concentration of tectonic activity into narrowing sections of the floor. This produced the most recently active inner troughs/graben some 25-35 km wide, generally lying off-centre within a depression about twice that width. These inner troughs tend to lie closer to the rift margin along which major downwarps are significant features and they are usually separated from spectacular single fault scarps on the opposite side of the rift by step-fault platforms, tilted blocks, or uparched portions of the floor which expose basement rocks as well as older volcanic and sedimentary successions. Only in the southern part of the rift does the inner trough run more or less axially.

Despite the emphasis placed here on the youthful character of the inner structures, it must be stressed that rift boundary faults were rejuvenated well into Pleistocene times by depression of the floor and uplift of the shoulders of the graben. As King (1978) has pointed out, the present morphology of the rift was produced by Quaternary faulting and downwarping on a scale which outpaced the rate of accumulation of volcanics and sediments within the depression.

Volcanics: general distribution and thickness

The volcanics are soda-rich and they embrace the full compositional range from basalts (both alkaline and transitional varieties) and nephelinites to trachytes, phonolites and alkali rhyolites.

An appreciation of the development of the main structural features in the Kenya rift is necessary in order to understand where and how the most complete volcanic successions are available for study. It has been shown, for instance, that volcanism as well as tectonic activity tended to become concentrated with time into narrow inner troughs within the rift. On the other hand, not all Quaternary volcanic events were restricted to the depression: voluminous eruptions also took place well beyond the structural confines of the rift.

Cenozoic volcanics are exposed along the entire length of the Kenya rift. At various stages in its evolution, the volcanic rocks and interbedded sediments filled the depression, and some flows spread across the flanking ground to coalesce with the products of eruptions at or far beyond the rift margins. As a result, Miocene to Holocene lavas and pyroclastics are now distributed in a meridional belt up to 550 km wide.

No boreholes drilled within the rift have gone deep enough to penetrate basement rocks: indeed, in only one area has drilling provided significant data bearing on the important question of total thickness of volcanics. This is in the central part of the rift where late Quaternary rhyolitic and trachytic lavas and pyroclastics are proved to be more than 1 km thick. Elsewhere, careful mapping has shown that the volcanics occupy the graben to depths exceeding 3 km.

In areas where it has been possible to compare sequences within the trough with those exposed along the margins, the latter are shown to be attenuated and incomplete. The results of these studies warn against speculation based solely on evidence displayed in rift boundary scarps and the flanking terrain, for the successions exposed there represent either episodes when the volcanics filled and overtopped a developing trough, or they belong to eruptive centres located on the rift shoulders. In the former case, correlation right across the rift may be regarded as a feasible proposition, though it presupposes that very fluid flows were erupted in sufficient volume to spread evenly and that no constraints existed on flow across the axial region of the rift. In fact, it has proved surprisingly difficult to make detailed correlations of this kind between sequences on opposite sides of the rift (King, 1978; Williams & Truckle, 1980). In the latter case, volcanism of quite different character on opposing shoulders can be expected, despite broad contemporaneity as indicated by isotopic age determinations. Truncation of major volcanoes by faults at the rift margin locally exposes successions up to 2 km thick.

Volcanic landforms and intrusive complexes

Kenya is a region of great contrasts in volcanic landforms, and some of these are probably unique. The most spectacular features are giant cones several kilometres high and with basal diameters up to 100 km. The most impressive examples (e.g. Mt. Kenya) are situated outside the rift depression. Miocene nephelinite volcanoes of the Kenya-Uganda border, some 200-300 km west of the rift, are sufficiently dissected to reveal intrusive alkaline complexes composed of ijolitic rocks and carbonataties (King et al., 1972). Less severe erosion of younger basaltic, trachytic and phonolitic central volcanoes, situated up to 100 km east of the rift, exposes no more than high-level plug and dike complexes.

Quaternary volcanism east of the rift produced predominantly basaltic linear multicentre ranges (Fig. 6). Each of these chains consists of several hundred cinder cones and/or maars on thick foundations of lavas formed by great upwellings along fissures which have trends oblique to those of the rift faults (Williams, 1978b).

Trachytic volcanoes within the rift are low-angle shield structures. The least dissected examples are caldera volcanoes, somewhat unevenly distributed in a chain along the inner trough. Calderas south of the equator are 10 km or more in diameter, and they are the likely sources for voluminous air-fall pyroclastics which mantle areas up to 100 km west of the axial line of the rift. In contrast, the volcanoes in a northern group are surmounted by calderas which are rarely more than 8 and generally less than 5 km across. They gave rise to much more restricted pyroclastics, but all display basalts which were mostly erupted from fissures aligned parallel to the rift faults (Williams, 1978b). Ejected blocks of syenite in the south, and syenite and gabbro in the north, provide some evidence of the nature of sub-volcanic complexes. The caldera volcanoes in northern Kenya are flanked by older, tilted, deeply eroded trachytic centres in which dike complexes are exposed (Webb & Weaver, 1975).

For descriptive purposes, attempts were made (Williams, 1970) to distinguish the products of central-vent activity from volumetrically more important 'plateau'/'flood' volcanics related to fissure and multicentre eruptions. This understandably led to gross oversimplification, yet it served to draw attention, for instance, to problems connected with the sources and origins of the extensive plateau phonolites and trachytes of central and southern Kenya.

More recently, the recognition of intrusive complexes in the rift floor suggests that at least some of the plateau phonolites came from large shield volcanoes (King & Chapman, 1972; Lippard, 1973), though the question of origin of large volumes of salic magma still remains somewhat speculative. Fissure sources cannot be dismissed for some of the plateau trachytes, but there is mounting evidence to indicate that other extensive trachytic and ignimbritic formations are associated with shield structures.

Infilling and overtopping of rift depressions

The distribution of volcanics of different compositions at various stages in the evolution of the Kenya rift is shown in Fig. 7, and no systematic commentary will be provided in this account. More details are readily accessible in several recent review articles (Williams, 1978a, 1978b; King, 1978) and in regional summaries referred to elsewhere in the text.

All the evidence assembled during recent studies shows that the oldest volcanic rocks now exposed within or close to the rift accumulated in broad inter-connected downwarps, destined later to become fault-bounded troughs. The early depressions developed on a surface of high relief carved in late Precambrian metamorphics.

The early history of subsidence is now well documented in a large area north of the equator because old sequences and structures can be studied within the confines of the rift and compared with those of the rift margin (Chapman et al., 1978). Exposures of basement rocks within the rift make this a particularly rewarding region for detailed investigation.

The rift depressions in Kenya probably developed earlier in the north than in the south, though it will be apparent from Fig. 7 that Quaternary formations extend across nearly the full width of the rift in southern Kenya, obscuring the initial stages of development. Confinement of young basalts and trachytes to the narrow Baringo-Suguta trough north of the equator was achieved by continuous downwarping, so that progressively older Tertiary volcanics and sediments are exposed towards the margins of the rift floor: a combination of faulting and arching reveals even basement rocks on the western side. The oldest volcanics in the north are basalts which have yielded dates in the range 16-20 m.y.; some may be as old as 30 m.y. They commonly rest on coarse sediments (the Turkana Grits) derived from basement rocks. These deposits are traditionally assigned to the Miocene, but the possibility that they (and consequently the Turkana depression) date back to the Lower Tertiary cannot be dismissed. The oldest dated volcanics in southern Kenya are melanephelinites and augitites with ages in the range 12-15 m.y. These infilled a pre-existing depression (Crossley, 1979).

Two periods of massive overspill of 'flood' volcanics are recognized: the first led to widespread overtopping of the rift by 'plateau' phonolites around 12-15 m.y. ago, and the second by trachytic lavas and ignimbrites some 2-6 m.y. ago (see Fig. 7).

The importance of pre-existing basement topography in controlling the distribution and thickness of flows spilling over from the rift is frequently underestimated and, at times, not even considered. The most common misconceptions arise because of failure to appreciate the superior resistance to erosion of lavas such as phonolites, as compared to basement gneisses and schists: yet this is a crucial issue which cannot be avoided in attempts to reconstruct the form of any sub-volcanic surface, or to estimate the original extent and volume of some of the older lava

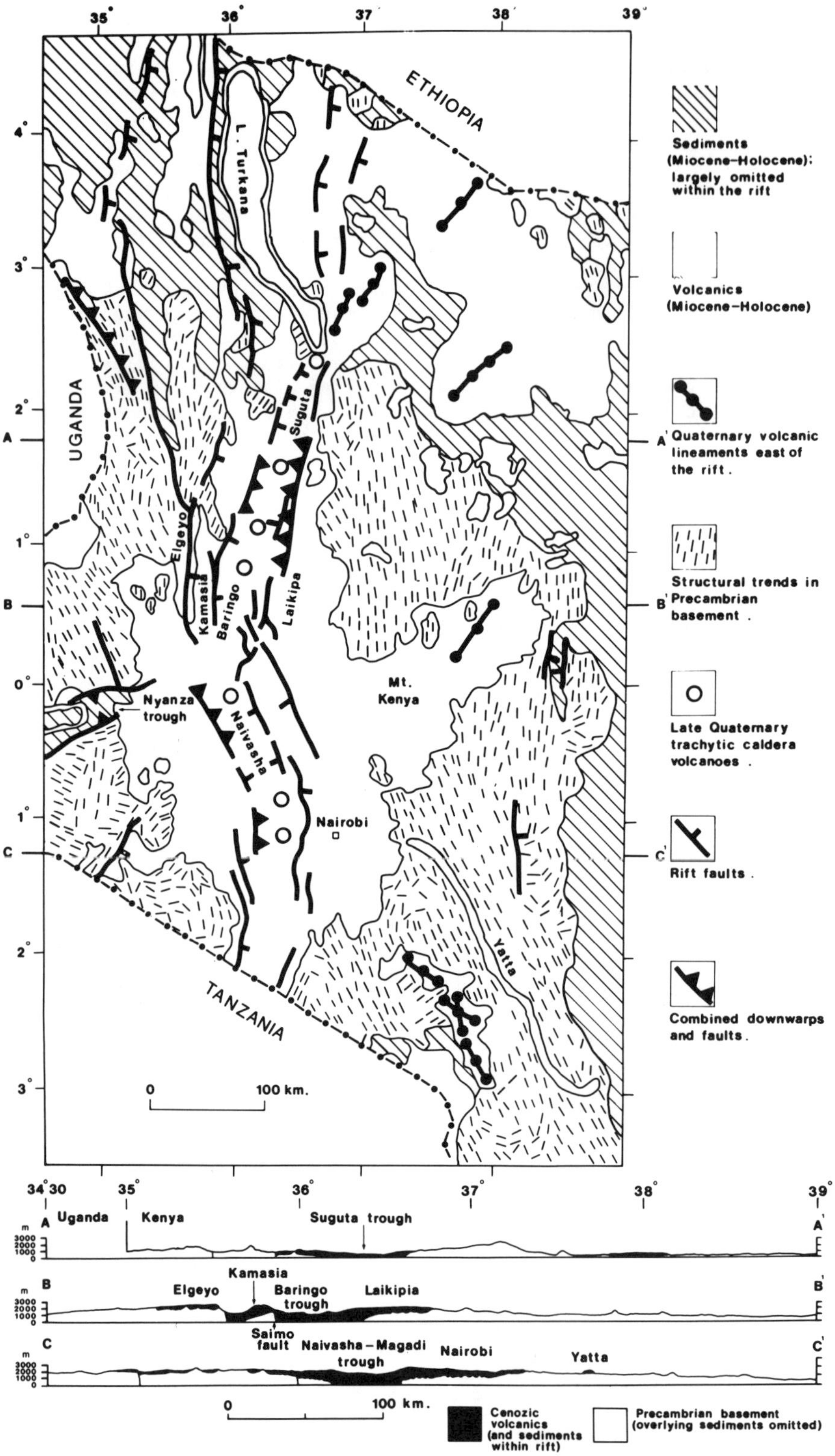

Fig. 6. The Kenya rift: simplified structural map and sections.

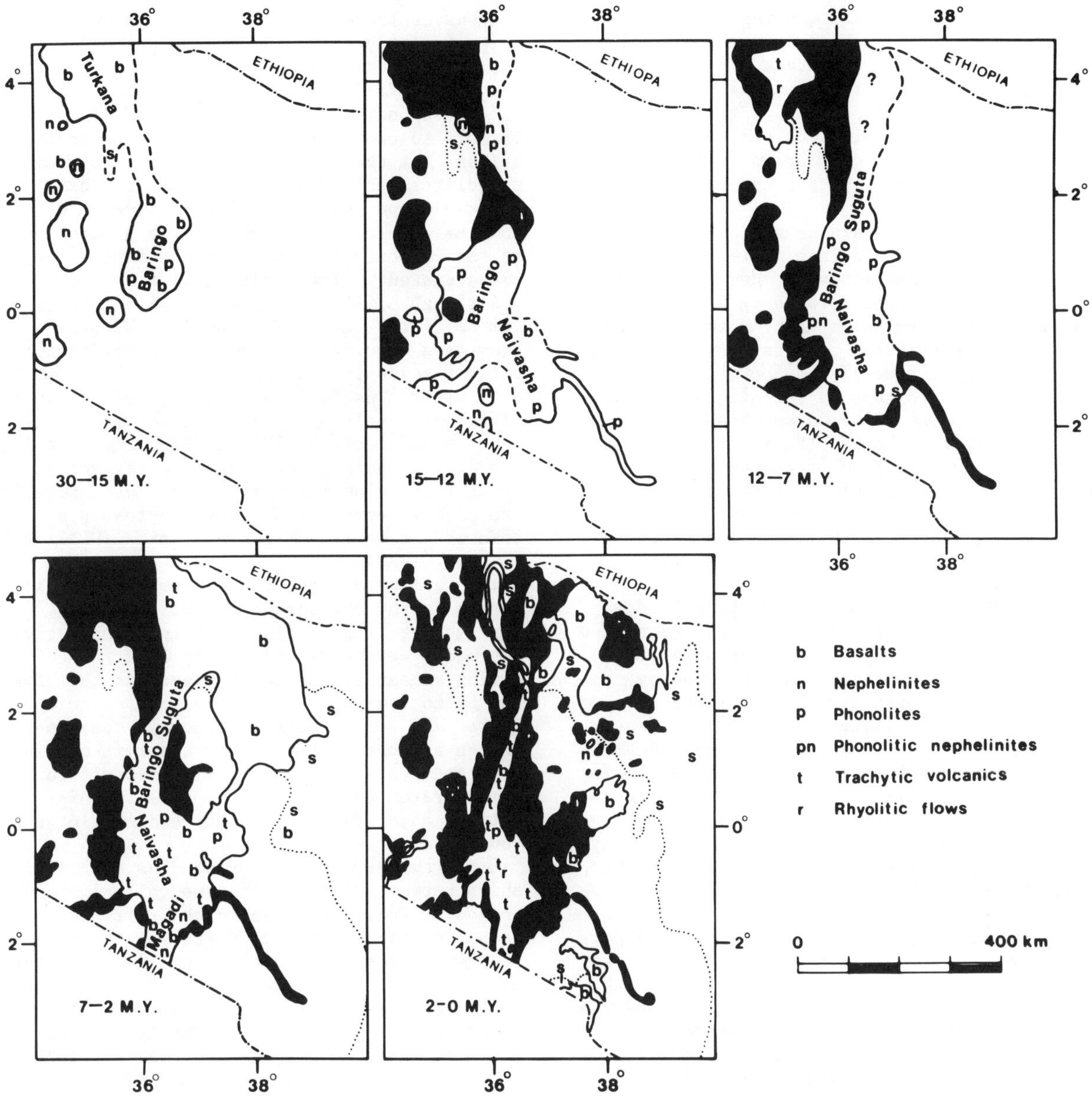

Fig. 7. Volcanic evolution of the Kenya rift.

sheets. The problems are best illustrated by considering the Miocene 'plateau' phonolites. Few deny the spectacular inversion of topography which left a 250 km-long 13 m.y. phonolite valley-flow (Fig. 6) now perched as a thin lava capping which preserved basement rocks of the Yatta Plateau well above the elevation of the adjacent plains in southeastern Kenya. Many, however, are reluctant to accept the effects of differential erosion elsewhere, and it is not widely appreciated that where phonolite sheets end at present-day rivers this does not necessarily infer any substantial reduction in the length of the flow; in many cases, this situation arises merely by erosion of a former basement barrier which terminated the flow (Williams, 1978a).

Relationships between volcanic and tectonic activity

Early reviews attempted to show that prominent episodes of volcanism were bracketed by major tectonic events such as faulting, uplift and downwarping. These claims tended to be based on the notion that a few widespread phases of faulting could be identified and dated; but only sporadic age determinations were available for the calibration of volcanic sequences proposed during rapid reconnaissance mapping. Recent reviews (King, 1978; Chapman & Brook, 1978) have criticized the frequently mistaken assumption that it was volcanism rather than tectonic activity which was nearly continuous during the evolution of the rift, and they point out that misleading conclusions are likely to be drawn in areas where long periods are simply not represented by any volcanic activity.

One of the first accounts examining the age and tectonic relationships of East African volcanic rocks was that of Kent (1944). He endorsed many of the earlier views put forward by Sikes (1939) in coming to the conclusion that 'rift faulting has repeatedly terminated periods of volcanic extrusion'. More recently, support for the assertion that 'simultaneous formation of shear fractures and ascent of magma up those fractures is mechanically improbable' (Baker at al., 1978) is considered to come from 'repeated episodes of flood volcanism and faulting' in the southern part of the Kenya rift. A more general statement based on this hypothesis (Williams and McBirney, 1979) implies that volcanism alternated with faulting throughout the Kenya rift.

Full and critical discussion of the evidence leading to these conclusions cannot be undertaken here, but one or two aspects deserve brief comment. The problems are best reviewed by dealing separately with examples from the northern part of the rift, where older sequences and structures have been studied in great detail, and from southern Kenya where the early infilling formations are effectively concealed within the trough by Pliocene-Quaternary volcanics.

North of the equator, recent work in the Elgeyo-Kamasia-Baringo area on the western side of the rift shows that both volcanic activity and closely-spaced faulting migrated inwards with time, resulting ultimately in confinement of Quaternary activity to a narrow inner trough situated closer to the eastern than the western rift margin (Chapman et al., 1978). Downwarping allowed the accumulation of very thick successions of Miocene phonolites and sediments (16-7 m.y.), but there is not much direct evidence for faulting before a first major episode about 7 m.y. ago which produced displacements up to about 2000 m on the Elgeyo boundary fault and along the parallel Saimo fracture zone situated within the depression some 30 km from the margin. The fact that the most voluminous phonolite eruptions preceded these movements by some 5 m.y. led Chapman and others to conclude that periods of maximum intensity of volcanic activity were not terminated by major faulting: there is no denial that some important eruptive episodes closed with phases of faulting. Indeed, evidence from the Pliocene record shows that significant fault movements took place, for instance, after the eruption of the Kaparaina basalts (5.5 - 4.5 m.y.), but before sedimentation in local basins (4 - 2 m.y.).

In several areas on the western side of the rift, major faulting about 7 m.y. ago coincided with a change from essentially basaltic and phonolitic to basaltic and trachytic volcanism. There was no corresponding simple change from, for example, fissure to central-vent activity. A traditional belief that the Miocene basalts and 'plateau' phonolites were erupted from fissures has been largely dispelled by the identification of complex shield structures in both groups. Most of the Pliocene trachytes in northern Kenya are also related to low-angle shields.

Although still largely unexplained, the compositional changes in the late Miocene provide a convincing example of a relationship between volcanic and tectonic activity, and they also offer a useful basis for an initial subdivision of volcanic sequences between the equator and latitude 1°30'N (King and Chapman, 1972). No such relationship is recognized in areas farther north, however, because phonolites are poorly represented and trachytes appear among some of the earliest rift volcanics (Williams and Truckle, 1980).

Uplift and major faulting occurred in the Kamasia region during the period 2 - 0.5 m.y. ago (Chapman et al., 1978). The most impressive movements were those resulting from the rejuvenation of earlier structures such as the Elgeyo and Saimo faults, whereas the most voluminous Quaternary volcanism took place in the inner trough at some distance from the rift margins. Understandably, it has proved difficult to demonstrate precise temporal relationships between major fault movements and many of the younger eruptive episodes.

More rewarding areas for studies of relationships between Plio-Pleistocene volcanism and phases of faulting close to the rift margins are found in southern Kenya, where some of the youngest volcanic formations extend right across the trough. On the western side of the rift, faulting preceded or accompanied volcanic eruptions, many of which evidently occurred close to the fracture zones (Crossley, 1979). This part of the rift depression was never completely infilled by flood volcanics, so that fault scarps commonly have a facing or ledge of younger lavas representing remnants of formations which were subsequently let down into the rift floor either by rejuvenation of the original fault, or by movement on a nearby parallel fracture stepping down towards the rift. The amount of scarp retreat which can be inferred in these examples gives some idea of the variable length of the interval between faulting and a subsequent phase of volcanic activity.

Early movements on major faults which define the western side of the rift in southern Kenya cannot be dated more precisely than middle to late Miocene (12-7 m.y.), but it seems likely that the displacements coincided with the beginning of a long period of basalt-trachyte volcanism. This followed, but did not replace, widespread strongly undersaturated volcanic activity (Crossley, 1979).

The eastern side of the depression in southern Kenya was overtopped by trachytes less than 2 m.y. ago. By careful correlation, these lavas (Limuru trachytes) have been traced from the shoulders of the rift, across a step-fault platform, to exposures in tilted horst blocks in the floor of the graben where they are ultimately concealed by younger volcanics. Baker and Mitchell (1976) concluded that the Limuru trachytes belong to the earlier of two main Plio-Pleistocene basalt-trachyte 'cycles', each of which was preceded and followed by important phases of rift faulting recognizable largely because 'volcanic activity ceased or was at a minimum during those time periods'. There are no exposed contacts between the Singaraini basalts (2.3 m.y.) and the Limuru trachytes (1.9 m.y.) so precise relationships remain speculative. It is clear, however, that faulting occurred between eruptions of the Ol Tepesi basalts (1.4 - 1.6 m.y.) and Plateau trachytes (1.2 - 0.9 m.y.) of the second 'cycle'; and faulting continued during accumulation of the Plateau trachytes, culminating in more significant movements at the close of that eruptive episode.

It is of interest to note that longer-term (Miocene-Quaternary) cyclic basalt-trachyte volcanism in northern Kenya was not neatly punctuated by episodes of faulting (Williams and Truckle, 1980). In that region, however, the trachytes are not flood lavas; they form discrete central volcanoes.

Volumes of igneous rocks

Several attempts have been made to estimate the volumes of volcanics associated with the Kenya rift, but no detailed discussions or maps are available in published form to show precisely what evidence was used. It is impossible, therefore, to assess the validity of assumptions demanded by such order-of-magnitude calculations. Another difficulty arises in comparing the estimates because one account (Williams, 1972), for instance, deals strictly with the volcanics in Kenya, whereas another (King, 1978) more realistically includes major central volcanoes in eastern Uganda, and some (but not all) the volcanics in northern Tanzania. Making adjustments for these differences in the regions selected (Table 3) leads to surprisingly close agreement in the figures quoted for the total volume of volcanics (King's estimate: 230,000 km^3. Williams' adjusted estimate: 226,000 km^3).

This is not to say that the two accounts are in complete harmony when it comes to examination of the volumes quoted for some central volcanoes, or the necessarily even more speculative figures

TABLE 3. Volumes of volcanics associated with the Kenya rift

(1) Kenya, E. Uganda and part of N. Tanzania (King, 1978)

	km^3
Basalts, mugearites	80,000
Nephelinites, basanites etc.	70,000
Phonolites	50,000
Trachytes, alkali rhyolites	30,000
Total	230,000

(2) Kenya (Williams, 1972)

	km^3
Fissure/multicenter basalts	60,000
Plateau phonolites	25,000
'Flood' trachytes, rhyolites	20,000
Miocene central volcanoes (nephelinites etc.)	5,000
Pliocene central volcanoes (all compositions)	31,000
Quaternary central volcanoes (basalt, trachyte)	3,000
Total	144,000

(3) Kenya, E. Uganda and part of N. Tanzania: figures from Williams (1972) modified to embrace region covered by King (1978)

	km^3
Miocene flood volcanics (Kenya)	60,000
central volcanoes (Kenya)	5,000
central volcanoes (E. Uganda)	55,000
Pliocene flood volcanics (Kenya)	25,000
central volcanoes (Kenya)	31,000
Quaternary flood volcanics (Kenya)	20,000
central volcanoes (Kenya)	3,000
N. Tanzania (part)	27,000
Total	226,000

derived for some groups of 'plateau' volcanics. For example, there are differences of opinion about the original volume of the Mt. Kenya volcanics; and King follows Lippard (1973) in taking the volume of the Miocene 'plateau' phonolites to be around 40,000 - 50,000 km^3. However, Lippard's revision of earlier estimates is itself based on a map which completely ignores thin yet extensive phonolites in parts of southwestern Kenya, including the only known remnants of a phonolite shield volcano outside the rift.

Lippard suggested that the present author underestimated the thickness of phonolites within the rift in arriving at a figure of 25,000 km^3 for the Miocene flows, and it is certainly true that the great thickness of rift infill which has been demonstrated locally could not have been anticipated simply from examination of condensed and incomplete sequences exposed on the rift shoulders. On the other hand, there is now some danger of over-correction based on assumptions that flood volcanics periodically inundated much larger areas of the rift floor than can be illustrated by slender, and frequently circumstantial, evidence.

Recent workers favour derivation of the phonolites from large shield-like structures in the rift floor. The thickest known sequence is in the Kamasia area at about $0^{o}30'$N where phonolite lavas make up some 1300 m of a succession 2000 m thick which includes mugearites, hawaiites, basanites, tephrites and sedimentary formations. The phonolites show substantial lateral variations in thickness as may be expected close to shield sources, and extrapolation into less well exposed regions presents obvious difficulties.

Problems of a different kind arise in estimating the volumes of 'flood' volcanics covering the plateaus flanking the rift, and a matter of great importance concerns the former extent of well-exposed flows. Some diagrams give the impression that there is ample justification for joining up all known outcrops of lavas like the Miocene phonolites simply on the basis of lithological similarities, or broad correlation from scattered age determinations. This approach sometimes leads to dubious conclusions because the effects of former topographic controls are neglected. Marked lateral variations in thickness do occur, both as a result of rugged buried topography as well as thickening associated with accumulation in broader downwarps than may be expected from the width of the present rift depression. Lippard (op. cit.) has shown, for instance, that Miocene phonolites are about 800 m thick in the Elgeyo escarpment, which now forms the margin of the rift west of the Kamasia area, whereas the average thickness across the flanking plateau is less than half this figure. In many other areas outside the rift, the plateau phonolites are less than 100 m thick.

Lippard evidently visualises the total area originally under a phonolite cover to be some 100,000 km^2. Tentative figures for areal distribution within and outside the rift are taken from his map, despite omissions mentioned above, and these are combined with ranges of average thickness in Table 4. It is clear that the accuracy of estimates of total volume hinges more on personal bias concerning matters such as the distribution of postulated concealed sources of phonolites in the rift floor, than on knowledge of the formations on the plateaus. Nevertheless, detailed studies of the latter continue to demonstrate some important principles.

Domal uplifts and tilts: evidence from the volcanics

Prolonged erosion of the late Precambrian/early Paleozoic Mozambique Belt in East Africa spanned some 400 m.y. and led to the development of several widespread erosion surfaces which were subsequently uplifted and dissected (Pulfrey, 1960; Saggerson and Baker, 1965). Full discussion of controversial matters such as the ages, original forms, correlation and history of deformation of these surfaces is out of place here. It is important to realise, however, that many problems which have arisen over concepts of domal uplifts, particularly disputes about the timing and amount of uplift, have yet to be resolved. Consequently, the distinction between rift-related and unrelated vertical movements remains unclear.

The difficulties are compounded in many cases simply because of failure to recognize the fundamental difference between crustal uparching prior to major faulting (doming) and later tilts associated with uplifts of the rift shoulders. The disposition of lava flows erupted on, or spilling on to, the flanks of the rift can be expected to provide crucial evidence in areas where it can be reasonably assumed that the volcanics were essentially horizontal at the time of emplacement. This is a more rewarding approach than one which relies on somewhat speculative reconstruction of the form of largely buried pre-volcanic surfaces.

Early reviews (e.g. Baker and Wohlenberg, 1971) related the evolution of the Kenya rift to three main phases of uparching established from studies of erosion surfaces. It was claimed that late Miocene domal uplift of some 300 m central Kenya culminated in major faulting in the Pliocene, and that further uplift of 1400 m occurred in late Pliocene to mid-Pleistocene times. On the other hand, King (1978) maintains that much of the present elevation of the basement along the rift zone was inherited from much earlier episodes of uplift. He claims that major uparching of a late Mesozoic surface was associated only in part with rift tectonics.

Chapman and others (1978) showed that the lava plateau flanking the western side of the rift north of the equator was tilted gently away from the depression in post-Miocene times. Understandably, it proved impossible to place an upper age limit on this episode of tilting, though it evidently took place after major faulting. In the same area, Jones and Lippard (1979) claim that sediments beneath Middle Miocene volcanics were deposited on deeply dissected terrain 'carved into the upwarped margin of the proto-Kenya rift'. This statement implies rift-related uparching prior to major faulting which took place around 7 m.y. ago, and before the back-tilting of lavas mentioned above. In the same article, reference to substantial uplift of the Kenya dome in the Miocene is evidently accepted from the earlier work of Saggerson and Baker (1965), despite the fact that the authors maintain that no doming of the basement surface can be detected in north-south sections along the rift flanks.

TABLE 4. Areal distribution, assumed average thicknesses, and total volume of Miocene 'plateau' phonolites in Kenya

Likely areal distribution (km^2)			Observed maximum aggregate thickness (m)	Assumed average thickness (m)	Volume (km^3)
Within rift		38,000	2,800	(a) 1,000	38,000
				(b) 500	19,000
Plateau west of rift	25,000	62,000	400	(a) 200	12,400
Plateau east of rift	37,000			(b) 100	6,200
Total volume: 25,000 - 50,000 km^3					

Pyroclastics are generally less rewarding than lavas or sediments in providing evidence of tilting, because few assumptions can be made about the original attitude of a cover which is frequently draped across the pre-existing topography. However, studies of the development of drainage patterns on thick volcanic ashes sometimes yield valuable information about surprisingly recent tilts associated with uprise of the rift shoulders (Williams, 1978b).

Discussion

These concluding remarks seek more to identify than to explain some of the broad similarities and striking contrasts in magmatic activity in the three regions selected for review. Neither the topics nor the discussions presented below are intended to offer exhaustive treatment.

It is instructive to refer not only to igneous activity, but also to some of the major events in the structural evolution of the rifts, in particular to those which bear directly on issues such as the distribution of volcanics. Similarly, it is appropriate to make reference to some aspects of the structural settings and early histories of the regions used for this study, if only to highlight conflicting views about the significance of domal uplifts, and problems in distinguishing between pre-rift and rift-related magmatism.

Pre-rift histories of the regions

Structures in basement formations undoubtedly exert some control at least on the orientations of segments of rift systems. This is most clearly illustrated by reference to the Cenozoic East African system in which two distinct rift branches developed by deflection around the ancient "Tanganyika shield'. It so happens that the two branches are marked by volcanism of quite different character, composition and intensity: but the reasons have yet to be fully explained.

The extent to which basement structures may account for different relationships between the constituent basins or troughs in a rift system can seldom be demonstrated with conviction. Consequently, the contrasting structural patterns in the Rio Grande (en echelon basins with intense magmatism along at least one prominent offset) and in Kenya (abrupt changes in the orientation of troughs arising from intersecting sets of faults) cannot be explained at present. Similarly, the explanation for a marked offset in late parallel troughs in the Oslo graben remains conjectural. Unlike the Rio Grande 'Jemez lineament', the offset in the Oslo rift did not have associated with it a noticeable concentration of igneous activity, though it may have dictated the distribution of some volcanic centres (Fig. 2).

No truly transverse structures exist in Kenya, though divergent volcanic lineaments occur east of the rift, and the Nyanza (Kavirondo) trough branches off on the western side (Fig. 6). These are very different features, however, and there are great contrasts in the compositions and ages of volcanics on either side of the rift. To the east, some basement influence on the orientations of linear basaltic ranges has been suggested but these chains of cones seem to be more closely related to structural culminations and depressions than to nearly meridional foliation trends in the basement rocks (Williams, 1977).

The interval between rifting and preceding orogenic movements deserves brief consideration, if only because of the fundamentally different kinds of igneous associations found in tensional and compressional environments (Martin and Piwinskii, 1972; Macdonald, 1975), and the difficulties likely to be encountered if the change-over was rapid. In this respect, the pre-rift history of East Africa stands apart from that of regions traversed by the Oslo and Rio Grande rifts.

Away from the coastal belt in Kenya, high-grade metamorphic rocks of the late Precambrian/early Paleozoic Mozambique Belt were subjected to prolonged erosion spanning some 400 m.y. before rift development commenced. There are few distracting occurrences of Paleozoic or Mesozoic igneous rocks to complicate the pattern of rift-related magmatism. On the other hand, the patchy preservation of sedimentary formations close to the rift zone makes it necessary to rely heavily on the volcanics in attempts to unravel the structural evolution of the region.

In contrast, a far less significant interval separated Cenozoic rifting and Laramide orogenic events (Upper Cretaceous-Lower Tertiary) along the Rio Grande. Rift-related igneous activity was but one stage in a complex sequence of tectono-magmatic episodes embracing subduction processes as well as intra-plate block faulting. There is still lack of agreement over the time of inception of rifting, though a better preserved sedimentary cover than that flanking the Kenya rift provides rewarding areas for study of basin development.

Igneous activity associated with the Oslo graben commenced in Upper Carboniferous or Lower Permian times and, in general, it has been possible to make a convincing distinction between rift-related episodes and those connected with Caledonian and early Carboniferous events. Investigations are hampered, however, by near-complete erosion of volcanics in areas flanking the graben; and, as in Kenya, sedimentary formations are sparse outside the rift.

Crustal swells and alkaline magmatism

Some workers (e.g. Le Bas, 1971) draw attention to the common occurrence of peralkaline igneous rocks and abundant alkaline basalts in continental rifts, stressing the association of various igneous provinces with crustal swells formed by dome-like uplift of large regions. Indeed, it has been suggested (Le Bas, op. cit.) that the swells have a more vital role in determining the distribution of peralkaline provinces than do the rift valleys which commonly transect these domal uplifts. If true, this assertion provides one possible explanation for the contrasting compositions of igneous suites found in the East African and Rio Grande rifts. But its proof depends largely on the validity of claims that pulses of massive domal uplift occurred in the former region, and that no doming can be demonstrated in the latter.

Examined in more detail, it soon becomes apparent that problems arise in making distinctions between genuine (i.e. pre-faulting) doming and features such as late uplifts of the shoulders of rifts, and between rift-related crustal swells and topographic highs inherited from earlier tectonic events: to demonstrate merely deformation and elevation of erosion surfaces is insufficient. Moreover, it is dangerous to rely heavily on conclusions reached in studies of paleorifts for these are the very situations where evidence outside the rift zones is scant, and where models of rift evolution sometimes have built-in notions of doming as a necessary prerequisite for graben formation and/or alkaline magmatism.

Duration of magmatism

It is not yet possible to define precisely the periods spanned by igneous activity in the three rifts selected for review. Problems range from those encountered in attempts to reconcile radiometric dates with field and paleontological evidence, to difficulties arising from inadequate knowledge of the extent and full structural history of each rift system. Nevertheless, there is some support for a notion that the main episodes of magmatism were spread over some 20-30 m.y. in each case.

Differences of opinion about the date of inception of rifting in the Rio Grande region leave conflicting views concerning the duration of rift-related igneous activity. A minimum figure is evidently 21 m.y. (though the generally accepted concept of a 'middle Miocene lull' in volcanism effectively reduces this figure to between 13 and 17 m.y.), and a maximum about 32 m.y.

The main problems in Kenya revolve around issues such as disputes over the age of early basalts in the north (certainly 16-20 m.y. old, but perhaps up to 30 m.y.); lack of exposure of deep infill to the south where no volcanic formations on the rift margin are older than 12-15 m.y.; and the significance and reliability of Oligocene to early Miocene dates (25-31 m.y.) from the older intrusive alkaline complexes of southeastern Uganda. All these figures must be viewed against claims that volcanic activity in the Ethiopian rift commenced some 40-48 m.y. ago (Davidson and Rex, 1980).

The results from paleorifts are even more problematical because of conflicting opinions about both the upper and lower limits of associated igneous activity. Early determinations on igneous rocks from the Oslo graben produced a very wide spread of ages (355-136 m.y.), with a cluster between about 280 and 240 m.y. Later, it was suggested that plutonic activity had a surprisingly restricted time range (less than 3-4 m.y.), but the most recent work extends the span of magmatic activity to at least 10 m.y.

TABLE 5. Compositions of igneous rocks in the Kenya, Oslo and Rio Grande rifts

	Volcanics	Vol. %	Exposed intrusives
KENYA	Alkali basalts, transitional basalts, trachybasalts; Basanites, tephrites; Melanephelinites, nephelinites, phonolitic nephelinites, carbonatites	55-65	Syenites, trachytes, nepheline syenites, phonolites, basalts, ijolites, carbonatites
	Phonolites	20	
	Trachyphonolites, trachytes, alkali rhyolites	15-25	
OSLO	Alkali basalts, tholeiites, trachybasalts; Basanites, nephelinites	55	Alkali granites, alkali syenites, nepheline syenites, monzonites, monzodiorites, diorites, gabbros, pyroxenites
	Rhomb porphyries (trachyandesites/trachytes)	35	
	Trachytes, rhyolites	10	
RIO GRANDE	Alkali basalts, basanites, tholeiites, basaltic andesites, andesites, dacites, rhyodacites, rhyolites		Granites, andesites, trachyandesites, basalts, rhyolites

(Sundvoll, 1978a) or 40 m.y. (Sundvoll, 1978b). Dates in the range 290-250 m.y. seem plausible from the evidence currently available. Developments of Mesozoic and Tertiary igneous rocks in this rift system, but beyond the confines of the Oslo graben proper, conceivably reflect renewed activity triggered off by events during the evolution of the Central graben of the North Sea, and early opening of the North Atlantic.

Compositions and volumes of igneous rocks

The broad compositional ranges of volcanic and plutonic rocks exposed in the Oslo, Rio Grande and Kenya rifts are summarized in Table 5, and only selected topics are taken for discussion here. Petrochemical comparisons are excluded to avoid likely repetition of data presented in another chapter.

The Kenya and Rio Grande rifts both contain abundant alkali basalts, but the tholeiites and calc-alkaline suites of the latter have no true counterparts in East Africa. Conversely, there are no equivalents in the Rio Grande of the strongly undersaturated rocks of the melanephelinite/nephelinite - phonolitic nephelinite - phonolite (- carbonatite) association found in western and southern Kenya. Even more striking is the lack of flood lavas in the Rio Grande corresponding to the voluminous 'plateau' phonolites of the Kenya rift. Nor can these salic volcanics be matched volume-wise in other parts of the East African rift system: indeed, they are unique in records of alkaline magmatism in any structural environment. Petrographic series embracing basalts (both alkalic and 'transitional'), trachybasalts, and either quartz trachytes (and locally alkali rhyolites) or trachyphonolites set the Kenya rift further apart from the Rio Grande in terms of igneous rock compositions.

It is worth noting that Barker (1977) has drawn attention to certain similarities between the volcanics of the alkaline provinces of Kenya and Trans-Pecos (west Texas). Despite, however, the close proximity of the latter to the Rio Grande, and temporal overlap in magmatism in the two regions, there are still differences of opinion over the significance of these features. Some regard the Trans-Pecos province as a continental rift; others relate the magmatism to subduction processes, or to ensialic back-arc spreading (see Barker, 1979).

Comparison of the Oslo and Kenya rift volcanics is superficially more attractive in seeking compositional similarities. Both rifts contain abundant alkali basalts, occurrences of strongly undersaturated mafic lavas, and a significant proportion of trachytes or rhyolites. Furthermore, the voluminous rhomb porphyries of the Oslo graben are reminiscent of the Kenya rift flood phonolites in characteristics such as their fluid behaviour and remarkable uniformity of composition and texture over large areas. However, petrographic comparisons between the flood volcanics in the two regions show less impressive similarities. The rhomb porphyries are subalkaline to weakly alkaline, and they are frequently quartz-bearing. Texturally, they closely resemble some of the Mt. Kenya and Mt. Kilimanjaro volcanics (major complexes outside the Kenya rift), but the East African lavas more frequently contain nepheline and/or analcime.

Attempts to estimate the volumes and proportions of igneous rocks in continental rifts are beset with familiar problems, and most of the figures are order-of-magnitude calculations based on inadequate data. In some active rifts (e.g. Rio Grande), even crude estimates are hard to come by because of the depth of burial of early formations. Though instructive in allowing examination of plutonic complexes, the deep levels of erosion in paleorifts generally make it necessary to rely on pre-conceived notions about the original distribution of volcanics in efforts to calculate total volumes of igneous rocks.

Despite these limitations, published figures do serve to emphasize a striking aspect of the volcanism in Kenya where the sheer volume of erupted material led to complete infilling and overtopping of depressions at two stages during rift development.

Relationships between igneous and tectonic activity

Modern investigations have gone a long way towards disproving once widely held views of rather simple relationships between volcanic events and major fault movements, and between igneous activity and pulses of uplift. In many cases, improved knowledge of local stratigraphies shows that displacements now observed along faults resulted from near-continuous movement spanning a long time range, rather than just a few short-lived episodes of intense activity separated by long periods of tectonic quiescence. Certain misconceptions arise through failure to recognize the almost instantaneous nature of individual volcanic events compared to the long periods represented by sedimentary formations of equivalent thickness (King, 1978; Chapman and Brook, 1978). In the absence of intercalated sediments, gaps in the accumulation of volcanics tend to go unreported resulting in over-simplification of the history of fault development.

Zones of strongest igneous activity seldom coincide with the main faults framing segments of rifts. Indeed, there is little evidence of any upwelling of magmas actually along the boundary faults in the Kenya and Rio Grande rifts; and even the deeper levels exposed in the Oslo graben fail to reveal marginal concentrations of dike swarms or sub-volcanic complexes.

Downwarping is locally more important than faulting in the formation of rift depressions. Where the early histories of the basins and troughs can be investigated, it is generally shown that the initial downwarps preceded major faulting. Moreover, the oldest infilling formations are usually sedimentary so it is clear that the first stages of subsidence, whether by warping or faulting or both, were not related to processes such as the evacuation of large magma chambers. The extent to which movement of magma towards the surface may have influenced later phases of subsidence is difficult to demonstrate but comparison with non-volcanic rifts, or segments of rifts, shows that such a process does not provide a fundamental mechanism for graben formation.

In parts of northern Kenya, a general correlation is found between volcanism and closely-spaced faulting, both having migrated towards the centre of the rift in the course of its evolution (Chapman et al., 1978). Equally striking, however, is the fact that a first episode of major faulting coincided with marked compositional changes in erupted material, phonolites giving way to members of alkali basalt-trachyte suites.

A combination of arching, downwarping and faulting of the rift floor in northern Kenya led to progressive narrowing of the active part of the depression, so that ultimately the Quaternary infill was concentrated into an inner trough locally less than 20 km across. This restriction in width of the active zone was accompanied by changes in the composition of basic lavas which became progressively less undersaturated with time. The gradation from Miocene basanites and alkali basalts to Quaternary transitional basalts is attributed to decreasing depth and/or greater degrees of partial melting beneath the rift (Lippard and Truckle, 1978).

The distribution of alkali olivine basalts and tholeiites in the Rio Grande rift is complicated, and the latter show no simple restriction to axial zones; transverse sheer structures like the Jemez lineament evidently play a strong part in determining the spatial distribution of magma types (Baldridge, 1979). Alkali olivine basalts and olivine tholeiites were erupted contemporaneously in Plio-Pleistocene fields along the Albuquerque basin, whereas the volcanics of equivalent age farther south are almost exclusively alkali basalts and basanites.

Demonstrated temporal trends in basic lava chemistry, and spatial patterns in the distribution of basalt types with respect to rift structures, thus show less impressive similarities between Kenya and the Rio Grande than are sometimes claimed (e.g. Lippard and Truckle, 1978). It may be noted that occurrences of both alkali olivine basalts and olivine tholeiites in the younger lava sequences within the Oslo depression (Ramberg and Larsen, 1978) are more reminiscent of the central Rio Grande than any part of the Kenya rift, though lack of preserved volcanics outside the Oslo graben precludes full comparison of patterns in basalt chemistry with those under investigation in active rift systems.

A tendency for strings of late-stage central volcanoes to develop more or less axially within segments of rifts is demonstrated by the distribution of Quaternary silicic caldera volcanoes along the inner troughs in central and northern Kenya, and by a rather more dispersed pattern of shields surmounted by cauldrons in the Oslo graben. Attention has been drawn to the regular spacing of many of these centers in attempts to relate their distribution to crustal or lithospheric thickness (Mohr and Wood, 1976; Ramberg and Larsen, 1978).

Detailed studies have provided little support for notions that phases of salic volcanism were accompanied by pulses of domal uplift, or that the most voluminous eruptions in rifts were terminated by major faulting. It has been shown, for instance, that the Miocene phonolites in Kenya accumulated, not on the flanks of a developing domal structure, but in broad steadily subsiding depressions. Moreover, the phonolite eruptions reached a climax some 5 m.y. before the commencement of widespread faulting (Chapman et al., 1978).

Conclusions

The highly selective nature of this review makes it unwise to offer profound concluding statements which may be quoted out of context and applied to all continental rifts. Nevertheless, despite the small sample chosen for discussion, it will be obvious that rifts have less in common in terms of magmatic activity than would be expected from cursory examination of the literature.

Though some relationships between volcanic and tectonic activity have been demonstrated, there is little justification for assuming that periods of maximum intensity of volcanism were always terminated by episodes of major faulting. Moreover, notions of simple relationships between volcanic events and pulses of domal uplift are not substantiated by the most recent studies of rift systems.

The igneous rocks encountered in continental rifts (frequently alkaline, but in some cases calc-alkaline or tholeiitic) occur also in other geological environments. Supposedly characteristic suites of alkaline rocks do not therefore provide reliable criteria for the identification of paleorifts.

Finally, it must be stressed that some continental rifts (or segments of rift systems) display little or no associated magmatism so that processes such as evacuation of large magma chambers cannot be entertained as a fundamental mechanism for graben formation. Indeed, it has already been pointed out (King, 1978) that in sectors of rift systems which approximate to simple graben the essential structures are similar regardless of the nature of the infill which may be predominantly sedimentary, volcanic or even water.

References

Accademia Nazionale dei Lincei. Geodynamic Evolution of the Afro-Arabic Rift System, Rome, 1980.

Aubele, J.C. Geology of Cerros del Rio volcanic field, in Guidebook to the Rio Grande rift in New Mexico and Colorado, edited by J.W. Hawley, New Mex. Bur. Mines Min. Res., Circ. 163, 198-201, 1978.

Bailey, D.K. Continental rifting and alkaline magmatism, in The Alkaline Rocks, edited by H. Sørensen, pp. 148-159, John Wiley & Sons, London, 1974.

Bailey, R.A., and R.L. Smith. Volcanic geology of the Jemez Mountains, New Mexico, in Guidebook to the Rio Grande rift in New Mexico and Colorado, edited by J.W. Hawley, New Mex. Bur. Mines Min. Res., Circ. 163, pp. 184-196, 1978.

Baker, B.H., R. Crossley, and G.G. Goles. Tectonic and magmatic evolution of the southern part of the Kenya rift valley, in Petrology and Geochemistry of Continental Rifts, edited by E.-R. Neumann and I.B. Ramberg, pp. 29-50, D. Reidel Publishing Co., Dordrecht, Holland, 1978.

Baker, B.H., and J.G. Mitchell. Volcanic stratigraphy and geochronology of the Kedong-Olorgesailie area and the evolution of the South Kenya rift valley, J. geol. Soc. Lond., 132, 467-484, 1976.

Baker, B.H., P.A. Mohr, and L.A.J. Williams. Geology of the Eastern Rift System of Africa, Geol. Soc. America, Special Paper 136, 67 pp., 1972.

Baker, B.H., L.A.J. Williams, J.A. Miller, and F.J. Fitch. Sequence and geochronology of the Kenya rift volcanics, Tectonophysics, 11, 191-215, 1971.

Baker, B.H., and J. Wohlenberg. Structure and evolution of the Kenya Rift Valley, Nature, 229, 538-542, 1971.

Baldridge, W.S. Petrology and petrogenesis of Plio-Pleistocene basaltic rocks from the central Rio Grande rift, New Mexico, and their relation to rift structure, in Rio Grande Rift: Tectonics and Magmatism, edited by R.E. Riecker, pp. 323-353, AGU, Washington, 1979.

Baltz, E.H. Resume of Rio Grande depression in north-central New Mexico, in Guidebook to the Rio Grande rift in New Mexico and Colorado, edited by J.W. Hawley, New Mex. Bur. Mines Min. Res., Circ. 163, pp. 210-228, 1978.

Barker, D.S. Northern Trans-Pecos magmatic province: introduction and comparison with the Kenya rift, Geol. Soc. America Bull., 88, 1421-1427, 1977.

Barker, D.S. Cenozoic magmatism in the Trans-Pecos Province: relation to the Rio Grande rift, in Rio Grande Rift: Tectonics and Magmatism, edited by R.E. Riecker, pp. 382-392, AGU, Washington, 1979.

Bishop, W.W. (editor). Geological Background to Fossil Man, 585 pp., Scottish Academic Press, Edinburgh, 1978.

Bishop, W.W., J.A. Miller, and F.J. Fitch. New potassium-argon age determinations relevant to the Miocene fossil mammal sequence in East Africa, Amer. J. Sci., 267, 669-699, 1969.

Chapin, C.E. Evolution of the Rio Grande rift - a summary, in Rio Grande Rift: Tectonics and Magmatism, edited by R.E. Riecker, pp. 1-5, AGU, Washington, 1979.

Chapman, G.R., and M. Brook. Chronostratigraphy of the Baringo Basin, Kenya, in Geological Background to Fossil Man, edited by W.W. Bishop, pp. 207-223, Scottish Academic Press, Edinburgh, 1978.

Chapman, G.R., S.J. Lippard, and J.E. Martyn. The stratigraphy and structure of the Kamasia Range, Kenya Rift Valley, J. geol. Soc. Lond., 135, 265-281, 1978.

Crossley, R. The Cenozoic stratigraphy and structure of the western part of the Rift Valley in southern Kenya, J. geol. Soc. Lond., 136, 393-405, 1979.

Davidson, A., and D.C. Rex. Age of volcanism and rifting in southwestern Ethiopia, Nature, 283, 657-658, 1980.

Dons, J.A., and B.T. Larsen (editors). The Oslo Paleorift, Norges geol. Unders., 337, 199 pp., 1978.

Eaton, G.P. A plate-tectonic model for late Cenozoic crustal spreading in the western United States, in Rio Grande Rift: Tectonics and Magmatism, edited by R.E. Riecker, pp. 7-32, AGU, Washington, 1979.

Elston, W.E. Rifting and volcanism in the New Mexico segment of the Basin and Range province, southwestern USA, in Petrology and Geochemistry of Continental Rifts, edited by E.-R. Neumann and I.B. Ramberg, pp. 79-86, D. Reidel Publishing Co., Dordrecht, Holland, 1978.

Elston, W.E., and T.J. Bornhost. The Rio Grande rift in context of regional post-40 m.y. volcanic and tectonic events, in Rio Grande Rift: Tectonics and Magmatism, edited by R.E. Riecker, pp. 416-438, AGU, Washington, 1979.

Girdler, R.W. (editor). East African Rifts, Tectonophysics, 15, No. 1/2 (special issue), 186 pp., 1972.

Harkin, D.A. The Rungwe volcanics at the northern end of Lake Nyasa, Geol. Surv. Tanganyika Mem., 2, 1960.

Hawley, J.W. (editor). Guidebook to the Rio Grande rift in New Mexico and Colorado, New Mex. Bur. Mines Min. Res., Circ. 163, 241 pp., 1978.

Hawley, J.W., and W.R. Seager. New Mexico-Texas State line to Elephant Butte Reservoir, in Guidebook to the Rio Grande rift in New Mexico and Colorado, edited by J.W. Hawley, New Mex. Bur. Mines Min. Res., Circ. 163, pp. 71-89,1978.

Henningsmoen, G. Sedimentary rocks associated with the Oslo region lavas, in The Oslo Paleorift, edited by J.A. Dons and B.T. Larsen, pp. 17-24, Norges geol. Unders., 337, 1978.

Holmes, A., and H.F. Harwood. The volcanic area of Bufumbira: The petrology of the volcanic field of Bufumbira, south-west Uganda, and other parts of the Birunga field, Geol. Surv. Uganda Mem., 3 (pt. 2), 300 pp., 1937.

Holtedahl, O. Norges geologi, Norges geol. Unders., 164, 1118 pp., 1953.

Jones, W.B., and S.J. Lippard. New age determinations and the geology of the Kenya Rift-Kavirondo Rift junction, W. Kenya, J. geol. Soc. Lond., 136, 693-704, 1979.

Kent, P.E. The age and tectonic relationships of East African volcanic rocks, Geol. Mag., 81, 15-27, 1944.

King, B.C. Vulcanicity and rift tectonics in East Africa, in African Magmatism and Tectonics, edited by T.N. Clifford and I.G. Gass, pp. 263-283, Oliver and Boyd, Edinburgh, 1970.

King, B.C. Structural and volcanic evolution of the Gregory Rift Valley, in Geological Background to Fossil Man, edited by W.W. Bishop, pp. 29-54, Scottish Academic Press, Edinburgh, 1978.

King, B.C., and G.R. Chapman. Volcanism of the Kenya rift valley, Phil. Trans. R. Soc. Lond., A271, 185-208, 1972.

King, B.C., M.J. Le Bas, and D.S. Sutherland. The history of the alkaline volcanoes and intrusive complexes of eastern Uganda and western Kenya, J. geol. Soc. Lond., 128, 173-205, 1972.

King, B.C., and L.A.J. Williams. The East African Rift System, in Geodynamics: Progress and Prospects, edited by C.L. Drake, pp. 63-74, AGU, Washington, 1976.

Lamb, C.M. Volcanism in the Naivasha area of Kenya (abstract), 19th ann. Rep. res. Inst. afr. Geol., Univ. Leeds, p.78, 1975.

Le Bas, M.J. Per-alkaline volcanism, crustal swelling and rifting, Nature Phys. Sci., 230, 85-87, 1971.

Lipman, P.W., B.R. Doe, C.E. Hedge, and T.A. Steven. Petrologic evolution of the San Juan volcanic field, southwestern Colorado: Pb and Sr isotope evidence, Geol. Soc. America Bull., 89, 59-82, 1978.

Lipman, P.W., and H.H. Mehnert. Late Cenozoic basaltic volcanism and development of the Rio Grande depression in the southern Rocky Mountains, Geol. Soc. America Mem., 144, 119-154, 1975.

Lipman, P.W., and H.H. Mehnert. The Taos Plateau volcanic field, northern Rio Grande rift, New Mexico, in Rio Grande Rift: Tectonics and Magmatism, edited by R.E. Riecker, pp. 289-311, AGU, Washington, 1979.

Lippard, S.J. The petrology of phonolites from the Kenya Rift, Lithos, 6, 217-234, 1973.

Lippard, S.J., and P.H. Truckle. Spatial and temporal variations in basalt geochemistry in N. Kenya rift, in Petrology and Geochemistry of Continental Rifts, edited by E.-R. Neumann and I.B. Ramberg, pp. 123-131, D. Reidel Publishing Co., Dordrecht, Holland, 1978.

Logatchev, N.A. (editor). Major Problems of Rift Genesis, USSR Acad. Sci., Siberian Branch, Novosibirsk, 1977 (in Russian).

Macdonald, R. Tectonic settings and magma associations, in Oversaturated Peralkaline Volcanic Rocks, edited by D.K. Bailey, F. Barberi and R. Macdonald, Bull. volcanol., 38 No. 3 (Special Issue), 575-593, 1975.

Macdonald, R., D.K. Bailey, and D.S. Sutherland. Oversaturated peralkaline glassy trachytes from Kenya, J. Petrology, 11, 507-517, 1970.

Manley, K. Stratigraphy and structure of the Espanola basin, Rio Grande rift, New Mexico, in Rio Grande Rift: Tectonic and Magmatism, edited by R.E. Riecker, pp. 71-86, AGU, Washington, 1979.

Martin, R.F., and A.J. Piwinskii. Magmatism and tectonic settings, J. geophys. Res., 77, 4966-4975, 1972.

McConnell, R.B. Geological development of the Rift System of Eastern Africa, Geol. Soc. America Bull., 83, 2549-2572, 1972.

McConnell, R.B. East African Rift System dynamics in view of Mesozoic apparent polar wander, J. geol. Soc. Lond., 134, 33-39, 1977.

Merla, G. Missions geologica nell'Ethiopia meridionale del C.N.R. 1959-1960, Notizie geomorfologiche, G. Geol. Bologna, 13, 1-56, 1963.

Merla, G., and E. Minucci. Missione geologica nel Tigrai, Reale Accad. Ital., Roma, 1, 363 pp., 1938.

Mohr, P., and P. Gouin. Ethiopian Rift System, in Geodynamics: Progress and Prospects, edited by C.L. Drake, pp. 81-87, AGU, Washington, 1976.

Mohr, P.A., and C.A. Wood. Volcano spacings and lithospheric attenuation in the Eastern Rift of Africa, Earth Planet. Sci. Letters, 33, 126-144, 1976.

Neumann, E.-R. Petrology of the plutonic rocks, in The Oslo Paleorift, edited by J.A. Dons and B.T. Larsen, pp. 25-34, Norges geol. Unders., 337, 1978.

Neumann, E.-R., and I.B. Ramberg (editors). Petrology and Geochemistry of Continental Rifts, D. Reidel Publishing Co., Dordrecht, Holland, 296 pp., 1978.

Oftedahl, C. Studies on the igneous rock complex of the Oslo region: XII, The lavas, Skr. Norske Vidensk.-Akad. i Oslo, Mat.-naturv. Kl. 1952, No.3, 1952.

Oftedahl, C. Cauldrons of the Permian Oslo rift, J.Volc. Geotherm. Res., 3, 343-371, 1978(a).

Oftedahl, C. Main geologic features of the Oslo graben, in Tectonics and Geophysics of Continental Rifts, edited by I.B. Ramberg and E.-R. Neumann, pp. 149-165, D. Reidel Publishing Co., Dordrecht, Holland, 1978 (b).

Pilger, A., and A. Rosler (editors). Afar between Continental and Oceanic Rifting, Schweizerbart, Stuttgart, 1976.

Pilger, A., and A. Rosler (editors). Afar Depression of Ethiopia, Schweizerbart, Stüttgart, 1975.

Pulfrey, W. Shape of the sub-Miocene erosion bevel in Kenya, Geol. Surv. Kenya Bull., 3, 1960.

Ramberg, I.B. Gravity interpretation of the Oslo graben and associated igneous rocks, Norges geol. Unders., 325, 194 pp., 1976.

Ramberg, I.B., and B.T. Larsen. Tectono-magmatic evolution, in The Oslo Paleorift, edited by J.A. Dons and B.T. Larsen, pp. 55-73, Norges geol. Unders., 337, 1978.

Ramberg, I.B., and E.-R. Neumann (editors). Tectonics and Geophysics of Continental Rifts, D. Reidel Publishing Co., Dordrecht, Holland, 444 pp., 1978.

Riecker, R.E. (editor). Rio Grande Rift: Tectonics and Magmatism, AGU, Washington, 438 pp., 1979.

Saggerson, E.P., and B.H. Baker. Post-Jurassic erosion-surfaces in eastern Kenya and their deformation in relation to rift structure, Quart. J. geol. Soc. Lond., 121, 51-72, 1965.

Seager, W.R., and Morgan P. Rio Grande rift in southern New Mexico, west Texas, and northern Chihuahua, in Rio Grande Rift: Tectonics and Magmatism, edited by R.E. Riecker, pp. 87-106, AGU, Washington, 1979.

Segalstad, T.V. Petrology of the Skien basaltic rocks and the early basaltic (B_1) volcanism of the Permian Oslo rift, in Petrology and Geochemistry of Continental Rifts, edited by E.-R. Neumann and I.B. Ramberg, pp. 209-216, D. Reidel Publishing Co., Dordrecht, Holland, 1978.

Sikes, H.L. Notes on the geology of the country surrounding Nairobi, Rep. geol. Surv. Kenya, 1939.

Smith, R.R. Early rift magmatism at Spanish Peaks, Colorado, in Rio Grande Rift: Tectonics and Magmatism, edited by R.E. Riecker, pp. 313-321, AGU, Washington, 1979.

Smith, R.L., R.A. Bailey, and C.S. Ross. Geologic map of the Jemez Mountains, New Mexico, U.S. Geol. Surv., Misc. Geol. Inv. Map I-571, scale 1:125000, 1970.

Sundvoll, B. Isotope- and trace-element chemistry, geochronology, in The Oslo Paleorift, edited by J.A. Dons and B.T. Larsen, pp. 35-40, Norges geol. Unders., 337, 1978 (a).

Sundvoll, B. Rb/Sr-relationship in the Oslo igneous rocks, in Petrology and Geochemistry of Continental Rifts, edited by E.-R. Neumann and I.B. Ramberg, pp. 181-184, D. Reidel Publishing Co., Dordrecht, Holland, 1978 (b).

Webb, P.K., and S.D. Weaver. Trachyte shield volcanoes: a new volcanic form from South Turkana, Kenya, Bull. Volcanol., 39, 1-19, 1975.

Tuttle, O.F., and J. Gittins (editors). Carbonatites, John Wiley and Sons, New York, 591 pp., 1966.

Tweto, O. The Rio Grande rift system in Colorado, in Rio Grande Rift: Tectonics and Magmatism, edited by R.E. Riecker, pp. 33-56, AGU, Washington, 1979.

Williams, H., and A.R. McBirney. Volcanology, Freeman, Cooper and Co., San Francisco, 397 pp., 1979.

Williams, L.A.J. Volcanic associations in the Gregory Rift Valley, East Africa, Nature, 224, 61-64, 1969.

Williams, L.A.J. The volcanics of the Gregory Rift Valley, East Africa, Bull. Volcanol., 34, 439-465, 1970.

Williams, L.A.J. The Kenya rift volcanics: a note on volumes and chemical composition, in R.W. Girdler (editor), East African Rifts, Tectonophysics, 15 (1/2), 83-96, 1972.

Williams, L.A.J. Some aspects of Cainozoic volcanism in Kenya, in Major Problems of Rift Genesis, edited by N.A. Logatchev, pp. 139-148, USSR Acad. Sci., Siberian Branch, Novosibirsk, 1977 (in Russian).

Williams, L.A.J. The Volcanological development of the Kenya rift, in Petrology and Geochemistry of Continental Rifts, edited by E.-R. Neumann and I.B. Ramberg, pp. 101-121, D. Reidel Publishing Co., Dordrecht, Holland, 1978 (a).

Williams, L.A.J. Character of Quaternary volcanism in the Gregory Rift Valley, in Geological Background to Fossil Man, edited by W.W. Bishop, pp. 55-69, Scottish Academic Press, Edinburgh, 1978 (b).

Williams, L.A.J., and P.H. Truckle. Volcanic sequences in the Kenya rift, in Geodynamic Evolution of the Afro-Arabic Rift System, pp. 123-132, Accademia Nazionale dei Lincei, Rome, 1980.

CHEMICAL ASPECTS OF RIFT MAGMATISM

Franco Barberi[1], Roberto Santacroce[1] and Jacques Varet[2]

1. Istituto di Mineralogia e Petrografia, University of Pisa, Italy;
2. Bureau de Recherches Geologiques et Minières, Orleans, France.

Introduction

Rifts are faults troughs occurring in places where the entire thickness of the lithosphere has ruptured under tension. They are characteristically associated to abnormally thin lithosphere, either in oceans, marginal basins and in continents. Oceanic rifts are sites of crustal separation, with formation of new crust and horizontal extension rate ranging from 1 to 10 cm/year (accreting plate boundaries). In continental rifts the horizontal extension rate is instead much lower and they may develop or not ("failed rifts") to form oceans; they represent therefore either "within plate" tectonic structures or embryonic plate boundaries.

Continental rifts can form in a variety of tectonic environments at different stages of lithosphere evolution. In the East African Rift System different evolution stages are observed that are usually considered to represent the typical sequence of rift development ("Wilson cycle", Dewey and Burke,1974). They include uplift and continental rifting on a variously thinned lithosphere, the development of a triple rift system and the continental break-up at the Afar-Red Sea-Gulf of Aden junction. The Rhinegraben and Baikal rift are instead considered a result of continental impact ("impactogene rifts"), while the active rifts of Rio Grande and the Basin and Range province in south-western USA and Mexico are "apparently products of tension related to the imperfection of the rotation of the North American plate past the Pacific plate along the San Andreas fault" (Burke, 1978).

Other fault trenches cutting the entire lithosphere occur both in ocean ("fracture zones") and in continents (e. g. the Aqaba-Dead Sea-Jordan trough). They represent transform plate boundaries and differ from typical rifts for the dominant shear character with only minor tension, and the lack of lithosphere thinning. Magmatic activity is associated to any major rift development stage, but with highly variable intensity, nature and time-space distribution. The aim of this paper is to give a brief comprehensive review of the volcanic activity related to the main types of rifts, with emphasis on its chemical aspects.

Ocean rifts

The very interesting magmatic and tectonic event occurring in the Krafla area in Iceland since 1975 (and that of Asal, eastern Afar, of late 1978) and the reconstruction of similar events in the past (Bjornsson et al., 1979) led to the conclusion that ocean rifting is a discontinuous rather than a continuous process, with an approx. 100 years period in the slowly spreading North Atlantic ridge (spreading rate of 1 cm/year or, better, 1m/century). Rifting phases, and related magmatic activity, should occur with a much higher frequency and rapidly spreading ridges, such as Juan de Fuca and East Pacific Rise. In any cases, there is not doubt that oceanic spreading is characterized by very intense magmatic activity with emplacement, at the crust of mid-ocean ridges, of very huge volumes of basaltic magmas, as injected dykes, sills and submarine (subaerial in very few cases) extrusives. Oceanic basalts certainly represent the Earth's most voluminous recent volcanic product. This fact and the recognition of their importance for the understanding of upper mantle nature, composition and evolution, has focused a great interest in the study of these rocks. An impressive amount of new data on ocean floor basalts has been collected in the last few years, mainly trough the extensive sampling provided by the Deep-Sea D-illing Project (DSDP), the FAMOUS Project and the International Phase of Ocean Drilling (IPOD).

Composition of oceanic basalts.

Chemical data on ocean floor basalts have been recently assembled, compared and discussed by Melson et al., (1976), Bryan et al., (1976) and Hart (1976). Average chemical composition for major elements of basalts and basaltic glasses related to spreading ridges in the three major oceans and the Red Sea are reported in Table 1 and compared with recent basalts from the emerged portion of Mid-Atlantic ridge in Iceland. It is apparent from Table 1 that these rocks are tholeiites from continents or mid-plate oceanic islands. Averages from the various oceans do not vary greatly, with the exception of the distinctly higher FeO and TiO_2 content of the average from the rapidly spreading East Pacific Rise. This and other compositional differences (e.g. in the normative An-Ol-Hy projection) in basalts sampled from slow and fast spreading ridges, led to the suggestion that basalt composition could be related to spreading rate (Bass, 1971). More detailed sampling of oceanic ridges has however shown a considerable diversity of composition (mostly emphasized by trace elements and isotopic ratios, but reflected also in major elements distribution) even in basalts colleceted from limited sectors (i.e. FAMOUS) or in single DSDP or IPOD drilling sites (Bryan et al., 1976; Wood et al.,1979); furthermore no apparent relation of basalt composition with age has been so far found.

During the last years, there has been a general agreement among geochemists and petrologists that a parti-

TABLE 1. AVERAGE COMPOSITION OF BASALTS AND BASALTIC GLASS FROM MID-OCEANIC RIDGES AND ICELAND.

	1	2	3	4	5	6	7	8
SiO_2	50.67	50.19	50.93	49.94	49.61	48.79	49.34	49.94
TiO_2	1.28	1.77	1.19	1.35	1.43	0.87	1.55	2.15
Al_2O_3	15.45	14.86	15.15	15.09	16.01	16.82	14.96	14.48
FeO*	9.67	11.33	10.32	10.85	11.49	9.18	11.51	12.67
MgO	8.05	7.10	7.69	7.77	7.84	8.33	7.44	6.37
CaO	11.72	11.44	11.84	11.90	11.32	13.66	12.35	11.07
Na_2O	2.51	2.66	2.32	2.56	2.76	2.07	2.33	2.54
K_2O	0.15	0.16	0.14	0.18	0.22	0.09	0.23	0.43

1, Atlantic, 155 glass analyses (Bryan et al., 1976); 2, East Pacific Rise, 38 glass analyses (Melson et al.,1976); 3, Indian, 12 glass analyses (Melson et al., 1976); 4, Red Sea central trough, 19 basalt analyses (Barberi et al., 1978); 5, average oceanic tholeiite (Cann,1971); 6,7 and 8, Groups 1 (11 samples), 2 (17 samples), and 3 (8 samples) of recent basalts from Iceland rift zone (Sigvaldason, 1974). FeO*, total iron as FeO.

cular type of basalt, called abyssal tholeiite, or Mid-Oceanic Ridge Basalt (MORB), or Large-Ion Lithophile (LIL) elements-depleted tholeiite, is characteristic among modern basalts erupted at mid-ocean spreading ridges. These MOR basalts have rather uniform geochemical characteristics, such as low content of LIL, a light REE depleted pattern (Fig.1), low $^{87}Sr/^{86}Sr$(0.7022-0.7028; Brooks et al., 1976), low $^{206}Pb/^{204}Pb$ (less than 18.8; Tatsumoto, 1978) and $^{143}Nd/^{144}Nd$ higher than 0.51305 (Bickford and Van Schmus, 1979). This LIL-depleted tholeiite dominates in "normal" sectors od mid-oceanic ridges, whereas in "anomalous" ridge segments, characterized by topographic highs (as in the proximity of island platform or chains) other types of tholeiites appear and may become dominant over MORB, such as at Mid-Atlantic Ridge near Azores and Iceland. These are geochemically transitional between MORB and oceanic island basalts, being characterized, with respect to MORB, by LIL enrichment with flat or light REE enriched patterns, and higher values of Sr and Pb isotopic ratios (this variability in composition implies that the use of discriminating diagrams to infer the geological setting of basalts, e.g. Pearce and Cann, 1973 may be misleading). Bryan et al.,(1976) divided the oceanic basalts into a dominant homogeneous group I made of LIL-depleted MORB and a LIL-enriched compositionally heterogenous group II, made of basalts from various oceanic geological settings (ridge, off-ridge, island-platforms, fracture zones, aseismic ridges). Some distintive major and trace element ratios of the two groups are reported in Table 2, where "critical" trace elements (Y, Zr, REE) are those appearing to be less sensitive to moderate sea water-alteration and metamorphism. The LIL-enriched basalts of group II are easily distinguished from MORB by their different REE patterns (Fig.1). Most of the North-Atlantic ridge basalts from FAMOUS area, leg 37 DSDP and leg 49 IPOD fall however in a intermediate category indicating that there is no sharp compositional boundary between the two groups, but most probably a

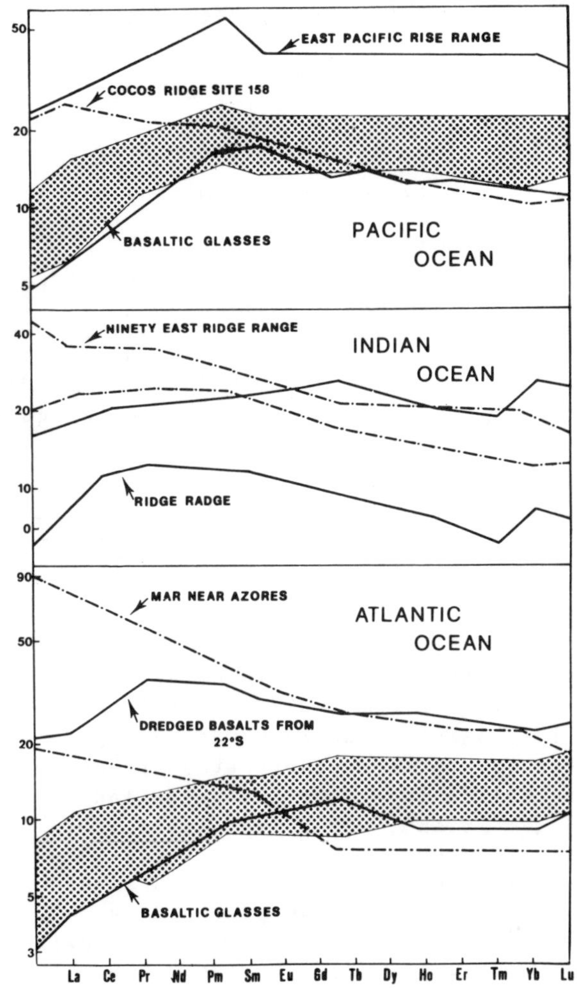

Fig.1 REE normalized to chondrites, patterns of oceanic basalts (after Bryan et al., 1976, simplified).

wide variety of compositions for prymary basalt liquids, which wre apparently not related to the tectonic setting but rather reflect mantle source heterogeneities (Wood et al., 1979).

Most of mid-oceanic ridges rocks fall in the basalt compositional field, and intermediate compositions (such as icelandites) are comparatively rare. Their abundance increases in "anomalous" oceanic areas, such as in the 90° East aseismic ridge, and particularly on oceanic islands located near the ridge crest, where rock-series, more or less complete, ranging from basalts to salitic products also occur. They have either a tholeiitic nature (tholeiite-icelan-

TABLE 2. MAJOR AND "CRITICAL" TRACE ELEMENTS RATIOS AND DISTRIBUTION IN OCEAN FLOOR BASALTS.

	FeO* /MgO	Ni	La/Sm	La/Yb	Yb	Zr
LIL-depleted basalts						
Atlantic (glasses)	0.81-1.7	80-200	0.57-0.95	0.38-1.06	1.7-3.85	35-150
Juan de Fuca (avg.)	1.8	77	1.03	1.00	4.71	-
Nazca Plate	1.20-1.83	60-103	0.65-1.01	0.76-1.00	2.63-5.32	81-178
Eastern Indian Ocean	0.83-1.40	58-140	0.76-1.01	0.51-0.86	2.0-2.9	28-85
LIL-enriched basalts						
Atlantic	0.84-1.55	58-150	1.27-4.97	1.41-13.55	1.8-2.9	35-100
Pacific (avg.)	1.29	57	2.05	3.33	2.2	127
Indian	0.85-1.15	110-210	1.00-3.85	1.17-5.21	2.29-3.0	114-170

Ranges of variations refer mostly to specific DSDP site averages. From Bryan et al., (1976), semplified.

dite-dacite-rhyodacite-rhyolite) as in Iceland, or a mildly alkalic affinity (mildly alkalic basalt-hawaiite-mugearite-benmoreite-oversaturated trachyte-peralkaline rhyolite; the so-called "Coombs" trend, Mihashiro, 1978) as in Easter, Ascension, Bouvet, Pitcairn islands (Baker, 1975). These rocks are usually interpreted as produced by shallow crystal fractionation of associated basalts (Carmichael, 1964; Baker, 1968 and 1969) although a genesis by partial remelting of older basalts under the high geothermal gradient of the rift zone has been proposed for icelandic acid rocks (Grönvold, 1976). Silica undersaturated alkalic volcanic rock series occur frequently in oceanic islands (e.g. Hawaii, Gough, Tristan and Cunha) (Mihashiro, 1978) but are not known from near ridge-crest environments. This indirectly confirms that undersaturated alkali basalts must be absent or extremely rare in oceanic spreading ridges. The few examples of alkalic basalts so far reported from the ocean floor seem actually related to a transform (fracture zone) environment (Thompson and Melson, 1972;Bonatti, 1973).

Petrogenesis of oceanic basalts.

Most of the oceanic tholeiites are saturated, at their low-pressure liquidi, with olivine, plagioclase and calcic pyroxene. This suggests that they have undergone some crystal fractionation before eruption on the ocean floor and therefore that they do not represent primary mantle melts. Several geochemical data (e.g. Ni, Cr, Co contents) indicate also a fractionated nature of oceanic basalts. Let use Ni as un example. Oceanic basalts usually have Ni content lower than 200 p.p.m. (see Table 2). The Ni content of primary mantle partial melts is instead estimated to range between 220 and 350 p.p.m. (Sun and Hanson, 1975) and these values decrease rapidly by shallow fractionation.

Is has also been shown that the considerable variation in major oxides concentration (i.e.Al_2O_3, total Fe,TiO_2) observed in ocean floor basalts from specific sites are due to, or can be explained by, a small scale crystal settling, as emphasized by great variability in phenocrysts distribution (Hart, 1976; Wood et al., 1979). Crystal fractionation alone cannot, however, explain the great geochemical differences reported for oceanic basalts, as emphasized mostly by trace elements and stable isotope geochemistry. Petrological studies of basalts from the FAMOUS area in the Mid-Atlantic ridge suggest that a process of dynamic partial melting from a

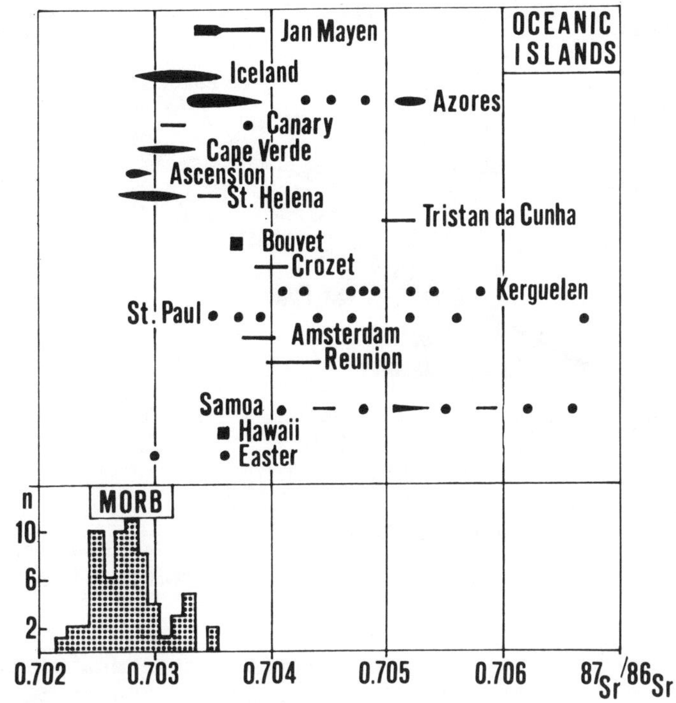

Fig.2 $^{87}Sr/^{86}Sr$ ratios of MORB and oceanic islands basalts(after Hoffmann and Hart, 1978)

homogenous mantle source accounts for observed variation (Langmuir et al., 1977).

However, there is by now general agreement among geochemists that the compositional gradient of oceanic basalts reflects heterogeneity in the mantle source, whereas controversy largely exists on both the number of the mantle sources and the relative role played in oceanic basalt petrogenesis by different processes like crystal fractionation, magma mixing, equilibrium or disequilibrium partial melting of the mantle source.

The best evidence for mantle heterogeneity come from radiogenic isotope data. Figs. 2 and 3 illustrate the systematic differences found in Sr and Pb isotopic ratios for MORB and oceanic island basalts. Island basalts have, with very few exceptions (St. Helena, Ascension), $^{87}Sr/^{86}Sr$ ratios higher than 0.7030, also when the island is located on or near the ridge crest, whereas MORB have lower values. This is commonly considered as suggesting that island basalts have a distinct mantle source from MORB and mantle is vertically heterogenous (Hofman and Hart, 1978). Attempts of explaining the Sr isotopic differences by disequilibrium partial melting of an homogenous source (O'Nions and Pankhurst, 1974) or by fractionation (O'Hara, 1973, 1975, 1977) have been later abandonnedby the same proponents (O'Nions et al., 1976) or are subjected to strong criticism by geochemists (Hofman and Hart, 1978). Also the whide variation in $^{206}Pb/^{204}Pb$ ratios is interpreted as indicative of isotopic heterogeneity in the mantle source(Tatsumoto, 1978). Isotopic ratios in oceanic basalts therefore indicate that source regions with different composition exist within the Earth's mantle. These compositional variations are due to enrichment and depletion events in the mantle, which have been generated over long periods of time (O'Nions et al.,1976). A study of basalts of the Iceland-Reykjanes region, and tentative applications of mixing equations indicate that basalts derived

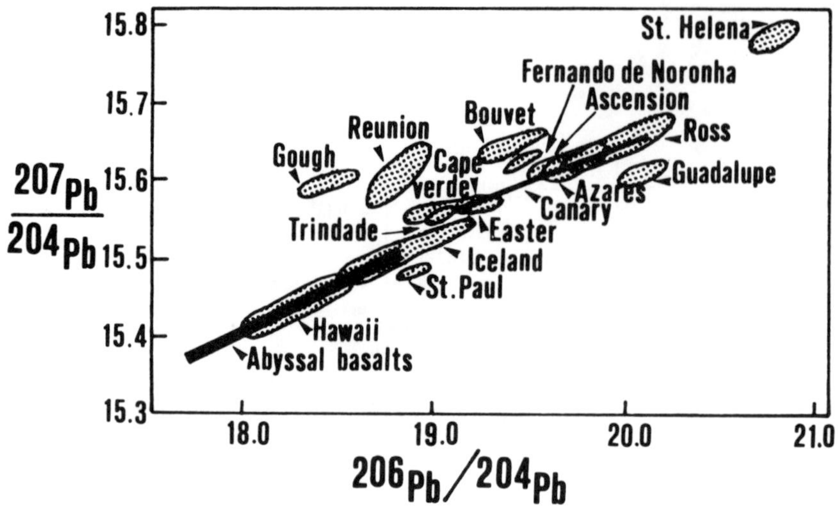

Fig.3 $^{207}PB/^{204}Pb$ vs. $^{206}Pb/^{204}Pb$ diagram for abyssal tholeiites and rocks from oceanic islands (after Tatsumoto, 1978)

from a multiplicity of sources with a similar history which have been isolated for hundreds of milions of years (Wood et al., 1979).

A two sources model with magma mixing, originally proposed by Schilling (1973 a) to account for the geochemical trend observed along the Reykjanes ridge approaching Iceland, became very popular and has been applied to many other sectors of oceanic ridges (Schilling, 1973 b, 1975; Anderson et al., 1975; White et al., 1975). Basically this model implies that ridges consist of "normal" segments, "anomalous hot spot" areas (oceanic island on the ridge crest or other topographically anomalous sector of MOR) and segments transitional between them (e.g. Reykjanes ridge). Basalts in the hot spot areas would be fed by LIL-undepleted deep source rising as mantle plumes or blobs, whereas a LIL-depleted shallower mantle source would feed the "normal"ridge segments, with mixing of the two sources occurring in the transitional areas. This model, originally based on REE distribution, then received support from isotopic studies (Hart et al., 1973; Sun et al., 1975). In recent years it has been however strongly criticized, both in light of new geochemical results from more extensive sampling of crucial areas and of the theoretical modeling of the mixing process (Langmuir et al., 1978). New data from Iceland (O'Nions et al., 1976; Sigvaldason et al.,1976) and mid-Atlantic ridge near Iceland and Azores (Wood et al., 1979) have actually shown that:
i) the geographic compositional trends are less evident and regular than previously reported;
ii) LIL-enriched and LIL-depleted basalts are simultaneously erupted from the same fissure swarm;
iii) LIL-depleted abyssal tholeiites are present, although volumetrically subordinated, even in "anomalous hot spot" areas and "transitional" ridge segments; geographic compositional differences approaching "hot spot" zones, seem then due to the variable amplitude (and abundance) of LIL-enriched basalt variation range, and not to the presence or absence of MORB.
Further difficulty to the "plume" model arises from the remark that the LIL-undepleted peridotite is denser than depleted peridotite and therefore should not tend to rise buoyantly in the mantle (O'Hara, 1975); from the observation that some "hot spot" are actually "hot lines" affecting the oceanic lithosphere for thousands of Km (Bonatti et al., 1977); from the sug-

gestion that undepleted mantle could lie above and not below depleted material (Hedge, 1978; Barberi et al., 1980 a).

The following picture therefore emerges at present for oceanic basalt petrogenesis:

- isotopic data suggest heterogeneity in the mantle source, both vertically and laterally. A simple two-source mixing model is however inadeguate to explain the great geochemical variation shown by oceanic basalts and basalts petrogenesis is probably a much more complex process. A LIL-depleted MORB mantle source seems however characteristically present in all spreading ridges;
- only in specific sectors (i.e. FAMOUS) of MOR, characterized by constancy of isotopic composition, basalts are possibly fed by a single homogenous source. Here the large variation in major and trace elements distribution can be explained by a combination of crystal fractionation, mixing of more primitive magmas injected into a magma chamber differentiating beneath the ridge, equilibrium partial melting or dynamic melting (O'Hara, 1977; Bryan and More, 1977; Langmuir et al., 1977; Wood et al., 1979). The actual occurrence, time persistence and petrogenetic importance of shallow magma chamber beneath the ridge is however still subject to controversy (e.g. Wood et al., 1979);
- notwithstanding the enormous effort and progress made in last years in the oceanic basalt study, their petrogenesis is open to further progress, that will probably mostly come from improvement of quantitative models of trace elements and radiogenic isotope behaviour in magmagenetic processes, together with the fine-scale geologically controlled submersible-dredging studies combined with drilling programmes.

Continental rifts

A review of the chemical aspects of continental rift magmatism is faced to several difficulties, arising from their large number, the relative great variability of their geophysical-geological features, including their age as well as the intensity and nature of associated magmatism, and finally the strongly inhomogeneous quantity and quality of available chemical data. Volcanic rocks of continental rifts have unfortunately not received the same attention that petrologists and geochemists devoted in the last years to oceanic basalts and island arc volcanics. With few remarkable exceptions, we still lack, therefore, of modern geochemical data (mostly of trace elements and radiogenic isotopes) on many significant sectors of the world continental rift system, and moreover also major elements data appear often qualitatively inhomogenous.

Our first choice in preparing this chapter has been to concentrate only on modern rifts, neglecting paleorifts of all ages, as a tribute of the "principle of actualism"(interested readers are referred to the two volumes of the Proceedings of the Symposium "Paleorift systems with emphasis on the Permian Oslo rift". NATO-ASI, 1978 (Eds. E.R. Neumann and I.B. Ramberg). Secondly we were obliged, in order to keep a minimum uniformity of exposure, to use mostly major elements data. The third choice has finally been that of selecting for a brief description examples of modern continental rifts covering as much as possible the category's spectrum. We thought that the "volcanicity" i.e. the frequence and volumetric importance of rift related volcanism was the best selection criterion, not only because obviously appropriate in a chapter on magmatism, but also because of our deep convinction that volcanism is an important tracer of geodynamic processes.

Low-volcanicity rifts

a) The Rhinegraben. According to stratigraphic evidences, the Rhinegraben began to form as down-warping in Middle Eocene whereas the step faulted structure initiated in Chattian and Aquitanian times (Wagner, 1953). Two stages in the rift development are recognized (Illies, 1978): the older structure had a main tectonic direc-

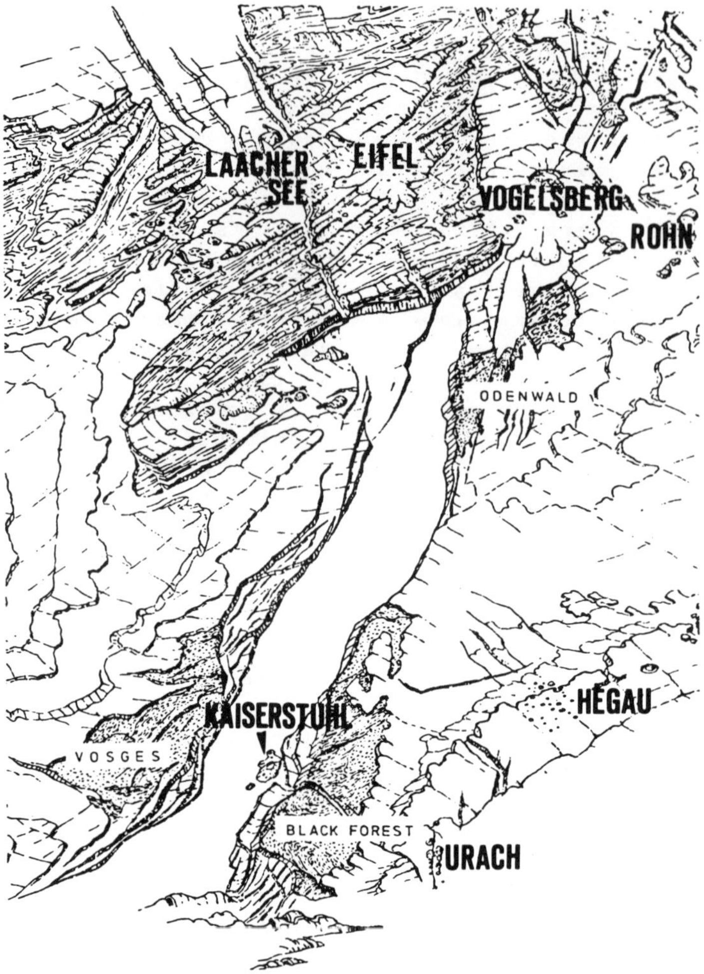

Fig.4 Block diagram of the Rhinegraben region (after Cloos, 1955, slightly modified)

tion of about 190° and probably ended in Lower Miocene. Graben tectonics died out in Middle Miocene to Lower Pliocene and resumed in Middle Pliocene, henceforth controlled by a sinistral shear motion. All Pleistocene to Recent active graben faults show a preferred strike direction of about 170°. The Paleogene tensional rift system is interpreted as an intraplate reaction to the contemporaneous main stage of Alpine collision. The Pliocene-Recent structure could be, on the contrary, an active subplate boundary determined by the coincidence of a pre-existing crustal anisotropy and the regional stress pattern. The Rhinegraben s.s. (between Basel and Frankfurt) is the only segment where the active structure coincides with the Paleogene one (Illies; 1978).

Volcanic rocks do not form a continuous cover and a central sector, more than 100 Km long,is free of volcanism. Moreover they occur along faults mostly on the highlands bordering the graben. The main graben floor is actually free of volcanics with the exception of Kaiserstuhl central volcano, which lies at the intersection with the Bonndorf graben fault system. Vogelsberg volcano is located at the northern prolongation of the graben, outside the rift zone (Fig.4). The Eifel volcanic complex is located at the intersection with NW faults, transversal respect to the main graben trend.

Volcanism initiated in Late Cre-

Table 3. Average major elements chemical compositions of basic rocks from Rhinegraben area

	Vogelsberg				Röhn				Laacher See			Daun Eifel	Kaiserstuhl	Hocheifel		Hegau
SiO_2	43.2	45.1	51.2	54.1	40.4	42.3	44.0	49.4	41.7	44.2	46.8	43.2	43.4	43.7	46.6	38.1
TiO_2	3.0	2.9	2.5	2.6	3.1	2.6	2.2	1.5	2.9	2.7	2.3	3.0	3.1	2.9	2.7	3.0
Al_2O_3	12.6	12.7	13.6	13.4	11.9	13.0	13.4	13.1	14.4	14.1	14.9	13.5	14.3	14.2	14.0	9.5
FeO°	10.2	11.1	10.5	10.0	13.9	11.2	11.8	11.6	10.5	10.2	10.7	10.2	12.4	10.4	10.3	13.2
MgO	14.0	12.1	9.2	7.3	12.8	13.3	12.3	8.3	8.3	9.4	9.0	8.8	7.9	10.6	10.3	15.3
CaO	9.9	11.1	7.9	7.2	12.7	12.5	11.5	10.8	13.4	12.2	10.4	15.0	13.5	12.9	11.1	15.3
Na_2O	4.0	2.7	3.5	3.8	3.4	2.8	2.7	3.1	4.0	3.2	4.1	2.7	3.0	3.5	3.1	2.9
K_2O	2.4	1.4	1.1	0.92	1.3	1.2	0.77	1.1	3.9	3.3	1.2	2.6	2.0	1.0	1.1	1.6
P_2O_5	0.68	0.72	0.46	0.50	0.33	0.54	0.86	0.76	0.7	0.5	0.5	0.76	0.4	0.6	0.5	0.8
MgN.	0.75	0.70	0.65	0.61	0.66	0.72	0.69	0.60	0.63	0.66	0.64	0.64	0.57	0.68	0.67	0.71
Rock type	BA	AB	OT	SB	NE	BA	AB	OT	NE	BA	AB	BA	BA	BA	AB	MA
N.of anal.	10	14	12	13	3	5	4	7	4	20	2	15	5	7	5	10

Analyses water-free to 100; FeO°=total iron as FeO; MgN.= Mg/(Mg+0.85Fe"tot); BA = basanite; AB = alkali-basalt; OT = olivine-tholeiite; SB = subalkaline basalt; NE = nephelinite; MA = melilite-ankaratrite. Source of data: Ficke,1961; Schorer,1970; Mortel,1971; Wimmenauer,1974; Gehnes and Wimmenauer, 1975; Duda and Schmincke,1978.

taceous (about 80 m.y.) and shows two peaks in intensity, in Middle-Late Eocene and in Pliocene-Quaternary, when major faulting took place, as indicated by the frequence of occurrencies and the volume of erupted products (Lippolt et al., 1974). The available K-Ar age data cover, however, almost continuously the time span from Late Cretaceous to Holocene. Neugebeuer (1978) describes an outward migration of volcanic activity with time from the crest of the initially uplifted area.

The magmatic association is extremely complex (Table 3) and includes many rocks varieties with subalkaline and strongly to moderately alkaline affinity (nephelinites, phonolites, carbonatites, basanites, alkaline and subalkaline basalts) with both sodic and potssic (leucitic) character (Mortel, 1971; Wimmenauer, 1974; Gehnes and Wimmenauer, 1975; Huckenholz and Noussinanos, 1977; Duda and Schmincke, 1978). The strongly alkalic and silica undersaturated basic rocks (basanites, nephelinites, ankaratrites) occur along the whole activity time span. Saturated and weakly undersaturated basic rocks seem instead confined to the last 20 m.y.. Poorly differentiated basic products largely prevail on evolved varieties (mostly trachytes and phonolites) in the Rhinegraben l.s. area; the latter are however widespread at central volcanoes (Kaiserstuhl, Vogelsberg) and elsewhere (Rohn, Laacher See, Hocheifel, Hegau).

b) The Baikal rift zone. The Baikal rift system extends for about 2500 Km mostly in a SW-Ne direction, at the connection between the Precambrian Siberian platform and the Sayan-Baikal fold belt. It is made by several different units comprising mountain ranges, volcanic and sedimentary plateaus and rift depressions.Such a complex structure results from an alongate uplift (Late Cenozoic-Paleocene) that assumed in Eocene a step faulted rift character (Logatchev and Florensov, 1978) in South Baikal area, and successively extended both westwards and northeastwards. Tectonic were continuous up to present and different phases of faulting have been recognized (Sherman, 1978). Two main

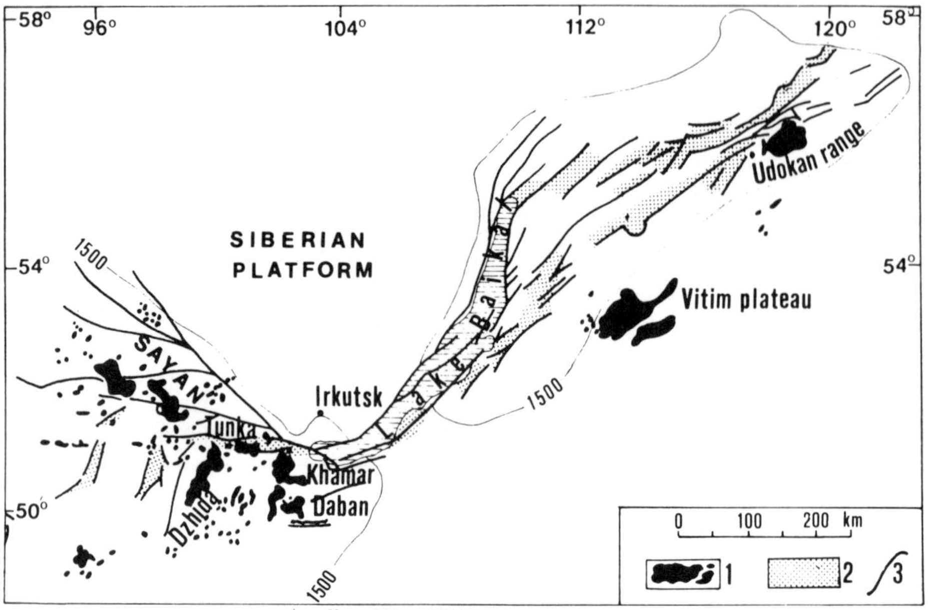

Fig.5 Structural sketch map of the Baikal rift zone. 1.Tertiary and Quaternary volcanics; 2.Tertiary and Quaternary sediments filling the major grabens; 3.Major faults. The 1500 m contour line indicates the limit of the uplifted area (after Kiselev et al., 1978, simplified)

stages in the rift zone evolution were recognized by the study of the thick Cenozoic sedimentary rift filling formations (Logatchev and Florensov,1978). The earlier and longer stage (Eocene-Early Pliocene) is characterized by moderate tectonic movements and mostly plastic deformation of the basement. During the second stage (Middle Pliocene to Recent) brittle fracturing with displacements along both normal and wrenchs faults dominates: the total extension is of few Km (possibly up to 10 Km in South Baikal rift segment; Logatchev and Florensov, 1978).

Volcanismrelated to the development of the rift zones has been active from Early Paleocene (possibly late Cretaceous) to Holocene. The greatest volumes of volcanic products were erupted during the first stage of the rift history and affected the regions located at both the western (Sayan, Khamar-Daban) and eastern (Udokan) ends of the Sayan-Baikal uplift. Volcanic activity was mainly fissural forming large basaltic fields with maximum thickness of 500-600 m. Cinder cones and associated lava flows generally characterize more evolved magmas (benmorites to trachytes) as in Udokan basin. Volcanism in the Baikal rift zone is however geographically discountinuous and is not associated with major faults and depression (Fig. 5). Only the Tunka rift valley is partly filled by basalts (Paleocene to Quaternary) which form interlayers whithin the Cenozoic sedimentary sequence.

Chemical data show that during the entire period of volcanic activity there has been a complex alternance of basaltic lavas with alkaline, transitional and tholeiitic character (Table 4). Alkaline and subalkaline lavas (olivine tholeiites and tholeiites) are scattered over the entire volcanic region, with the remarkable exception of Tunka district where tholeiites largely predominate. According to Kiselev et al., (1978) slightly differentiated lavas (ha-

Table 4. Average major elements chemical compositions of basic rocks from the Baikal Rift Zone.

	Kitoy river		Oka river	Tunka		Khamar-Daban		Dzhida	Khubsugul		Vitim		Udokan	
SiO_2	48.7	46.7	49.3	47.9	47.5	49.2	47.0	48.3	50.3	48.1	49.9	46.1	50.1	46.7
TiO_2	2.2	2.5	2.6	2.4	2.1	2.9	2.9	2.0	2.3	2.5	2.2	2.5	1.6	2.9
Al_2O_3	15.3	15.2	15.0	16.5	16.5	15.4	14.8	14.8	15.6	15.7	14.8	14.6	15.4	14.6
FeO°	12.1	12.4	11.5	11.6	11.6	11.5	12.1	13.3	10.7	10.9	11.6	12.8	11.4	13.6
MgO	9.0	9.2	7.7	8.5	7.9	7.8	8.4	8.3	6.7	8.0	7.5	8.3	7.5	10.0
CaO	9.2	9.2	8.0	8.9	8.4	7.9	8.4	8.5	7.8	8.0	8.3	8.1	9.2	9.4
Na_2O	1.8	2.9	3.3	2.1	3.6	3.2	4.0	2.4	3.3	3.6	3.8	4.6	3.2	3.9
K_2O	0.92	1.4	1.9	1.3	1.7	1.5	2.0	1.5	2.4	2.2	1.4	2.4	0.96	1.9
P_2O_5	0.56	0.23	0.60	0.53	0.36	0.46	0.36	0.71	0.66	0.81	0.35	0.56	0.30	0.35
MgN.	0.61	0.61	0.60	0.61	0.59	0.59	0.60	0.57	0.57	0.61	0.58	0.58	0.58	0.61
Rock type	SB	OT-AB	OT-AB	OT	AB	OT	AB-BA	OT	OT	AB	OT	BA	OT	AB-BA
N.of anal.	14	4	22	32	8	10	7	3	14	8	24	34	3	34

Keys as in table 3. Source of data: Kiselev et al.,1978.

waiites) are quite common. More evolved products (mugearites to trachytes) are restricted to Udokan Quaternary district. Strongly undersaturated alkaline rocks are scarce: some basanites, nephelinites and ne-rich hawaiites are reported from the Dzhida river basin, Vitim plateau and Udokan basin. As a whole, alkali basalts (and hawaiites) and subalkaline basalts are interminled with no apparent preferential distribution either in time or space, but with tholeiites dominating in the only volcanic district directly related to a rift valley structure (Tunka). Strongly alkalic products on the contrary seem to occur only in Quaternary times, during the second stage of the Baikal rift zone evolution.

c) Western Branch of the East African rift system. In Eastern Africa the rift system extends some 3400 Km from Gulf of Zula (Red Sea) in the North to Zambesi river in the South. The zone of rift fractures separates Danakil Alps, Somalia, Eastern Kenya and Tanzania and North Mozambique stable blocks from the rest of Africa. Lake Victoria block (Tanzania, Uganda, West Kenya, Rwanda, Burundi) separates the Eastern rift branch from the shorter Western one; to the South the two arms (Lake Rukwa and Great Ruaha rifts) join to form the single Malavi rift (Rungwe triple junction of Burke and Whiteman, 1972) (Fig.6). Frequence, intensity and nature of volcanism is strikingly different in the two main rift branches and only the Western branch displays the features of a low-volcanicity rift.

Western rift affects the region of uplifted plateaus related to the regional updoming of Eastern Africa, probably initiated in Late Cretaceous. The initial downwarp of the rift occurred in Early Miocene as indicated by the oldest sediments of the Lake Albert (Hopwood and Lepersonne, 1953, in King, 1970). Major faulting episodes have Pleistocenic age, post-dating the formation of the main erosion surface of the rift. Fault scarp, possibly indicating a minor Miocene episode, have been reported (Cahen, 1954 in Baker, 1971) but step faulting does not seem to exceed 5.0 m.y. in age (Burke and Whiteman, 1972).

Volcanism occur discontinuously in four zones located at the culmination of the major uplifts and in the proximity of offsets of rift arms

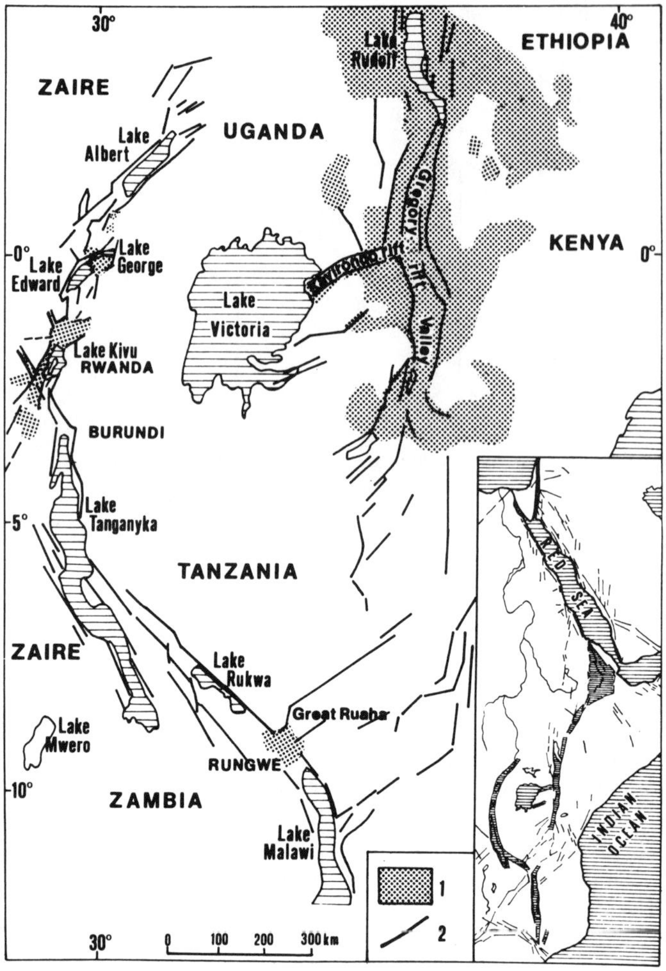

Fig.6 Structural sketch map of the southern termination of the East African rift system. Major faults (1) and Tertiary and Quaternary volcanic areas (2) are indicated (after King, 1970, slightly modified)

junction: Rungwe, South Lake Kivu, Virunga and Toro-Ankole (Fig.7). In the Rungwe volcanic region alkali basalts, nephelinites, phonolites and trachytes (both lavas and pyroclastics) were erupted in Pliocene-Quaternary times and evidences exist for an historic pyroclastic eruption (Harkin, 1960). Cretaceous carbonatitic intrusions underlye recent volcanics. Between Rungwe area and Lake Kivu the Lake Tanganyka (Nyasa)-Lake Rukwa graben system is free of volcanism. The South Kivu volcanic area lyes at the crossing between NW-SE ("Tanganyka") and NE-SW ("Albert") tectonic trends. Two volcanic zones occur in the area: the Mwenga-Kamituga lava field located to the SW of Lake Kivu, out of the intersection with Tanganyka tectonic trend, and the Tshibinda zone on the Southwestern border of Lake Kivu. In both zones volcanism in mainly fissural, with clear alignement of volcanic vents along NE-SW directions. Basic lavas are the most abundant erupted products: they include basanites, alkalibasalts, quartz-tholeiites and (prevailing) transitional olivine-tholeiites. Intermediate products, although scarce, are present (ferrobasalts, dark trachytes) as well as oversaturated trachytes (Haute Ruzizi trachytes; Denaeyer and Schnellinck, 1965) the latter being

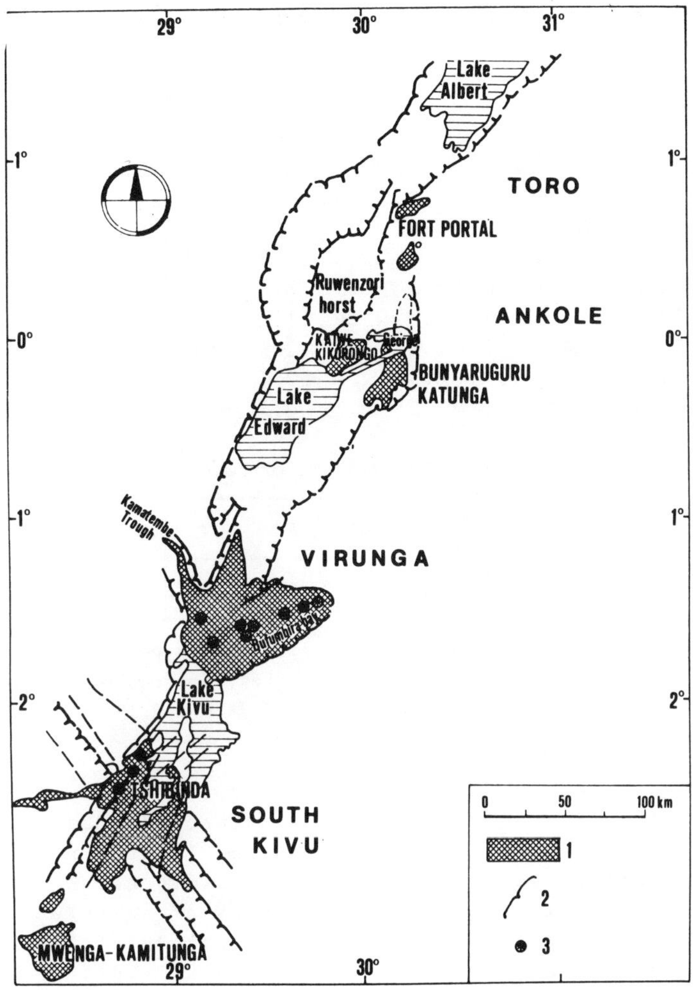

Fig.7 Western branch of the East Africa rift system. 1.Tertiary and Quaternary volcanic areas; 2.Major faults; 3.Major central volcanoes (after Guibert, 1978, modified)

limited to the South of Tshibinda area. The age of volcanism is poorly known: it seems to range from Late Cretaceous (70-90 m.y.) to present with probable gaps. Activity possibly increased in Lower to Middle Miocene and in Pliocene to Quaternary times. An eruption is recorded in the last century (Guibert, 1977 a, 1977 b, 1978).

The Northern border of Lake Kivu (the culmination of the floor of the Western Rift) is occupied by the Virunga volcanic range, at the intersection with the NW-SE trending Kamatembe trough. In this region eight major central volcanic complexes were built up in Plio-Quaternary times (Guibert et al.,1975) along a ENE-WSW trend ("Bufumbire bay" trend). With the exception of the deeply eroded Mikeno volcano, all other seven volcanoes must be regarded as active, although only Nyamalagira, Visoke and Nyragongo have a record of historic activity. Strongly undersaturated alkaline lavas (kalsilite and melitite bearing nephelinites to nepheline leucitites) characterize Nyragongo activity and are generally considered as typical of the whole Virunga range. These views are actually somewhat contradicted of available data indicating an high variability of erupted lavas as shown in Table 5. By the way it has to be noted that the number

Table 5. Average major elements chemical compositions of basic rocks from the Western African Rift.

	South Kivu			Virunga					Toro-Ankole			
SiO_2	50.6	47.8	45.9	38.9	38.9	42.9	45.6	47.3	40.2	41.3	38.2	40.3
TiO_2	1.7	1.8	2.0	1.9	3.3	2.9	2.9	2.2	4.5	4.6	5.1	5.0
Al_2O_3	15.3	14.5	15.2	9.9	12.8	12.7	13.6	12.7	8.1	8.1	8.6	10.2
FeO°	11.2	11.2	10.7	9.8	12.3	11.8	11.6	11.9	11.0	11.2	12.0	12.9
MgO	7.0	8.7	9.5	17.6	8.0	9.3	8.8	9.2	15.3	15.5	12.6	8.8
CaO	10.4	11.5	11.5	15.9	15.7	13.6	11.8	9.9	12.8	12.0	16.5	15.6
Na_2O	2.6	2.6	3.1	2.6	3.8	3.0	2.5	2.0	1.2	1.8	1.7	2.6
K_2O	0.51	1.0	1.1	1.8	3.8	2.8	2.4	1.9	6.1	4.7	4.1	3.5
P_2O_5	0.56	0.66	0.83	1.1	1.2	0.75	0.54	0.52	0.68	0.62	1.0	0.85
MgN.	0.57	0.62	0.65	0.79	0.58	0.62	0.62	0.62	0.75	0.75	0.69	0.59
Rock type	SB	OT	AB	NA	NE	BA	AB	OT	MP	UG	KT	KA
N.of anal.	13	17	15	6	16	34	41	12	8	10	8	17

Keys as in table 3. NA = nepheline-ankaratrites; KA = potassic ankaratrites; KT = katungites; UG = ugandites; MP = maphurites. Sources of data: Denaeyer and Schellinck,1965; Denaeyer,1972; Guibert,1977a,b and 1978.

of analyses available for the various rock-types does not reflect the actual volume proportions of erupted products, because of the large amount of samples related to Nyragongo and Nyamalagira historic eruptions.

North of Virunga range, the rift structure mantains a NE trend in the Lake Edward graben. Then an offset zone is observed; the Semliki river-lake Albert graben running to the West of the Ruwensori horst block is free of volcanism, whereas volcanic activity has affected the offset zone to the NE of Lake Edward in Bunyaruguru-Katunga and Katwe-Kikorongo areas (Fig.7). The last volcanic zone occurs more to the North, outside the rift structure, in Fort Poral region.In the three areas (Toro-Ankole province) pyroclastic products are abundant and several explosion craters are reported. Plagioclase free, higly undersaturated alkali potassic mafic products are typical of the Southern volcanic areas of the province, while carbonatitic products (lavas and tuffs) characterize Fort Portal area. To our knowledge no radiometric age determinations of Toro-Ankole volcanic rocks exist. According to Holmes and Harwood (1932) they should be Pliocene to Pleistocene in age, but historic activity has been also reported. Averages of selected chemical analyses of basic rocks from Western rift are reported in table 5; they show a clear increase in the potassic character of the mafic alkalic lavas northwards from South Kivu to Toro-Ankole.

d) <u>Continental rifts related to transform plate boundary: the Dead Sea rift</u>. With This category we refer to fault trenches cutting the entire lithosphere, and therefore displaying a rift-type morphology but characterized, with respect to "typical" rifts, by the predominance of a shear rather than tensional component and corresponding to transform (and not accreting) plate boundaries. The Dead Sea rift is reported as an example of this kind of structure in a continental environment; its oceanic equivalent is to be found in the so-called "fracture zones" offsetting mid-oceanic ridges.

The Gulf of Aqaba-Dead Sea-Jordan Valley rift represent the North-western transform boundary of the Arabian plate connecting the Red Sea (accreting) and Zagros (consuming) boundaries. Here volcanic activity occurs in distinct locations with respect to the transform plate boundary (Fig.8):

- within the narrow Dead Sea rift zone, either in few and very small outcrops and as part of larger volcanic fields, extending within the Arabian plate far from the rift (e.g. As Shamah lava field extending from Lake Tiberias to Eastern Jordan, Syria and Saudi Arabia);
- on both sides of the rift margin, at distances from few to more than 100 Km from the rift axis.

The volcanics products were emitted through the main fissure systems trending E-W and NW-SE that were active respectively in Early Miocene and Pliocene-Holocene. Available age data indicate the intermittent character of volcanic activity and a remarkable correspondance with the main eruptive cycles at the accreting Afar boundary (Barberi et al.,1980b). Volcanics consist of basic alkalic lavas with varying degree of silica undersaturation ranging from alkali-basalts to basanites and nepheline-basanites, with minor hawaiitic intermediates. As a whole, there is apparently a tendency towards an increase of alkalinity with time.

Tectonic studies have shown that the Aqaba-Jordan Valley rift is characterized by segments with a pure strike motion alternating with others with an extensional component (Freund,1965). Only the latter are affected by volcanism and it can be said that the occurrence of basaltic volcanism is by itself an indication that the tectonics that affected the lithosphere in that area had an extensional component large enough to produce gaping fissures allowing melts to rise from the upper mantle. The rather high "primitive" character of all basic lavas in the region indicates that magmas rised rapidely to the surface from the mantle source region, without strong differentiation during the ascent.The moderate to strong alkaline character of the lavas furtherly indicates low degree of mantle partial melting possibly related to the low rate of extension.

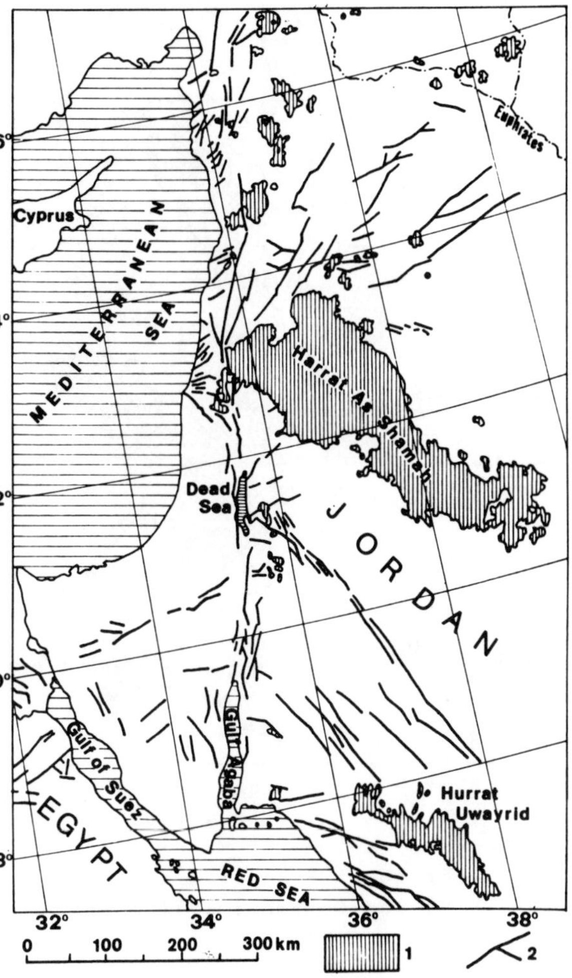

Fig.8 Gulf of Aqaba - Dead Sea - Jordan Valley transform system. 1. Tertiary and Quaternary volcanic areas; 2. Major faults.

High volcanicity rifts: the East African Rift System

A very high volcanicity characterizes the eastern branch of the East African Rift System from Kenya to its northern termination in Afar, where a complex junction with Red Sea and Gulf of Aden oceanic rift arms occurs. The Eastern Rift is not continous from Kenya to Ethiopia: in lakes Stephanies and Rudolph area the rift structural features attenuate to form the 300 Km wide Gofa "basin and range" province (Moore and Davidson, 1978) separating the Kenyan (or Gregory) from the Ethiopian rift.Although forming a single major structure affetting the

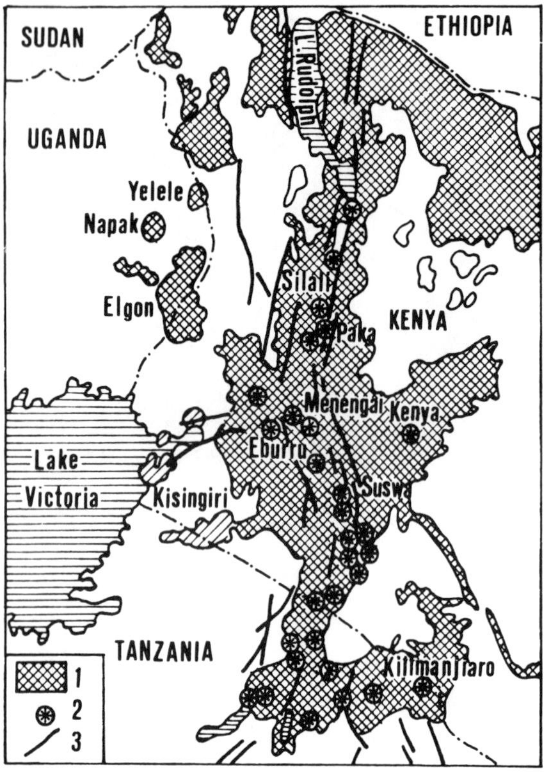

Fig.9 Sketch map of the Gregory rift valley. 1. Tertiary and Quaternary volcanics; 2. Major central volcanoes; 3.Major faults (after King and Chapman, 1972,modified).

Earth lithosphere, the Kenyan, Ethiopian and Afar rift sectors displaysome variations, that in our opinion reflect different stages of advancement in the rifting process leading to the continental break-up.

a) The Kenya rift. The Kenya (or Gregory) Rift Valley is a complex graben structure some 750 Km long with an average width of 60 Km (Fig.9). It is characterized by numerous subparallel normal faults forming the step-faulted margin of the rift valley and by small scale horst and graben affecting its floor. In Northern Kenya the graben zone is only the youngest part of a broader zone of rift fracturing (Turkana depression), while in Southern Kenya a single deep graben occur up to North of the Equator, where the East-West trending Kavirondo rift joins the main rift valley (Nakuru triple junction of Burke and Whiteman, 1972). More to the South, the Kenya rift changes in character, forming a number of NE-SW trending depression upfaulted on their Western margins, South-East of Kilimanjiaro, the NW-SE trending Pangani graben occurs constituting another supposed rift junction. In Tanzania the Gregory rift ceases to be a continuous chain of graben and becomes a series of widely spaced antithetic fault blocks downstepping eastwards which are free of volcanism and join the Western rift in Rungwe area.

The Kenya rift area has been marked by continuous volcanic activity from Oligocene-Miocene times to the present day. The volcanic region is broadly centered on the Kenya "dome", a topographic culmination resulting from a regional uplift occurred between the end of Cretaceous and Upper Miocene times. The rift valley s.s. crosses the higher parts of this dome.

The tectono-magmatic evolution of the area has been the object of several papers (Baker and Wolhemberg, 1971; Baker et al., 1971, 1977; King and Chapman, 1972; Logatchev, 1976; Chapman et al.,1978; Williams, 1978; Crossley, 1979). According to Baker and Wolhemberg (1971), the rift appears to have initiated in Late Oligocene-Early Miocene as a downwarp with only minor faults in the Turkana area. An upwarped region along the Kenya-Uganda border was the site of Lower to Middle Miocene nephelinitic central volcanism, contemporaneous to alkaline products that partly filled the Turkana depression. In Late Miocene the central rift area was slightly uplifted and was the site of fissure phonolite eruptions. The first extensive rift faulting episode developped in Early Pliocene producing a meridional asymmetrical trough faulted mainly in its western side. In the floor of the central part of this trough, a minor phonolite and trachyte volcanism occurred in Middle Pliocene, followed by much more voluminous basalt eruptions along nearly the whole lenght of the trough. In Late Pliocene time massive eruptions of trachytic ignimbrites occurred in the Naivasha sector of

Table 6. Average major elements chemical compositions of basic rocks from the Gregory Rift System.

	Napak		Moroto		Lake Rudolph	various	Silali	Chyulu
SiO_2	43.1	39.8	44.6	46.7	45.5	47.6	48.1	45.3
TiO_2	1.4	2.0	2.4	2.3	2.2	2.5	2.8	3.0
Al_2O_3	9.1	12.5	10.3	15.2	16.1	15.0	15.9	11.3
FeO°	12.6	14.7	11.5	14.7	9.3	12.2	11.8	13.3
MgO	18.1	7.0	10.8	7.8	8.9	6.3	5.8	11.5
CaO	11.0	14.6	14.6	8.6	11.8	11.7	11.1	10.8
Na_2O	2.6	5.5	3.9	3.3	3.7	3.0	2.6	2.9
K_2O	1.3	2.7	1.1	0.96	1.8	1.0	0.95	1.1
P_2O_5	0.40	0.89	0.61	0.40	0.46	0.47	0.73	0.65
MgN.	0.75	0.50	0.65	0.53	0.67	0.52	0.51	0.65
Rock type	NA	NE	BA-AB	OT	AB-BA	AB	OT	AB
N.of anal.	10	5	3	3	5	23	41	12

Keys as in tables 3 and 5. Sources of data: Varne,1968; King and Sutherland,1966; Brown and Carmichael,1969;1971; Goles,1975; McCall and Hornung,1972; Baker et al.,1977.

the rift floor locally filling the rift and overflowing its flanks. At the same time the last and major uplift of the marginal plateaus accompained the first of several phases of graben faulting which created the structure of the present Gregory Rift. The developing graben was partly filled by Lower to Middle Pleistocene flood trachytes. In Later Quaternary trachyte, basalt-trachyte and phonolite caldera volcanoes built up axially in the floor of the inner graben, while in areas well to the East of the rift, extensive linear multicentre basaltic eruptions occurred.

Several stages can be therefore distinguished in the volcanic development of the Kenya rift. Williams(1978) recognized five main volcanic cycles: 30-15 m.y., 15-12 m.y., 12-7 m.y., 7-2 m.y., and 2-0 m.y., in rough agreement with the six stages of Logatchev et al., (1972) (23-16; 13.5-12; 10-5; 5-2; 2-0.7; 0.7-0).

In general terms (King and Chapman, 1972) the earliest and latest volcanic activity occurred respectively on the western and eastern flanks of the rift; tectonic and volcanic event were initiated earlier in the northern part of the rift and progressed southwards. Flood eruptions were characterized by great volumes of uniforme compositions, but show a general trend from undersaturated to oversaturated types in the course of time. The Quaternary central volcanoes of the axial part of the rift show great diversities of composition, suggesting differentiation in local reservoirs. Average composition for basic and salic volcanics are reported in Tables 6 and 7.

Lippard and Truckle (1978) recognized two trends in geochemistry of basic rocks from North Kenya rift:

- increasing undersaturation at any time away from the rift axis;
- decreasing undersaturation with time leading to the development of transitional basalts in the Quaternary along the axis of the rift.

This has been recently confirmed by Truckle and Williams (1979) that have also shown that whereas basic lavas erupted to the West of the rift are strongly undersaturated and display relatively limited ranges in composition, the magmatism to the East of the rift

Table 7. Average major elements chemical compositions of salic rocks from the Gregory Rift System.

	1	2	3	4	5	6	7
	"plateau phonolites"	"Kenya phonolites"	Suswa		Silali	Oldoinyo nyokie	Eburru
SiO_2	57.0	60.3	57.8	60.4	61.3	63.5	70.0
TiO_2	0.55	0.67	1.1	0.74	0.65	0.88	0.53
Al_2O_3	20.7	17.2	17.1	16.4	16.5	13.5	9.3
FeO°	4.6	6.9	7.5	7.3	8.3	8.1	7.8
MgO	0.53	0.58	1.0	0.55	0.43	0.37	0.14
CaO	1.5	1.5	2.7	1.8	1.5	1.5	0.49
Na_2O	8.6	7.1	7.4	7.2	5.7	6.4	6.9
K_2O	6.2	5.4	4.8	5.1	5.0	5.3	4.5
AI	1.01	1.02	1.02	1.06	0.90	1.20	1.74
KI	0.32	0.33	0.30	0.32	0.37	0.35	0.30
Rock type	PH	PHT	PH	PHT	T	PT	P
N.of anal.	14	10	7	6	60	19	1

AI = (Na+K)/Al mol.; KI = K/(Na+K) mol.;T=trachytes;PT=pantelleritic trachytes; P=pantellerite; PH=phonolites; PHT=trachyte-phonolites. Sources of data: Lippard,1973 (av.1,2); Baker,1977 (6); Nash et al., 1969 (3,4); McCall and Hornung,1972 (5); Weaver et al.,1972 (7).

has a much greater range of variations from nephelinites to mildly alkaline compositions.

The same geographical and temporal trend seems also reflected in the nature of salic volcanic rocks that tend to become silica oversaturated with time (transition, with overlapping, from phonolites to trachytes to peralkaline rhyolites).

b) The main Ethiopian rift.The main Ethiopian rift begins at about 6° Lat. N (Lake Chamo) and runs up to approx. 11° Lat. N, where it widens into the Afar triangle (Fig. 10). The rift valley is tectonically characterized by en echelon swarms of NNE trending faults. These are normal faults, gaping and emissive fissures that fed either basalts and silicic products (mainly peralkaline rhyolites and oversaturated trachytes) (Baker et al.,1972).

The tectonic style changes at about 9° Lat.N, where the scarp of the Somalian plateau turns to a E-W direction. South of this latitude, in the Lakes region, the active part of the rift is marked by en echelon swarms of open fissures with associated volcanic activity (both fissural and through central vents) (Di Paola, 1972). No clear axial graben is observed and silicic volcanic products are widespread. In the northern part, the rift is instead marked by a well defined linear graben of NNE direction. Tectonics and volcanism are concentrated along the axis of the graben, that is characterized by fissural basaltic eruptions as well as by intermediate and silicic products mostly related to central volcanoes. These volcanoes are regularly spaced (55 Km) (Mohr and Wood, 1976) and their position is apparently controlled by the occurrence of transverse tectonic lineaments. The early phases of tectonics and volcanism on the main Ethiopian rift are poorly known, mostly because of the widespread Plio-Quaternary cover that prevents direct observation of the older

Fig.10 Geological sketch map of the Ethiopian rift valley. 1. Crystalline basement; 2.Pre-Pliocene volcanics; 3. Miocene and Quaternary volcanics; 4. Quaternary sediments; 5. Major central volcanoes; 6.Major faults (after Merla et al.,1979, geological map, redrawn).

products. The early volcanic episode in the area dates back to Eocene, but it seems limited to the central part of the Ethiopian plateau, where the so-called Ashangi basalts were erupted (Zanettin et al., 1979). Volcanism became widespread in Late Oligocene-Early Miocene (30-25 m.y.) on all the continental blocks bordering the Ethiopian, Afar, Red Sea and Gulf of Aden rifts. Evidence have been reported for an early Miocene initiation of rifting in the Afar triple junction (Barberi et al., 1975 a; Arnò et al., 1980). The oldest dates so far reported for the main Ethiopian rifts refer to Middle Miocene (13-10 m.y.) rhyolites outcropping on the rift margins (Zanettin et al., 1979) but also in this sector most probably rifting dates back to early Miocene. Silicic volcanic activity in the Ethiopian rifts shows two intensity peaks, respectively in Middle Miocene (13.5-11.5 m.y.) and Plio-Quaternary (5.5-1.5 m.y.). It is important to note the time correspondance with volcanologically similar events in Kenya, but characterized by different composition of erupted products. The Middle Miocene event produced flood phonolites in Kenya (15-12 m.y.) and alkaline and peralkaline rhyolites in the Ethiopian rift (13.5-11.5 m.y.) and in Afar (14-9 m.y.); the Plio-Quaternary event produced trachytes in Kenya (6-2 m.y.), and still oversaturated silicics, mostly peralkaline, in the Ethiopian rift (5.0-1.5). A change is observed in Afar where peralkaline as well as subalkaline rhyolites occur at that time (4.0-0.5 m.y.), but within a volcanic sequence (Afar Stratoid Series) The associated basic rocks are also less silica undersaturated in Ethiopia being dominated by alcalic to mildly alcalic basalts. Strongly alkalic varieties (i. e. nephelinites) are absent in Ethiopia and basanites, phonolites and trachytes are reported only from limited sectors of the Ethiopian plateau whereas they lack completely in both main Ethiopian rift and Afar. Ethiopian plateau volcanics therefore include some undersaturated alkalic,basic and salic,varieties that are not found within the rift, buth there is a large overlapping of composition between rift and off-rift rocks in the dominant moderately to mildly alkalic compositional field (see Tables 8 and 9).

c) Transition from continental to oceanic rift: the Afar triangle. The Afar rift has an exceptional geological interest being one of the few areas in the world where the transition from continental rifting to continental break-up

Table 8. Ethiopian plateau and rift basic rocks - Average major elements chemical compositions

	1	2	3	4	5	6	7	8	9	10
SiO_2	47.7	52.5	50.1	50.0	45.5	47.9	43.0	47.2	47.6	48.2
TiO_2	4.9	2.3	3.0	2.6	2.7	2.6	3.6	1.7	2.1	2.4
Al_2O_3	11.9	14.8	13.0	13.4	13.8	15.3	13.4	16.4	17.0	15.5
FeO°	14.2	11.5	11.6	11.7	11.9	12.2	14.3	11.3	10.3	12.1
MgO	6.4	5.3	7.9	7.6	11.4	6.3	9.0	9.8	8.0	6.2
CaO	10.5	9.6	10.3	10.7	9.9	10.0	11.1	9.3	10.8	10.6
Na_2O	2.5	2.9	2.6	2.5	2.9	3.4	3.4	3.1	3.0	3.3
K_2O	1.0	0.61	0.93	0.88	0.98	1.2	0.83	1.1	0.72	0.93
P_2O_5	0.59	0.22	0.40	0.43	0.57	0.74	1.2	0.10	0.35	0.43
MgN.	0.49	0.49	0.59	0.58	0.67	0.52	0.57	0.65	0.62	0.52
rock type	AB	SB	OT	OT	AB	AB	BA	AB	OT-AB	AB
anal.	4	5	10	10	1	6	2	5	8	3
form.	Ashangi		Aiba	Alaji		Termaber		Injibara	Tullu Moje	Various
	PLATEAU								RIFT	

Keys as in table 3. Source of data: Piccirillo et al., 1979 (averages 1,2,6,7); Zanettin et al., 1976 (3,4,5); Treuil and Varet, 1973 (8); Di Paola unpublished data (9); Di Paola, 1972; Weaver et al., 1972; Brotzu et al., 1974 (10).

can be observed. It has been intensively studied in the last ten years mainly through the effort of a joint french-italian research project. These studies (e.g. Barberi et al., 1975 a; Barberi and Varet, 1977 and 1978) allowed the formulation of a two-stage evolution model for Afar rift: a long initial stage of continental rifting initiated in early Miocene and lasted about 20 m.y., followed by a stage (4m.y. to present) during which continental break-up occurred and the proto-oceanic structures of Afar were formed.

The continental rifting stage can be studied only on the rift margins where Miocene rocks outcrop, whereas the inner part of the rift is almost entirely covered by recent volcanics emplaced in the 4 m.y. (Fig.11). The early stage is characterized (Barberi et al., 1975 a) by phases of volcanic paroxism occurred approximately at 25-16 m.y.; (mainly basalts), at 14-10 m.y. (mainly flood rhyolites) and at 8-6.5 m.y.(mainly basalts), probably corresponding to similar events in Kenyan and Ethiopian rifts. These early volcanic events are still little known petrologically and geochemically. Basalts have however a mostly mildly alkaline character and Middle-Miocene rhyolites are similar to the coheval Ethiopian rhyolites being mostly represented by alkaline and peralkaline (comenditic) varieties. During Pliocene, approximately 4-5 m.y. ago, the Afar rift changes character and its volcanic and tectonic behaviour is different from the rest of the East African rift valley. The first significant episode is the emplacement of the Afar Stratoid Series, a dominantly basaltic flood formations emplaced in the last 4 m.y. through fissures parallel to the regional structural trend, that changes from NNW to WNW southwards. Rhyolites occur mainly in the upper part of the series, either as ignimbrites and flows apparen-

Table 9. Ethiopian plateau and rift salic rocks - Average major elements chemical compositions

	1	2	3	4	5	6	7	8	9	10	11	12
SiO_2	74.6	75.1	72.4	74.9	75.1	65.8	64.1	75.0	58.5	73.6	71.5	67.1
TiO_2	0.45	0.51	1.1	0.32	0.37	0.61	0.12	0.13	0.33	0.35	0.34	0.26
Al_2O_3	13.2	12.0	9.2	12.9	12.0	15.9	17.9	12.9	20.9	8.8	13.1	15.8
FeO°	1.8	2.3	7.0	2.6	2.8	4.8	4.8	2.4	3.4	6.6	3.9	4.0
MgO	0.28	0.27	0.41	0.06	0.18	0.54	0.19	0.03	0.52	0.02	0.17	0.14
CaO	0.28	0.18	0.20	0.13	0.18	0.35	0.85	0.24	2.7	0.29	0.65	0.69
Na_2O	4.8	4.5	3.7	4.3	4.4	6.7	7.0	4.7	7.0	5.9	6.0	6.6
K_2O	4.5	4.9	5.4	4.6	5.0	5.1	4.7	4.6	6.3	4.2	4.2	5.2
rock type	AR	C	P	AR	C	CT	T	AR	PH	P	C	CT
$\frac{(Na+K)}{Al}$	0.96	1.06	1.35	0.94	1.05	1.04	0.90	0.98	0.88	1.62	1.11	1.04
$\frac{K}{(Na+K)}$	0.39	0.41	0.49	0.41	0.43	0.33	0.31	0.39	0.37	0.32	0.32	0.34
N of anal.	7	11	1	4	5	4	8	3	4	13	11	1
	Oligocene Alaji			Miocene Alaji		Termaber	Adua-Axum		Termaber † Injibara	Ethiopian Rift (Tullu Mojé)		Ganjuli

AR= alkali rhyolite; C= comendite; P= pantellerite; CT=comenditic trachyte; PH= phonolite; T=trachyte; Source of data: Zanettin et al.,1976 (1,2,3,4,5,6); unpublished data (7); unpublished data and Mohr, 1970 (8); Zanettin et al., 1976 and Treuil and Varet, 1973 (9); Di Paola, unpublished data (10,11); Zanettin et al., 1978 (12).

tly related to fissure activity and (mostly) as flows, domes, and pyroclastic rocks related to central volcanoes. These dominantly rhyolitic centers characteristically occur in the intersection of different fracture trends. The basic lavas are dominated by slightly differentiated mildly alkaline basalts and hawaiites, but few samples of alkali basalts and quartz tholeiites have been also found. Rhyolites show a rather range of variation from the prevailing peralkaline varieties to subaluminous types. Intermediate rock-types also occur both as fissural flows and in the central volcanoes. In the fissural sequence they are however quantitatively strongly subordinated to both basalts and rhyolites. This volcanic series has been interpreted as oceanic crust produced during Pliocene Afar spreading or alternatively as related to a continental rifting phase (see discussion in Barberi and Varet, 1977).

Since 1.0 m.y. Afar is characterized by the development of volcano-tectonic structures interpreted as proto-oceanic plate boundaries, of accreting and transform type. The first corresponds to the so-called axial volcanic ranges, that is to active volcanic structures built up by frequent fissural eruptions of basalts from a small central graben (1 to 2 Km across) running parallel to the regional tectonic trend. Parallel normal faults affect both sides of the graben for a width of as much as 15 Km. Their age, as well as the age of basalts, decreases symmetrically from the borders towards the axis of the graben. This fissural stage is present in all axial ranges of Afar and only basalts with minor iron-rich intermediate lavas occur. In some axial ranges a volcano-magmatic evolution is observed leading to shield or more complex central volcanoes where complete differentiation series from basalt to rhyolite occur. The axial range basalts are mostly mildly alkalic. Low-K varieties, approaching MORB composition, also occur in some cases (e.g. Asal rift in Gibuti Republic;

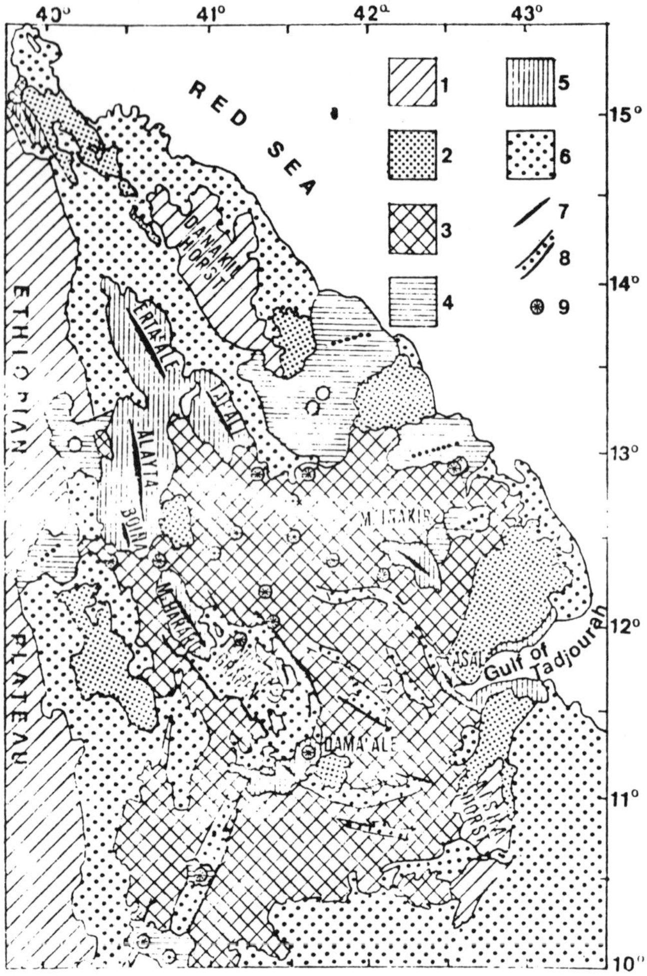

Fig.11 Geological sketch map of the Afar Rift. 1.Pre-Miocene basement; 2. Miocene volcanic formation of the rift margin ("Mabla" and "Dalha" series); 3.Pliocene to Quaternary Stratoid Series; 4. Quaternary transverse ranges; 5. Quaternary axial ranges; 6.Tertiary and Quaternary sediments; 7. Present oceanic spreading axes; 8. Major graben structures; 9. Major silicic centres at the top of the Stratoid Series (redrawn simplified after Barberi et al., 1971 and CNR-CNRS, 1975, 1:500.000 geological maps).

Richard, 1979). Differences are observed both geographically and in time within the same axial range (e.g. Erta'Ale) that only rarely can be explained by equilibrium partial melting and largely reflect heterogeneity of the primary source (Barberi et al., 1980 a). The differentiation series shows chemical features clearly controlled by the nature of the basaltic parent magma. Where basalts are slightly more alkalic, the series ends with mildly peralkaline pantellerites, whereas slightly peralkaline comendites are associated with less alkalic basalts and non peralkaline rhyolites are found when basalts approach a tholeiitic composition (Barberi et al., 1975 b; Ferrara and Santacroce, 1980) (Tables 10 and 11).

Transverse volcanic structures (Barberi et al., 1974 a) occur mostly on the rift margins and appear aligned with offsets of axial ranges and with offsets of the continental blocks bordering the Afar depression. They are of late Quaternary age but apparently located along prerift lines of tectonic weakness. Volcanism has a fissural character but without displaying neither the intensity nor the tensional features (open fissures, axial graben) of the axial ranges. Finally they are characterized by undersaturated alkali basaltic magmas.

A comprehensive glance on continental rift volcanism

The description previously reported, although necessarely short and possibly oversimplified, shows some relevant aspects of continental rift volcanism that certainly have important genetic and structural implications. Among the most interesting questions are the following:

-which is the reason for the strongly different volcanic and magmatic behaviour shown by the two kinds of continental rifts that we called respectively high and low-volcanicity;

-which is the significance of the spatio-temporal variation in the nature of the volcanic products observed in the East African high volcanicity rift branch and which the origin of the huge flood salic volcanics characteristically erupted within the high-volcanicity continental rifts.

We shall try now to give some answers to these problems, with the purpose of contributing to the discus-

Table 10. Afar rift basic rocks - Average major elements chemical compositions.

	Asal		Boina	Assab Range	Stratoid Series		Dalha Series	
SiO_2	49.9	46.9	47.7	47.1	48.6	48.0	48.3	47.6
TiO_2	1.0	2.6	2.4	2.5	2.9	2.9	1.6	3.1
Al_2O_3	15.4	14.6	15.5	15.6	15.4	15.6	15.4	14.2
FeO°	10.1	13.4	11.9	10.5	11.9	11.7	12.0	13.5
MgO	9.5	7.9	7.5	8.8	6.4	6.3	7.9	6.5
CaO	12.6	10.9	10.7	10.6	10.5	10.8	11.4	10.9
Na_2O	2.1	2.8	2.8	3.1	2.9	3.2	2.7	3.1
K_2O	0.07	0.51	0.93	1.1	0.62	0.67	0.49	0.72
P_2O_5	0.07	0.27	0.40	0.41	0.48	0.55	0.25	0.39
Mg N.	0.65	0.54	0.57	0.64	0.53	0.53	0.58	0.49
rock type	SB	OT	OT-AB	AB	OT	AB	OT	AB
N. of anal.	10	9	4	7	9	2	2	2

Keys as in Table 3. Source of data: Richard, 1979; Barberi et al., 1975 b, 1974 a; De Fino et al., 1973.

sion on the rift magmatism genesis and significance, well aware of the fact that in some cases we will not be able to go much beyond the presentation of reasonable working hypotheses.

Differences between high and low-volcanicity rifts.

Data relative to volumes of erupted products, ranges of variation of the K/(Na+K) ratio in the basic rocks, frequency of basic, intermediate of salic rocks and frequency of the various types of basic rocks, based on published analytical data classified according to the simple alkalies-silica diagram of Cox et al. (1979) are reported in Figs. 12 and 13. They have a more qualitative than quantitative meaning and further work might change the single figures, but it is unliked that it will affect the general picture. These data and previous descriptions show clearly that low-volcanicity rifts (e.g. Rhinegraben, Baikal, West African) are characterized by:
-comparatively much lower volumes of erupted products;
- a space discontinuity in the volcanic activity along the rift;
-much wider variation of the nature of the erupted basic lavas, as emphasized by the higher ranges of K/(Na+K) and by REE distribution pattern (Fig. 14); (note the inverse relation between volume of volcanic products and width of the alkalis ratio variation range shown by Fig.12);
-little quantity of salic products with unimodal distribution of rock-types, contrasting with the bimodal distribution (basic and salic peaks) typical of high-volcanicity continental rifts. Furthermore the distribution of basic rocks varieties shows that strongly alkalic and undersaturated rocks (nephelinites, basanites) prevail in low-volcanicity rifts and there is a tendency for slightly undersaturated olivine-tholeiitic types to become more abundant when the volume of erupted products increases.

The rate of extension is lower in low than in high-volcanicity rifts and this certainly corresponds to a different degree of continental lithosphere attenuation by rifting, as shown by the different crustal thickness and the

Table 11. Afar silicic rocks - Average major elements chemical compositions

	1	2	3	4	5	6	7	8	9	10
SiO_2	72.4	75.2	72.5	73.6	73.6	71.1	72.4	64.1	64.9	71.4
TiO_2	0.31	0.14	0.36	0.29	0.25	0.43	0.36	0.51	0.57	0.22
Al_2O_3	13.0	12.0	9.7	11.0	12.9	12.1	10.0	15.8	14.1	12.1
FeO°	3.6	2.2	6.6	3.9	3.4	4.4	6.2	5.3	5.8	4.0
MgO	0.30	0.19	0.10	0.11	0.16	0.30	<0.01	0.77	0.41	0.12
CaO	1.6	0.59	0.39	0.54	1.1	1.4	0.37	3.6	1.8	0.58
Na_2O	4.8	4.9	5.7	5.5	5.0	6.0	5.9	5.8	7.0	6.1
K_2O	3.9	4.7	4.4	4.9	3.6	4.2	4.5	3.9	5.2	5.3
rock type	AR	C	P	C	AR	C	P	T	PT	C
AI	0.94	1.10	1.45	1.30	0.93	1.19	1.47	0.87	1.22	1.30
KI	0.35	0.38	0.34	0.37	0.32	0.32	0.33	0.31	0.33	0.36
N of anal.	4	6	8	3	9	5	7	1	1	2
	Afar Stratoid Series			Ribta	Dama' Ale	Tat' Ali	Boina	Dubbi	Ma' Alalta	
					Axial Ranges			Marginal Units		

Keys as in Table 7 and 9. Source of data: Barberi et al., 1974 b and unpublished data (1,2,3,8,9,10); Richard, 1979 (4); Ferrara and Santacroce, 1980 (5); unpublished data (6); Barberi et al., 1975 b (7).

kind of crust-mantle transition, suggesting the occurrence, at variable depth beneath the rift, of "anomalous" soft and hot mantle material (Fig.15). In other words, the two types of rifts differ greatly in the degree of asthenosphere penetration into the lithosphere, and low-volcanicity rifts are probably affected only by lithosphere fracturing with limited asthenosphere uprising and consequently by limited extension, all these features being reflected in their "cooler" nature (scarce volcanism, low heat flow).

These quite different geodynamic conditions can have a relevant influence on the magmagenetic process. In high-volcanicity rifts, basic magmas are most obviously generated by partial melting of the asthenosphere intruded underneath the rift axis; they rise up rapidely because of the high extension rate and their composition should be mostly controlled by the nature of the asthenospheric source and the mechanism and intensity of the partial melting process. Their compositional differences are rather limited and their significance will be discussed in the next chapter.

In low-volcanicity rifts, the process could be completely different. The slow rate of extension should induce only limited melting of the deep seated asthenosphere with production of strongly alkalic underasturated basic magmas. In the same time the pressure gradient created by lithosphere fracturing, could activate a flux of volatile elements from the asthenosphere. This flux could modify by metasomatism the overlying mantle and also locally changes the fluid phase conditions, both in pressure and composition. The simultaneous generation of magmas from variously metasomatized mantle portions, possibly under differing H_2O and CO_2 conditions,could explain the great variability in the composition of low-volcanicity rift basic magmas, also when erupted in close proximity. It is worthwile to recall that metasomatism by mantle degassing has been frequently invoked in recent years to explain the genesis of alkaline rocks of continental rifts (e.g. Bailey,1972 and 1978). Such a model would also account for the relatively large variation observed in the $^{87}Sr/^{86}Sr$ ratios of low-volcanicity basic magmas (Fig.16), whereas the high isotopic values found in the salic rocks of the same environment are more probably due to the interaction with the continental crust (Bell and Powell, 1969, 1970; Hoefs and Wedepohl, 1968).

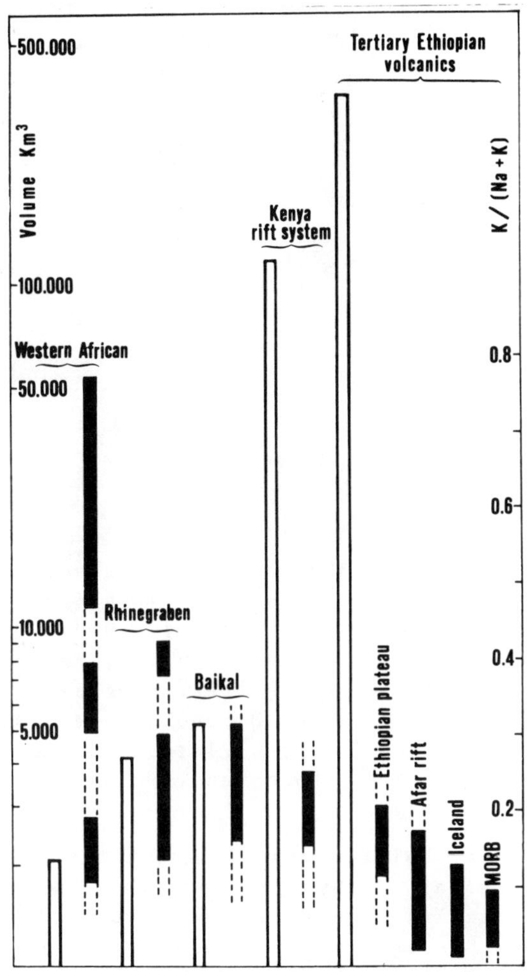

Fig.12 Volumes of erupted volcanics products (in black) and range of K/(Na+K) ratios of basic (MgO 5%) rocks from different rift zones. Volume estimates are from Logatchev and Florensov, 1978 (Baikal), Baker et al., 1972 (Kenya and Ethiopia)and have been roughly evaluated for Rhinegraben and Western African Rift, considering surface extension and the few informations on lava piles thickness found in the literature.

The significance of the spatio-temporal variations in East African rift magmatism.

It is apparent from the previous description that significant chemical variations are observed in both time and space in the high-volcanicity East African rift system. Summarizing they are the following:

- a decrease in alkalinity and silica undersaturation with time and in space in Kenya proceeding from the bordering plateau toward the rift axis and northward along the rift in recent times;
- similar differences, although less evident, are observed in Ethiopia, emphasized by the occurrence of strongly alkalic undersaturated varieties both basic and salic (basanites and trachiphonolites) only in the plateau. These alkalic rocks are however scarce and there is a large compositional overlap between plateau and rift volcanism, as shown by the work of Zanettin et al., (1976) and Piccirillo et al.,(1979);
- Kenya and Ethiopia rifts show great differences in the magmatic association, but on overall similitude in the kind of volcanic activity. In fact both display a strong bimodal basic-salic distribution (Fig.13) and the timing of main volcanic events is grossly the same in light of available geochronological data.

An important aspect is the existence of a close relation between the nature of basic and salic rocks erupted in the same rift sector in the same time span. This is well shown by the diagrams of Fig.17 where basic and salic volcanic association on the East African rift system are compared. Strongly undersaturated alkalic phonolitic rocks occur in fact in close association with nephelinites and basanites, and the same do trachytes with alkali basalts, peralkaline rhyolites with mildly alkalic

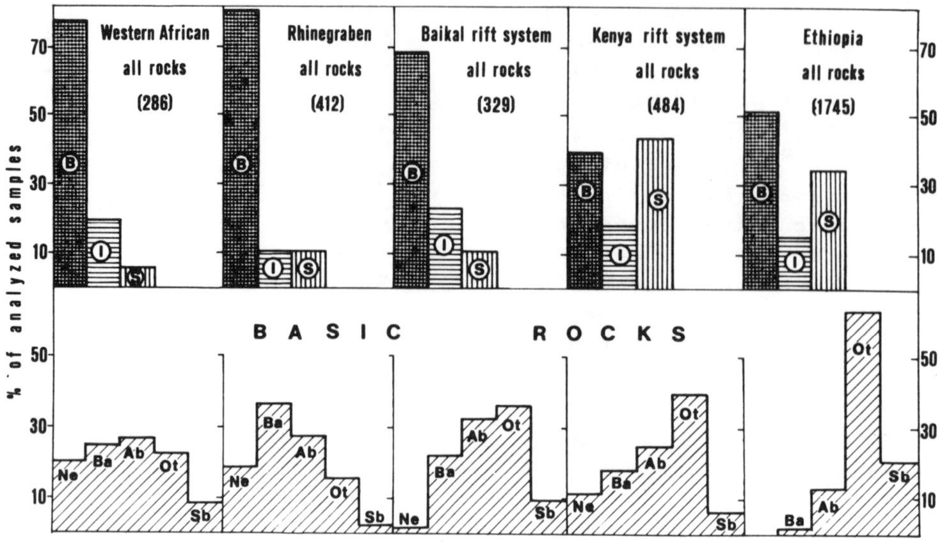

Fig.13Frequency histograms of available analyses of volcanic rocks. Rocks were classified according to alkalies vs. silica plot of Cox et al.(1979). B=basic; I=intermediate; S=salic; Ne=nephelinites; Ba=basanites; Ab=alkali basalts; OT=olivine-tholeiites; SB=subalkaline (normative qz) basalts. 1.Lavas from Nyragongo and Nyamalagira recent eruptions are excluded; 2.Plateau volcanics are included; 3.Includes Afar rift, Ethiopian rift and plateau volcanics.

(olivine tholeiitic) basalts, and finally subalkaline rhyolites and subalkaline basalts. The first two occur only in the Kenyan rift and the third is somewhat more abundant in Afar. Fig.17 shows also the more limited range of variation of Ethiopian and Afar rift rocks with respect to Kenyan ones. It is important to note that the composition of the salic products correspond always to that of the potential final liquids that would be produced by fractionation of the associated basic magmas. And this holds not only for the almost complete rock-sequences erupted at central vents, but also for the fissural flood salic volcanics, where the strong bimodal basic-salic distribution and the lack or paucity of intermediate rocks represent difficulties for the application of current fractionation genetic models. In our opinion these data speak in favour of a close genetic link between basic and salic rocks of the whole East African rift system. An origin by crystal-liquid fractionation at shallow depth has been proved in a number of rift central volcanoes (Barberi et al., 1975 b; Weaver et al., 1972; Nash et al., 1969; Ferrara and Santacroce; 1980; Brown and Carmichael, 1969 and 1971) and examples of the salic rocks sequences and related parent magmas at upper crustal levels are indicated in Fig.17. Some data in favour of a similar fractionation origin for a basalt-trachyte fissural association in Kenya rift have been published by Baker et al.(1977). Baker et al. (1978) presented a tentative model for the magmatic evolution of South Kenya rift, in which the time compositional evolution from nephelinites to mildly alkalic basalts is explained by an increase in the degree of partial melting of an uprising mantle body. The uprising magmas rested in shallow reservoirs formed by dilation and injection of the crust; here they underwent low pressure fractionation producing the huge volumes in light salic magmas erupted at the surface.

We think that a model of this kind can be extrapolated to the whole high-volcanicity East African continental

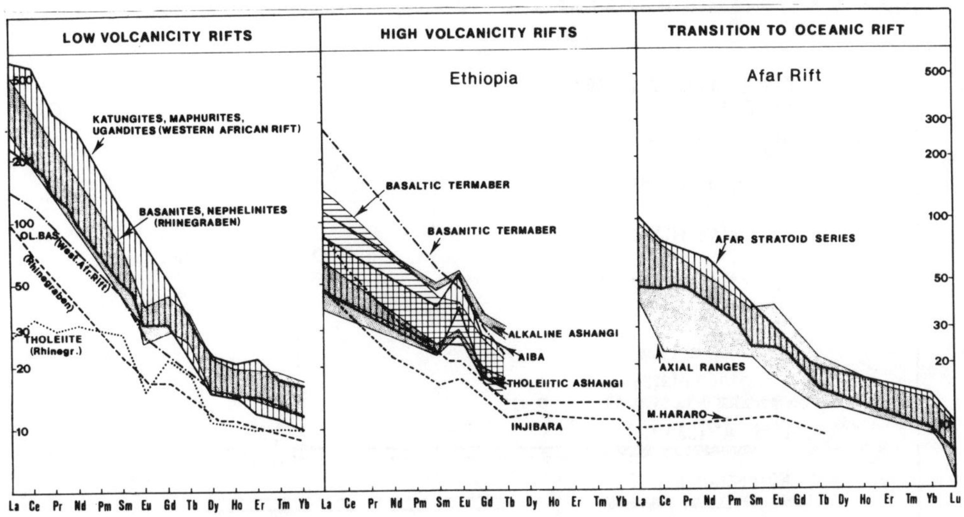

Fig.14REE normalized to chondrites patterns for basic volcanic rocks of continental rifts and Afar. Source of data: Western African rift: Beloussov et al.(1974); Rhinegraben: Herrmann (1968); Ethiopian plateau: Piccirillo et al.(1979) and Treuil and Joron (1976); Afar Rift: Barberi et al. (1975 b) and Treuil and Joron (1976)

rift. Following a model proposed by Marinelli (1975) for other volcanic environments, we suggest that block tilting which commonly occur in continental rifts as a consequence of extension, may create tectonic traps along the rift axis where basic magmas rest and differentiate creating salic light re-

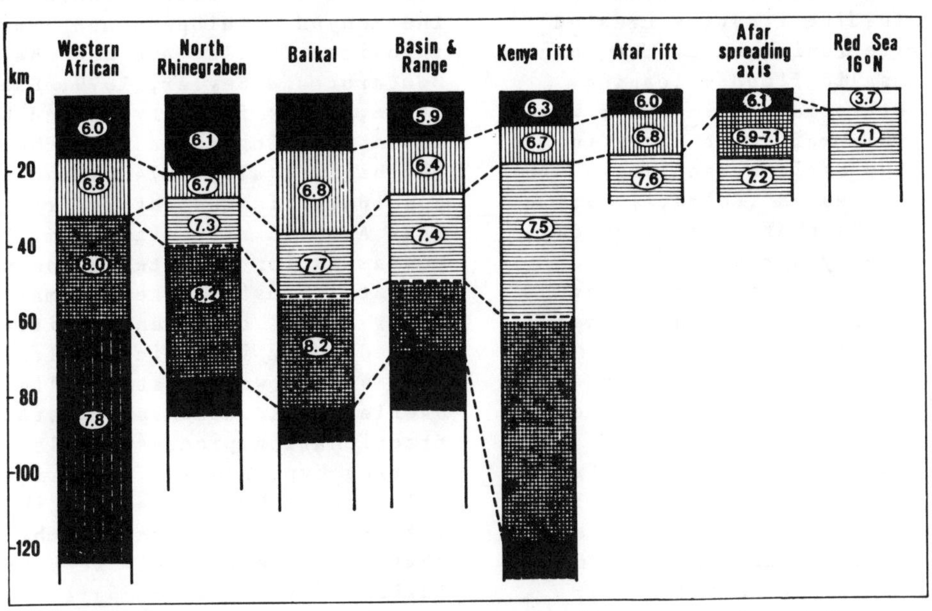

Fig.15Seismic crustal sections of some rift zones. 1 and 2. Crust; 3.Upper mantle layer with anomalous low P-velocity (7.1 - 7.7 Km/sec); 4. Mantle with normal P-velocity (8.0 - 8.2 Km/sec); 5.Low velocity zone (after Puzyrev et al.,1978; Berckhemer et al., 1975; Ruegg, 1975)

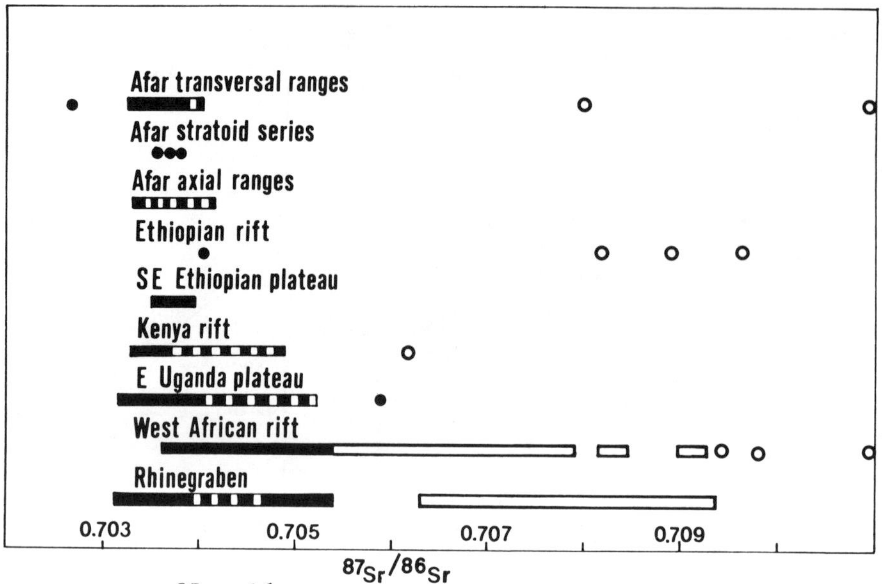

Fig.16 Range of $^{87}Sr/^{86}Sr$ ratios in continental rifts and Afar. Basic rocks are in black; salic and intermediate in white. Source of data: Afar: Barberi et al.(1970, 1980 a); Ferrara and Santacroce (1980); Ethiopian rift:Spies(1975); Barberi et al.(1980 b); SE ethiopian Plateau:Barbieri et al(1976); Kenya: Rock(1976); East Uganda: Bell and Powell (1970); Rock(1976); West African rift:Bell and Powell (1969) Rhinegraben: Hoefs and Wedepohl (1968)

sidual liquids that are erupted at the surface along feeding fissures created in a following tectonic phase. The changing nature of salic flood volcanics from Kenya to Ethiopia and in Kenya with time, would simply reflect changes in the nature of the basic parent magma. It has been also suggested (Goles, 1975; Baker et al., 1978) that salic liquids could preferentially uprise simply because of their low density, producing the bimodal distribution of rock-types tipically found in high-volcanicity rifts. This is certainly a possible eruptive mechanism but the apparent close timing of similar volcanic events all along the East African rift system from Kenya to Afar and the fact that the last Plio-Quaternary salic flood eruptions in Kenya and Ethiopia rifts occur simultaneously with a mostly basaltic fissural volcanism in the fasterly expanding Afar rift (stratoid series) seems to speak in favour of an overall tectonic control, e.g. a spreading phase.

An origin by crustal anatexis, proposed by several authors to explain the Kenyan, Ethiopian and Afar fissural rhyolites (Bailey and Macdonald, 1970; Boucarut and Seyler, 1979; Kazmin and Seife, 1979) is contradicted by the available petrological and geochemical data (Barberi et al., 1974 b) and fails when considered in the entire framework of East African rift magmatism. Process of contamination by interaction with continental crustal material may locally occur in all continental rifts, as suggested by Sr isotopic results (Fig.16). They seem however usually limited and overlapped to dominant crystal-liquid fractionation processes. Also the interesting hypothesis of an origin of the salic rocks by partial melting of the basalts buried underneath the rift floor, that could tentatively be exported from Iceland (Gronvold, 1976) to high-volcanicity continental rifts, hits apparently against some difficulties. In fact if the heat flow is high enough to melt the buried basaltic layers, even more so it should melt the underlying

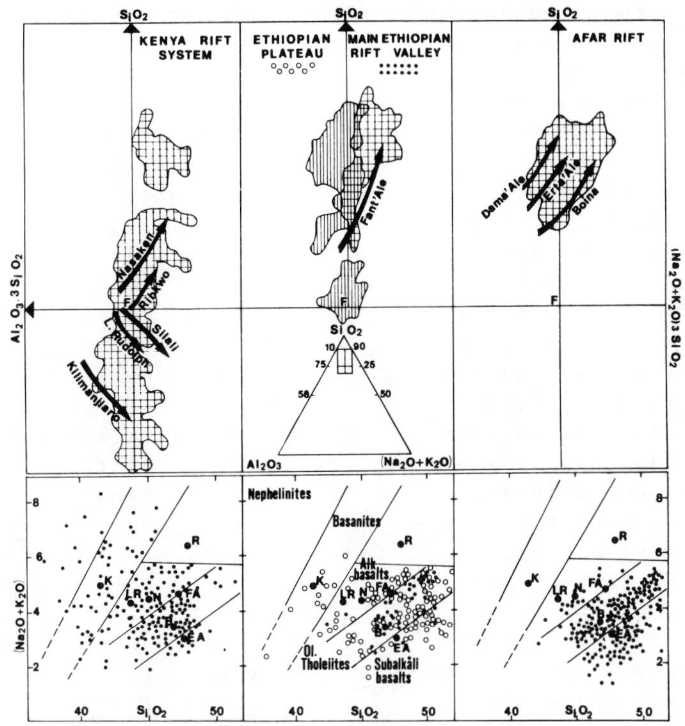

Fig.17Variation ranges of salic rocks of Kenya, Ethiopia and Afar rifts in the SiO_2 - Al_2O_3 - (Na_2+K_2O) molecular triangle (upper part). Arrows indicate rock-sequences (fractionation trends) erupted at central volcanoes. F is the alkali feldspar point. Central vertical line separates the peralkaline (right) from the non peralkaline field. Available analyses of basic rocks from the same areas are plotted in a silica-alkalies diagram in the lower diagram. Fields of different rock types after Cox et al. (1979). Large dots are basic rocks considered as parental magma of the differentiation series shown by arrows in upper diagram. K= Kilimanjiaro; L.R.=Lake Rudolph; N=Nasaken; R=Ribkwo; FA=Fant'Ale; B=Boina; EA=Erta'Ale.

continental crust producing silicic magmas that instead lack in important sectors of continental rifts. Furthermore such a genetic mechanism could not explain the close magmatic affinity of salic and basic rocks preserved also in rift sectors characterized, as in Kenya, by marked compositional changes with time and where the deeply buried old basalts could then be incompatible with the composition of the salic rocks varieties erupted at the surface.

It remains to be explained the reason of the compositional variation in the basic magmas observed along the East African rift. There is apparently a relation between the volume of erupted basic lavas and their composition, also in high-volcanicity rifts (Fig. 12), in the sense that mildly alkalic basalts tend to prevail when the volcanicity increases. Such a tendency could be extrapolated from continental to oceanic rifts, where the less alkalic MORB certainly represent by far the most voluminous rift volcanic product. The volumes of erupted basic magma can in turn be considered an expression of the intensity of the melting process affecting the asthenosphere diapir under the rift axis. Following this way of reasoning, the overall controlling parameter of rift magmatism would be the rate of extension, controlled by asthenosphere uprising dynamics. Ethiopia continental block would have been affected since the beginning by a faster rate of asthenosphere uprising, with the extension rate and simultaneous eruption, almost since the early stages, of large volumes of mostly mildly alkalic basalts produced by high degree of partial melting. The rate of extension (rate of asthenosphere uprising) was lower in the Kenya rift and initial magmas include abundant strongly alkalic varieties produced by smaller degree of melting and higher depth. Further uprising of the asthenosphere produces volcanic pulses, with changes in both the volume and the nature of erupted products. The latters are more evident in Kenya than in Ethiopia rift because of the initially lower extension rate of the former and the highest volcanicity is observed in Afar and coincides with the transition to the continental break-up. This is a possible general qualitative mechanism for rifting and related magmatic processes and cannot obviously explain the small scale geochemical variations observed within specific sectors of both oceanic and continental rifts. Recent studies (Treuil and Joron, 1976) have shown important geochemical differences among recent Afar basalts that can be only in little part explained by partial melting of an homogenous mantle source (Barberi et al., 1980 a). Also these data, as those from oceanic basalts, therefore point to the existence of geochemical heterogeneities in the uprising asthenosphere, and the same most probably holds also for continental rifts where geochemical data are scanty. The heterogeneity of the source, the dynamics of the melting process and of melt uprising are certainly all important factors controlling the fine composition of the rift basic magmas. We remain however convinced that the major chemical differences in rift magmatism associations, and more generally the volcanic behaviour of rifts, are overall controlled by the dynamics of asthenosphere uprising.

References

Anderson R.N., Clague D.A., Klitgord K. D., Marshall M., and Nishimori P.K. (1975). Magnetic and petrologic variations along the Galapagos spreading center and their relation to the Galapagos melting anomaly. Geol. Soc.Am.Bull., 86, 683.

Arnõ V., Bakashwin M.A., Bakor A.Y.; Barberi F., Basahel A., Di Paola G. M., Ferrara G., Gazzaz M.A., Giuliani A., Heikel M., Marinelli G., Nassief A.O., Rosi M. and Santacroce R. (1980). Recent volcanism within the Arabian plate: preliminary data from harrats Hadan and Nawasif - Al Buqum. In: Geodynamic Evolution of the Afro-Arabian rift system. Acc. Naz. Lincei, Roma (in press.).

Baker B.H. (1971). Explanatory note on the structure of the southern part of the African rift system. In: Tectonics of Africa, UNESCO, Earth sciences, 6,543-548.

Baker B.H.(1977). Geology and Geochemistry of the Ol Doinyo Nyokie trachyte ignimbrite vent complex, South Kenya rift valley. Bull.Volcanol., 39, 3, 420-440.

Baker B.H., Crossley R. and Goles G.G. (1978). Tectonic and magmatic evolution of the southern part of the Kenya rift valley. Neumann and Ramberg (eds) Petrology and Geochemistry of Continental Rifts. Reidel Publ.Co., Dordrecht, Holland, 29-50.

Baker B.H., Goles G.G., Leeman W.P. and Lindstrom M.M. (1977). Geochemistry and Petrogenesis of a basalt-benmoreite-trachyte suite from the southern part of the Gregory Rift, Kenya. Contr. Miner. Petrol., 64, 303-332.

Baker B.H., Mohr P.A., and Williams L. A.J. (1972). Geology of the Eastern Rift System. Geol. Soc. America Sp. Paper, 136.

Baker B.H., Williams L.A.J., Miller J.A. and Fitch F.J. (1971). Sequence and geochronology of the Kenya rift volcanics. Tectonophysics, 11, 191-215.

Baker B.H. and Wohlemberg J. (1971). Structure and evolution of the Kenya rift valley. Nature, 229, 538-542.

Baker I. (1968). Intermediate oceanic volcanic rocks and the "Daly gap". Earth Planet. Sci.Lett., 4, 103-106.

Baker I. (1969). Petrology of the volcanic rocks of Saint Helena island South Atlantic. Geol.Soc. Am. Bull., 80, 1283-1310.

Baker P.E. (1975). Peralkaline acid volcanic rocks of oceanic islands. Bull. Volcan., 38, 737-754.

Bailey D.K. (1972). Uplift, rifting and magmatism in continental plates. J. Earth Sci., Leeds, 8, 225-239.

Bailey D.K. (1978). Continental rifting and mantle degassing. In:Neumann and Ramberg (eds), Petrology and Geochemistry of continental rifts, 1-15, Reidel Publ.Co., Dordrecht, Holland.

Bailey D.K. and Macdonald R. (1970). Petrochemical variations among mildly peralkaline (comendite) obsidians from Oceans and Continents. Contr. Mineral. Petrol., 28, 340-351.

Barberi F., Bonatti E., Marinelli G.and Varet J. (1974 a). Transverse tectonics during the split of a continent: data from Afar Rift. Tectonophysics, 23, 17-29.

Barberi F., Borsi S., Ferrara G., Marinelli G. and Varet J. (1970). Relations between tectonics and magmatology in the northern Danakil Depression (Ethiopia). Phil. Trans. Roy.Soc. Lond., A 267, 293-311.

Barberi F., Cioni R. and Santacroce R. (1978). Upper Miocene submarine basalts from the western margin of the Red Sea (Secca Fawn-1 hole, Massawa Channel). Geol.Rundschau, 67, 2, 598-614.

Barberi F., Capaldi G., Gasparini P., Marinelli G., Santacroce R., Scandone R., Treuil M; and Varet J. (1980a). Miocene to Quaternary basalts of Jordan: a paleomagnetic, geochronological and petrochemical study. In: International Symposium on Geodynamic Evolution of the Afro-Arabian Rift System. Roma, Acc. Naz. Lincei (in press).

Barberi F., Civetta L.and Varet J. (1980 b). Strontium isotopic composition of Afar volcanics and its implications for mantle evolution. Earth Planet. Sci. Lett. (in press).

Barberi F., Ferrara G., Santacroce R., Treuil M. and Varet J. (1975 b). A transitional basalt-pantellerite sequence of fractional crystallization, the Boina centre (Afar rift, Ethiopia). J. Petrology, 16, 22-56;

Barberi F., Ferrara G.,Santacroce R. and Varet J. (1975 a). Structural evolution of the Afar triple junction. In: Pilger and Rosler (eds):Afar Depression of Ethiopia, 1, 38-54; Stuttgart (E. Schweizerbart'sche Verlagsbuchhandlung).

Barberi F., Giglia G., Marinelli G., Santacroce R., Tazieff H. and Varet J. (1971). Geological map of the Danakil Depression (Northern Afar, Ethiopia). 1:500.000. CNR-CNRS. Geotechnip. La Celle-Saint-Cloud.

Barberi F., Santacroce R. and Varet J. (1974 b). Silicic peralkaline volcanic rocks of the Afar depression (Ethiopia). Bull.Volcanol.,38,755-790.

Barberi F. and Varet J. (1977). Volcanism of Afar: Small scale plate tectonics implications. Geol. Soc. Am. Bull., 88, 1251-1266.

Barberi F. and Varet J. (1978). The Afar Rift Junction. In: Neumann and Ramberg (eds), Petrology and Geochemistry of Continental Rifts, 55-70, Reidel Publ.Co., Dordrecht, Holland,

Barbieri M., Brotzu P.,Morbidelli L., Penta A., Piccirillo E.M. and Traversa G. (1976). Trace elements and $^{87}Sr/^{86}Sr$ ratios of the basic stratoid volcanism in the South-Eastern Ethiopian plateau. Periodico Mineralogia, 45, 129-145, Roma.

Bass M.N. (1971). Variable abyssal basalt populations and their relation to sea-floor spreading rates. Earth Planet. Sci. Lett., 11, 18-22.

Bell K. and Powell J.L. (1969). Strontium isotopic studies of alkalic rocks: the potassium-rich lavas of the Birunga and Toro-Ankole regions. East and Central Equatorial Africa. J. Petrology, 10.

Bell K. and Powell J.L. (1970). Strontium isotopic studies of alkalic rocks: the alkaline complexes of Eastern Uganda. Bull.Geol.Soc.Am.,81.

Beloussov V.V., Gerasimovsky V.I.,

Goryatchev A.V., Dobrovolsky VV., Kapitsa A.P., Logatchev N.A., Milanovsky E.E., Poliakov A.I., Rykunov L.N. and Sedov V.V. (1974). East African Rift System. Results of researches on the International Geophysical Projects, Vol. III, Nauka Publ. House, Moscow.

Berckhemer H., Baier B., Bartelsen H., Behle A., Burkhardt H. Gebrande H., Makris J., Menzel H., Miller H. and Vees R. (1975). Deep seismic soundings in the Afar region and on the highland of Ethiopia, In: Pilger and Rossler (eds), Afar Depression of Ethiopia, Vol.2, 89-107. E. Schweizerbart'sche Verlagsbuchhandlung, Stuttgart.

Bickford M.E.and Van Schmus W.R. (1979). Geochronology and radiogenic isotope research. Review Geophys.Space Phys., 17, n.4;, 824-839.

Bjornsson A., Johnsen G., Sigurdsson S. and Thorbergsson G. (1979). Rifting of the plate boundary in North Iceland in 1975-1978. J. Geophys. Res., 84,86, 3029-3038.

Bonatti E. (1973). Origin of offsets of the Mid-Atlantic Ridge in fracture zones. J. Geology, 81,144-156.

Bonatti E., Harrison C.G.A., Fisher D. E., Honnorez J., Schilling J.G., Stepp J.J. and Zentilli M. (1977). Easter volcanic chain (South-East Pacific): a mantle hot line. J. Geophys. Res., 82, 2457-2478.

Boucarut M.and Seyler M. (1979). Volcanic formations of the Republic of Djibouti: geochemical and petrological data. In: Geodynamic evolution of the Afro-Arabian rift system (Abstract). Acc.Naz.Lincei, Roma.

Brooks C., Hart S.R., Hofmann A. and James D.E. (1976). Rb-Sr mantle isochrons from oceanic regions. Earth Planet.Sci.Lett., 32, 51-61.

Brotzu P., Morbidelli L;, Piccirillo E. M. and Traversa G. (1974). Petrographycal and chemical researches on the stratoid volcanics on the Main Ethiopian Rift. Mem.Museo Tridentino Sci. Naturali,20, 101-122, Trento (Italy).

Brown F.H. and Carmichael I.S.E. (1969). Quaternary volcanoes of the Lake Rudolph region: 1. The basanite-tephrite series of the Korath range. Lithos, 2, 239-260.

Brown F.H. and Carmichael I.S.E. (1971). Quaternary volcanoes of the Lake Rudolph region; 2. The lavas of North Island, South Island and the Barrier. Lithos, 4, 305-323.

Bryan W.B. and Moore J.G. (1977). Compositional variations of young basalts in the Mid-Atlantic-Ridge rift valley near 36° 49'N. Geol.Soc.Am.Bull., 88, 556-570.

Bryan W.B., Thompson G., Frey F.A. and Dickey J.S. (1976). Inferred geologic setting and differentiation in basalts from the Deep-Sea Drilling Project. J. Geophys.Res., 81, 23, 4285-4304.

Burke K. (1978). Evolution of continental rift system in the light of plate tectonics. In:Ramberg and Neumann (eds), Tectonics and Geophysics of Continental Rifts, 1-9, Reidel Publ. Co., Dordrecht, Holland.

Burke K. and Whiteman A.J. (1972). Uplift rifting and the break-up of Africa. In: Tarling and Runcorn (eds), Implications of continental drift to the Earth Sciences. Vol.2, 735-755. Academic Press, London.

Cahen L. (1954). Geologie du Congo Belge - Liège - Belgique.

Cann J.R. (1971). Major element variations in ocean-floor basalts. Phil. Trans. Roy. Soc. London, Ser.A, 268, 295-505.

Carmichael I.S.E. (1964). The petrology of Thingmuli, a Tertiary volcano in eastern Iceland. Journ. Petrol.,5, 435-460.

Chapman G.R., Lippard S.J. and Martyn J. E. (1978). The stratigraphy and structure of the Kamasia range, Kenya rift valley. J.Geol.Soc.London, 135, 265-281.

Cloos H. (1955). Geologische Strukturkarte der Mittelgebirge. Geol. Rundsch., 44, 480-481.

Cox K.G., Bell J.D. and Pankhurst R.J. (1979): The interpretation of Igneous Rocks. G. Allen and Unwin, London.

CNRS (France) - CNR (Italy) (1975). Geological map of Central and Southern

Afar (Ethiopia and TFAI), 1:500.000. Geotechnip. La Celle-Saint-Cloud.

Crossley R. (1979). The Cenozoic stratigraphy and structure of the western part of the Rift Valley in Southern Kenya. J.Geol.Soc.London, 136,393-405.

De Fino M., La Volpe L. and Lirer L. (1973). Volcanology and Petrology of the Assab Range (Ethiopia). Bull. Volcanol.,37,1, 1-16.

Denaeyer M.E. (1972). Les laves du fossé tentonique de l'Afrique centrale. Ann. Musée Roy.Afr.Centr.Tervuren. Ser. in 8.

Denaeyer M.E. and Schellinck F. (1965). Recueil d'analyses des laves du fossée tectonique de l'Afrique Centrale. Ann.Musée Roy. Afr.Centr.Tervuren, Ser.in 8,n.49.

Dewey J.F. and Burke K. (1974). Geology, 2,57.

Di Paola G.M. (1972). The Ethiopian Rift Valley (between 7°00' and 8°40' lat. North). Bull.Volcanol., 36, 4, 517-560.

Duda A. and Schmincke H.U. (1978). Quaternary basanites, melitite nephelinites and tephrites from the Laacher See area (Germany). N. Jb. Miner. Abh., 132,1, 1-33.

Ferrara G. and Santacroce R. (1980). Dama'ale: a tholeiitic volcano in Southern Afar(Ethiopia). In:Geodynamic Evolution of the Afro-Arabian system. Acc. Naz. Lincei, Roma (in press).

Ficke B. (1961). Petrologische Untersuchungen an tertiären basaltischen bis phonolitischen Vulkaniten der Rhön. Tschermaks miner.petrogr.Mitt., 3,7, 335-436.

Freund R. (1965). A model of the structural development of Israel and adjacent areas since Upper Cretaceous times. Geol.Mag., 102,3, 189-205.

Gehnes P. and Wimmenauer W. (1975). Geochemical studies on igneous rocks of the Rhine graben region (Germany). N. Jb. Miner.Mon., 49-56.

Goles G.G. (1975). Basalts of unusual composition from the Chyulu Hills, Kenya. Lithos,8, 47-58.

Goles G.G. (1975). Some constraints on the origin of phonolites from the Gregory Rift, Kenya, and inferences concerning basaltic magmas in the Rift System. Lithos, 9, 1-8.

Grönvold K. (1976). Variation in magma types in North Iceland. Bull. Soc. Geol. France, 7,18,4.

Guibert P. (1977a). Contribution à l'étude du volcanisme du Kivu sud (Zaire). I: la chaine volcanique Tshibinda - Kalehe. Arch.Sci.Genève, 30,1, 15-27

Guibert P. (1977 b). Contribution a l'étude du volcanisme du Kivu sud (Zaire). II: les epanchements basaltiques anciens et recents de l'ile Idjwi. Arch. Sci.Genève, 30,1,29-43.

Guibert P. (1978). Contribution a l'étude du volcanisme de la chaine des Virunga (Rep. du Zaire). Le volcan Mikeno. Unpubl.Th., Université de Genève.

Guibert P., Delaloye M. and Hunziker J. (1975). Contribution a l'étude du volcan Mikeno, chaine des Virunga (Zaire). I:données geochronometriques K/Ar. II:données isotopiques Rb/Sr. C.R.S.P.H.N. Genève, 10,1, 57-66.

Harkin D.A. (1960). The Rungwue volcanics at the northern end of lake Nyassa. Mem.Geol.Survey Tanganyka, 2.

Hart S.R. (1976). Chemical variance in deep ocean basalts. Initial Report DPSP, 34, 301-335.

Hart S.R., Schilling J.G. and Powell J. L. (1973). Basalts from Iceland and along the Reykjanes Ridge. Nature, 242, 565-575.

Hedge C.E. (1978). Strontium isotopes in basalts from the Pacific Ocean basin. Earth Planet. Sci.Lett., 38, 88-94.

Hermann A.G. (1968). Die Verteilung der Lanthanides in basaltischen Gesteinen. Contr.Mineral.Petrol., 17, 275-314.

Hoefs J. and Wedepohl K.H. (1968). Strontium isotope studies on young volcanic rocks from Germany and Italy. Contr.Mineral.Petrol., 19, 328-338.

Hofmann A.W. and Hart S.R. (1978). An assessment of local and regional isotopic equilibrium in the mantle. Earth Planet.Sci.Lett., 38, 44-62.

Holmes A. and Harwood H.F. (1932). Petrology of the volcanic fields east

and south-east of Ruwenzori, Uganda, Ql.J.Geol.Soc.London, 88,370-442.
Hopwood A.T. and Le Personne J. (1953). Presence de formation d'age Miocene inferieur dans le fossé tectonique du lac Albert e de la basse Semliki. Ann.Soc.Belge geol., 77, 83-113.
Huckenholz H.G. and Noussinanos T. (1977). Evaluation of temperature and pressure conditions in alkalibasalts and their peridotite xenoliths in NE Bavaria, Western Germany. N.Jb. Miner.Abh., 129, 2, 139-159.
Illies J.H. (1978). Two stages Rhinegraben rifting. In Neumann and Ramberg (eds). Tectonics and Geophysics of continental rifts. Reidel Publ.Co., Dordrecht,Holland.
Kazmin V. and Seife M.B. (1979). Evolution of the Northern part of the Ethiopian rift. In:Geodynamic evolution of the Afro-Arabian rift system. (Abstract). Accad.Naz.Lincei, Roma.
King B.C. (1970).Volcanicity and rift tectonics in East Africa). In:Clifford and Gass (eds). African Magmatism & Tectonics. Oliver & Boyd, Edinburgh, 263-283.
King B.C. and Chapman G.R. (1972). Volcanism of the Kenya Rift Valley. Phil.Trans.Roy.Soc.London, A 271, 185-208.
King B.C. and Sutherland D.S. (1966). The carbonatite complexes of eastern Uganda. In: Carbonatites, 73-126. J.Wiley & Sons Ltd, New York and London.
Kiselev A.I., Golovko H.A. and Medvedev M.E. (1978). Petrochemistry of Cenozoic basalts and associated rocks in the Baikal rift zone. Tectonophysics, 45, 49-59.
Langmuir C.H., Bender J.F., Bence A.E., Hanson G.N. and Taylor S.R. (1977). Petrogenesis of basalts from the FAMOUS area: Mid-Atlantic Ridge. Earth Planet.Sci.Lett., 36, 133-156.
Langmuir C.H., Vocke R.D. and Hanson G. N. (1978). A general mixing equation with application to Icelandic basalts Earth Planet.Sci.Lett., 37, 380-392.
Lippard S.J. (1973). The petrology of phonolites from the Kenya rift. Lithos, 6, 217-234.
Lippard S.J. and Truckle P.H. (1978). Spatial and temporal variations in basalt geochemistry in the N. Kenya rift. In: Neumann and Ramberg (eds), Petrology and Geochemistry of Continental Rifts, 123-131, D. Reidel Publ.Co., Dordrecht, Holland.
Lippolt H.J., Todt W. and Horn P.(1974). Apparent Potassium-Argon ages of Lower Tertiary Rhine Graben volcanics. In: Illies and Fuchs (eds). Approaches to Taphrogenesis, 213-221. E. Schweizerbart'sche Verlagsbuchhandlung, Stuttgart.
Logatchev N.A. (1976). Magmatism-tectonics relationship during the Kenya rift development and the prime cause of strongly and mildly alkaline volcanic suite formation. In: Pilger and Rosler (eds). Afar between Continental and Oceanic Rifting, 96-106, E.Schweizerbart'sche Verlagsbuchhandlung, Stuttgart.
Logatchev N.A., Beloussov V.V. and Milanovsky E.E. (1972). East African rift development. Tectonophysics,15,71-81.
Logatchev N.A. and Florensov N.A. (1978). The Baikal system of rift valleys. Tectonophysics, 45, 1-13.
Marinelli G. (1975). Magma evolution in Italy. Earth Sci.Soc., Libyan Arab Rep.
Mc Call G.J.H. and Hornung G. (1972). A geochemical study of Silali volcano, Kenya, with special reference to the origin og the intermediate-acid eruptive of the central rift valley. Tectonophysics, 15, 97-113.
Melson W.G., Vallier T.L., WrightT.L., Byerly G. and Nelen J. (1976). Chemical diversity of abyssal volcanic glass erupted along Pacific, Atlantic and Indian ocean sea-floor spreading centers. In: "The Geophysics of the Pacific Ocean Basin". Geophys.Monogr. Sci., 19, AGU, Washington D.C.
Merla G., Abbate E., Azzaroli A., Brum P., Canuti P., Fazzuoli M., Sagri M. and Tacconi P.(1979). A geological map of Ethiopia and Somalia (1973), 1:2.000.000. Consiglio Nazionale delle Ricerche, Italy, Firenze.
Mihashiro A. (1978). Nature of alkalic volcanic rock series. Contrib.Mineral. Petrol., 66, 91-104.
Mohr P.A. (1970). Catalogue of chemical

analyses of rocks from the intersection of the African, Gulf of Aden and Red Sea Rift Systems. Smithsonian Contr.Earth Sci.,2.

Mohr P.A. and Wood C.A. (1976). Volcano spacings and lithospheric attenuation in the Eastern rift of Africa. Earth Planet.Sci.Lett., 33, 126-144.

Moore J.M. and Davidson A. (1978). Rift structure in Southern Ethiopia. Tectonophysics, 46, 159-173.

Mortel H. (1971). Foide und Zeolithe basaltischer Gesteine des Vogelsberges. N.Jb.Miner.Abh.,115,1, 54-97.

Nash W.P., Carmichael I.S.E. and Johnson R.W. (1969).The mineralogy and petrology of Mount Suswa, Kenya. J. Petrology, 10, 3, 409-439.

Neugebauer H.J. (1978). Crustal doming and the mechanism of rifting. Part 1: Rift formation. Tectonophysics, 45, 159-186.

O'Hara M.J. (1973). Non primary magmas and dubious mantle plume beneath Iceland. Nature, 243, 507.

O'Hara M.J. (1975). Is there an Icelandic mantle plume? Nature, 253,708.

O'Hara M.J. (1977). Geochemical evolution during fractional crystallization of a periodically refilled magma chamber. Nature, 266, 503-507.

O'Nions R.K. and Pankhurst R.J. (1974). Petrogenetic significance of isotope and trace element variations in volcanic rocks from the Mid-Atlantic. J. Petrology, 15, 603.

O'Nions R.K., Pankhurst R.J. and Gronvold K. (1976). Nature and development of basalt magma sources beneath Iceland and the Reykjanes Ridge. J. Petrology, 17, 315-338.

Pearce J.A. and Cann J.R. (1973). Tectonic settings of basic volcanic rocks determined using trace element analyses. Earth Planet.Sci.Lett., 19, 290-300.

Piccirillo E.M., Justin Visentin E., Zanettin B., Joron J.L. and Treuil M. (1979). Geodynamic evolution from plateau to rift: major and trace element geochemistry of the Central Eastern Ethiopian Plateau volcanics. N.Jb.Geol.Palaon.Abh., 158,2, 139-179.

Puzyrev N.N., Mandelbaum M.M., Krylov S.V., Mishenkin B.P., Petrik G.V. and Krupskaya G.V. (1978). Deep structure of the Baikal and other continental rift zones from seismic data. Tectonophysics, 45, 15-22.

Richard O. (1979). Etude de la transition dorsale oceanique-rift emergé: le golfe de Tadjourah (Republique de Djibouti). Unpubl.Theses, University of Paris Sud.

Rock N.M.S. (1976). The comparative Strontium isotopic composition of alkaline rocks: new data from Sothern Portugal and East Africa. Contrib. Mineral. Petrol. , 56, 205-228.

Ruegg J.C. (1975). Main results about the crustal and upper mantle structure of the Djibouti region (T.F.A.I). In: Pilger and Rosler (eds), Afar Depression of Ethiopia, Vol.2, 121-134. E. Schweizerbart'sche Verlagsbuchhandlung, Stuttgart.

Schilling J.G. (1973 a). Iceland mantle plume: geochemical study of Reykjanes ridge. Nature, 242, 565-571.

Schilling J.G. (1975). Afar mantle plume: rare earth evidence. Nature Phys.Sci., 242, 2-5

Schilling J.G. (1975). Azores mantle blob: rare earth evidence. Earth Planet. Sci. Lett., 25, 103-115.

Schorer G. (1970). Die Pyroxene tertiarer Vulkanite des Vogelsberges. Chem. d. Erde., 29, 69-138.

Sherman S.I. (1978). Faults of the Baikal rift zone. Tectonophysics, 45, 31-39.

Sigvaldason G.E. (1974). Basalts from the centre of the assumed Icelandic mantle plume. J. Petrology, 15, 497-524.

Sigvaldason G.E., Steinthorsson S., Imsland P. and Oskarsson N. (1976). The similtaneous production of basalts enriched and depleted in large lithopile thace ions (LIL) within the same fissure swarms in Iceland. Bull.Soc.Geol. France, 18, 863-867.

Spies O. (1975). Inclusion-bearing Pliocene peralkaline silicic ignimbrites from South-West of Mt. Ayelu, Southern Afar (Ethiopia). In: Pilger and

Rosler (eds), Afar Depression of Ethiopia, Vol.I, 334-340. E.Schweizerbart'sche Verlagsbuchhandlung, Stuttgart.

Sun S.S. and Hanson G. (1975). Evolution of the mantle. Geology, 3, 297.

Sun S.S., Tatsumoto M. and Schilling J. G. (1975). Mantle plume mixing along the Reykjanes Ridge axis: lead isotopic evidence. Science, 190, 143-147.

Tatsumoto M.(1978). Isotopic composition of lead in oceanic basalt and its implication to mantle evolution. Earth Planet.Sci.Lett., 38, 63-87.

Thompson G. and Melson W.G. (1972). The petrology of oceanic crust across fractures zones in the Atlantic ocean: evidence of a new kind of sea-floor spreading. J. Geol.,80, 526-538.

Treuil M. and Joron J.L. (1976). Etude geochimie des elements en traces dans le magmatisme de l'Afar. Implications petrogenetiques et comparaison avec le magmatisme de l'Islande et de la dorsale medio-Atlantique. In: Pilger and Rosler (eds):Afar depression of Ethiopia, 2, 26-79. Stuttgart (E. Schweizerbart'sche Verlagsbuchhandlung).

Treuil M. and Varet J. (1973). Critères pétrologiques géochimiques et structuraux de la genèse et de la differentiation des magmas basaltiques: example de l'Afar. Unpubl. Thèse, University of Paris Sud, Orsay (France).

Truckle P.H. and Williams L.A.J. (1979). Variations of basic magma petrology of the Kenya rift valley. In: Geodynamic Evolution of the Afro-Arabian Rift System. Roma, Acc. Naz. Lincei (Abstract).

Varne R. (1968). The petrology of Moroto mountain, Eastern Uganda and the origin of nephelinites. J. Petrology, 9, 2, 169-190.

Wagner W. (1953). Das Gebiet Rheintalgrabens wahrend des Mesozoikums und zu Beginn seiner tertiaren Ablagerungen. Notizbl.Hess.L. Amt Bodenforsch., 81, 164-194.

Weaver S.D., Sceal J.S.C. and Gibson I.J. (1972). Trace-element data relevant to the origin of trachytic and pantelleritic lavas in East African Rift System. Contr.Mineral. Petrol., 36, 181-194.

White W., Hart S.R., Schilling J.G. (1975). Geochemistry of the Azores and the Mid-Atlantic Ridge 29° N to 60° N. Carnegie Inst. Washington, Yearb. 74, 224.

Williams L.A.J. (1969). Geochemistry and Petrogenesis of the Kilimanjaro volcanic rocks of the Amboseli area, Kenya. Bull. Volcanol., 33, 3, 862-888.

Williams L.A.J. (1978). The volcanological development of the Kenya rift. In:Neumann and Ramberg (eds), Petrology and Geochemistry of Continental Rifts, 101-121. D. Reidel Publ. Co., Dordrecht, Holland.

Wimmenauer W. (1974). The alkaline province of central Europe and France. In:Sorensen (ed). The alkaline rocks, 238-271, Wiley & Sons, London.

Wood D.A., Tarney J., Varet J., Saunders A.D., Bougault H., Joron J.L., Treuil M. and Cann J.R. (1979). Geochemistry of basalts drilled in the North Atlantic by IPOD Leg 49: implications for mantle heterogeneity. Earth Planet.Sci.Lett., 42, 77-97.

Zanettin B., Gregnanin A., Justin Visentin E., Mezzacasa G. and Piccirillo E.M. (1976). New chemical analyses of the Tertiary volcanics of Central Ethiopian Plateau. Consiglio Nazionale delle Ricerche, Padova.

Zanettin B., Justin Visentin E., Nicoletti M. and Petrucciani C. (1978). The evolution of the Chencha escarpment and the Ganjuli graben (Lake Abaya) in the Southern Ethiopian rift. N. Jb. Geol. Palaont. Mh.,8, 473-490.

Zanettin B., Justin Visentin E., Nicoletti M. and Piccirillo E.M. (1979). Correlation among Ethiopian volcanic formations with special reference to the chronological and stratigraphical problems of the trap series. In: Geodynamic evolution of the Afro-Arabian rift system. Abstract. Acc. Naz.Lincei, Roma.

METALLOGENESIS IN RELATION TO RIFTING

Frederick J. Sawkins

Department of Geology and Geophysics
University of Minnesota, Minneapolis, MN 55455 USA

Abstract. The application of certain criteria, notably mafic-felsic bimodalism in igneous rocks and thick clastic sedimentary sequences, to many geologic terrains permits identification of past rifting or extensional tectonic events, despite subsequent structural and metamorphic overprinting in many instances.

The metal deposits that are found in many such terrains represent a well-defined spectrum of ore types that is distinctive from most metal deposits generated at convergent plate boundaries. Oceanic rifting (spreading) tends to generate massive sulfide deposits of Cyprus-type. The early stages of continental rifting events can lead to formation of tin and molybdenum deposits associated with granitic intrusives and subsequently, important stratiform and lesser hydrothermal copper deposits. In addition, some large Paleozoic and Proterozoic massive sulfide deposits can be inferred to have formed in rift-related environments. Similar considerations apply to certain copper-nickel deposits in layered mafic complexes and some carbonate-hosted lead-zinc deposits.

Clearly the magmatic, tectonic and sedimentation events that accompany rifting processes provide fertile ground for the generation of certain metal deposit types. The association in time of formation of many such deposits with the breakup of Pangea, and widespread late Proterozoic rifting events adds credence to the metallogenic significance of rifting.

Introduction

In recent years considerable attention has been focused on the geology and geophysics of active rift systems. The relationship between metallogenesis and rifting is obscured by the fact that few examples of ore deposition can be observed in modern rift systems. Despite this, it is abundantly clear that metallogenesis is an important facet of rifting activity, as several workers have noted (Russell, 1973; Sillitoe, 1972a; Vokes, 1973; Guild, 1974; Sawkins, 1976a).

Any fruitful evaluation of the spectrum of possible metal deposit types related to rifting environments must consider paleo-rifting in a broad rather than narrow context. Accordingly, in this paper we shall consider metallogenesis as it relates to all facets of extensional tectonics, rather than confine ourselves to more restricted concepts of rift valley formation. Metallogenesis can result from both oceanic and continental rifting, but the latter is more significant in terms of major ore deposition events preserved in the geologic record. Significant insights into the mechanisms by which continental rifts develop are detailed by Burke and Dewey (1973), and Dewey and Burke (1974) have emphasized continental rifting as a necessary initial step in the operation of Wilson Cycles. Not all rifting events lead to ocean opening events, however.

The full operation of Wilson Cycle, or ensialic orogenic events (see later) leads to considerable tectonic complexity, and because of this, many initial rift related igneous and sedimentary sequences tend to become obscured by later structural and metamorphic overprinting.

Recognition of Rifting Events in the Geologic Record

As Burke (1978) has observed, rifting involving rupture of the entire thickness of the lithosphere can take place in a wide variety of plate tectonic settings. However, on the basis of current world tectonic activity, four types of extensional environments are of possible significance in terms of metallogenesis:

(1) Rifting along oceanic spreading centers.
(2) Hotspot induced rifting of a formerly stable continental interior.
(3) Rifting behind convergent plate boundaries (back arc basins).
(4) Impact rifting adjacent to continental collision mountain belts.

For most of the geologic record evidence for paleo-rifting events must be sought within the fabric of the continents. In certain instances paleorifts are identifiable as discreet, morphologic entities (Hoffman, 1973), but in most cases evidence for past rifting events must be sought by evaluation of disrupted meta-igneous and metasedimentary sequences in structurally complex terrains.

The outstanding petrochemical signature of rift-related magmatism is mafic-felsic bimodalism (Martin and Piiwinski, 1972), typically manifest as (meta) basalt-rhyolite assemblages or their plutonic equivalents. In addition, peralkaline igneous rocks, anorogenic granitic ring complexes, and continental basaltic magmatism in the form of plateau basalts, dike swarms, or layered mafic complexes are all viewed as indicative of subcontinental hotspot activity, that could be rifting-related (Sawkins, 1976b).

Sedimentary sequences indicative of extensional tectonics are most readily identified in terms of their aggregate thickness, protracted development and the importance of feldspathic arenites within them, especially if an intracontinental setting for sedimentation is indicated. Critical to this argument is the concept of isostasy, for the accumulation of thick (>5,000 m) sedimentary sequences in intracontinental troughs is simply not permitted by isostatic constraints, unless extensional attenuation of the continental crust occurs.

Clearly it is hazardous to infer ancient rifting events on the basis of any one of the above criteria, but in many instances combinations of these criteria can be identified in geologic terrains inferred to contain rifting related ore deposits.

Spectrum of Metal Deposit Types Related to Rifting

The most pragmatic approach to the subdivision of metal deposits into distinctive types involves primarily a consideration of the lithologic and tectonic setting in which they are generated and secondarily a consideration of their geometric relationship to their immediate host-rock, and their metal composition. Of all these parameters only the tectonic setting requires inference, and the requisite clues can generally be read from the broad lithologic setting.

Cyprus-type deposits

The massive sulfide deposits that form during oceanic rifting at divergent plate margins (Sillitoe, 1972a; Rona, 1978) occur within highly distinctive lithologic environments: the upper portions of ophiolite complexes. Considerable attention has been focused on such deposits (Constantinou and Govett, 1972; Spooner and Bray, 1977), and their occurrence in areas outside of the Mediterranean area noted (Sillitoe, 1972a; Sawkins, 1976c).

The accessibility of Cyprus-type deposits to the exploration geologist and miner, however, requires rather special tectonic events (obduction of oceanic crust). These events necessitate the operation of plate convergent mechanisms and only minute fractions of oceanic crust survive the subduction process and become embedded in the fabric of the continents.

Recent work in the eastern Pacific (Francheteau et al., 1979) has demonstrated that the processes that lead to the formation of Cyprus-type deposits are probably widespread along many oceanic spreading systems.

Hotspot related tin deposits

Burke and Dewey (1973) elucidated the sequence of events, through which continental rifting and breakup occur (Figure 1) and their scheme provides a framework within which the spectrum of metal deposits related to continental rifting can be addressed.

The initial step involves the influence of a mantle hotspot on the upper-most portion of overlying continental lithosphere. This concentration of thermal energy on the continental crust leads to crustal melting and the emplacement within the upper crust of anorogenic granites rich in lithophile elements. As a result tin deposits can form.

Examples of such tin deposits include those of the Nigeria (Jos Plateau), Southwest Africa, Brazil (Rondonia) and the Bushveld Igneous Complex, South Africa (see Sillitoe, 1974; Sawkins, 1976a). It is worth noting that the potential for economic concentration of other lithophile elements, notably uranium, also exists in this tectono-magmatic environment.

Hydrothermal copper deposits

The overwhelming majority of hydrothermal copper deposits (e.g., porphyry coppers) are generated within convergent plate boundary environments (Sillitoe, 1972b), but certain significant copper deposits of hydrothermal origin can form in rift settings. The prime examples are the Messina deposits in northern South Africa, which lie within the failed arm of the lower Limpopo triple junction.

Recent work on these breccia pipe deposits (Sawkins, 1977; Sawkins and Rye, 1979) has emphasized their relationship to rift tectonics and indicated copper-rich rift basalts as the source of the copper they contain. The hydrothermal copper-bearing breccia pipes of the Tribag Mine, Ontario (Norman, 1977) relate to Keweenawan rifting, although in this instance a relationship to felsic intrusives controlled by rift structures is indicated.

Stratiform copper deposits

Copper deposits in which the ore minerals are confined to specific shale or sandstone horizons represent an important facet of rift metallogeny. Most prominent are the ores of the Zambian copper belt (Mendelsohn, 1961), but examples are widespread (Table 1). Brown (1971, 1978) has argued that textures and mineralogic zoning patterns in many such deposits indicate that metallization occurred subsequent to initial accumulation of the host sediments.

Sawkins (1976a) noted that copper deposits, mainly of stratiform type, are particularly prominent in rift environments and suggested that a logical source for the copper was rift basalts.

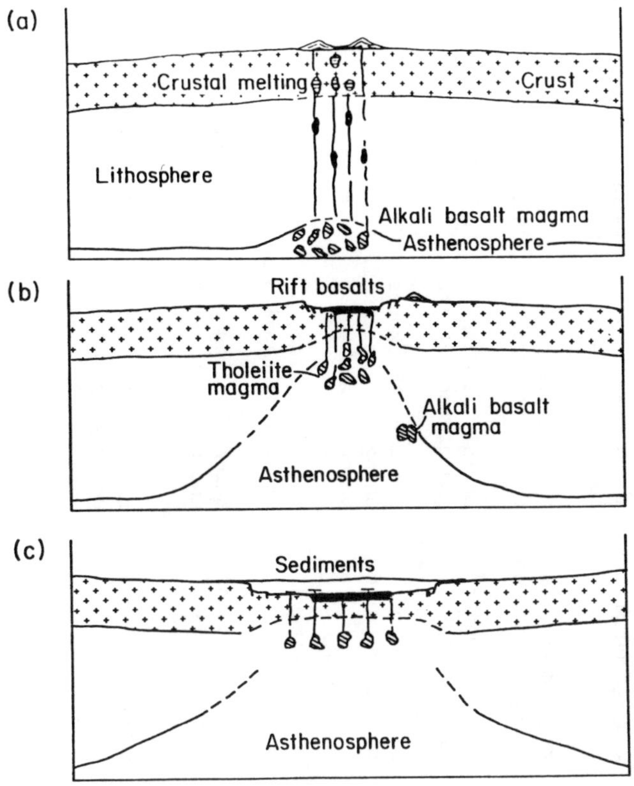

Fig. 1. Generalized diagrams to show sequence of events typical of continental rifting episodes.

Such continental basalts exhibit a marked enrichment in copper with respect of mid-ocean ridge basalts (see Sawkins, 1976a; Table 2), and a significant amount of this copper presumably occurs as discrete sulfide blebs that are easily leached by heated oxygenated waters circulating within the volcanic-sedimentary sequences that fill the rift through (Figure 2).

The marked concentration of stratiform copper deposits in sedimentary sequences laid down during late Proterozoic time is noteworthy (see Table 1), and can be related to widespread rifting events that affected the continents at this stage of earth history (Sawkins, 1976b).

Native copper deposits in basalts

Only a single economically important example of native copper deposits in rift-related basaltic rocks is known, but the deposits of the Keweenawan basalts in Upper Michigan serve to emphasize the copper-rich nature of such basalts (White, W.S., 1968). Isotopic studies on these deposits and the enclosing basalts (Livnat et al., 1976) indicate that large volumes of heated groundwater circulated through the rocks and that these were important in the deposition of the native copper in porous flow tops.

These observations serve to emphasize that the combination of thick volcanic and sedimentary sequences and high heat flow that characterize development of many rifts make such tectonic features particularly fertile ground for the generation of metal deposits of various types.

Magmatic Cu-Ni deposits

The relationship between rift-type tectonics and the emplacement of layered mafic complexes is particularly clear in some instances. For example, the Skaergaard Complex, Greenland relates to North Atlantic rifting and the Duluth Complex, Minnesota relates to Keweenawan rifting (Chase and Gilmer, 1973). It can be argued (Sawkins, 1976a) that layered mafic igneous complexes formed in intracontinental settings or during continental rifting episodes are typical manifestations of mantle hotspot activity, and thus are properly viewed within the context of rift tectonics.

Several layered mafic intrusives contain significant Cu-Ni sulfide mineralization. The most important of these is the Sudbury Irruptive, and although many subscribe to the idea it was initiated by a meteor impact (French, 1970), it none the less represents a major mantle melting event. The Duluth Complex (Weiblen and Morey, 1975) contains very large low grade copper-nickel resources and is unequivocally related to Keweenawan rifting.

Although the Bushveld Igneous Complex does not contain major copper-nickel deposits, it is an important repository of magmatic ores (chrome, platinum, iron, vanadium: Willemse, 1969). Early events associated with the Bushveld Complex; initial doming, volcanism, Cu-bearing alkali magmatism (Palabora) and aulacogen formation (see Hunter, 1976; Jansen, 1975) all suggest hotspot activity.

Molybdenum deposits

Many small molybdenum vein deposits occur within the confines of the Oslo Rift in Norway (Ihlen and Vokes, 1977) and larger disseminated molybdenum occurrences are now under investigation there.

More significantly, the post-Laramide zone of porphyry molybdenum deposits that stretches from Colorado through New Mexico into Texas (Guild, 1978; Sharp, 1979) exhibits a spatial relationship to the Rio Grande rift zone that cannot be coincidental. Fluorite is a common accessory mineral in molybdenum deposits and Van Alstine (1976) has demonstrated the strong relationship between many fluorite deposits and the Rio Grande and other young rift systems.

Massive sulfide deposits in relation to rifting

That massive sulfide deposits can indeed form in rift environments is indicated by the sulfide layers present below the brine pools in the axial rift of the Red Sea (Degens and Ross, 1969). It could be argued that these concentrations of base metal sulfides are forming in an oceanic environment, and are more properly viewed as a variant of Cyprus-type deposits. Given this approach the question "Does continental rifting (as opposed to oceanic rifting) lead to the formation of certain massive sulfide deposits?" remains.

TABLE 1. Stratiform Copper Deposits

Deposit or district	Age	Host rocks	Lithology	Evidence for rifting	Reference
Basal Cretaceous, Angola	Cretaceous	Cuvo-formation	mudstone and siltstone	association with South Atlantic rifting	Van Eden, 1978
Mansfeld copper deposits, Central Europe	Permian	Kupferschiefer	black shale	basin and range type faulting, basalt-rhyolite (see reference)	Lorenz and Nicholls,
Aynak copper deposits, northeastern Afgahanistan	Infracambrian	?	metasandstone metadolomite	association with formation of Tethys Ocean	Sillitoe, 1978
White Pine, Michigan, U.S.A.	10 b.y.	Nonesuch Shale	gray-maroon shale	Keweenawan basaltic magmatism	White, D.E., 1968
Belt copper prospects Montana, U.S.A.	late Proterzoic	Revett Formation	white quartzite and siltite	thick sedimentary sequence, basaltic magmatism	Harrison, 1972
Seal Lake Labrador, Canada	late Proterzoic	Adeline Island Formation	gray-green shales	thick clastics, associated basalts	Gandhi and Brown, 1975
Coppermine River N.W.T. Canada	late Proterzoic	Rae Group	gray-green glauconitic sandstone	thick clastics, associated basalts	Kindle, 1972
Redstone Copper Belt N.W.T. Canada	late Proterzoic	Redstone River Formation	siltstones and carbonates	normal faulting, associated basalts	Ruell, 1978
Zambian Copper Belt Central Africa	late Proterzoic	Ore shale Lower Roan Formation	argillite and impure dolomite	sediment thickness changes, time equiv. Bukoban basalts	Mendelsohn, 1961
Witvlei and Klein Aus mines, Namibia	late Proterzoic	Wituki Formation Nosib group	greenish-grey argillite	thick clastics, underlying basalts	Toens, 1975
Adelaidean copper province, South Australia	late Proterzoic	lower Willouran Adelaidean System	dolo-siltstones carbonaceous dolo-arenites	thick clastics underlying basalts	Rowlands, 1973
Kilembe Mine Uganda	early Proterzoic	Kilembe Series	meta-siltites	thick meta-clastics, underlying meta-basalts	Davis, 1967

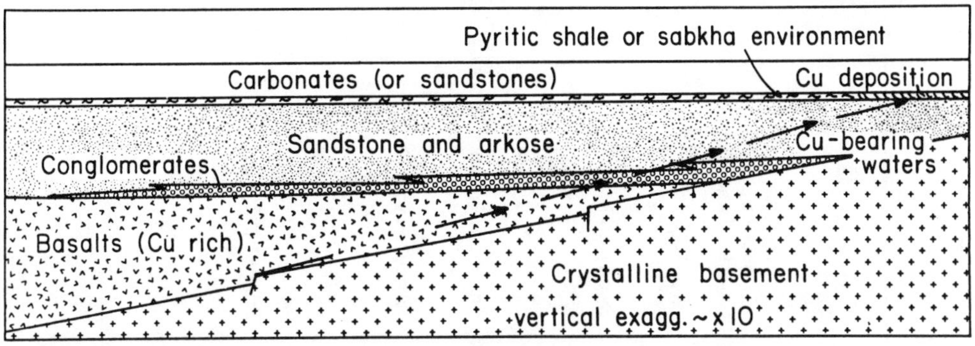

Fig. 2. Sketch to illustrate the concept of formation of stratiform copper deposits. View represents a longitudinal section along a continental rift.

It has been previously argued (Sawkins, 1976a, 1976c) that a number of important massive sulfide deposits occur within lithologic sequences, both volcanic and non-volcanic, that provide evidence for extensional rather than the more familiar (e.g. Kuroko-type deposits) plate convergent tectonic settings. The majority of such deposits (Table 2) occur within structurally complex terrains and a case for their original formation within extensional tectonic enviornments can only be made by identifying certain of the lithologic and petrochemical criteria for rifting discussed in an earlier section of this paper.

Massive sulfide deposits inferred to be rifting-related fall into two distinctive types: the so-called shale-hosted deposits (see Table 2) and those associated with felsic volcanics in terrains dominated by bimodal felsic-mafic volcanics. Such bimodalism has been recently recognized in the Tetagouche Group volcanics associated with the New Brunswick massive sulfide province (Harley, 1979) and the volcanics associated with the important southern Iberian massive sulfide province (Munha, 1979).

The massive sulfide deposits that occur within essentially non-volcanic lithologic sequences (see Table 2) have remained enigmatic in terms of their genesis. Nevertheless there are strong indications that deposits such as Sullivan, B.C., Mt. Isa, N.S.W. and Rammelsberg and Meggen in West Germany formed in extensional tectonic environments. In particular, the great thicknesses of enclosing sedimentary successions, and evidence of broadly contemporaneous basaltic magmatism (see Sawkins, 1976a, 1976c) support this view.

In recent years a whole new province of Paleozoic shale-hosted massive sulfide deposits has been found in northwest Canada and northern Alaska (Dawson, 1977; Nokelberg et al., 1979). In many instances the enclosing lithologies are black shales and cherts indicative of starved basin conditions similar to those of the 'Bathyl Lull' during Devonian rifting events and massive sulfide formation in western Europe (Ziegler, 1978; Sawkins and Burke, 1980).

Rift-related massive sulfide deposits are of considerable importance in economic terms for many contain tens or even in some cases hundreds of millions of tons of high grade base metal ore. The source of the metals in such deposits and their precise mode of genesis remains enigmatic, but the recognition that they form in extensional tectonic environments represents a significant step to a more complete understanding of such deposits.

Inclusion of the Broken Hill lead-zinc deposits in New South Wales in this category obviously involves a degree of geopoetic license, but the latest work on these deposits and their high grade enclosing metamorphic rocks (Laing et al., 1978) suggests original lithologies involving tholeiitic basalts, thick clastic sedimentary sequences, and possibly a depositional framework that involved normal (i.e. tensional) faulting. Similar considerations apply to the massive sulfide province discovered within the last decade in the northern Cape, South Africa (McMillan, 1976; Rozendaal, 1978).

Carbonate-hosted lead-zinc deposits in relation to rifting

Lead-zinc deposits in carbonate rocks of intracontinental setting represent important sources of these metals. The so-called Mississippi Valley-type deposits of North America, and elsewhere, the Alpine-type deposits of Europe and North Africa, and Proterozoic carbonate-hosted lead-zinc deposits in Africa and Australia are all characteristized by the absence of a definitive time-space relationship to igneous events.

Fluid inclusion studies of many of these deposits indicate ore deposition by saline water comparable to connate brines at temperatures typically ranging from 70-150°C (Heyl, 1969). The position of many such deposits over, or on the flanks of, basement highs adjacent to broad intracratonic basins, taken in conjunction with the fluid inclusion data, suggests that the migration of basinal brines to basin edges is an important factor in the genesis of such deposits (Hanor, 1979). Based on the above it could be assumed that there is no compelling reason to explore possible interrelationships between carbonate-hosted lead-zinc deposits and rift-type tectonics.

TABLE 2. Rift-related Massive Sulfide Deposits

Mine or deposit	Age	Host rock Lithology	Composition	Evidence for rifting	References
Rammelsberg, W. Germany	Devonian	slates	Pb, Zn, Cu, Ba	starved basin setting, thick section	Kraume *et al.*, 1955 Sawkins and Burke, 1980
Meggen, W. Germany	Devonian	slates	Pb, Zn, Ba	starved basin setting, thick section	Zimmermann and Amstutz, 1971 Sawkins and Burke, 1980
McMillan Pass area N.W.T. Canada	Upper Devonian-Mississippian	Carbonaceous shale	Pb, Zn, Ba	synsedimentary graben structure	Dawson, 1977
Wulik and Red Dog deposits, Western Brooks Range, Alaska	Mississippian	Carbonaceous shale and chert	Pb, Zn, Ba	starved basin setting	Nokelberg *et al.*, 1979
Otjihase Matchless Belt, Namibia	late Proterozoic	Biotite schist	Cu, Zn	very thick sequence	Anhaeasser and Button, 1974
Ducktown, Tennessee	late Proterozoic	Biotite schist	Cu, Zn	very thick sequence, basaltic magmatism	Magee, 1968
Sullivan B.C. Canada	Proterozoic	argillites	Pb, Zn	very thick section, basaltic magmatism	Freeze, 1966
Mt. Isa Queensland, Australia	Proterozoic	shales	Pb, Zn, Cu	very thick section, basaltic magmatism	Dunnet, 1976
McArthur River N.T., Australia	Proterozoic	shales	Pb, Zn	thick section, synsedimentary faulting	Walker *et al.*, 1977
Broken Hill	Proterozoic	gneisses	Pb, Zn	thick section, meta-sediments, meta-basaltic rocks	Laing *et al.*, 1978

Despite these considerations, carbonate-hosted lead-zinc deposits exhibiting a relationship to rift environments do occur. For example, small lead-zinc deposits occur in Miocene sediments surrounding the Red Sea (Hamilton, 1976) and similar deposits are found in Cretaceous sediments of the Benue Trough (Farrington, 1952).

The lead-zinc deposits that occur within the Triassic carbonate sequences of the Eastern Alps (Maucher and Schneider, 1967) may well relate to a period of rifting associated with a well-documented westward migration of the Tethyan Ocean during the Jurassic (Dewey et al., 1973). Furthermore, the Mississippi Valley deposits of southern Illinois and Kentucky exhibit a space-time relationship to alkaline mafic igneous activity of hotspot affinities (Heyl, 1972).

The Irish lead-zinc province contains an important suite of deposits in Carboniferous carbonate rocks and Russell (1973, 1976) has argued persuasively for their formation within an extensional tectonic setting. Similarly (Sawkins, 1966), noted a possible relationship between the formation of the Mississippi Valley-type deposits of the North Pennines and contemporaneous rifting events in the Midland Valley of Scotland (Francis, 1968).

Carbonate-hosted lead-zinc deposits of Proterozoic age include those of the McArthur River area, Queensland (Walker et al., 1977). Recent work on these deposits (Rye and Williams, 1978)

TABLE 3. Relationship of various metal deposits to past rifting events.

Rifting Event	Deposit Type	Examples
Breakup of Pangea	Stratiform Cu	Angola sed. Cu Kupferschiefer
	Cu breccia pipes	Messina, S. Africa
	Carbonate hosted Pb-Zn	Benue Trough deposits Red Sea margins Pennines, England
	Crustal Granite Sn	Jos Plateau tin district
Devonian rifting Europe	Shale hosted massive sulfide	Rammelsberg, W. Germany Meggen, W. Germany
	Volcanogenic massive sulfide	Rio Tinto district, Spain
Late Proterozoic rifting	Stratiform Cu	Zambian deposits Namibian deposits White Pine, U.S.A. Adelaidean deposits Seal Lake, Canada Coppermine River, Canada Redstone River, Canada
	Cu breccia pipes	Tribag, Ontario, Canada
	Magmatic Cu-Ni	Duluth Complex
	Shale hosted massive sulfide	Ducktown, Tennessee Otjihase, Namibia
	Carbonate hosted Pb-Zn	Tsumeb, Namibia
	Crustal granite Sn	Rhondonia deposits, Brazil
Mid Proterozoic rifting	Shale hosted massive sulfide	Mt. Isa, Australia McArthur River, Australia Sullivan, Canada Broken Hill, Australia
	Carbonate hosted Pb-Zn	McArthur River area, Australia

indicates that they are of Mississippi Valley-type. Sediment thicknesses and faults in this essentially undeformed terrain provide clear indications of a rifting environment. Much the same considerations apply to the lead-zinc (+ copper) deposits of the Tsumeb area, Namibia that occur within the thick Otavi Dolomites (Haughton, 1969).

In conclusion, the relationship between intracratonic carbonate-hosted lead-zinc deposits, especially Mississippi Valley-type deposits, and extensional tectonics is at best a subtle one, although a few relatively clear examples of such a relationship can be identified.

Some additional possible examples of metallogenesis and rifting

The banded iron formations of the Hamersley Range, western Australia represent an important example of the unique world-wide episode of iron deposition that occurred approximately 2 billion years ago. There is geophysical, stratigraphic and petrochemical (basalt-rhyolite bimodalism) evidence that the Hammersley Basin represents a rift trough between the adjacent Yilgarn and Pilbara shield areas (Trendall and Blockley, 1970; Horwitz and Smith, 1976).

The sedimentation events associated with other major iron deposits of this general period such as those in southern Africa, the Labrador Trough, and the Superior region of the U.S.A. may also relate to rifting events, but this is not clear at present.

Clear evidence does exist, however, for a rift related origin for the early Proterozoic Huronian successions of southern Canada that contain the important uranium deposits of the Elliot Lake region near their base (Frarey and Roscoe, 1970). Here initial mafic-felsic bimodal volcanism, great thickness, and contemporaneous normal faulting all suggest a tectonic environment characterized by rifting. Much the same considerations apply to the essentially time equivalent succession of the Witwatersrand basin, host for important gold and uranium deposits (Pretorius, 1975).

Finally, a possible connection between rifting and certain high grade metamorphic belts deserves consideration. Examples include the Grenville Province, the Limpopo and Northern Cape Belts in southern Africa, and the Willyama and Gawler Block terrains in Australia. The upper amphibolite to granulite grade metamorphism that characterizes such belts renders interpretation of the original stratigraphy obscure at best and uncertainty surrounds their origin. Nevertheless, the massive sulfide deposits that have been discovered in these belts (Stamatelopolau-Seymour and Mclean, 1977; McMillan, 1976; Laing et al., 1978) appear to parallel less metamorphosed massive sulfide deposits of more definite rift association. Also, the copper-nickel deposits at Pikwe-Selebe in the Limpopo Belt (Gordon, 1973) occur in association with what appears to be a typical, albeit highly metamorphosed, rift-related layered mafic complex.

Discussion

In the foregoing sections I have attempted to demonstrate that a clear relationship between certain metal deposits and rifting events can be demonstrated, and that an association of many others with rifting can be implied. Important in this regard is the well-defined spectrum of metal deposits associated with rifting events and the recurrence of certain deposit types at different times in the geologic past (Table 3).

One point central to any understanding of the metal deposits patterns associated with Mesozoic and Cenozoic rifting events, especially the breakup of Pangea, is that most stratabound deposits formed at this time now lie deeply buried beneath thick prisms of miogeoclinal sediments (Fig. 3). Only after these sequences are involved in collisional orogeny and subsequent erosion are some of the ore deposits they contain likely to approach the surface. Thus, apart from the Red Sea sulfide muds, virtually all known, rift-related, massive sulfide and stratiform copper deposits occur within Paleozoic or older terrains that have suffered a major orogenic event (see Tables 1 and 2).

A noteworthy exception to this is the Keweenawan rifting event (Chase and Gilmer, 1973). For some reason neither Wilson Cycle nor ensialic orogeny followed this rifting episode, which appears simply to have aborted after some tens of kilometers of plate separation.

The age distribution of rift-related deposits is of some interest. Metal deposits of this general type are essentially absent from Archean terrains, presumably because this was a period of continent building, and continental rifting had to await the

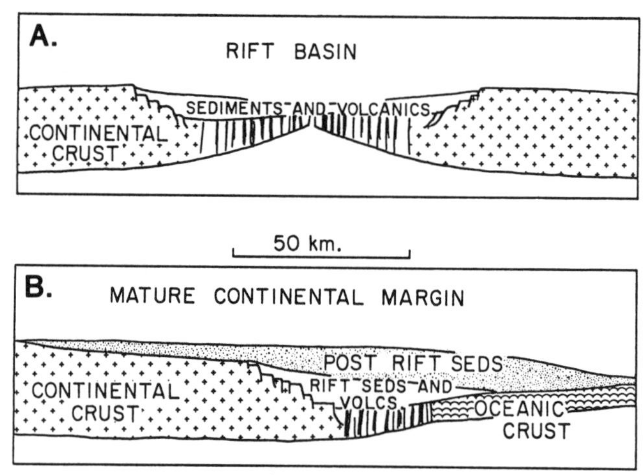

Fig. 3. Sketch to illustrate how rift sediments and associated volcanics (A) tend to become deeply buried by miogeoclinal sediment wedges, (B) after an ocean-opening event. Only subsequent tectonic (collisional) reactivation of such a continental margin will exhume the rift sequences and any metal deposits they contain.

completion of this stage of earth evolution. During the Proterozoic however, especially during later Proterozoic time, evidence for continental rifting is widespread (Sawkins, 1976c) and this is clearly reflected in the age distribution of metal deposits with rifting affinities (see Tables 1 and 2). Reasons for the paucity of certain important types of rift-related metal deposits during later Phanerozoic time were discussed earlier in this section.

Earlier reference was made to the controversy surrounding Wilson Cycle versus ensialic orogeny with respect to many older orogenic belts. It is inappropriate to review here the voluminous literature surrounding this important dichotomy in our understanding of past orogenic events, but from the perspective of rift-related metal deposits and their associated rocks, it appears that both phenomena are initiated by rifting events. Clearly, much remains to be learned about the precise mechanisms involved, but the recognition of precursor rifting as a common denominator is considered significant.

REFERENCES

Anhaeusser, C.R. and Button, A., A review of southern African deposits - their position in time and space, Econ. Geol. Research Unit, Univ. of the Witwatersrand, Inform. Circ., 85, 48, 1974.

Brown, A.C., Zoning in the White Pine copper deposit, Ontonagon County, Michigan, Econ. Geol., 65, 543-573, 1971.

Brown, A.C., Evidence for post-sedimentary origin of stratiform copper deposits (abstract), Int. Assoc. Genesis of Ore Deposits Meeting, Snowbird, Utah, 61, 1978.

Burke, K., Continental Rifts (abstract), 1978 Inter. Symp on the Rio Grande Rift, Santa Fe, N.M., 21-22, 1978.

Burke, K. and Dewey, J.F., Plume-generated triple junctions: Key indicators in applying plate tectonics to old rocks, Jour. Geology, 86, 406-433, 1973.

Chase, G.C. and Gilmer, T.H., Precambrian plate tectonics: The midcontinent gravity high, Earth Planet. Sci. Letters, 21, 7078, 1973.

Constantinou, G. and Govett, G., Genesis of sulfide deposits, ochre and number of Cyprus, Trans. I.M.M., 81, B34-46, 1972.

Davis, G.R., Aspects of the metamorphosed sulfide ores at Kilembe, Uganda: Sedimentary ores, Proc. 15th Inter-University Geol. Cong. Spec. Publ. No. 1, Univ. of Leicester, 273-296, 1967.

Dawson, M.K., Regional metallogeny of the Northern Cordillera, Geol. Surv. Canada, 77-1A, 1-4, 1977.

Degens, E.T. and Ross, D.A., Hot brines and recent heavy metal deposits in the Red Sea, New York, Springer-Verlag, 600, 1969.

Dewey, J.F. and Burke, K., Hotspots and continental breakup: Implications for continental orogeny, Geology, 2, 57-60, 1974.

Dewey, J.F., Pitman, W.C., Ryan, W. and Bonin, J., Plate tectonics and the evolution of the Alpine System, Bull. Geol. Soc. Am., 84, 3137-3180, 1973.

Dunnet, D., Some aspects of the Panatarctic craton margin in Australia, Ohil. Trans. R. Soc. Lond. Ser. A, 273, 641-654, 1976.

Farrington, J.L., A preliminary description of the Nigerian lead-zinc field, Econ. Geol., 47, 583-608, 1952.

Francheteau, J. and others, Massive deep-sea sulfide ore deposits discovered on the East Pacific Rise, Nature, 277, 523-528, 1979.

Francis, E.H., Review of Carboniferous-Permian volcanicity in Scotland, Geol, Rundschau, 57, 219-228, 1968.

Frarey, M.J. and Roscoe, S.M., The Huronian Supergroup north of Lake Huron, Geol. Survey Canada Paper 70-40, 143-158, 1970.

Freeze, A.C., On the origin of the Sullivan orebody, Kimberley, B.C., Can. Inst. Min. and Metall. Spec. Vol. 8, 263-294, 1966.

French, B.M., Possible relationship between meteorite impact and igneous petrogenesis as indicated by the Sudbury structure, Ontario, Canada, Bull. Volcanol., 34, 466-517, 1970.

Gandhi, S.S. and Brown, A.C., Cupriferrous shales of the Adeline Island Formation, Seal Lake Group, Labrador, Econ. Geol., 70, 145-163, 1975.

Gordon, P.S.L., The Selebi-Pikwe nickel-copper deposits, Botswana, in: Granites, gneisses and related rocks: Geol. Soc. South Africa Spec. Publ. 3, 167-187, 1973.

Guild, P.W., Metallogenesis in the western United States, J. Geol. Soc. Lond., 135, 355-376, 1978.

Hamilton, L., Mineralization of a trailing plate margin: The Red Sea, in: 25th Internat. Geol. Congress, Australia Abstracts, 3, 738-739, 1976.

Hanor, J.S., Sedimentary genesis of hydrothermal fluids, in: Geochemistry of Hydrothermal Ore Deposits, H.L. Barnes, ed., 2nd ed., New York, John Wiley and Sons, chapter 4, 1979.

Harley, D.N., A mineralized Ordovician resurgent caldera complex in the Bathurst-Newcastle Mining District, New Brunswick, Canada, Econ. Geol., 74, 786-796, 1972.

Harrison, J.E., Precambrian belt basin of northwestern United States: it geometry, sedimentation and copper occurrences, Geol. Soc. Am. Bull., 83, 1215-1240, 1972.

Haughton, S.H., Geological history of Southern Africa, Geol. Soc. of South Africa, 535, 1969.

Heyl, A.V., Some aspects of genesis of zinc-lead-barite-fluorite deposits in the Mississippi Valley, U.S.A., Trans. Inst. Mining and Metall., 78, B148-160, 1969.

Heyl, A.V., The 38th parallel lineament and its relation to ore deposits, Econ. Geol., 67, 879-894, 1972.

Hoffman, P., Evolution of an early Proterozoic continental margin: the Coronaton Geosyncline and associated aulacogens of the northwestern Canadian Shield, Phil. Trans. R. Soc. Lond., A273, 547-581, 1973.

Horwitz, R.C. and Smith, R.E., Bridging the Yilgarn and Pilbara Blocks, western Australian Shield (abstract), I.G.C., Sydney, 12, 1976.

Hunter, D.R., Some enigmas of the Bushveld Complex, Econ. Geol., 71, 229-248, 1976.

Ihlen, P.M. and Vokes, F.M., Metallogeny associated with the Oslo Rifting: the Oslo Paleorift. Part 2, Guide to the excursions, NATO ad. study inst., Oslo, 147-174, 1977.

Jansen, H., The Southpansberg Trough (Northern Transvaal) - An aulacogen, Trans. Geol. Soc. S. Africa, 78, 129-136, 1975.

Kindle, E.E., Classification and description of copper deposits, Coppermine River area, District of Mackenzie, Geol. Surv. Canada Bull., 214, 198, 1972.

Kraume, E., Dahlgrun, F., Ramdohr, P. and Wilke, A., Die Erzlager des Rammelsberges bei Goslar: Hannover, Monographien der Deutschen Blei-Zink Erzlagerstatten, 4, 394, 1955.

Laing, W.P., Marjoribanks, R.W. and Rutland, R.W., Structure of the Broken Hill Mine area and its significance for the genesis of the orebodies, Econ. Geol., 72, 1112-1136, 1978.

Livnat, A., Rye, R.O. and Kelley, W.C., Stable isotope and fluid inclusion studies of the Keweenawan Copper District, Northern Michigan (abstract), Geol. Soc. Amer. Ann. Mtgs., Denver, 1980-1981, 1976.

Lorenz, V. and Nicholls, I.A., The Permocarboniferous Basin and Range Province of Europe. An application of plate tectonics, in: The Continental Permian in Central, West, and South Europe, H. Falke, ed., D. Reidel Publ. Co., Bordrecht-Hollad, 313-342, 1976.

Magee, M., Geology and ore deposits of the Ducktown District, Tennessee: Ore deposits of the United States, 1933-1967, A.I.M.E. Graton-Sales, I, 207-241, 1968.

Martin, R.F. and Piwinskii, A.J., Magmatism and tectonic setting, Jour. Geophys. Res., 77, 4966-4975, 1972.

Maucher, A. and Schneider, H.J., The Alpine lead-zinc ores, in: Econ. Geol. Monograph No. 3, Genesis of stratiform lead-zinc-barite-fluorite deposits, 71-89, 1967.

McMillan, M.D., Two different models for copper-lead-zinc-mineralization in Southern Africa, Inter. Geol. Congress Abst., Sydney, 175, 1976.

Mendelsohn, F., The geology of the Northern Rhodesian Copper Belt, MacDonald and Co., Ltd., London, 523, 1961.

Munha, J., Blue amphiboles, metamorphic regime and plate tectonic modelling in the Iberian Pyrite Belt, Contrib. Min. Pet., 69, 279-290, 1979.

Nokelbeg, W.J., Plahuta, J.T., Lange, I.M. and Grybeck, D., Volcanogenic zinc-lead-barite deposits in pelagic rocks of late Paleozoic and early Mesozoic age, Northwest Brooks Range, Alaska (abstract), Geol. Soc. America Ann. Mtg., San Diego, 487-488, 1979.

Norman, D.I., Geology and geochemistry of Tribag Mine, Batchawana Bay, Ontario, unpubl. Ph.D. thesis, University of Minnesota, 257pp, 1977.

Pretorius, D.A., The depositional environment of the Witwatersrand goldfields: a chronological review of speculations and observations, Minerals Sci. Engin., 7, 18-47, 1975.

Rona, P.H., Criteria for recognition of hydrothermal mineral deposits in oceanic crust, Econ. Geol., 73, 135-160, 1978.

Rowlands, N.J., The Adelaidean system of South Australia: A review of its sedimentation, tectonics and copper occurrences, Belt Symposium Volume, Idaho Bureau of Mines and Geology, 80-112, 1973.

Rozendaal, A., The Gamsberg Zinc Deposit, South Africa: A banded stratiform base metal sulfide ore deposits (abstract), 5th IAGOD Symposium, Snowbird, Utah, 169, 1978.

Ruelle, J.C., Depositional environment and genesis of stratiform copper deposits of the Redstone Copper Belt, Mackenzie Mountains, N.W.T. (abstract), Geol. Soc. Amer. Ann. Mtgs., Toronto, 482-483, 1978.

Russell, M.J., Base-metal mineralization in Ireland and Scotland and formation of the Rockall Trough, in: Implications of Continental Drift to the Earth Sciences, S.K. Runcorn, ed., Academic Press, London and New York, 1, 577-594, 1973.

Rye, D.M. and Williams, N., Stable isotope geochemistry of the McArthur River ore deposits: An epigenetic sedimentary-type Pb-Zn deposits (abstract), Geol. Soc. Amer. Ann. Mtgs., Toronto, 483-484, 1978.

Sawkins, F.J., Ore genesis in the north Pennine orefield in the light of fluid inclusions studies, Econ. Geol., 61, 385-401, 1966.

Sawkins, F.J., Metal deposits related to intracontinental hotspot and rifting environments, Jour. Geol., 84, 653-671, 1976a.

Sawkins, F.J., Widespread continental rifting: some consideration of timing and mechanism, Geology, 4, 427-430, 1976b.

Sawkins, F.J., Massive sulfide deposits, in relation to geotectonics, in: Metallogenesis and Plate Tectonics, Spec. Pub. No. 14, Geol. Assoc. Canada, 221-240, 1976c.

Sawkins, F.J., Fluid inclusion studies of the Messina copper deposits, Transvaal, South Africa, Econ. Geol., 72, 619-631, 1977.

Sawkins, F.J. and Rye, R.O., Additional geochemical data on the Messina copper deposits, Transvaal, South Africa, Econ. Geol., 74, 684-689, 1979.

Sawkins, F.J. and Burke, K., Extensional tectonics and mid-Paleozoic massive sulfide occurrences in Europe, Geol. Rundschau, 69, 349-360, 1980.

Sharp, J.E., Cave Peak, a molybdenum-mineralized breccia pipe complex in Culberson County, Texas, Econ. Geol., 74, 517-534, 1979.

Sillitoe, R.H., Formation of certain massive sulfide deposits in sites of seafloor spreading, I.M.M. Trans., 81, B141-B148, 1972a.

Sillitoe, R.H., A plate tectonic model for the origin of porphyry copper deposits, Econ. Geol., 67, 184-197, 1972b.

Sillitoe, R.H., Tin mineralization above mantle hotspots, Nature, 248, 497-499, 1974.

Sillitoe, R.H., Stratabound ore deposits related to infracambrian rifting along Northern Gonwanaland (abstract), 1978 IABOD Meetings, Snowbird, Utah, 175, 1978.

Spooner, E.T. and Bray, C.J., Hydrothermal fluids of seawater salinity in ophiolitic sulfide ore deposits in Cyprus, Nature, 266, 808-812, 1977.

Stamatelopolou-Seymorcer, K. and Maclean, W.H., The geochemistry of possible metavolcanic rocks and their relationship to mineralization at Montauban-Les-Mines, Quebec, Can. Jour. E. Sci., 14, 2440-2453, 1977.

Toens, P.D., The geology of part of the southern foreland of the Damara Orogenic Belt in S.W.A. and Botswana, Geol. Rundschau, 64, 175-192, 1975.

Trendall, A.F. and Blockley, J.G., The iron formations of the Precambrian Hamersley Group, Western Australia, Bull. 119, Geol. Surv. W. Aust., 366, 1970.

Van Alstine, R.E., Continental rifts and lineaments associated with major fluorspar districts, Econ. Geol., 71, 977-987, 1976.

Van Eden, J.G., Stratiform copper and zinc mineralization in the Cretaceous of Angola, Econ. Geol., 73, 1154-1160, 1978.

Vokes, F.M., Metallogeny possibly related to continental break-up in southwest Scandinavia, in: Implications of Continental Drift to the Earth Sciences, 1, D.H. Tarling and S.K. Runcorn (eds.), Academic Press, London, 573-579, 1973.

Walker, R.N., Logan, R.G. and Binnekamp, J.G., Recent geological advances concerning the H.Y.C. and associated deposits, MacArthur River, N.T., Jour. Geol. Soc. Australia, 24, 365-380, 1977.

Weiblen, P.W. and Morey, G.B., The Duluth Complex - a petrologic and tectonic summary, Minn. Geol. Survey Reprint Series 28, 28, 1975.

White, D.E., Environments of generation of some base metal ore deposits, Econ. Geol., 63, 301-335, 1968.

White, W.S., The native copper deposits of northern Michigan, in: Ore Deposits of the United States, 1933-1967 (Graton-Sales Volume), New York Am. Inst. Mining Metall. Engineers, 303-325, 1968.

Willemse, J., The geology of the Bushveld Igneous Complex, the largest repository of magmatic ore deposits in the world, Econ. Geol. Monograph No. 4, Magmatic Ore Deposits, 1-22, 1969.

Ziegler, P.A., North-western Europe: Tectonic and basin development, Geologie en Mijnbouw, 57, 589-626, 1978.

Zimmerman, R.A. and Amstutz, G.C., The Meggen pyrite-barite deposit, Internat. Sed. Cong., 8th Heidelberg, 254-261, 1971.

TERRESTRIAL ENERGY CURRENTS AND TRANSFER IN ICELAND

Gunnar Bodvarsson

School of Oceanography and Department of Mathematics, Oregon State University, Corvallis, Oregon 97331

Abstract. Present estimates of terrestrial energy currents through the surface of Iceland indicate the following components: total heat conduction 15 GW, time averaged advection by lava 7 GW, advection by thermal waters 8.5 GW and mechanical power expended for intruding magma 0.3 GW. The conduction flow per unit surface area is 150 mW/m^2 and the total energy flow about 300 mW/m^2. Compared to other zones of crustal strain, the conduction flow is not particularly large but the currents due to volcanism and thermal activity are quite anomalous. There are strong indications that the abnormal heat currents through the surface of Iceland are caused by a more or less continuous upflow of magma along the global rift zone traversing the island. The magma that is generated in the upper mantle is partially extruded by volcanism and partially intruded into a surface zone of 20 to 30 km thickness. Heat released by the intrusions is the cause of the supernormal component of the conduction current and of the extensive high-temperature geothermal activity in Iceland. The overall heat input per unit length rift axis below Iceland appears to be representative of the global oceanic rift system.

Acknowledgements. The writer is indebted to the two reviewers, Dr. Arthur H. Lachenbruch, research geophysicist, U.S. Geological Survey, Menlo Park, California, and Professor Sveinbjorn Bjornsson, University of Iceland, and to Professor Shaul Levi, Oregon State University, for their very helpful comments on this paper. This work was supported by the National Science Foundation under Grant No. EAR 77-23938.

(1) Introduction

The evolution of the earth is dominated by thermal processes that are manifested at the surface by plate motion, volcanism, thermal and tectonic activity. These phenomena are clearly related to global zones of crustal strain or extension and, in particular, to the ocean rift system. For a better understanding of the evolution of the earth, we are therefore interested in the energetics of the rift system.

Iceland is a conspicuous part of the North Atlantic Ridge that has received a great deal of attention during the past few decades. A considerable amount of geological and geophysical work has been carried out on the island where there are now over 100 boreholes of depth in excess of one km that are available for sampling and logging. The data material is quite substantial.

The present paper has been written for the purpose of reviewing and discussing some important aspects of the energetics of the Iceland rift system and its implications with regard to global thermal processes. Our attention will focus on relevant processes in the crust and upper mantle down to the depths of the order of a few tens of kilometers. Because of the local importance, particular attention will be paid to rock/water heat transfer in the crust. The main results can be displayed in a simple energy current graph.

For a comprehensive review of the geology and geophysics of Iceland, we refer to the interesting review of Palmason and Saemundsson (1974). Our data sources will be quoted in the course of the discussion.

(2) Surface Energy Currents and Flows

A preliminary investigation of terrestrial energy currents in Iceland was carried out by Bodvarsson (1954) almost three decades ago. Although the results obtained then were based on very incomplete field material, an overall picture emerged that still appears to be reasonable. The total heat current though the surface of Iceland (103,000 km^2) was estimated at about 32 GW consisting of a conduction current of 20 GW, average rate of advection by surface lava of 8 GW and rate of advection by thermal water of about 4 GW. The average integrated flow through the surface was estimated at 310 mW/m^2 that is about 5 times the global average conduction flow of 60 mW/m^2.

The quantity and quality of field data have improved very considerably during the last two decades and many of the previous estimates have been revised and improved by Palmason (1974) and Palmason and Saemundsson (1979). In particular, the number of boreholes available for temperature measurements has increased greatly. Interesting samples of borehole temperature profiles were published by Palmason (1974, see Figure 1) and the inferred temperature gradients in available boreholes are shown on a recent map prepared by Palmason and Saemundsson (1979, see Figure 2). The number of data points in Figure 2 is 48 and the average gradient 86°C/km. Estimating the average conductivity of the tholeiitic basalt formations of Iceland at 1.7 W/m°C (Palmason and Saemundsson, 1979) we obtain an average conduction flow of almost 150 mW/m^2 and a total estimated conduction current of 15 GW. Considering that based on the global average, the normal conduction current through the surface of Iceland would be 6 GW, we find an abnormal conduction component of 9 GW. Using more recent data on the average rate of production of surface volcanics, Palmason and Saemundsson (1979) have revised the rate of advection by surface lava to 6 GW, but they retain the estimate of Bodvarsson (1954) of 4 GW for the rate of advection by thermal waters. (Figs. 1 and 2).

In the present discussion, we will adopt the above estimates of Palmason and Saemundsson (1979) for the conduction current and the average mass extrusion of subaerial lava of 0.045 km^3/year = 1.4 m^3/s. Assuming a density of 3×10^3 kg/m^3 and a total sensible heat content of 1.6×10^6 J/kg, the

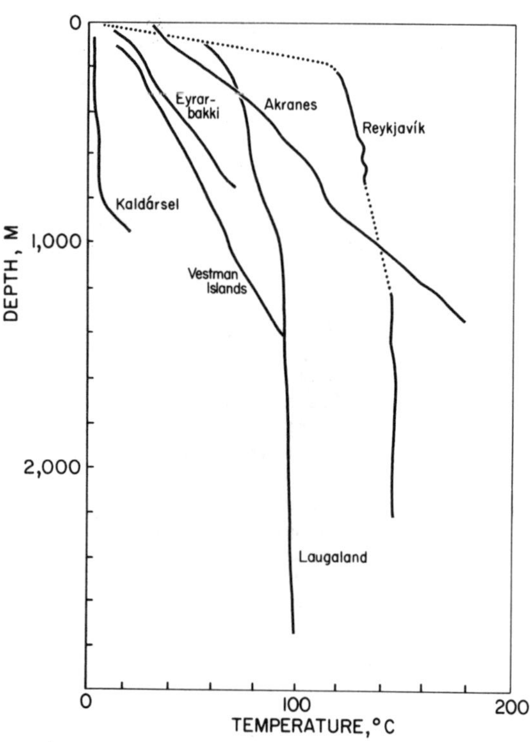

Fig. 1. A few samples of temperature profiles in boreholes in Iceland (from Palmason, 1974).

average advection by surface lava is 7 GW. Moreover, because of recent data by Bjornsson et al. (1980) on the output of the Grimsvotn caldera, we will raise the estimate of the total advection by thermal waters to 8.5 GW (see also section (4) below). It must be stated that this estimate is more uncertain than the two results adopted above.

For an overview it is useful to enter the present estimates into Table I. Columns (a) and (b) give respectively the currents and flows per unit surface area of Iceland and following Palmason (1974), the third column indicates the specific currents per unit length ridge axis (300 km) traversing Iceland.

We are now interested in downard continuing the above data to obtain the currents at subsurface levels. For this purpose the various modes of current interactions due to hydrothermal and magmatic processes in the crust and the upper mantle have to be estimated. It is reasonable to assume that the hydrothermal processes are largely confined to the uppermost few kilometers, viz., mainly to seismic layeres 1 and 2 of Iceland (Palmason, 1971). Below this zone, the magmatic processes will dominate the energy transfer picture.

(3) Rift Zone Models & Heat Advection by Magma

The results of Bodvarsson (1954) already indicated that heat advection by ascending magma provides the most plausible and acceptable explanation for the abnormally large total heat current through the surface of Iceland. The present thermal situation can be accounted for by assuming that magma generated in the upper mantle is intruded into a surface zone extending down to a depth of a few tens of kilometers where it forms various types of intrusions that dissipate heat to the surrounding rock. Assuming episodes of uniform intrusion rates and distribution, Bodvarsson (1954) concluded that in order to maintain present surface currents, the magma had to be intruded at a rate four times the present rate of lava extrusion by subaerial volcanism. Some later investigations have resulted in rather similar conclusions. It would therefore appear quite evident that the problem of the sources of the energy currents will have to be considered in the broader context of regional magma generation and rift zone thermomechanics.

A very considerable amount of work has been devoted to the discussion of rift zone models and to the problems of the North Atlantic rift, in particular. In an early attempt to describe the evolution of the local structure of the Iceland rift zone, Bodvarsson and Walker (1964) suggested the dike-injection model shown in Figure 3a. An interesting combined lava-subsidence/dike-injection model has been developed by Palmason (1973, 1980a). The model shown in Figure 3b is consistent with many field observations. In this context, it is of interest to note that some questions have been raised as to the

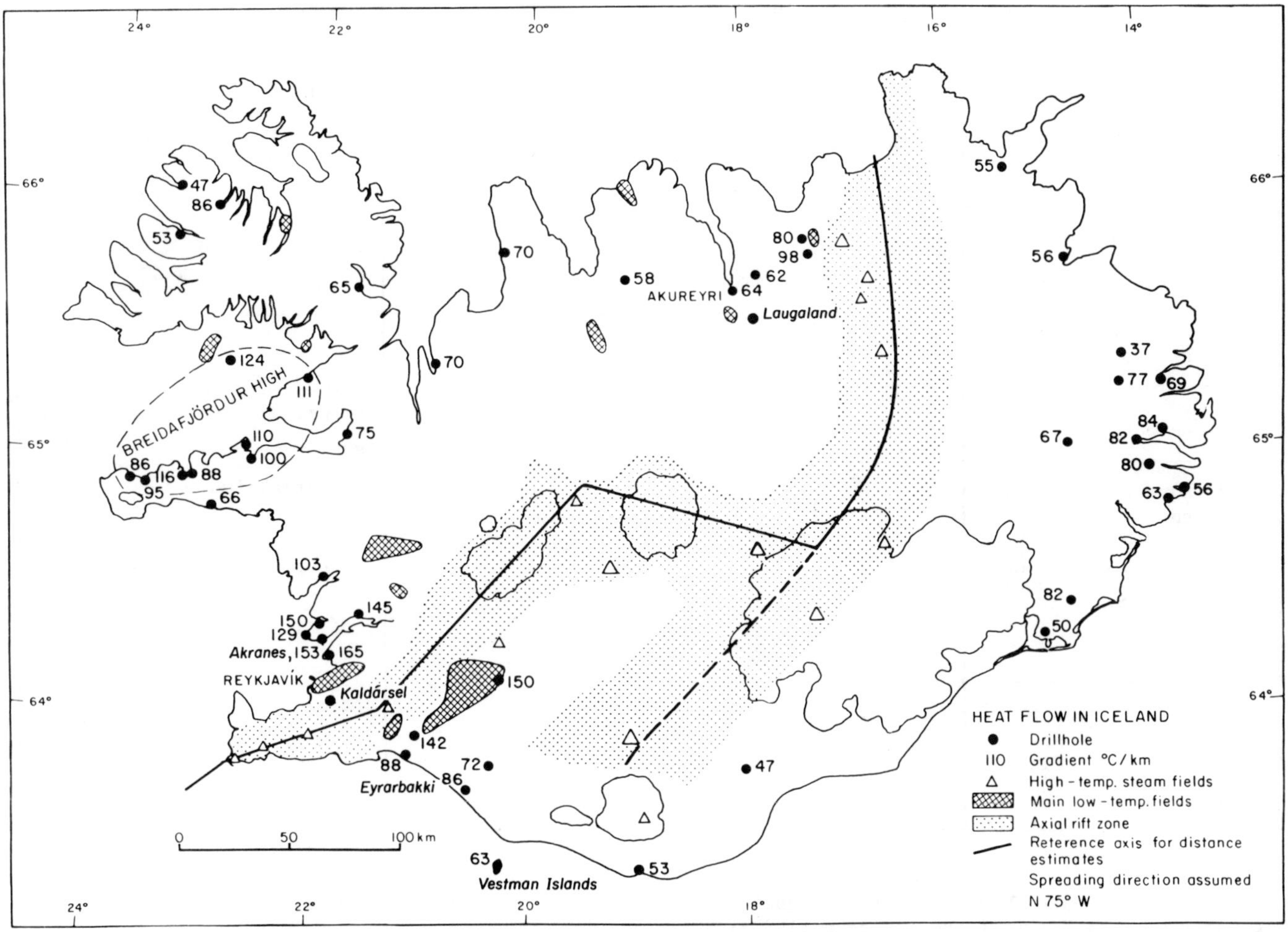

Fig. 2. Locations of boreholes used to determine thermal gradients in Iceland. The main centers of geothermal activity are shown on the map (from Palmason and Saemundsson, 1979).

permanence of the Iceland rift zone (Saemundsson, 1974 and 1979) and Palmason and Saemundsson (1979) find evidence for an extinct rift axis in the extreme west of Iceland. (Figs. 3a and 3b)

The major features of current rift zone or diverging plate models have been discussed by Bottinga and Allegre (1976). They consider the three cases shown in Figures 4a to 4c, viz., the dike-intrusion model of McKenzie (1967) and the asthenospheric upwelling models of Parker and Oldenburg (1973) and Langseth et al. (1966). The more mechanical aspects of the principal spreading center and rift zone processes have been discussed in considerable detail by Lachenbruch (1973, 1976). (Figs. 4 and 5)

The results of Lachenbruch (1978/1979) and Lachenbruch and Sass (1978) on the thermal effects of tectonic extension in the Great Basin of North America appear to be relevant to rift zone processes and to the thermomechanics of Iceland, in particular. They conclude that the high conduction heat flow in extending regions can be accounted for by mainly three processes as sketched in Figures 5a to 5c, viz., (1) lithosphere stretching combined with solid-phase accretion at the base, (2) lithosphere stretching combined with underplating by liquid basalt and (3) extension by distributed dike injection similar to the model used for Iceland by Bodvarsson and Walker (1964). All three processes may possibly be in progress in Iceland. However, considering the substantial volcanism and outpouring of lava, it would appear that the solid phase accretion model displayed in Figure 5a would be less relevant to the case of Iceland. On the other hand, basaltic dikes are numerous and are therefore contributing to the heat input. Moreover, recent magnetotelluric work by Beblo and Bjornsson (1980) indicates the presence of a few kilometers thick high electrical conductivity layer in the 10 to 20 km depth zone. Having a conductivity of around 0.07 mho/m, the layer is possibly semi-liquid and thus lends some support for underplating in the upper mantle under

TABLE 1 ESTIMATES OF ENERGY CURRENTS AND FLOWS

		Iceland (area = 10^5 km^2)			Earth (area = 5 x 10^8 km^2)	
		(a)	(b)	(c)	(d)	(e)
	Component	Current GW	Flow mW/m^2	Current per unit length of ridge axis (300 km) MW/km	Current GW	Column (c) x 50,000 GW
1	Total conduction	15	150		30,000	
2	Conduction in excess of normal	9	90	30		1,500
3	Advection by lava	7	70	23	subaerial 30 submarine 300	
4	Advection by thermal waters	8.5	85	28	subaerial 100 submarine ?	
5	Mechanical power (upper rift zone)	0.3	3	1	seismic power 20	50
6	Total	30.8	308	82		4,100
7	Estimated heat release by sea floor spreading of one cm/year			30		

Iceland. It is also to be noted that magnetotelluric work by Hermance and Pedersen (1980) provides evidence for a similar situation in the Rio Grande rift of North America. Underplating may thus be a common situation in rift zone areas and may in the case of Iceland take place in conjunction with dike-injection in any of the models displayed in Figures 3 to 5. There is consequently clear evidence for a substantial heat advection by magma in the crust of Iceland.

To estimate the mass flow of magma required to maintain the supernormal conduction current of heat in Iceland, we assume that the average intrusion depth is of the order of 10 km where the formation temperature is probably 500°C to 600°C. Relative to a total heat of fusion of $1.6x10^6$J/kg, the possible average total heat release by the cooling intrusions will thus be 10^6 J/kg and to sustain a heat current of 9 GW, a total average mass upflow of 9,000 kg/s will be required. Since some of the magma is likely to be injected as dikes, the magma/rock heat transfer can be quite rapid. Elementary heat conduction theory (Carslaw and Jaeger, 1959) indicates that the 90% cooling time of a dike of a width w is of the order of $w^2/40$ years. A dike of w = 10 m will lose its sensible heat in a few years and there are thus no difficulties in accounting for the required magma/rock heat transfer.

(4) Hydrothermal Processes

There exists a considerable literature on thermal activity and hydrothermal processes in Iceland. During the first half of this century, important contributions were made by Thorkelsson (1941) and Einarsson (1942). For the present purpose, an overview given by Bodvarsson (1964) is an useful reference. For interesting results on the tectonic, volcanic and hydrothermal activity in the Krafla area in Northern Iceland we refer to Bjornsson et al. (1979).

Thermal activity in Iceland falls into two main groups (Bodvarsson, 1964). First, the high-temperature activity that has a base temperature mostly in excess of 200°C and is associated with the present rift zone system. As shown in Figure 2 there are a total of 16 locations of high-temperature activity distributed along both rift axes and clearly associated with centers of very recent and/or present volcanism. The field setting indicates that the high-temperature areas represent convective hot-spots at and around magmatic intrusions as is most obvious in the Krafla area in the North (Bjornsson et al., 1979).

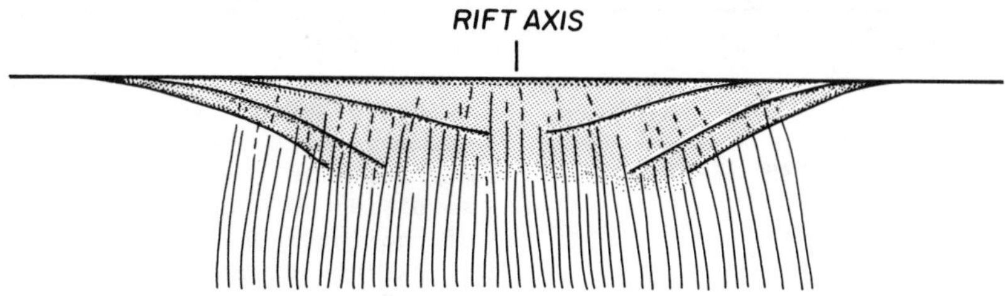

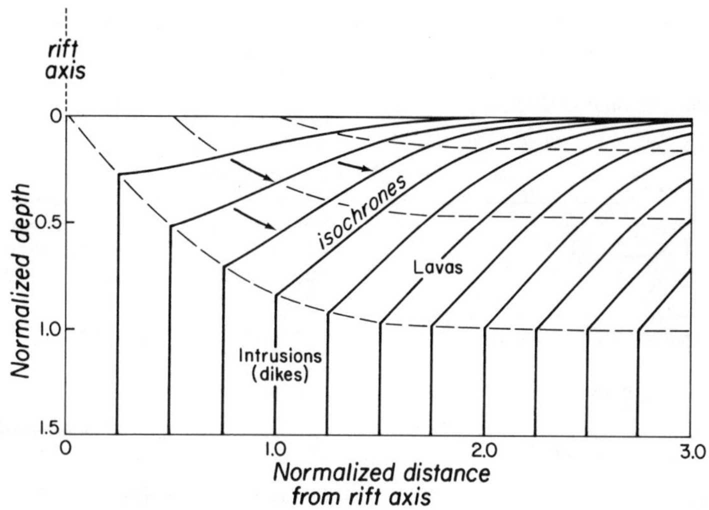

Fig. 3a. Dike-injection model of Bodvarsson and Walker (1964). Fig. 3b. Combined lava-subsidence/dike-injection model of Palmason (1973, 1980).

Second, the low-temperature activity that is distributed over the Tertiary and early Quaternary flood basalt areas in the western half of the country. The location of some of the main areas of low-temperature activity are indicated in Figure 2. Recent isotope work by Arnasson (1976) shows that in most cases, the recharge of these systems is located near to the center of the country, and the low-temperature springs are therefore outlets for quite extended hydrological circulation systems.

As of now, no reliable data are available on the depths of the hydrothermal circulation systems. In the Reykavik and Akureyri areas, boreholes have produced thermal water from depths of about 3 km. It is clear that the seismic layers 1 and 2 (Båth, 1960; Palmason, 1971) are permeable to water, and microearthquake studies (Ward and Bjornsson, 1971) indicate that water may also penetrate into the upper parts of layer 3. The circulation base of many of the major systems may thus be at depths of 3 to 5 km, and possibly deeper. The total duration of individual hydrothermal systems is another problem on which we have little data. In view of the difficulties associated with long term energy supply, the present writer generally favors hydrothermal age estimates of the order of a few 10^3 rather than a few 10^4 or 10^5 years. Moreover, it is fairly evident that the long periods of Pleistocene glaciation of Iceland have interferred greatly with subsurface water circulation and most of the present system may therefore have been active only during the post-Glacial period of 8,000 to 10,000 years.

Based on the results of Bodvarsson (1954) and considering recent investigations of Bjornsson et al. (1980), we estimate the total rate of heat dissipation by the high-temperature activity at 8 GW. Retaining the figure of Bodvarsson (1954) for the low-temperature activity of 0.5 GW we obtain the total rate of heat dissipation by thermal activity in Iceland of 8.5 GW as already stated in section (2) above.

Quantitative investigations of the source mechanisms of the high-temperature activity (Bodvarsson, 1951; White, 1968) indicate that it is difficult to account for the localized rock/water heat transfer unless the water actually penetrates into the intrusions that are in the process of cooling. This contention has been supported by later results by Bjornsson et al. (1980). The rock/water heat transfer in the case of the low-temperature activity appears to be less constrained. In view of the importance of hydrothermal heat advection for the local upper crustal temperature field and, in particular, because of possible interference with borehole

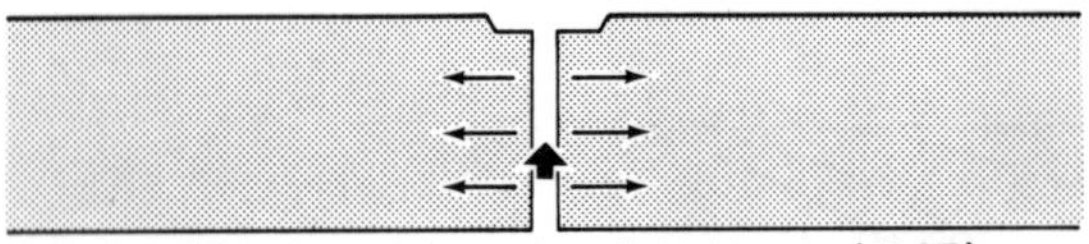

Fig 4a. Dike-intrusion model of McKenzie (1967)

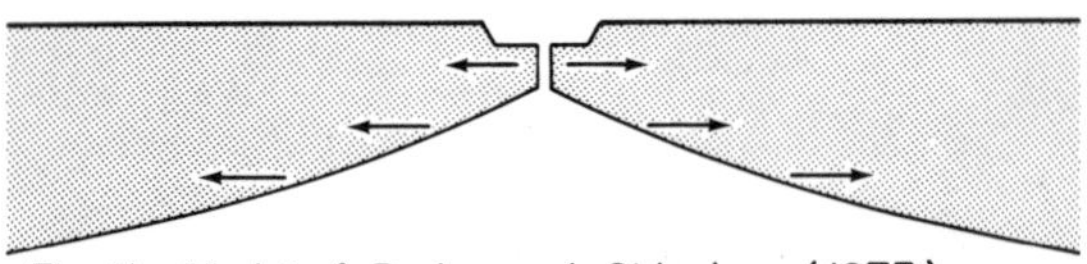

Fig 4b. Model of Parker and Oldenburg (1973)

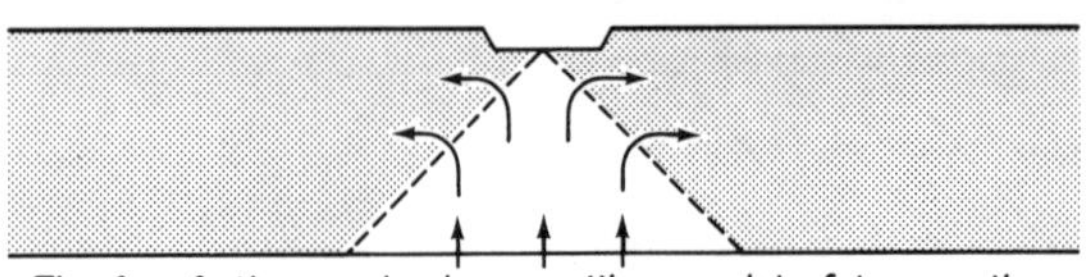

Fig 4c. Asthenospheric upwelling model of Langseth et al. (1966)

Fig. 4. Rift zone models.

temperature data, we will discuss the heat transfer processes involved in a little more detail in the following section.

(5) Hydrothermal Heat Transfer

The relatively low heat conductivity of common rock places severe limitations on the possible rate of heat transfer across given rock/water surfaces. In order to evaluate the hydrothermal perturbation of the general subsurface temperature field, we are therefore interested in order of magnitude estimates of the rock/water surfaces required to maintain the two main types of the thermal activity in Iceland.

Rock/water heat transfer can take place in two types of flow regimes, namely, in forced flow where pressure gradients of hydraulic origin dominate and in free convective flow that is dominated by thermal buoyancy. While the forced flow mode is likely to be relevant to the low-temperature systems, the latter type of flow is of particular interest in situations where fairly large temperature differentials are involved such that thermal contraction of the rock combined with fluid convection in fractures contributes to the widening or even opening of new fractures. This process by which cold water can penetrate downward into hot rock is likely to be of particular importance in the high temperature environment in and around intrusions. We will refer to the phenomenon as convective downward migration. There is some literature on the subject. Bodvarsson and Lowell (1972) have discussed some simple fracture models that are relevant to the penetration of seawater into the ocean floor. A considerably more elaborate investigation of the sea floor processes has been carried out by Lister (1974).

In estimating possible rates of heat transfer in hydrothermal systems it is convenient to commence with the case of convective downward migration in the high-temperature environment involving magmatic intrusions in the process of cooling. Obviously, we are here dealing with an extremely complex problem of elastomechanical/hydrodynamical interaction that defies any exact mathematical description. The situation is greatly compounded by the fact that natural rock formations are usually far from being homogeneous/isotropic as is being assumed in most theoretical modeling. Lister (1976) has extended his analysis for the sea floor (Lister, 1974) to high-temperature geothermal systems and presents some estimates of process rates. Because of the general interest of convective downward migration of fractures, we will here offer an alternative approach (Bodvarsson, 1979) that describes the phenomena from a somewhat different vantage point.

Consider a situation where there is a large upper crustal magmatic intrusion in the process

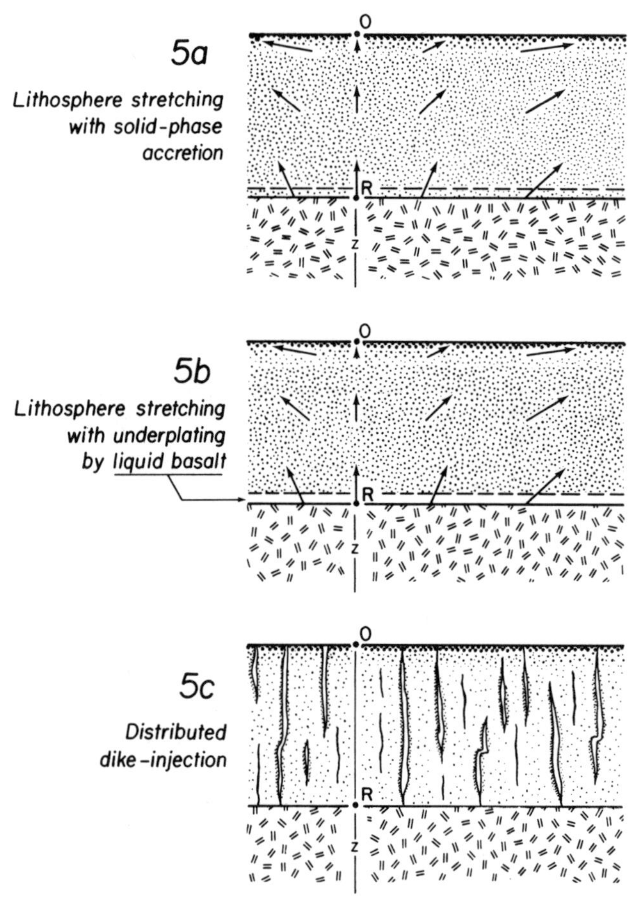

Fig. 5. Tectonic extension models of Lachenbruch (1978/79) and Lachenbruch and Sass (1978) for the Great Basin of North America.

of cooling and solidification. Water is penetrating into the solid section through a system of parallel vertical fractures that migrate downward by the process described above. Cold water of surface origin is piped by natural channels down to the fracture system where it sinks to the bottom of the fractures, cools the adjacent formations and thereby generates the thermal contraction that is required to migrate the system downward. The water heated in the process rises up through the fracture system. On the basis of the model, we take that there is a steady temperature differential of ΔT between the water at the bottom of the fracture system and the adjacent formations.

For analytical convenience we consider now a circular fracture of diameter D in common igneous rock that is closed because of a contact pressure p_c. Forcing water into this fracture at a pressure $p>p_c$ will open it such that the width at the center is (Sneddon and Lowengrub, 1969)

$$w = 1.5\times10^{-11}\ (p-p_c)D, \quad \text{SI} \qquad (1)$$

Much of the same effect can be achieved by a thermoelastic contraction of a volume of the surrounding rock such that the resulting tension across the fracture become sufficiently large to relieve the contact pressure p_c. To estimate the necessary magnitude of ΔT for this purpose, we assume stress isotropicity and the following simple relation

$$\Delta T > p_c/\alpha k \qquad (2)$$

where α is the thermal expansivity and k the bulk modulus of the rock. Clearly, to obtain the required effect the contraction, and thereby the temperature reduction, has to affect a certain volume containing the fracture. An estimate of the linear dimension of this volume is given by the diameter D.

A system of a number of parallel long fractures in the process of downward migration at a rate v can under rather general circumstances be modeled as a cold temperature front moving downward with the velocity v relative to the rock. On a one-dimensional heat conduction theory (Carslaw and Jaeger, 1959, page 396), the temperature differential ahead of the front is given by

$$T = \Delta T \exp(-vx/a) \qquad (3)$$

where x is the distance from the front and a is the thermal diffusivity of the rock. The expression on the right of (3) defines the penetration depth $d = a/v$, that is, the depth to where the amplitude of the front has decreased to 1/e of its full value.

The critical assumption to be made here is to identify the penetration depth d with the linear dimensions D resulting in the estimate

$$v \sim a/D \qquad (4)$$

To proceed with a numerical investigation, we take the possible situation at the depth of 3 km, for example. Let $\alpha = 2\times10^{-5}/°C$ and $k = 5\times10^{10}$Pa. The hydrostatic pressure at the depth of 3 km is about 3×10^7Pa and the average minimum horizontal stress will be of the order of 5×10^7Pa. The values will be lower in regions under strain and we can take that the possible values for the contact pressure on vertical planes of discontinuity is then of the order of 10^7Pa. On these premises equation (2) yields the estimate $\Delta T>10°C$. This is a small value showing that temperature differentials of a few tens of degrees C should be sufficient for convective downward migration.

Values for the parameter D are much more uncertain. Estimates of the order of a few tens of meters would appear reasonable. Inserting $a = 10^{-6}m^2/s$ and values D = 10 to 100 m in equation (4) gives then the estimate of v = 0.3 to 3 m/year. This result is roughly of the same order as the results of Lister (1976).

Having obtained the migration rate, it is a simple matter to estimate the average rate of rock/water heat transfer. Consider a homogeneous/isotropic heat conducting half-space that at time t = 0 has a uniform temperature T_o and hence a uniform heat content per unit volume h_o. Let the surface temperature from t = 0 on be zero. A simple exercise in heat conduction theory (Carslaw and Jaeger, 1959, page 62) reveals that the rate of heat transfer per unit surface area out of the solid is $(h_o/2)(a/t)^{1/2}$. This result will apply to a vertical fracture that is in the process of downward migration at a constant rate. Hence, at a distance x from the tip of the fracture, the transfer rate is $(h_o/2)(av/x)^{1/2}$. Assuming that the fracture has penetrated a distance H into the hot solid, we find by an integration over both surfaces that the average heat transfer rate per unit fracture length is

$$Q = 2h_o(avH)^{1/2} \qquad (5)$$

and thereby the average rate of transfer per unit area

$$q = 2h_o(av/H)^{1/2} \qquad (6)$$

Assuming penetration of surface water into an intrusion that initially is at the melting temperature, a reasonable value for the sensible heat content per unit volume is $h_o = 3\times10^9 J/m^3$. Moreover, using the average estimate v = 1 m/yr = 3×10^{-8}m/s and a penetration of H = 1 km, we find a rate of about $Q = 3\times10^4$W/m and $q = 30W/m^2$. On this basis we estimate that to maintain the present heat dissipation of the high-temperature activity in Iceland of 8GW, a total length of fractures of about 300 km and a total fracture area of 300 km^2 would be required. Considering that the 16 high-temperature areas cover a total surface area of a few 10^2 km^2 these results are not unreasonable.

Turning to the low-temperature activity, we

find some evidence for large scale convection and thereby the possibility for the convective downward migration of fractures. For example, the temperature profile measured in the Reykjavik borehole in Figure 1 above, displays a very steep temperature gradient near the surface but only a relatively moderate increase in temperature in the depth interval from 0.2 to 2.2 km. The situation at Laugaland in the North is quite similar. Moreover, the Kaldarsel borehole in the same figure is almost isothermal at about 5°C down to a depth of 0.8 km. Obviously cold water is being channeled downward in this location. Further data of this nature are available. The isotope results of Arnason (1976) have, on the other hand, indicated considerable recharge distances for the main low-temperature areas. Clearly, the water can also take up heat in forced horizontal flow either through localized channels or through the extensive sheet-like opening that are known to exist along the contacts of lava beds in flood-basalt areas. The situation with regard to the major low-temperature systems may thus be as sketched in Figure 6.

Rock/water heat transfer in forced sheet-like flow has been investigated at some length by Bodvarsson (1974, 1978). The principal results can be stated very briefly as follows. The basic model is an open flat flow-sheet of a very small aperture bounded by two homogeneous/isotropic heat conducting half-spaces that have a uniform initial temperature T_o. Starting at time $t = 0$ water at temperature $T_1<T_o$ enters the sheet along one boundary and is recovered along another boundary with a higher temperature T_2. The flow through the sheet is assumed to be uniform and unidirectional. Let T_{av} be the average of T_2 during the time interval t and q_{av} be the average rate of heat transfer per unit sheet area (both sides included). The principal result is that

$$q_{av} = \kappa_{av}\nabla T_m \qquad (7)$$

where ∇T_m is the log-mean temperature difference defined by

$$\nabla T_m = (T_{av}-T_1)/\ln[(T_o-T_1)/(T_o-T_{av})] \qquad (8)$$

and κ_{av} is the average heat transfer coefficient that can in the case of common igneous rock be estimated on the basis of the graphs obtained

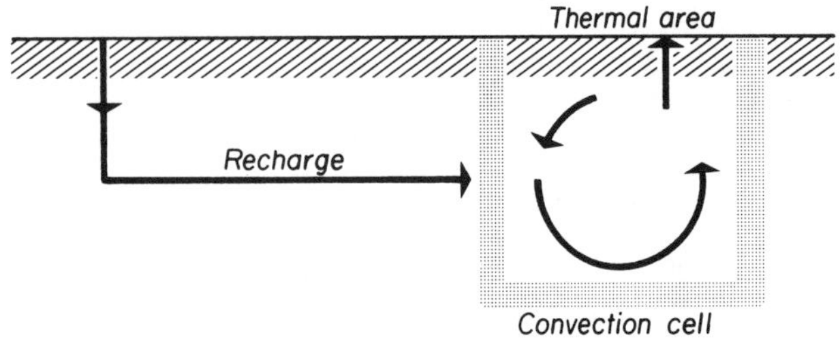

Fig. 6. Flow graph for a low-temperature system in Iceland.

by Bodvarsson (1974). Assuming mafic rock and system ages of 10^4 years, we find that approximately $\kappa_{av} \sim 5x10^{-3} W/°Cm^2$.

Heat uptake in forced flow through discrete channels of relatively small cross-flow dimensions is another possibility that should be considered. For the purpose of estimating channel lengths, we can resort to the case of cylindrical channels of a radius r and proceed in much the same way as above. Inserting the proper heat transfer coefficient in equation (7) then gives the heat transfer per unit length of the channel. Assuming again a constant flow of duration t and that the Fourier number $N = at/r^2 > 10^3$, we find on the basis of results of Carslaw and Jaeger (1959, page 334) that

$$\kappa_{av} = 4\pi k/\ln(1.2N) \qquad (9)$$

where k is the heat conductivity.

It is instructive to estimate the total heat transfer surfaces or channel lengths required for the low-temperature activity on both the free convective and forced sheet-flow models. In the case of the free convective cells such as in the Reykjavik system, the average rock/water temperature difference may be of the order of 100°C and the relevant sensible heat content per unit volume of rock then $h_o = 3x10^8 J/m^3$. Assuming the same value as above for the velocity v and the penetration depth H yields the estimates $Q=3x10^3$ W/m and $q = 3W/m^2$. Thus, to maintain the low-temperature activity dissipating a total of 0.5 GW a total fracture length of about 200 km and fracture area of 200 km^2 are needed.

Turning then to a forced flow regime in a single flow-sheet where, for example, $T_o = 150°C$, $T_1 = 5°C$, $T_{av} = 100°C$ we obtain $\Delta T_m = 90°C$ and on the basis of $\kappa_{av} = 5x10^{-3} W/°C\ m^2$ a value of $q = 0.45\ W/m^2$. On these premises the total sheet heat transfer area required for the low-temperature activity is estimated at about $10^3\ km^2$.

The assumption of only one flow-sheet is really irrelevant. The sheet can be partitioned into a number of smaller sheets and provided the assumptions with regard to the input/output temperatures remain unchanged, the final estimate of the total area is the same.

For the case of forced channel flow a reasonable radius is r = 1 m and since for $t = 10^4$ years, the Fourier number N turns out to be in the range 10^5 to 10^6, and a value of $\kappa_{av} \sim 1.2$ k is obtained. In the case of flood basalts k = 1.7 W/°C m is a good value and thus $\kappa_{av} \sim 2$ W/°Cm. Using the above figure of $\Delta T_m = 90°C$ yields then an estimate of $q \sim 180 W/m$. To maintain the heat output of the low-temperature activity of 0.5 GW a total channel length of 2,800 km would be required. Since equation (9) applies to a single channel, this estimate can be applied to a network of channels when the individual channel spacing is a few 10^2 m or more.

Clearly, the above results are only rough order magnitude estimates that have to be viewed with a considerable amount of caution. However, the rock/water heat transfer areas required for both types of thermal activity appear plausible and when compared with the total surface area of Iceland, the impression is that the rock/water heat transfer is a rather local process that is unlikely to interfere greatly with the overall subsurface temperature field. Excepting a few local anomalies such as in the region around Reykjavik in the South-west, it would thus appear that the temperature gradient data of Palmason and Saemundsson (1979) give a fairly representative picture of the subsurface conduction temperature field in the crust of Iceland.

(6) Mechanical Work

Some mechanical energy is also released in the rift zone of Iceland. Based on the concepts discussed in section (3) above, we take that this energy is expended for plate motion away from the local rift zone and that the motional engine is in part driven by the buoyancy of the upwelling magma. A rough estimate of the mechanical power released in the upper few tens of kilometers can be obtained in the following way. Assuming again that on the average the heat released per unit mass by the intrusions is 10^6 J/kg (see section (3) above), we find that to sustain the supernormal conduction heat current of 9 GW and the high-temperature thermal activity of 8 GW, an average of 17,000 kg/s or about 6 m^3/s of magma is required. The order of magnitude of the overpressure of the upflowing magma in crustal and upper mantle locations can be inferred on the basis of observation on the height of erupting vents and lava fountains. Considering kinetic energy and frictional pressure losses, values of the order of 10^8 Pa appear reasonable estimates. Somewhat lower values are, on the other hand, indicated by the mechanism of the inflation/deflation processes at the Krafla caldera in northern Iceland (Bjornsson et al., 1979). Adopting a figure of 5×10^7Pa, we obtain a total mechanical power for intruding magma of $6 \times 5 \times 10^7$ W = 0.3 GW. This figure is entered in (5.a) of Table I.

Clearly, the above figure represents only a fraction of the total mechanical power expended for plate dynamics along the Iceland rift zone. In particular, power for plate motion is also provided by the general convective upflow of mantle material below the rift zone. The upper section of this broad plume can be assumed to diverge away from the rift zone and thereby provide viscous drag acting along the lower boundary of the lithosphere. Lachenbruch (personal communication) has pointed out that the total pressure force involved can be estimated on the basis of the gravitation potential of the ridge material relative to the deep sea floor. However, since no data is available on the material current involved, no estimate can be made of the total power expended.

(7) The Energy Current Graph

It is helpful to portray the above results in a compact form in the time-averaged current graph shown in Fig. 7. Using the above energy transfer estimates, the surface current picture can be continued downward into the lower crust and upper mantle below Iceland. The graph is based on the assumption that the hydrothermal activity is confined to a zone extending from the surface down to a depth of about 5 km. Little magma/conduction transfer is assumed to take place in this zone, and the transfer data furnished in the graph are obtained with the help of present observations (Bjornsson et al., 1979) in the North of the rift zone. The episodic intrusive activity at the Krafla system is taken as a model for 10 such events during 10^3 years and the result is averaged out in space and time. The advection current of the low-temperature activity draws on the thermal capacitance of crustal rock and is therefore not a part of the conduction current.

(8) Iceland in the Global Setting

To put the above data into a proper perspective, it is of interest to compare the estimated energy currents in Iceland with results from other zones of crustal strain and with global figures.

Turning first to the conduction heat flow on and along the Mid-Atlantic Ridge (Lee and Uyeda, 1965) we find that the Iceland data as implied by the temperature gradients displayed in Fig. 2 are of a similar magnitude but somewhat more evenly distributed. In part, this may result from the complications due to the presence of two or possibly three rift axis in Iceland.

There are a number of continental zones of crustal strain. The two main zones in the coterminous United States, the Great Basin and the Rio Grande Rift zone, provide interesting and in many ways representative data. The Great Basin is an area of about 500,000 km^2 with an average conduction heat flow of 92 mW/m^2 (Sass et al., 1980) and including a subregion of about 100,000 km^2, the Battle Mountain High, where the values rise to 100 to 150 mW/m^2. Conductive heat flow in an approximately 100 km wide strip along the Rio Grande rift is in the range of 100 to 120 mW/m^2. The situation at the Rhine graben in Germany is rather similar (Cermak and Ryback, 1979). Comparing these results with the figures for Iceland in (1.b) of Table I indicates that Iceland has a similar or only a slightly higher conduction heat flow level than continental zones of crustal strain.

The situation with regard to volcanic and thermal activity is obviously quite different. Although, the data on the average global heat advection by such outlets are very incomplete, there is little doubt that Iceland accounts for a noticeable fraction of the total rate of energy release. Verhoogen (1946), estimated the total

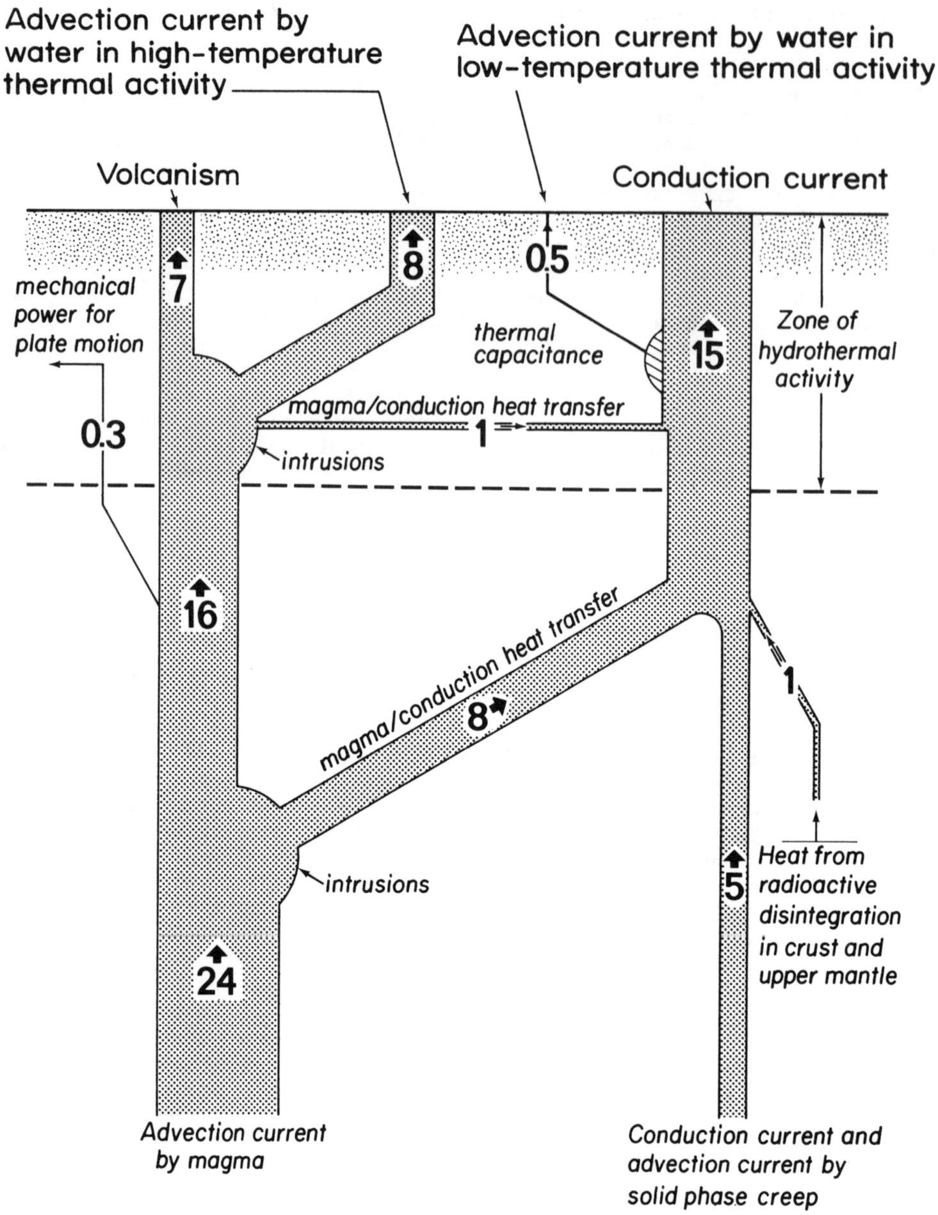

Fig. 7. Energy current graph for Iceland. The figures indicate current magnitude in GW = 10^9W.

global production of subaerial basaltic lava during post-Cambrian time (5×10^8 years) at $3\times10^7 km^3$. Assuming a heat release of $1.6\times10^6 J/Kg$ this corresponds to an average rate of heat advection by magma of about 10 GW. Since the figure does neither include gas, tefra nor silicic volcanism we estimate the average total post-Cambrian heat advection current by subaerial lava to be 30 GW. Assuming that the uppermost one kilometer of the sea floor has been formed by submarine volcanism during the last 2×10^8 years, the corresponding figure for submarine volcanism is found to be 300 GW. These figures are entered in (3.d) of Table I.

Very little data is available on the present global rate of heat advection by thermal activity. It is quite obvious that the principal surface activity is concentrated in very few areas such as Iceland; Yellowstone Park, Wyoming, U.S.A.; Central American volcanic belt, Taupo graben in New Zealand, similar areas in Indonesia, Philippines and Japan, etc. The present writer estimates that there are about 10 regions of comparable activity as Iceland. The global rate of advection by subaerial thermal activity may thus be of the order of 100 GW as entered into (4.d) of Table I. No data is available on submarine thermal activity.

A few additional data have been entered in Table I. First, the estimate of Kanamori (1977) of the average global rate of release of mechanical energy by seismic processes during the past 7 decades has been entered in (5.d). Moreover, estimating the total global length of extension or rift axes at $5x10^4$ km, the figures in (2.c), (5.c) and (6.c) have been multiplied by $5x10^4$ and entered in column (e). The resulting figures indicate the global energy release by the axis processes if the figures for Iceland are taken as representative averages. Finally, we estimate the heat release by sea floor spreading in Iceland on the basis of the model of Parker and Oldenburg in Figure 4b above. Assuming a spreading velocity of one cm/year and including only material accumulation from the surface down to depth of 15 km, that is, to the bottom of the inferred underplating layer of Beblo and Bjornsson (1980), we obtain the figure of 30 MW/km as entered in row (7).

Table I assembles our present estimates and other relevant data. This material prompts the following comments.

(1) The present coastal contours of Iceland are really irrelevant to the energy current picture. In particular, the conduction flow anomaly is likely to extend beyond the present boundary out on the continental shelf. The figures in (2.c) and (2.e) are therefore to be taken as underestimates, possibly by a factor of two. Moreover, (6.c) would then increase to 112 MW/k and (6.e) to 5,600 GW.

(2) Although the results in Table I are rough estimates that have been obtained from a number of mostly unrelated field data, the figures, nevertheless, reflect a rather satisfactory degree of consistency. It is interesting that (2.c) and (7.c) turn out to be equal.

(3) While, as already stated, the average conduction heat flow in Iceland is not much in excess of the values measured in other zones of crustal strain, and only 2.5 times the global average, Iceland accounts in our estimate for about 1/5 of the average global rate of advection by subaerial lava and about 1/10 of the global rate of advection by subaerial thermal activity. It would therefore appear that the excessive volcanism and thermal activity are rather the result of local conditions such as low crustal stresses and high crustal permeability than the result of a very large input of heat.

(4) It is noteable that the two figures for the mechanical power, (5.d) and (5.e) that are based on completely different sets of data material turn out to be of equal magnitude. Both represent fractions of the total power expended for global crustal dynamics. The seismic power represents a fraction that is being converted into strain energy and subsequently released as seismic energy.

(5) It is quite interesting that the figures in (6.3), even if it is raised to 5,600 GW, is only about 1/5 of the total global conduction current in (1.d). In other words, the estimated total supernormal rate of heat release at the global rift system is not a large fraction of the global conduction current. This is surprising in view of the fact that the rift processes appear to dominate the global crustal dynamics.

(6) Lachenbruch (personal communication) has pointed out a useful alternative to entry (7.c). Observations of ocean topography, heat flow, and hydrothermal loss can be interpreted in terms of a variety of simple models of plate evolution (e.g., Sclater et al., 1980, Table 2). On this basis, we obtain an estimate of the sum of the conduction and average hydrothermal advection flows of 475 $t^{-1/2}$ mW/m^2 where t is age of the sea floor in m.y. Hence, by a simple integration we obtain for the average rate of heat loss in a region 300 km wide centered on a ridge with spreading (half) rate of 1 cm/year the flow $473(2/15)15^{1/2}$ = 244 mW/m^2. The corresponding value from Table I is 238 mW/m^2 or 298 mW/m^2 depending on whether one includes advection by subaerial lava. In either case, the result says that heat loss from Iceland is typical of heat loss from any other area "15 m.y. wide" centered on a spreading center.

References

Arnason, B., Groundwater systems in Iceland traced by deuterium, Societas Scientiarium Islandica, Reykjavik, 1976.

Båth, M., Crustal structure of Iceland, J. Geophys. Res., 65, no. 6, pp. 1797-1807, 1960.

Beblo, M. and A. Bjornsson, A model of electrical resistivity beneath NE-Iceland, correlation with temperature, J. Geophys., 47, 184-190, 1980.

Bjornsson, A., Johnson, G., Sigurdsson, S., Thorbergsson, G. and E. Tryggvason, Rifting of the plate boundary in North-Iceland 1975-1978, J. Geophys. Res., 84, 3029-3038, 1979.

Bjornsson, H., Bjornsson, S. and T. Sigurgeirsson, Geothermal effects of water penetrating into hot rock boundaries of magma, submitted to Science, 1980.

Bodvarsson, G., Report on the Hengill Thermal Area, J. Eng. Assoc., Iceland, 36, pp. 1-49, 1951.

Bodvarsson, G., Terrestrial heat balance in Iceland, J. Eng. Assoc. Iceland, 39, 69-76, 1954.

Bodvarsson, G., Physical characteristics of natural heat sources in Iceland, Geothermal Energy, 1-Proc. UN Conf. on New Sources of Energy, Rome, 1961, 2, New York, 1964.

Bodvarsson, G., Geothermal resource energetics, Geothermics, 3(3), 82-92, 1974.

Bodvarsson, G., Convection and Thermoelastic Effects in Narrow Vertical Fracture Spaces with Emphasis on Analytical Techniques, Final Report for U.S.G.S., 111 pp., 1978.

Bodvarsson, G., Elastomechanical phenomena and the fluid conductivity of deep geothermal reservoirs and source regions, Workshop on Geothermal Reservoir Engineering, December 1979,

Stanford University, Stanford, California, 1979.
Bodvarsson, G. and G.P.L. Walker, Crustal drift in Iceland, Geophys. J.R. Astr. Soc., 8, 285-300, 1964.
Bodvarsson, G. and R.P. Lowell, Ocean-floor heat flow and the circulation of interstitial waters, J. Geophys. Res., 77, 4472-4475, 1972.
Bottinga, Y. and C. Allegre, Geophysical, petrological and geochemical models of the oceanic lithosphere, Tectonophysics, 32, 9-59, 1976.
Carslaw, H.W. and J.C. Jaeger, Conduction of Heat in Solids, 2nd ed., Claredon Press, Oxford 496 pp., 1959.
Cermak, V. and L. Ryback (editors), Terrestrial Heat Flow in Europe, Springer-Verlag, Berlin, 1979.
Einarsson, T., Ueber das Wesen der heissen Quellen Islands, Societas Scientiarum Islandica, Reykjavik, 1942.
Hermance, J.F. and J. Pedersen, Deep structure of the Rio Grande rift: a magnetotelluric interpretation, J. Geophys. Res., 85, 3899-3912, 1980.
Kanamori, H., The energy release in great earthquakes, J. Geophys. Res., 82, 2981-2987, 1977.
Lachenbruch, A.H., A simple model for oceanic spreading centers, J. Geophys. Res., 78, 3395-3417, 1973.
Lachenbruch, A.H., Dynamics of a passive spreading center, J. Geophys. Res., 81, 1883-1902, 1976.
Lachenbruch, A.H., Heat flow in the Basin and Range province and thermal effects of tectonic extension, P.A. Geoph., vol. 117, Birkhauser Verlag, Basel, pp. 34-50, 1979/1980.
Lachenbruch, A.H. and J.H. Sass, Models of an extending lithosphere and heat flow in the Basin and Range province, in Geol. Soc. Amer. Memoir 152, pp. 209-250, 1978.
Langseth Jr., M.G., Le Pichon, X., and M. Ewing, Crustal structure of mid-ocean ridges, J. Geophys. Res., 71, 5321-5355, 1966.
Lee, W.H.K. and S. Uyeda, Review of heat flow data, in Terrestrial Heat Flow (W.H.K. Lee, ed.), Geophys. Monogr. Am. Geophys. Union, 8, 87-190, 1965.
Lister, C.R.B., On the penetration of water into hot rock, Geophys. J.R. Astr. Soc., 44, 508-521, 1974.
Lister, C.R.B., Qualitative theory on the deep end of geothermal systems, Proceedings, 2nd UN Symposium on the Development and Use of Geothermal Resources, San Francisco, CA, p. 459-463, 1976.
McKenzie, I.D., Some remarks on heat flow and gravity anomalies, J. Geophys. Res., 72, 6261-6273, 1967.
Palmason, G., Crustal Structure of Iceland from Explosion Seismology, Soc. Sci. Islandica, Publ. 40, 187 p., 1971.
Palmason, G., Kinematics and heat flow in a volcanic rift zone, with application to Iceland, Geophys. J. R. Astr. Soc., 33, 451-481, 1973.
Palmason, G., Heat flow and hydrothermal activity in Iceland, in Geodynamics of Iceland and the North Atlantic Area, (ed. L. Kristjansson) D. Reidel Publishing Co., Hordrecht-Holland, 297-307, 1974.
Palmason, G., A continuum model of crustal generation in Iceland; Kinematic aspects, J. Geophys., 47, 7-18, 1980a
Palmason, G., Crustal rifting and related thermomechanical processes in the lithosphere beneath Iceland, Geologiche Rundschau (in press), 1980b.
Palmason, G. and K. Saemundsson, Iceland in relation to the Mid-Atlantic Ridge, Ann. Rev. Earth Planet. Sci., 2, 25-50, 1974.
Palmason, G. and K. Saemundsson, Summary of conductive heat flow in Iceland, in Terrestrial Heat Flow in Europe (V. Cermak and L. Rybach, eds.), Springer-Verlag, p. 218-220, 1979.
Parker, R. and D. Oldenburg, Thermal model of ocean ridge, Nature, 242, 137-139, 1973.
Saemundsson, K., Evolution of the axial rifting zone in northern Iceland and the Tjornes fracture zone, Geol. Soc. Am. Bull., 85, 495-504, 1974.
Saemundsson, K., Outline of the geology of Iceland, Jokull, 29, 7-28, 1979.
Sass, J.H., Blackwell, D.D., Chapman, D.S., Costain, J.K., Decker, E.R., Lawver, L.A. and C.A. Swanberg, Heat flow from the crust of the United States, in Physical Properties of Rocks and Minerals, (Y.S. Touloukian, W.R. Judd and R.F. Roy, eds.), McGraw Hill (in press), 1980.
Sclater, J.G., Jaupart, C. and D. Galson, The heat flow through oceanic and continental crust and the heat loss of the earth, Rev. Geophys. & Space Phys., 18(1), pp. 269-311, 1980.
Sneddon, I.N. and M. Lowengrub, Crack Problems in the Classical Theory of Elasticity, John Wiley and Sons, Inc., New York, 207 pp., 1969.
Thorkelsson, T., On Thermal Activity in Iceland, Reykjavik, 1941.
Verhoogen, J., Volcanic Heat, Amer. J. Sci., 745-771, 1946.
Ward, P.L. and S. Bjornsson, Microearthquakes, swarms and the geothermal areas of Iceland, J. Geophys. Res., 76, 3953-3982, 1971.
White, D.E., Hydrology, activity and heat flow of the Steamboat Springs Thermal System, Washoe County, Nevada, U.S. Geological Survey, Professional Paper 458G, 109 p., 1968.

MASS DISPLACEMENTS AND GEOTHERMICS WITHIN THE UPPER MANTLE OF THE RHINEGRABEN RIFT SYSTEM: A MODEL TO INTERPRET DATA FROM GRAVIMETRY AND SEISMOLOGY.*

D. Werner,[1)] H.-G. Kahle,[2)] J. Ansorge[1)] and St. Mueller[1)]

1) Institut für Geophysik, ETH-Hönggerberg, CH-8093 Zürich, Switzerland

2) Institut für Geodäsie und Photogrammetrie, ETH-Hönggerberg, CH-8093 Zürich, Switzerland

Abstract. In this paper a deep-seated geothermal anomaly beneath the Rhinegraben rift system is considered in order to resolve the discrepancy between the crustal structure derived by seismological methods and the observed gravity anomalies. The crustal structure is mainly characterized by a pronounced elevation of the M-discontinuity. The minimum crustal thickness of 24 km is encountered in the southern part of the graben. The cumulative gravity effect of the topography of the M-discontinuity and other crustal influences would result in a significant Bouguer anomaly of the order of 100 mgal (1 mgal = 10^{-5} ms^{-2}). Since this expected gravity anomaly is not observed the existence of a compensating mass has to be postulated. An attempt is made to relate this effect of compensation to a thermally induced negative density change. A corresponding geothermal anomaly has been designed by solving the heat transfer equation for a moving medium. The resulting temperature field is based on a kinematic model in which hot mantle material is upwelling. The mass displacement is in the order of 1 mm/year and the duration of this process is assumed to be 45 m.y. which corresponds approximately to the age of the Rhinegraben. In addition to accounting for the gravity discrepancy our model can also explain the thinning of the lithosphere underneath the Rhinegraben.

1. Introduction

The Rhinegraben between Basel and Frankfurt is the most pronounced surface expression of an extended continental rift system transecting the crust and upper mantle of Central Europe. In this paper an attempt is made to arrive at a kinematic model for the mass displacements associated with the rifting process by combining data from gravimetry, seismology and geothermics. A similar problem has been discussed by Lachenbruch [1976] and Lachenbruch and Sass [1978]. The detailed history of the graben formation had to be simplified in our model. It is mainly based on the postulate that a deep-seated thermal anomaly is present in the Rhinegraben rift which extends down into the asthenosphere. The temperature distribution of this anomaly is calculated by using the following geophysical temperature indicators:

The observed Bouguer gravity anomalies are not concordant with results from deep seismic soundings. The density distribution in the crust and uppermost mantle derived from seismological results leads to a calculated positive gravity anomaly which exceeds the observed gravity field significantly. This discrepancy reaches values up to +100 mgal (= 10^{-3} ms^{-2}) in the southern part of the Rhinegraben. It is mainly caused by the updoming of the crust-mantle boundary. This discrepancy can be explained if the temperature dependence of density is taken into account.

A second information about the temperature field can be obtained from the topography of the boundary between lithosphere and asthenosphere [Werner, 1977]. Recent studies of seismic surface-wave dispersion in Central Europe [Sprecher, 1976; Panza, Mueller and Calcagnile, 1980] have shown that the thickness of the lithosphere beneath the Rhinegraben area is considerably decreased as compared to the regions around the rift system.

Furthermore the relationship between electrical conductivity and temperature may serve as a temperature indicator. Magnetotelluric investi-

*) Contribution No. 232 of the Institute of Geophysics, ETH-Hönggerberg, CH-8093 Zürich, Switzerland
Separata No. 12 of the Institute of Geodesy and Photogrammetry, ETH-Hönggerberg, CH-8093 Zürich, Switzerland

gations have revealed a zone of increased electrical conductivity under the Rhinegraben [Reitmayr, 1975]. Again this observation can be interpreted by a deep-seated thermal anomaly. Observed geomagnetic anomalies also seem to indicate a positive correlation with the proposed thermal model [Bosum et al., 1977].

These geophysical findings taken independently from each other demonstrate that it is reasonable to postulate the presence of an important thermal anomaly in the Rhinegraben rift system. In the following section the kinematic model and the calculation of the geothermal anomaly are described. This discussion is followed by a review of seismological findings (section 3) and the presentation of the gravity problem mentioned above.

2. Kinematics and geothermics - derivation of a model

In order to find an explanation for the genesis of the postulated geothermal anomaly we consider a large-scale, long-time convection process, i.e. we assume that hot mantle material rises from the asthenosphere beneath the graben (Figure 1). Such a kinematic model is in accordance with the concept for the evolution of rift systems [see e.g. Mueller, 1978; Illies, 1978].

For the mass displacement field (field of the velocity vector $\vec{v}$) a mathematical expression has to be found which satisfies the following conditions:

A. The velocity $\vec{v}$ must be zero at the earth's surface, and the crust should not be significantly deformed.

B. $\vec{v}$ is nearly zero at the base of the asthenosphere (depth about 250 km).

C. The field $\vec{v}$ has to describe the upwelling mantle material beneath the graben in such a way that the resulting temperature field (and the corresponding density field) compensates the missing gravity anomaly.

D. The simplified law of mass conservation $\operatorname{div} \vec{v} = 0$ has to be satisfied.

A suitable formula for such a mass displacement field $\vec{v} = (v_x, v_z)$ - we consider only the two-dimensional case - can be written as follows [Werner and Kahle, 1980]:

$$v_x = 2 \sum_n u_n a_n (z - b_n z^3) \operatorname{atan} (a_n x) \exp (-b_n z^2)$$
$$v_z = \sum u_n a_n^2 z^2 (1 + a_n^2 x^2)^{-1} \exp (-b_n z^2) \qquad (1)$$

where
x, z = horizontal and vertical coordinates,
u_n, a_n, b_n are constants.

The field is symmetrical with respect to the graben axis at $x = 0$ (see Figure 1, left part). The velocity field is assumed to be constant during the evolution of the rift system, i.e. since about 45 million years [Illies, 1974]. We are aware that this assumption can approximate the rifting process only to a certain degree. According to Illies [1978] the rift system came to a standstill at the end of Miocene. It is presently not an active rift, as deduced from geological and tectonic evidence. It has preserved the structural features of a relatively early stage in the evolutionary process of transformation from continental to oceanic lithosphere [Mueller, 1978].

In our relatively simple model based on a pure kinematic supposition, the material of the crust, the lower lithosphere and the asthenosphere is assumed to be uniform. The model cannot explain the driving mechanism of the rifting process. Furthermore the influence of volcanic processes is not considered, i.e. the existence of the Kaiserstuhl volcano in the southern part of the graben is not included in our model. Details in the history of the rift formation are also neglected. We are only interested in the present temperature distribution in order to calculate the thermally induced density effect. Therefore it would be sufficient to assume an appropriate temperature field without defining its cause. To be more realistic our kinematic model has been developed in such a way that the deep-seated temperature anomaly (Figure 1, right part) can be attributed to a plausible upper mantle process, i.e. it is possible to calculate a corresponding temperature field from the mass displacement field. For this purpose we consider heat transport in a moving medium which consists of two processes: convection and thermal conduction. In this case the following differential equation has to be solved:

$$\vec{v} \operatorname{grad} T + \frac{\partial T}{\partial t} = \kappa \nabla^2 T + \frac{A}{c\rho} \qquad (2)$$

with:
$T = T(x,z,t)$ = temperature,
t = time,
$\vec{v} = \vec{v}(x,z)$ = mass velocity vector,
κ = thermal diffusivity,
$A = A(z)$ = radiogenic heat production,
c = specific heat,
ρ = density.

Our model originates from a formerly stationary temperature field. It is assumed that no geothermal disturbances existed before the initiation of the process. The radiogenic heat sources are confined to the earth's crust. According to the kinematic model the temperature calculations are two-dimensional and symmetrical with respect to the graben axis at $x = 0$.

The model calculation was based on a finite difference method [Werner, 1975]. A solution of this problem after Werner and Kahle [1980] is shown in Figure 1, right part. The isotherms re-

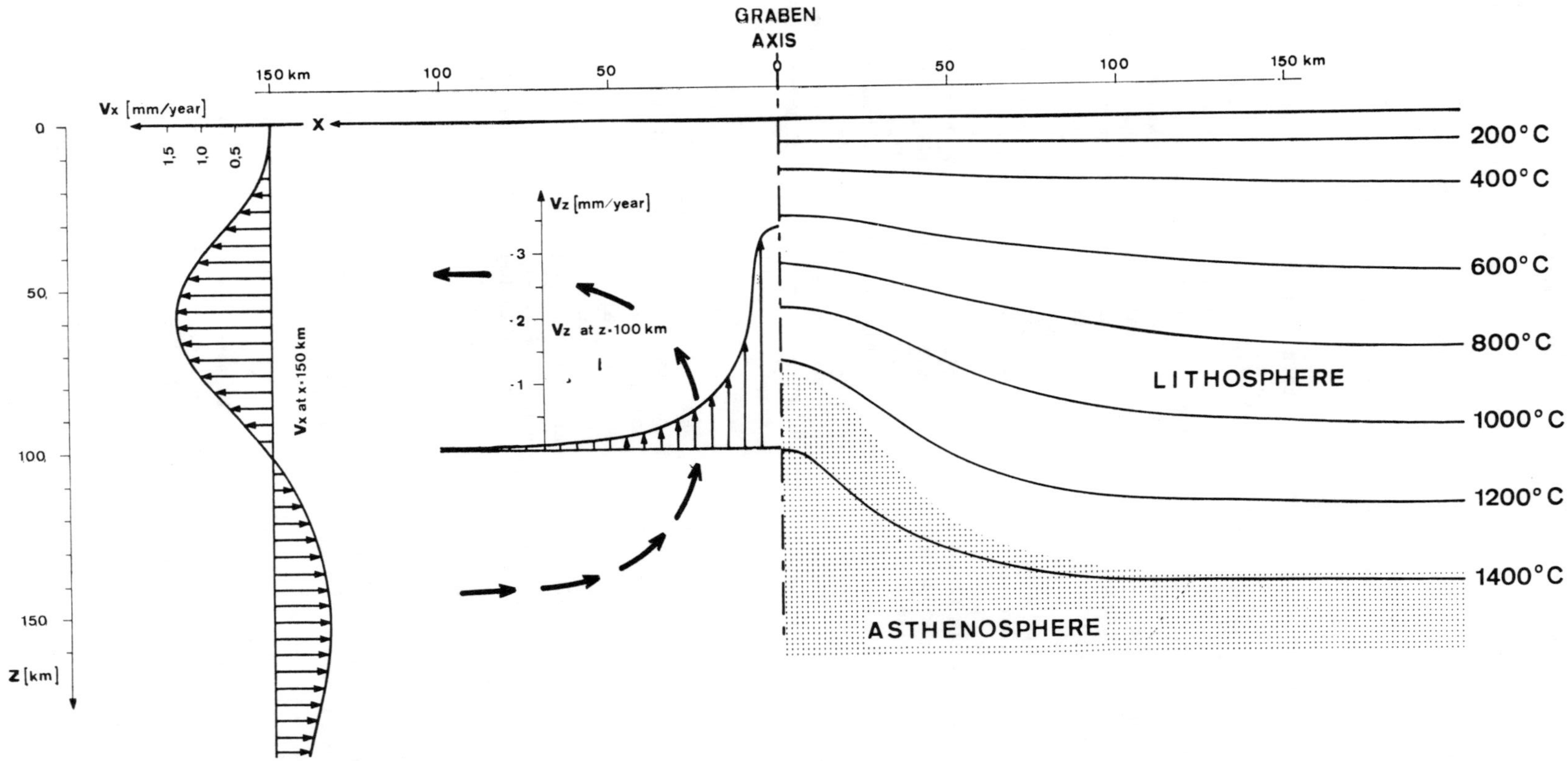

Fig. 1: Model of material and heat transport in the uppermost mantle

Left: The velocity field $\vec{v} = (v_x, v_z)$ for the upwelling mantle material under the rift zone which produces a thermal anomaly.

Right: The temperature field after 45 m.y. caused by convection and thermal conduction; the indicated boundary between lithosphere and asthenosphere (stippled area) depending on temperature and pressure is taken from Werner and Kahle [1980].

present the temperature field T(x,z,t) with t = 45 million years. The result obtained satisfies mainly the gravity discrepancy and the thinning of the lithosphere. The present-day boundary between lithosphere and asthenosphere based on the temperature model [see Werner and Kahle, 1980] is also indicated in Figure 1. The thinned lithosphere derived from our temperature model agrees with seismological results, as will be described in the next section.

3. Seismological findings

During the past ten years a detailed study of all available seismic refraction data has led to a regional picture of the crustal structure in the Rhinegraben area [Rhinegraben Research Group, 1974; Edel et al., 1975; Meissner et al., 1974, 1976; Prodehl et al., 1976]. The published interpretations show that the average P velocity in the upper and middle crust is relatively low with strong indications for the existence of a pronounced zone of low velocity within that part of the crust. A second low-velocity zone apparently lies immediately above the crust-mantle boundary superimposed by a pronounced fine-grade lamination [Mueller, 1977]. Reflection seismic measurements [Dohr, 1970, 1972] showed that the laminated transition structure seems to exist between the bottom of the second low-velocity zone and the top of the anomalous mantle under the rift [Mueller et al., 1973]. This fine-grade lamination may be due to numerous basic intrusions into the lower crust.

All these features can be considered as typical for the initial taphrogenic phase in the rifting and break-up of a continental plate with the subsequent progressive "oceanization" of the attenuated lithosphere. The Rhinegraben zone forms the western part of a larger area (Southern Germany) in which convincing evidence of a pronounced seismic velocity anisotropy within the uppermost mantle has been found [Bamford, 1976, 1977; Fuchs, 1975, 1977]. The maximum P_n velocity of 8.43 km/s has been observed in the direction N 20°E which is roughly the strike of the graben axis, in contrast to 7.77 km/s perpendicular to this direction. Observations along several long-range seismic profiles provided some insight into the velocity-depth distribution of P-waves in the upper part of the mantle [Ansorge, Bonjer & Emter, 1979]. Beneath the Rhinegraben and to the east of it the P-wave velocity increases from 8.0 km/s at the crust-mantle boundary to 8.6 km/s at a depth of 47 km with an average velocity of 8.4 km/s for this depth interval. No such high velocities have been observed in the uppermost mantle to the west of the Rhinegraben rift system.

Dispersion measurements on teleseismic Rayleigh waves across numerous two-station profiles in the European-Mediterranean area have recently been inverted uniformly to shear velocity-depth distributions based on a novel scheme of regionalization [Panza, Mueller & Calcagnile, 1980]. As a result of this inversion large lateral heterogeneities became apparent within the upper mantle of Europe. One of the more conspicuous features which could be outlined by this procedure is the so-called "Central European Rift System" extending from the North Sea to the Golfe du Lion. The Rhinegraben as the prominent segment of this continental rift system is characterized by a minimum lithospheric thickness of about 50 km in the Kaiserstuhl region with sizably lowered velocities both in the lithosphere and the asthenosphere. The generalized cross-section through the lithosphere-asthenosphere system in the central part of the Rhinegraben (Figure 1, on the right) based on our thermal model agrees with the result from the analysis of surface-wave observations presented by Sprecher [1976].

The increasing number of crustal seismic refraction profiles in Central Europe [see e.g. Giese, 1976] has also aided in elucidating the deep structure under the Rhinegraben rift system [Rhinegraben Research Group, 1974; Edel et al., 1975; Prodehl et al., 1976]. In Figure 2 a regional map of the crustal thickness has been compiled using all the available information from explosion seismology. The most conspicuous feature in the contour map of Figure 2 depicting the relief of the crust-mantle boundary is the mantle upwarp centered on the southern Rhinegraben with a depth of only 24 km in the Kaiserstuhl region west of Freiburg/Breisgau. The crustal thickness increases from 24 to 28 km towards the north-northeast along the strike of the Rhinegraben axis and falls off to about 30 km on both flanks of the rift system. Towards the southeast the crust-mantle boundary dips very rapidly as the Alps in Switzerland are approached.

The seismological findings described are much more complex than our relatively simple geothermal model. In our calculations only two important facts from seismology are taken into account, namely: The topography of the M-discontinuity (Figure 2), and the seismic velocity structure of the lower crust according to the models of Edel et al. [1975].

4. The gravity problem

In many cases the interpretation of regional Bouguer gravity anomalies has revealed a striking correlation between gravity data and the seismically determined depth of the crust-mantle boundary (M-discontinuity). It has been customary to explain this observation by a pronounced density contrast at the M-discontinuity: An elevated upper mantle causes a positive anomaly (see, e.g. oceanic areas), whereas negative anomalies are

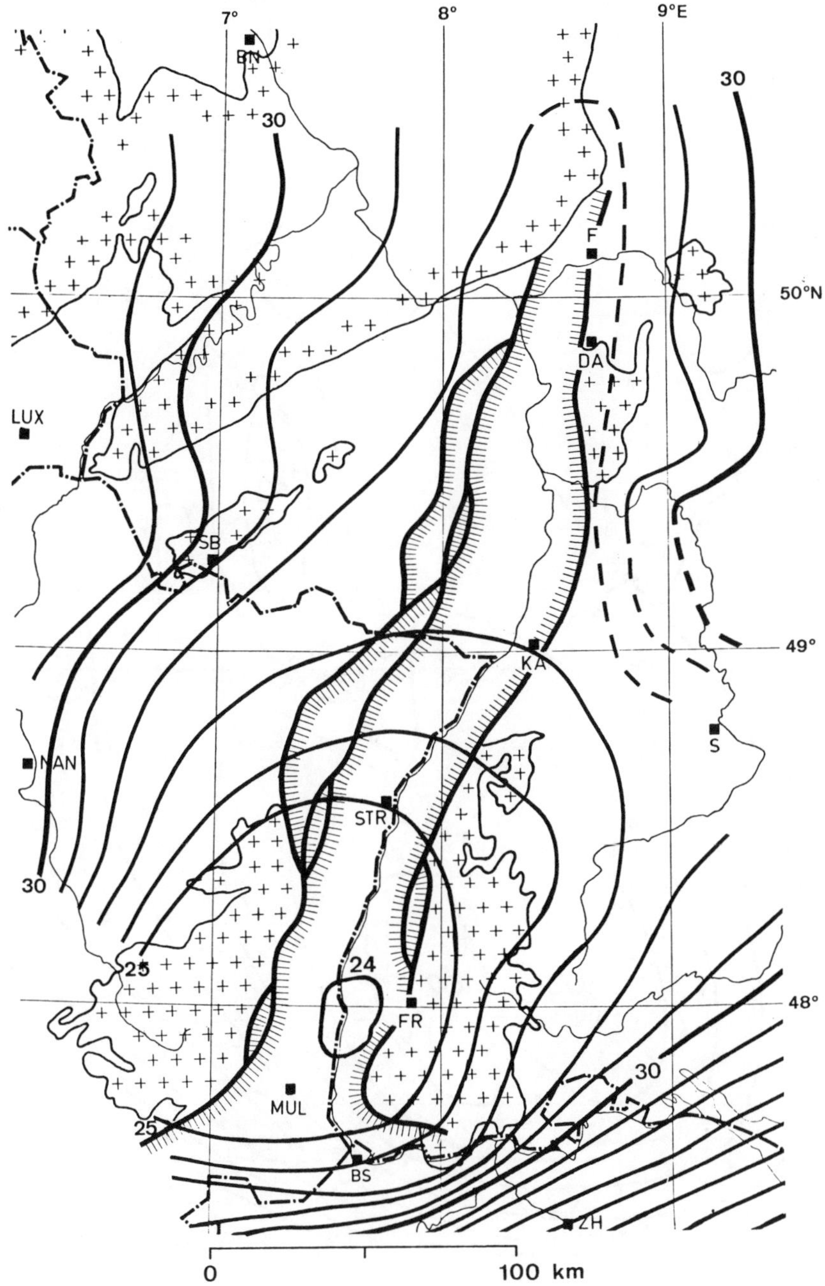

Fig. 2: Contour map of the M-discontinuity after Bamford [1976], Edel et al. [1975], Giese et al. [1973], Giese [1976], Kahle et al. [1976a, b; 1980], Meissner et al. [1974, 1976], Mueller et al. [1976].

BN = Bonn, F= Frankfurt, DA = Darmstadt, LUX = Luxembourg, SB = Saarbrücken, KA = Karlsruhe, S = Stuttgart, NAN = Nancy, STR = Strasbourg, FR = Freiburg i.Br., MUL = Mulhouse, BS = Basel, ZH = Zürich

The updoming crust-mantle boundary in the southern Rhinegraben should produce a significant gravity effect which is not observed (see Figure 4). Crossed areas indicate Hercynian crystalline massifs: The Black Forest and the Vosges Mountains east and west of the graben, respectively. The Kaiserstuhl volcano lies inside the 24 km depth-contour.

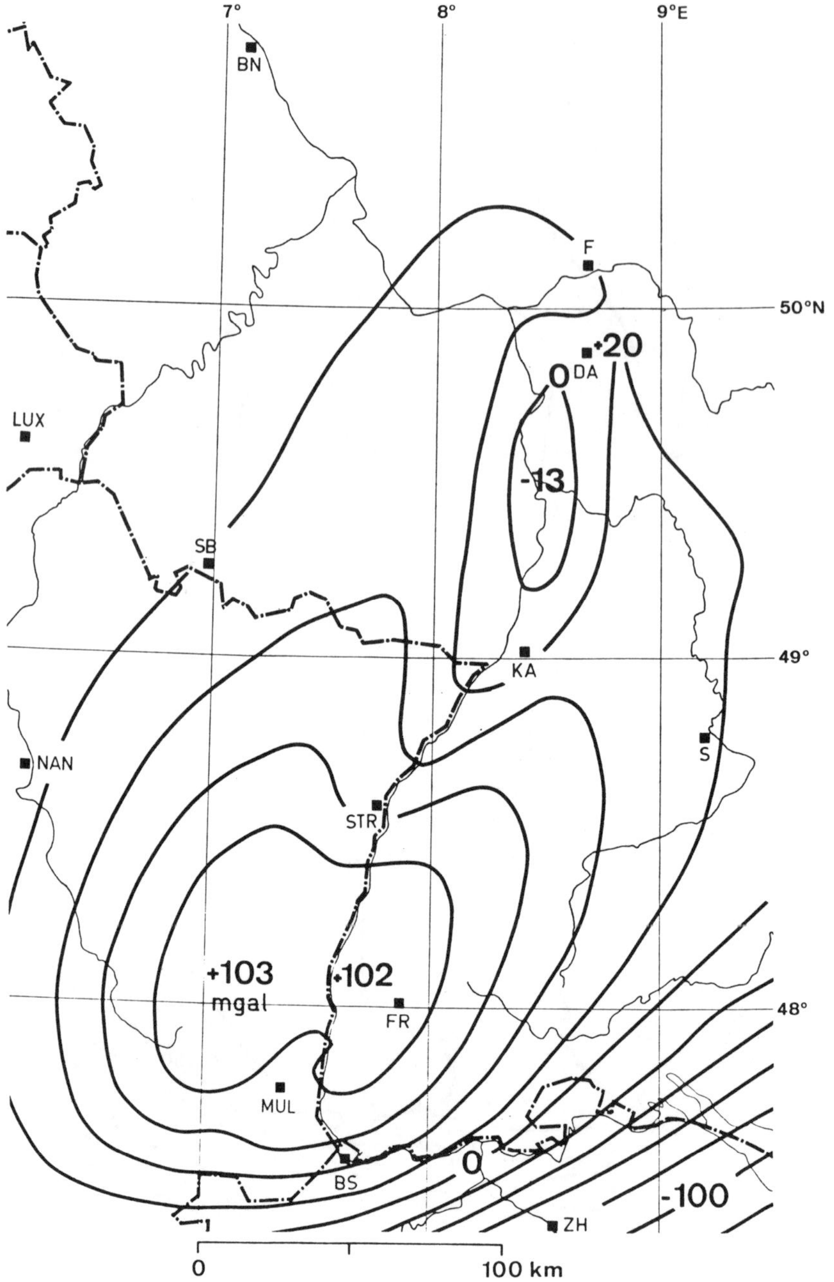

Fig. 3: Calculated gravity anomalies based on seismic and geological data after Kahle and Werner [1980].

most often found over areas associated with crustal thickening (see, e.g. the Alps [Kahle et al., 1976 a,b]). This empiric relationship is significantly violated in the vicinity of the Rhinegraben where the first case (uplifted upper mantle) is encountered (see Figure 2). A comparison with the observed Bouguer anomalies (Figure 4) demonstrates quite clearly that the relief of the M-discontinuity is not expressed in the gravity field. Figure 3 shows the calculated gravity anomaly derived from seismic and geological data. This map is the result of a three-dimensional model calculation taking into account the following effects: (a.) the upwarping of the M-discontinuity, (b.) the higher density in the lower crust, (c.) the root of the Alps, (d.) the

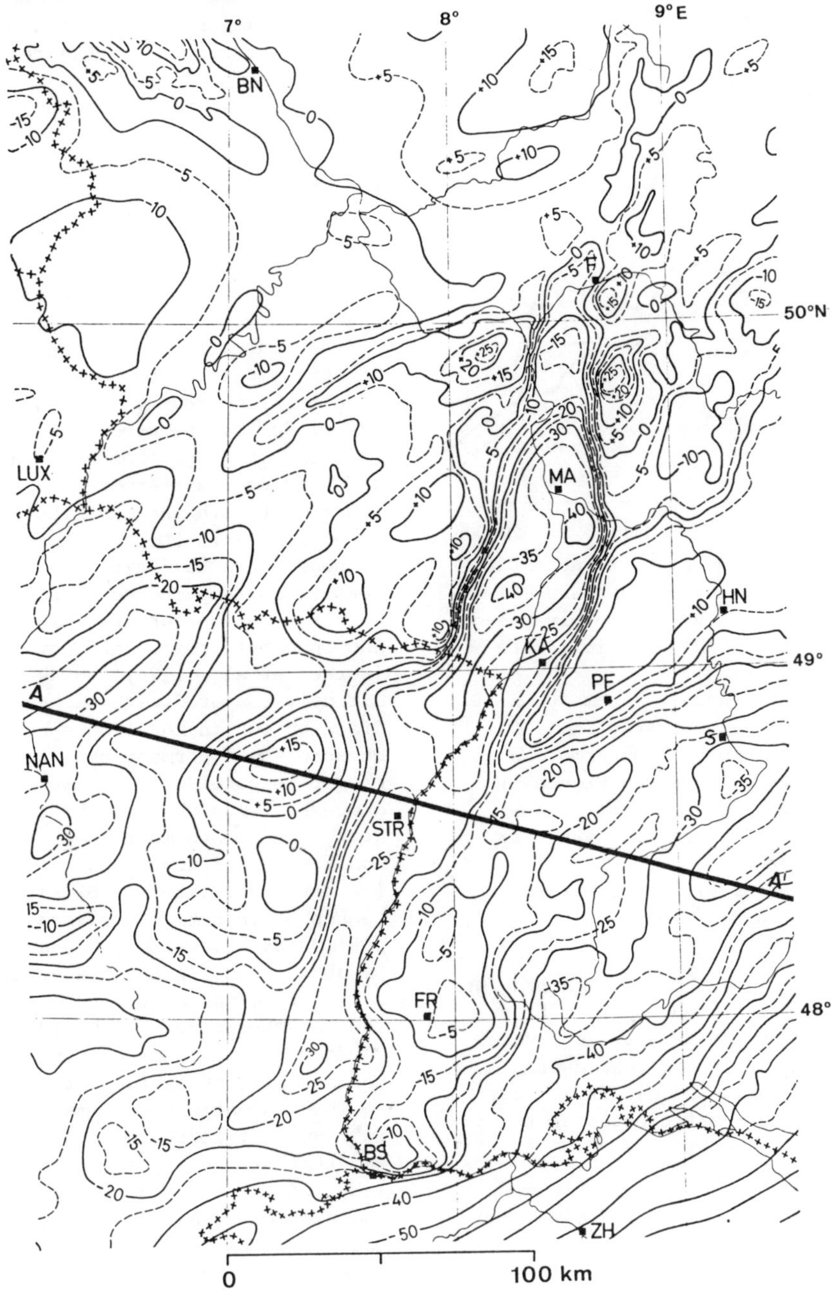

Fig. 4: Observed Bouguer anomalies in the same area (see Figure 2 and Figure 3) after Coron [1958], Gerke [1957], Kahle et al. [1976b], Lecolazet [1970], Mueller, and Rybach [1974]. Values in mgal.

Tertiary sediments in the Rhinegraben and in the Molasse basin, and (e.) the Mesozoic sediments in the vicinity of the graben.
The difference between the calculated and the observed gravity field can be seen by comparing Figure 3 and Figure 4.

This discrepancy has been envisioned already by a number of authors [see, e.g. Mueller et al., 1967; Fuchs, 1974; The Rhinegraben Research Group, 1974; Prodehl et al., 1976] and has led to some speculations on the nature of the compensating masses.

A quantitative approach to the compensation problem introducing a thermally induced density

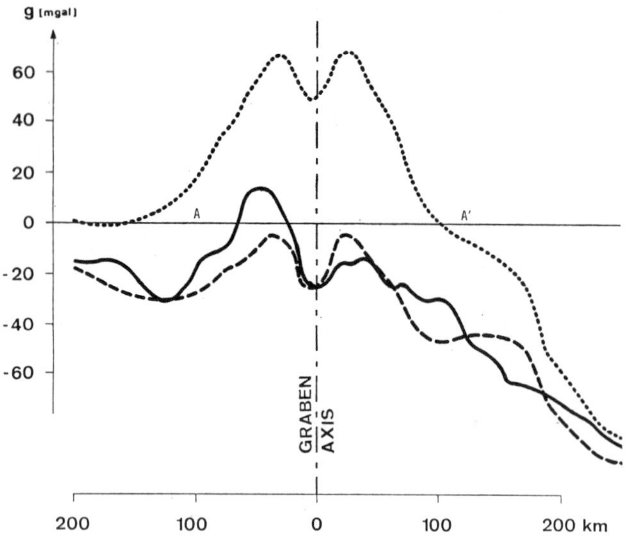

Fig. 5: Profile A - A' (see Figure 4) across the Rhinegraben rift system near Strasbourg (STR) - Dotted line: Calculated gravity anomaly based on seismic and geological data (see Figure 3); Dashed line: Calculated gravity anomaly including the thermally induced density effect taken from the temperature model in Figure 1; Solid line: Observed Bouguer anomaly (see Figure 4). It can be seen that the thermal density effect provides an explanation for the gravity discrepancy.

diminution provides a reasonable explanation for the observed discrepancy. Based on the effect of thermal volume expansion the density is lowered by elevated temperatures. In our case a simple linear relationship between the density change $\Delta\rho$ and lateral temperature differences ΔT is used:

$$\Delta\rho = -\alpha \, \rho_0 \, \Delta T, \quad (3)$$

where $\alpha = 3 \cdot 10^{-5}$ deg^{-1} [see, e.g. Forsyth and Press, 1971] is the thermal volume expansion coefficient, $\Delta T(x,z,t) = T(x,z,t) - T(\infty,z,t)$, and $\rho_0 = 3.30$ g cm^{-3} is the density for $\Delta T = 0$. The temperature field $T(x,z,t)$ is shown in Figure 1; $T(\infty,z,t)$ corresponds to the undisturbed field. By applying equation (3) the anomalous density distribution $\Delta\rho$ is obtained. Details about the model calculations are described by Werner and Kahle [1980] and Kahle and Werner [1980]. The gravity effects of the thermally induced density model have been subtracted from the calculated gravity effects which are caused by the elevated upper mantle and the dense lower crust. For demonstration we consider profile A - A' (compare Figure 4), perpendicular to the graben axis.

This profile crosses the graben near Strasbourg (France) passing through a region north of the minimum depth of the crust-mantle boundary. It is seen in Figure 5 that the thermal density effect can explain the gravity discrepancy described above. Both the amplitudes as well as the lateral extension of the observed regional gravity anomaly can be accounted for by this model.

5. Conclusions

The kinematic and geothermal model described in this paper can be taken as a useful concept in attacking gravity problems. At the same time, the model can be used to explain the observed thinning of the lithosphere in the Rhinegraben rift system. Furthermore the model seems to be concordant with the distribution of electrical conductivity in the uppermost mantle of the Rhinegraben area [Reitmayr, 1975]. In this context it is tempting to compare the present model with thermal measurements in boreholes penetrating into the sedimentary cover of the Rhinegraben [Werner et al., 1978]. However, they cannot be directly compared with the calculated temperature distribution in the lithosphere and asthenosphere because the uppermost crust of the Rhinegraben proper is considerably disturbed by hydrothermal processes. The observed high temperatures in the graben sediments must be explained by a special model of water upwelling from greater depth [Werner and Parini, 1980; Parini, 1981]. Processes of this kind are associated with time periods of about 100'000 years and depth ranges of about 6 km. In this paper we have considered a history of 45 m. y. and a depth range of 250 km. Therefore, we should not combine near-surface thermal observations with the long-term thermal history of the rift system.

The kinematic model (mass displacement field in Figure 1, left-hand side) is chosen in such a way that it can explain both the gravity discrepancy and the thinned lithosphere using geothermal arguments. At the same time it is a plausible model for the rift formation requiring a relatively narrow zone of upwelling mantle material.

Nevertheless our kinematic model can only be considered as a first approximation. It is obviously a simplified assumption that the mass displacement has been constant since 45 m. y. and was zero during the previous time. There are geological and tectonic indications that the horizontal displacements - perpendicular to the graben axis - came to a standstill about 10 m. y. ago [Illies, 1978]. Using this information an improved time - dependent model could be constructed, in which the mass velocities are slightly increased but zero during the last 10 m. y. Such a model leads to a temperature field at present which is nearly the same as sketched in Figure 1. Since the mass displacement field within the upper mantle is actually not known, any similar kinematic models as the one described here can be used in order to solve the gravity problem.

Naturally, the question arises if our Rhinegraben model can also be used to explain other graben systems. A similar gravity problem has to be considered, for instance, in the Gulf of Aden. The geophysical characteristics of this rift system can be explained in the same manner if the mass displacement field for this particular region [Kahle and Werner, 1979; Werner and Kahle, 1981] is taken into account.

Acknowledgments. The authors express particularly thanks to Miss A. Steiniger for competent and painstaking preparation of the "camera-ready" typescript.

References

Ansorge, J., K.P. Bonjer, and D. Emter, Structure of the uppermost mantle from long-range seismic observations in Southern Germany and the Rhinegraben area, Tectonophysics, 56, 31-48, 1979.

Bamford, D., An updated time-term interpretation of P_n-data from quarry blasts and explosions in western Germany, in Explosion Seismology in Central Europe, edited by P. Giese, C. Prodehl, and A. Stein, 215-220, Springer, Berlin, Heidelberg, New York, 1976.

Bamford, D., P_n velocity anisotropy in a continental upper mantle, Geophys. J. R. astron. Soc., 49, 29-48, 1977.

Bosum, W., A. Hahn, E.G. Kind, and R. Pucher, Geomagnetic anomalies in geothermal areas - Rhinegraben and Urach area, in Seminar on geothermal energy, Commission of the European Communities, Brussels, Vol. 1, 277-295, 1977.

Coron, S., Anomalies de Bouguer (1:1 Mill.), Feuille de Paris, Bureau Gravimétrique International, Paris, 1958.

Dohr, G., Reflexionsseismische Messungen im Oberrheingraben mit digitaler Aufzeichnungstechnik und Bearbeitung, in Graben Problems, edited by J.H. Illies and St. Mueller, Schweizerbart, Stuttgart, 207-218, 1970.

Dohr, G., Reflexionsseismische Tiefensondierung, J. Geophys., 38, 193-220, 1972.

Edel, J.B., K. Fuchs, C. Gelbke, and C. Prodehl, Deep structure of the southern Rhinegraben area from seismic-refraction investigations, J. Geophys., 41, 333-356, 1975.

Forsyth, D.W., and F. Press, Geophysical tests of petrological models of the spreading lithosphere, J. Geophys. Res., 76, 7963-7979, 1971.

Fuchs, K., Geophysical contributions to taphrogenesis, in Approaches to Taphrogenesis, edited by J.H. Illies and K. Fuchs, Schweizerbart, Stuttgart, 420-432, 1974.

Fuchs, K., Seismische Anisotropie des oberen Erdmantels und Intraplatten-Tektonik, Geol. Rundschau, 64, 700-716, 1975.

Fuchs, K., Seismic anisotropy of the subcrustal lithosphere as evidence for dynamical processes in the upper mantle, Geophys. J. R. astron. Soc., 49, 167-179, 1977.

Gerke, K., Die Karte der Bouguer-Isanomalen 1:1'000'000 von Westdeutschland, Angewandte Geodäsie, 46, 1957.

Giese, P., C. Morelli, and L. Steinmetz, Main features of crustal structure in western and southern Europe based on data of explosion seismology, Tectonophysics, 20, 367-379, 1973.

Giese, P., Main features of crustal structure in western Germany, in Explosion Seismology in Central Europe, edited by P. Giese, C. Prodehl, and A. Stein, Springer, Berlin, Heidelberg, New York, 195-214, 1976.

Illies, J.H., Taphrogenesis and plate tectonics, in Approaches to Taphrogenesis, edited by J.H. Illies and K. Fuchs, Schweizerbart, Stuttgart, 433-460, 1974.

Illies, J.H., Two stages Rhinegraben rifting in Tectonics and Geophysics of Continental Rifts, edited by I.B. Ramberg and E.-R. Neumann, Reidel, Dordrecht, 63-71, 1978.

Kahle, H.-G., E. Klingelé, St. Mueller, and R. Egloff, The variation of crustal thickness across the Swiss Alps based on gravity and explosion seismic data, Pure Appl. Geophys. 114, 479-494, 1976a.

Kahle, H.-G., E. Klingelé, St. Mueller, and R. Egloff, Gravimetrie, Sprengseismik und Krustenmächtigkeit entlang der Schweizer Geotraverse, Schweiz. Min. Petr. Mitt., 56, 679-684, 1976b.

Kahle, H.-G., and D. Werner, Golf von Aden, Geophysikalisches Modell für ein Riftsystem, paper presented at the 39th Annual Meeting of the German Geophysical Society, Kiel, 1979.

Kahle, H.-G., St. Mueller, E. Klingelé, R. Egloff, and E. Kissling, Recent dynamics, crustal structure and gravity in the Alps, in Earth Rheology, Isostasy, and Eustasy, edited by N.-A. Moerner, Wiley & Sons, Chichester, New York, Brisbane, Toronto, 377-388, 1980.

Kahle, H.-G., and D. Werner, A geophysical study of the Rhinegraben, Part II: Gravity anomalies and geothermal implications, Geophys. J. R. astron. Soc., 62, 631-647, 1980.

Lachenbruch, A.H., Dynamics of a passive spreading center, J. Geophys. Res., 81, 1883-1902, 1976.

Lachenbruch, A.H., and J.H. Sass, Models of an extending lithosphere and heat flow in the Basin and Range province, in Cenozoic tectonics and regional geophysics of the western Cordillera, edited by R.B. Smith, and G.P. Eaton, The Geological Soc. of America, Memoir 152, 209-250, 1978.

Lecolazet, R., La carte gravimétrique de l'Alsace, in Graben Problems, edited by J.H. Illies, and St. Mueller, Schweizerbart, Stuttgart, 233-234, 1970.

Meissner, R., and U. Vetter, The northern end of the Rhinegraben due to some geophysical measurements, In Approaches to Taphrogenesis, Schweizerbart, Stuttgart, 236-243, 1974.

Meissner, R., H. Berckhemer, and A. Glocke, Results from deep-seismic sounding in the Rhine-Main area, In Explosion Seismology in Central Europe, edited by P. Giese, C. Prodehl, and A. Stein, Springer, Berlin, Heidelberg, New York, 303-312, 1976.

Mueller, St., E. Peterschmitt, K. Fuchs, and J. Ansorge, The rift structure of the crust and upper mantle beneath the Rhinegraben, in The Rhinegraben Progress Report 1967, edited by J.P. Rothé, and K. Sauer, Abh. Geol. Landesamt Baden-Wuerttemberg, 6, 108-113, 1967.

Mueller, St., E. Peterschmitt, K. Fuchs, D. Emter, and J. Ansorge, Crustal structure of the Rhinegraben area, in The Structure of the Earth's Crust Based on Seismic Data, edited by St. Mueller, Tectonophysics, 20, 381-392, 1973.

Mueller, St., and L. Rybach, Crustal dynamics in the central part of the Rhinegraben, in Approaches to Taphrogenesis, edited by J.H. Illies, and K. Fuchs, Schweizerbart, Stuttgart, 379-388, 1974.

Mueller, St., R. Egloff, und J. Ansorge, Die Struktur des tieferen Untergrundes entlang der Schweizer Geotraverse, Schweiz. Min. Petr. Mitt., 56, 685-692, 1976.

Mueller, St., A new model of the continental crust, In the Earth's Crust, edited by J.G. Heacock, American Geophysical Union, Monograph 20, 289-317, 1977.

Mueller, St., Evolution of the earth's crust, In Tectonics and Geophysics of Continental Rifts, edited by I.B. Ramberg, and E.-R. Neumann, Reidel, Dordrecht, 11-28, 1978.

Panza, G.F., St. Mueller, and G. Calcagnile, The gross features of the lithosphere - asthenosphere system in Europe from seismic surface waves and body waves, Pure Appl. Geophys., 118, 1209-1213, 1980.

Parini, M., Geothermische Untersuchungen im Oelfeld Landau/Pfalz (Oberrheingraben, Deutschland) Ph.D. thesis, ETH Zürich, 1981.

Prodehl, C., J. Ansorge, J.B. Edel, D. Emter, K. Fuchs, St. Mueller, and E. Peterschmitt, Explosion seismology research in the central and southern Rhinegraben - A case history, in Explosion Seismology in Central Europe, edited by P. Giese, C. Prodehl, and A. Stein, Springer, Berlin, Heidelberg, New York, 313-328, 1976.

Reitmayr, G., An anomaly of the upper mantle below the Rhinegraben, studied by the inductive response of natural magnetic fields, J. Geophys., 41, 651-658, 1975.

Rhinegraben Research Group for Explosion Seismology, The Rhinegraben rift system: deep structure and tectonic features. The 1972 seismic refraction experiment in the Rhinegraben - first results, In Approaches to Taphrogenesis, edited by J.H. Illies, and K. Fuchs, Schweizerbart, Stuttgart, 122-137, 1974.

Sprecher, C., Die Struktur des oberen Erdmantels in Zentraleuropa aus Dispersionsmessungen an Rayleigh-Wellen, Ph.D. thesis, ETH Zürich, 1976.

Werner, D., Probleme der Geothermik am Beispiel des Rheingrabens, Ph.D. thesis, University of Karlsruhe, 1975.

Werner, D., Erdwärme und Lithosphäre im Raum Südwestdeutschland/Schweiz, Naturwissenschaften, 64, 626-631, 1977.

Werner, D., K. Fuchs, M. Parini, G. Sattel, G. Schroth, and F. Doebl, Exploration and interpretation of the strongly anomalous temperature field underneath the Rhinegraben, Publication of the European Communities, Brussels, 1978.

Werner, D., and H.-G. Kahle, A geophysical study of the Rhinegraben, Part I: Kinematics and geothermics, Geophys. J. R. astron. Soc., 62, 617-629, 1980.

Werner, D., and M. Parini, The geothermal anomaly of Landau/Pfalz - an attempt of interpretation, J. Geophys., 48, 28-33, 1980.

Werner, D., and H.-G. Kahle, Geothermics and gravity in continental and oceanic rift systems, J. Volcan. geotherm. Res., 1981, in press.

MUSINGS ON CONTINENTAL RIFTS

Paul Mohr

Department of Geology, University College, Galway, Ireland

Abstract. A review of our understanding of continental rift valleys, with a critical and only slightly jaundiced look at current ideas, is divided amongst the following headings:
Doming, volcanism and rifting,
Episodic development of rifts,
Continental rifts and antecedent structures,
Petrogenetic aspects of rift volcanism:
Wet and dry rifts,
Chemistry,
Daly and other gaps,
Mechanisms of continental rifting.
Some would conclude that much work, more money, and even more travel to exotic places are required to come up with answers.

Introduction

Sufficient unto rift valleys are the existing models thereof. Today's plethora of more or less scientific speculation on continental rifts forcibly encourages us to pause and gain a momentary perspective. We need to sift and digest observations old and new from the many rifts of the world, whilst a drastic winnowing of premature hypothesising should rapidly take care of itself. In a sense, we need to go back to square two and re-examine well-established facts bearing on fundamental aspects of rift genesis.

But first, what is a rift valley? Precise definitions have a habit of becoming embarrassing or irrelevant as the phenomenon being defined becomes better known. Definitions can also be persistently misused to the extent that tradition enshrines an oddball distortion of the original (the word 'mile' is a tart example). With these things in mind, I defer to Elie de Beaumont and J.W. Gregory, and utilise morphology and surface structure alone to define:

Graben: a simple, parallel-sided, downfaulted valley, with a width as little as a few hundred metres or as much as, say, ten kilometres. By simple is meant that a single style of deformation, or none, affects the graben floor. At the larger end of the scale, the term graben overlaps with the term

Rift valley (or just 'rift'): a parallel-sided, downfaulted valley, tens of kilometres in width and at least a few hundred kilometres in length. A rift valley may be simply a large graben, but in most cases the rift floor manifests a complex of smaller-scale graben, horst, warped blocks and important single faults. Graben and rift valleys are commonly asymmetric about their longitudinal axis, and where a faulted margin faces a warped margin, a so-called half graben is developed.

Rift system: a set of rift valleys united through closely parallel, en-echelon or anastomosing links. The structural genesis of a rift system, its constituent rift valleys and their constituent graben and horst is included under the term rifting, synonymous with the more imposing term taphrogenesis. Note that the term rift valley as defined here implies no particular sub-surface processes to cause the required fault-bounded subsidence. Rifting may well occur by more than one process. Nevertheless, mid-oceanic "rifts" are not correctly so named, as their floors are generated in situ rather than by subsidence of pre-existing rocks.

This essay touches briefly on certain fundamental aspects of continental rifts. Briefly, because consider that from a single NATO Symposium on continental rifts held in Oslo in 1977, two excellent volumes have emanated that amount to 59 articles and 740 pages. Even so, they neglect topics such as geophysics and precise geodesy. The approach here is cross-disciplinary, with the intention of provoking questions rather than supplying answers to:

1. relationships between doming, volcanism and rifting - chicken-and-egg or cause-and-effect?
2. episodic development of rifts - time-scale and magnitudes.
3. antecedent structures and rift orientation - grace or predestination?
4. petrogenetic aspects of rift volcanism - 'wet' and 'dry' rifts.
5. mechanism of crustal attenuation - 'active' or 'passive' rifts?

Perspective is needed for a subject that is rapidly becoming minutely detailed on narrow salients, with many synthetic models being based on insufficient understanding of the limitations imposed by existing data from neighboring salients.

Doming, volcanism and rifting

Is there a general pattern, in time and space, to crustal uplift, volcanism and rifting in the evolution of continental rifts? Despite widespread acceptance of one such postulated pattern, the interrelationships of these three phenomena remain surprisingly loosely determined.

Our present generalisations seem to stem from the results of the classic clay-cake modelling experiments of H. Cloos (1939) and E. Cloos (1968): arch-like uplift leads to crustal tension, fracture and extension; the facilitation of magmatic egress by trans-crustal fractures; and keystone-collapse of the crest of the uplift. Hans Cloos' arch became Holmes' (1945) and Quennell's (1960) swell and, via the Kenya Geological Survey, Gass' (1970) dome. This 'dominant hypothesis' has gained increasing acceptance concurrent with semantic evolution. Dainelli (1943), Le Bas (1971), Burke and Whiteman (1973), Gass (1973) and Bailey (1978), amongst others, have emphasised a sequence of domal uplift, alkaline volcanism upon the dome, and then crestal rifting across the dome. This sequence is now being assumed generally applicable to continental rifts, without proper recall to some long-standing observations. Thus Gass (1973) states: "rifts are secondary features, consequent on formation of the primary structures - the crustal domes". Ramberg and Spjeldnaes (1978) are induced to postulate domal uplift for pre-Cainozoic rift systems, even though "direct evidence is lacking" owing to subsequent denudation.

In light of plate tectonics, and heat of mantle hot-spots, Burke and Whiteman (1973) applied the dominant hypothesis to 16 localities within the African plate where oval-shaped doming of the basement (typically 300 x 100 x 1km) is associated with a total of 29 "rifts". They considered these domes to result from prolonged impingement of deep mantle 'plumes' on the underside of the African lithosphere, which has been stationary relative to these plumes since:

1. 25 m.y. ago, according with the positions of Atlantic volcanoes relative to the mid-Atlantic Ridge (Burke and Wilson, 1972), or
2. 40 m.y. ago, according with paleomagnetic evidence, and a presumed 15 m.y. thermal time-constant for the African lithosphere (Briden and Gass, 1974).

Burke and Whiteman consider that about 5 m.y. after initiation of continental rifting, triple rift (rrr) junctions develop. They identify 11 in Africa, 7 of them along the Eastern Rift of Tanzania, Kenya and Ethiopia. These rrr junctions can further develop into rrR, rRR, or RRR junctions as successive transition is made from continental rifting to crustal spreading of oceanic type. Yet several objections can be made to various aspects of this hypothesis.

Even a cursory review of the literature concerning each of Burke and Whiteman's 11 disparate rrr junctions shows that no consensus emerges on a geologically established succession of tectonic and magmatic events. There are enough individual exceptions to the dominant hypothesis to render it valueless. Taking the East African rift system, the continental rift system par excellence, as an example, we can ask which came first, domal uplift or trough-like subsidence of a 'proto-rift depression'? It is important to introduce a distinction here between regional domal uplift and shoulder uplift localised along the rift margins. Shoulder uplift is invariably a corollary of graben subsidence, whatever may be the mechanism of this subsidence, and contributes strongly to the high topography bordering the Eastern and Western Rifts of Africa. Indeed, these two rifts run the eastern and western <u>perimeters</u> of the East African uplift.

Baker et al. (1972), briefly reviewing observations on timing of domal uplift and rifting in the Eastern Rift System of Africa, conclude: "the rift zones were subsiding from the very beginning, and were not first uplifted on the crest of a dome". Evidence includes the thickening of early sedimentary and volcanic formations towards the future domal axis (Fuchs, 1939; Jepsen and Athearn, 1962; Hutchinson and Engels, 1970). King (1978a,b) concurs that, for the Kenya sector of the Eastern Rift, the first tectonic phenomenon was a broad downwarping, and notes that subsequent domal uplift in Kenya has been overestimated by previous workers impressed with the superimposed pile of late-Cainozoic volcanics. Indeed, King goes so far as to claim that domal uplift has been "local and trivial", and Kenyan rift relief has been sufficiently "insignificant" throughout most of its evolution to permit two major episodes of overspill of lavas and sediments beyond the proto-rift depression. So extreme a case can be disputed, but its basis finds support from other continental rifts. Williams (1978), in endorsing King's view on the small magnitude of domal uplift in Kenya, notes that further north in Ethiopia there has been no simple doming, but a complex mosaic of block uplifts superimposed either with shoulder uplift or marginal downwarping (Baker et al. 1972). Furthermore, the Ethiopian rift is positioned markedly off-axis on the eastern side of the Ethiopian uplift, in a manner which contradicts any simple mechanical relationship between crustal doming and rifting.

In the case of the Rhine graben, Illies (1978) has recently de-emphasised domal uplift within the following evolutionary context:

1. broad, gentle crustal subsidence in mid-Jurassic (Bathonian-Bajocian); this is redolent of the Afar story, but in each case a genetic connexion with subsequent Cainozoic events has yet to be proven.

2. undersaturated mafic lavas penetrated the platform crust along the future site of the Rhine graben (mid-Cretaceous).
3. subsidence with resulting brackish lagoonal sedimentation (mid-Eocene).
4. first climax of graben faulting and shoulder upwarping (upper Eocene-lower Oligocene); this episode was contemporaneous with the culmination of the Alpine compressive movements.
5. second climax (upper Pliocene-Quaternary).

Graben subsidence and faulting (episodes 4 & 5) coincided with and were roughly proportional to uplift magnitudes, and progressed northwards with time.

In a review of the Baikal rift, Logatchev (1978) emphasises a two-stage development:

1. Formation of a broad, shallow depression. Slow subsidence kept pace with supply of sediments from gently uplifted margins that lacked significant faulting (mid Eocene-lower Pliocene). Maximum sedimentary fill reached 5 to 7km thickness.
2. Vertical movements quickened by a factor of 6-10 in late Pliocene-Quaternary times, associated with rifting, shoulder uplift and crustal attenuation. Coarse clastic sedimentation, totalling 1 to 1.2km thickness, failed to keep pace with subsidence.

The timing of domal as distinct from shoulder uplift in this sequence remains unclear. Logatchev et al. (1978) identify the Sayan-Baikal uplift through its expression on the elevation profile of the Cretaceous-Paleogene erosion surface, but do not specify its temporal evolution. There is an unusual spatial relationship whereby deepest rift subsidence coincides with minimal uplift at the transverse Selenga saddle, whereas where uplift has been greatest the Baikal gräben branch and dissipate. This appears in contradiction to the direct relationship between degree of uplift and <u>relative</u> rift subsidence observed in the Eastern Rift of Africa; in eastern Africa, however, the rift floors and shoulders bear massive volcanic accumulations buttressed with near-surface basaltic intrusions. The greatest <u>absolute</u> rift subsidence exposed in Africa occurs in sectors of the Western Rift where, as at Baikal, domal uplift has been weak and irregular.

Recent reviews of the evolution of the Rio Grande rift, southwestern U.S.A., have been given by Chapin and Seager (1975) and Chapin (1979). The latter article states: "Rifting began between 32 and 27 m.y. ago... By 26 m.y., the crust along the developing rift has sagged sufficiently to form broad, shallow basins.... Preservation within rift basins of complete pre/early-rift volcanic sections eliminates regional doming as an early rift process" (Chapin, 1979, p.1). Uplift associated with the Rio Grande rift was concentrated in the 7-4 m.y. interval, though it has continued up to the present at a reduced rate. Maximum uplift now totals 1100m. In the case of the similar sized Permian Oslo graben, Ramberg and Larsen (1978) note that an early Paleozoic precursor to the rifting took the form of a 120km-wide, shallow sedimentary trough in which up to 2km of sediment accumulated.

The South Atlantic rifting apart of Africa and South America, initiated during the early Cretaceous, was foreshadowed by basin subsidence as early as the Carboniferous, according to Marsh (1973). In the North Atlantic, deep seismic reflexion profiles of the northern continental margin of the Bay of Biscay (rifted open in late Jurassic and early Cretaceous times) clearly reveal two important relationships (de Charpal et al. 1978):

1. rifting occurred along the line of a pre-existing Permo-Triassic epicontinental sedimentary basin, and
2. lack of erosion of the axis of the Biscay rift system, before or during rifting, suggests "that the rifting was not preceded by and thus is not a consequence of domal uplift".

The early Eocene splitting of Greenland from Europe was not accompanied by major uplift until post-early Oligoene time (Faller & Soper, 1979). Blundell (1978) explicitly dissents from the doming, volcanism, rifting sequence for the Proterozoic Gardar Igneous Complex of SW Greenland, but the time resolution is coarse by Cainozoic standards.

On the other hand, specific objection to precursory crustal sagging comes from Lowell <u>et al.</u> (1975), using the Red Sea rifting as an example. Putting aside Hutchinson and Engel's (1972) evidence for initial downwarping at the southern end of the Red Sea, Lowell claims that the first stage in Red Sea structural evolution was regional <u>arching</u> accompanied by volcanism, because:

1. Younger sediments dip with successively more gentle dip off the eastern flanks of the 2000 x 500km arch of Precambrian rocks (But this tells us nothing about what was happening on the site of the Red Sea).
2. No marine sediments occur indicative of any proto-Red Sea basin, only Oligocene continental sediments and interbedded basalt lavas along the Red Sea margins. (But Maestrichtian marine sediments of the central Red Sea sector are confined/preserved within the present Red Sea basin - Whiteman 1968. The Oligocene sediments and lavas referred to by Lowell are essentially restricted to the present Red Sea basin, just as Oligocene-Miocene marine and continental sediments in Somalia are restricted to the margin of the present Gulf of Aden basin).
3. Normal faults bounding the Red Sea basin are parallel to the coastlines, not perpendicular to the direction of the plate separation (This argues for possible basement structural control, not against

the thesis of an initial crustal sag. In any case, Whiteman (1968) adduces evidence that marginal warping rather than faulting dominates the Red Sea margins).
Lowell concludes that Red Sea rifting was immediately consequent on arching, but it remains doubtful that he has a clear distinction between precursory crustal subsidence and true rifting.

Fischer (1975, p.58) has stated the general case for the proto-rift hypothesis: "Whereas Cloos (1939) and others saw the uplift as the primary feature, and the rift as a secondary one, we now view the rift as a feature of lithospheric magnitude, and the marginal uplifts as a secondarily induced response to the heat flow along the rift". Indeed, Sengör and Burke (1978), following Neumann and Ramberg (1978), have discreetly and prudently moved away from the doming, volcanism, rifting sequence of the 'dominant' hypothesis, as field evidence begins to reimpose itself. Furthermore, to see rifting merely as the result of crustal doming is to ascribe an independent origin to the various sectors of the African and other continental rifts, as indeed is claimed by Le Bas (1971) and Gass (1975). The near-parallelism of rifts within a given rift system must then be ascribed to control by basement structural grain rather than a regional stress field, a question addressed below.

It is not my purpose to argue against the Cloos-Burke-Whiteman concept in all instances. Small radial graben may well develop on small domal uplifts. But the larger, true rift systems of the world have been presaged for as much as 50 m.y. by broad, shallow crustal subsidence. Stresses responsible for rifting are not solely or even largely those produced by regional uplift; rather, there has to have been active horizontal lithospheric separation (Searle, 1970; Illies, 1970; Artemjev and Artyushkov, 1971; Baker et al., 1972; de Charpal, 1978). The amount of horizontal stretching produced by 2000m domal uplift of a 1000km-wide planar medium is only about 600m. The observed separation in Ethiopia, on fault planes alone, is several kilometres. Separation for various rifts can also be estimated from gravity profile data:

Baikal	c. 10km maximum (Logatchev and Florensov, 1978).
Kenya	c. 8-15km maximum (Baker and Wohlenberg, 1971; Swain, 1976; Fairhead, 1976; King, 1978). But Searle (1970) proposes 25km.
Ethiopia	c. 40km in north, 25km in south (Le Pichon and Francheteau, 1978), but 65km in north according to McKenzie et al. (1970). Gravity profiles concur with lowest estimates (Searle and Gouin, 1972).
Rio Grande	50-70km in central sector, by means of basaltic dike injection (Chapin, 1979); but other estimates aired at the 1978 Rio Grande Symposium were c. 15km (L. Russell) and 5-8km (I.B. Ramberg).

How does volcanism fit into the rifting-uplift scenario? Does uplift, where it occurs at all, take place in one or multiple stages? To attempt an answer to these questions, I will focus more sharply on the prime example of a large and active rift system, that of eastern Africa.

Episodic development of rifts

A coarsely defined two-stage development of the Baikal and Rhine rifts has already been mentioned, development that has quickened and gained in magnitude with time. A similar phenomenon in the Rio Grande rift, where a marked 'mid-Miocene lull' in magmatism occurred between 20 and 13 m.y. (Chapin, 1979), may in part be related to the evolution of the Pacific-American plate boundary (Coney & Reynolds, 1977; Eaton, 1979). In the Eastern Rift of Africa, episodic uplift during the Cainozoic has been of successively greater magnitude, culminating in graben faulting associated with the latest, Pliocene-Pleistocene uplift episode (Baker et al., 1972).

In the Kenya rift sector of the Eastern Rift, a gentle arch, established at the time of Gondwana disruption (late Triassic), was rejuvenated during the early Tertiary before initiation of the proto-rift depression (Williams, 1978). From then on, the sequence of volcanotectonic events has expressed an accelerating sequence, with major change occurring about 7 m.y. ago:

50-32 m.y.	basalts of the southwest Ethiopian border region.
21-14 m.y.	Turkana flood basalts erupted around northern termination of the present rift; otherwise the Kenya rift remained magmatically quiescent during this interval.
15-11 m.y.	gentle regional uplift around shallow axial depression (maximum uplift 700m in west, 300m in east); phonolite flood lavas overflow this developing depression.
10±2 m.y.	margin faulting and shoulder uplift on western side of rift depression along virtually its entire length, forming an asymmetric 'half' graben.
c. 7 m.y.	previous alkaline basalt-phonolite volcanic association gives way to mildly alkaline basalt-mugearite-trachyte association, manifestation of an increase in magmatic silica-saturation.
7-5 m.y.	alkali trachyte central volcanism; mafic intrusions in lower crust.
6-2 m.y.	trachyte flood lavas overflow developing trough, which deepens further at:
4 m.y.	regional uplift (maximum of 1500m in central Kenya) accompanied by

	lithospheric attenuation and graben faulting.
3-2 m.y.	weakly alkaline basaltic dikes injected into rift floor.
2-1.6 m.y.	faulting of rift margins and especially inner graben, with further shoulder uplift; rapid accompanying eruption of flood trachytes.
1.6-0.9 m.y.	further faulting-with-volcanism cycles.
0.5 m.y.	grid faulting of rift floor.

Even allowing for the greater resolution of younger events overprinted on older ones, both quickening and spatial narrowing of the tectonic process are evident in the Kenya rift, as though crustal attenuation were approaching some critical value at which a new phenomenon (crustal spreading?) might take over. Also noteworthy is the coincidence of profuse silicic volcanism with episodes of rift subsidence.

Further north, in Afar, there has been "a history of continuous plate separation during which the availability of basalt magma has fluctuated markedly" (Black et al., 1975). Black et al. consider that when injection of basaltic dikes in Afar was intensely active, extensional faulting was weakly or not active, and vice-versa. Baker et al. (1978) concur: "The (Kenyan) rift is the result of mantle diapirism, crustal updoming and faulting, formation of crustal magma reservoirs, and graben formation, followed by repeated episodes of <u>alternating</u> volcanism $[\sigma_z < \sigma_x]$ and faulting $[\sigma_z > \sigma_x]$". The alternations necessarily occur, according to these authors, because "there is no (crustal stress) condition under which magma can ascend, and normal faulting can occur simulataneously". However, no indication is given by either set of authors as to how field observations can resolve significant ($> 10^{5-6}$yrs) alternations from short-term interdigitations of only years or even hours. Also, a distinction must be made between magmatic ascent due to dike injection in attenuating crust, and focussed ascent through upper crust that collapses into large silicic magma chambers.

Black et al. (1975) additionally observe that the intervals between dike and faulting episodes in SE Afar have shortened with time. This parallels the quickening episodicity of Kenyan rift development, succinctly explained by Baker et al. (1978) as follows: When an asthenolith rises through the lithosphere, approaching closer to the surface, uplift becomes progressively shorter in wavelength and bigger in amplitude. The degree of extensional faulting increases with time as the elastic crust thins; the thinner this crust, the lower the yield stress and the faster the strain-rate.

The detailed regional history of Afar shows resemblances to that of Kenya, allowing for a more rapid crustal extension compared with the Kenyan and the poorly known Ethiopian rift valleys. The radiometric datings of Zanettin and Justin-Visentin (1973), Barberi et al. (1975) and Black et al. (1975) for Afar can be summarised as:

? 45-38 m.y.	earliest Trap Series flood basalts on Ethiopian plateau.
?	faulting and warping/tilting in western Afar.
34-19 m.y. (climax at c. 26)	main Trap Series flood basalts, erupted from fissure swarms along warped and attenuated continental crustal margins, progressing southward with time and becoming more alkaline; peralkaline granite intrusions both within and marginal to Afar.
16 m.y.	faulting in E. Afar.
14-10 m.y.	rhyolite lavas and ignimbrites in E. and S. Afar; uplift of plateaux at about this time; lesser episode of marginal dike injection.
9 m.y.	faulting in E. and S. Afar.
8-6.3 m.y.	flood basalts in E. Afar, where stress-field changes from Red Sea to Gulf of Tajura provenance.
4.3-0.6 m.y.	flood basalts covered most of Afar; voluminous ignimbrites in Ethiopian rift.
1.2-0.0 m.y.	uplift of plateaux; Afar belt faulting, marginal graben, Wonji fault belt; quasi-oceanic axial range basalts (still active); trachyte and alkali rhyolite calderas along Ethiopian rift axis.

Comparison of this table with that for the Kenya rift reveals a measure of synchroneity for both tectonic and volcanic events. This would not be unexpected as a corollary of global reorganisations of plate motions, and finds support in worldwide continental rift correlations that may however be spurious or coincidental at the present level of ignorance. It is nevertheless intriguing that the Rio Grande rift began its development 32 to 27 m.y. ago, became established through basin aggregation and unification at 26 m.y., experienced a gap in volcanic activity between 20 and 13 m.y., akin to Afar and Kenya, and experienced a culmination in tectonic activity (subsidence together with margin uplift) between about 7 and 4 m.y. ago (Chapin and Seager, 1975; Chapin, 1979).

Bott (1976), through analysing simple schemes for graben development, concludes that faulting ought indeed to occur in episodes, as observed in Africa. Extension of the lower, ductile crust proceeds regularly (assuming that stress is maintained at the yield-point) but, in the brittle upper crust, stress gradually builds up until crustal rupture and stress-drop occur, after which stress increase recommences. This concept can be combined with the upward-migrating asthenolith concept of Baker et al. (1978), but one feature remains unexplained. Within the episodic, alternating and quickening volcano-tectonic evolution of continental rifts, the <u>volumes</u> of volcanic material have tended to decrease with time, following on a massive initial episode (Baker et al., 1972; Williams, 1972; Logatchev, 1978). This observation cannot be ameliorated on grounds that younger magmatism might involve progressively more intrusive activity at the expense of volcanism. The mid-Tertiary Ethiopian magmas rose to high structural levels through a thicker and lighter-density upper crust than that of the present rift, producing huge stacks of flood basalts and massive intrusive gabbros (Levitte, 1970). The early high-tide of volcanism in some continental rifts remains unexplained.

Continental rifts and antecedent structures

In considering the geographic plans of the world's continental rift systems, we note that the majority (eg. Africa, Rhine, Rio Grande, Oslo, North Sea, Dead Sea) trend within 30° of N-S. Exceptions include the Baikal and Narmada rift systems.

In their meridional bias, continental rifts imitate major plate boundaries, many of which are related to poles of rotation located close to the poles of the Earth's rotational axis (Burke and Whiteman, 1973). The implication is that rift plan is governed by fundamental plate tectonic factors, rather than by any pre-existing 'grain' in uplifted and stretched crust. On the other hand, if some continental rifts have been generated by tensional stresses behind continent-continent impact zones (Tapponnier and Molnar, 1976), it would be merely fortuitous if those stresses happened to act perpendicularly to pre-existing crustal grain.

But what if crustal anisotropy itself shows a generally preferred direction on planet Earth? Morris and Tanczyk (1978) consider that, from Archaean times on, near N-S structures have been generated more commonly than near E-W structures. According to these authors, Precambrian dikes in Canada, for example, are markedly biased to meridional orientations, and more dikes were generated in low rather than high latitudes (prior to subsequent continental driftings).

The frequently observed parallelism of Cainozoic rift faulting and Precambrian crustal 'grain' has led to two opposing theses:

1. the parallelism is coincidental, even allowing for any N-S bias in Earth structural orientations. The newer structures are merely taking advantage of pre-existing grain (King, 1970), which for significant distances is run oblique to, or jumped across, in response to a differently oriented stress field.
2. the parallelism expresses a genetic connexion, such that Cainozoic continental rifting is merely the latest tectonic phenomenon expressed along fundamental lithospheric sutures that have been active since the beginning of Earth history. The same thermal sources provided energy for forming Archaean mobile belts as for the recent production of rift valleys (McConnell, 1972, 1974, 1978).

Originally, McConnell proposed that the controlling antecedent structures are orogenic belts, better termed "ensialic fold belts" for the southern and central regions of Africa. However, rift valleys are observed to cut across these, or to terminate against them. Thus McConnell has more recently (1974) substituted the concept of "taphrogenic lineaments"; these are characterised by deep erosional exposure of high-grade metamorphic rocks that have been subjected to intense mylonitisation, cataclasis and shearing (though any lateral displacements have been small, according to McConnell). The zones have been uplifted as a crustal response to a localised, steep geothermal gradient. "Persistent" (J. Watterson) and "resurgent" (E.S. Hills) Precambrian lineaments, many dating back to 2.7 G.a. in the late Archaean, can purportedly be overprinted and obscured by subsequent orogenic activity at higher crustal levels. According to McConnell, this explains observed sectors of non-parallelism between major rift faults and Precambrian structures at the present surface level in Africa. Obviously, there is the danger of a circular argument here.

Despite the often close and even uncanny coincidence of rift and basement trends at the continental and local extremes of scale (> 1000km and < 10km), there can be gross disparities on the intermediate scale. Hepworth (1978) succinctly enquires how there can be any genetic connexion between such disparate structures as:

1. Proterozoic mobile ensialic belts, up to 300km wide, with a multiphase history of thrusting, metamorphism and migmatisation, and
2. late-Phanerozoic vertical tectonics, arching, and tensional fracturing affecting a width approximating the thickness (c. 50km) of cratonised crust/lithosphere.

King's (1970) analogy of timber splitting in response to oblique cross-cuts may be appropriate: the crustal grain is followed where mechanically convenient, and cut across elsewhere. More recently, King (1978) has been converted to recognising a common parallelism of not only Cainozoic but also Mesozoic and Karroo rifts with Precambrian structures in Africa, with one of two implications:

1. The source of the stress-field responsible for rifting lies in the crust [lithosphere]. Any subcrustal processes are essentially irrelevant, unless
2. there has been a fixed relationship between crust and a sub-crustal [sublithospheric] stress source during the last 2.5 G.a.

However, King holds to the observation that only those Karroo rifts were rejuvenated in the Cainozoic which had a N-S orientation (i.e. rejuvenation of old lines of weakness is not inevitable and automatic), thus implying changing stress fields.

The first alternative is endorsed by Sykes (1978) who considers that, during continental fragmentation, rifting usually follows the youngest zone of previous orogenesis (cf. McConnell!) and tends to avoid old cratonic areas. Sykes further considers that the positions of rift junctions are controlled by a pre-existing crustal mosaic rather than any active motive force involving, say, mantle plumes. Thirdly, the volcanic 'hot spots' of rift systems, purportedly coinciding with ancient crustal nodes, are passive centres rather than expressions of deep mantle perturbations - the spatial coincidences are too great for Sykes to find the mantle hypothesis acceptable.

King's (1978) second alternative raises a fundamental problem, addressed by Hepworth as follows: "It is very difficult to suppose that mantle cells or convection currents remained stationary with respect to the lithosphere during the Proterozoic and Phanerozoic, which is the implication of a 'taphrogenic lineament' capable of controlling the rift system". This difficulty is mitigated if continental lithosphere has carried deep mantle roots during the drifts of geological history (Jordan, 1978).

Three likely possibilities emerge regarding antecedent structural control on rifting:

1. McConnell's 'fixist, resurgence hypothesis', requiring permanent coupling of lithosphere and upper mantle.
2. Sykes' 'passive hypothesis', whereby rift stresses originate wholly in the crust.
3. King's (initial, 1970) 'mobilist hypothesis', rifting occurring as a response to new, sub-lithospheric processes.

Let us look briefly at the rifts themselves, first in Africa. If rifts are demonstrably an intraplate response to major processes at the boundaries of their plate, then thesis 1. is badly weakened, and perhaps 3. also.

Three major episodes of rifting have occurred in eastern Africa, according to Sowerbutts (1972):

(1)	Triassic-Jurassic:	N-S rifting in SE Africa, related to initial breakup of Gondwana (Africa-Antarctica separation).
(11)	Cretaceous:	NE-SW rifting in "Central" Africa, related to separation of Madagascar from Africa (Karroo rifts).
(111)	Cainozoic:	NNE-SSW rifting in E. Africa, related to separation of Arabia from Africa.

According to Sowerbutts (and here he is in contradiction with Dixey, 1956), there has been little or no rejuvenation of pre-existing rifts in subsequent rift episodes. For example, the Cainozoic Malawi rift cuts across and ignores the Ruvuma and other Karroo rifts. This weakens McConnell's thesis (#1). Let us scrutinise three of McConnell's proposed taphrogenic lineaments:

(a) the Ubendides of SW Tanzania, a "mobile" belt of intense late-Archean deformation, run northwest from the Livingston Mts. via the Mbeya triple-junction, the Rukwa rift and the south-central sector of the Tanganyika rift to the rim of the Congo basin. The Ubendide belt was activated c. 2.7 G.a. ago, and reactivated 2.0-1.8 G.a. ago. The extraordinarily linear cataclasite zones of the Ubendide belt (Brown, 1962) are faithfully followed by the rift-margin faults for more than 400km through the Rukwa sector, with a linearity quite atypical of rift faults elsewhere in eastern Africa. Yet the evident fact remains that the Western Rift escapes both north and south from the Ubendides via en-echelon transpositions i.e. rift structures show an independence from Basement structures on a regional (intermediate) scale.

(b) The Aswa mylonite zone marks the northern end of the Western Rift at the Uganda-Sudan border. It is a Proterozoic belt of deep crustal deformation running NW-SE, parallel to the Ubendide belt for 200km across the Uganda Shield. Yet the Western Rift, far from following this powerful lineament, terminates against it.

(c) A continent-spanning alignment of mafic intrusive complexes has been identified in Africa by Vail (1977). Twenty-five such complexes are aligned on a 3600km-long segment of a great circle trending NNE between latitudes 33º and 7ºN. As these intrusives were emplaced during the interval 2.6-1.6(?) G.a., the line would certainly qualify as persistent! Vail's line traverses a variety of tectonic domains old and new, including cratons, mobile belts and rift valleys. But the African Rift System itself follows this line for little more than a few (possibly) coincidental) hundred kilometres. Despite McConnell's (1978) claim that the line expresses a taphrogenic lineament, Vail himself explicitly stated that "None of the rift [nor mobile belt] movements seem to have displaced the complexes, nor do the latter appear to have had any control on the direction or style of the valley grabens".

Turning to Sykes' hypothesis (#2), we can note that, in Africa at any rate, 'hot spots' manifested in domal or swell uplift accompanied by profuse volcanism are not sited on any obvious, ancient structural nodes. Ethiopia and Kenya come immediately to mind (see McConnell, 1972, cf. Baker et al., 1972). What of other continental rift settings?

For the Baikal rift system, Logatchev and Florensov (1978) emphasise that the Archean boundary between the Siberian Platform and the Sayan-Baikal mobile belt has played a decisive role throughout the entire geological history of the area. In its central sector, the Baikal rift system is essentially tied to this ancient boundary. Yet the Sayan-Baikal uplift-rifting is recognised as a new and unique phenomenon following on the chain of forerunning events. Late Mesozoic depressions situated somewhat to the east of the Baikal zone were not true rifts, but possible proto-rift precursors. Logatchev and Florensov conclude non-commitally that "on the whole, orientations of young rift faults depend on Precambrian structures, but cross-cutting also occurs".

The Paleocene rifting of Greenland from Scandinavia (Larsen, 1978) occurred near the time of sea-floor magnetic anomaly no. 24 (c. 54 M.a.). It was preceded by crustal injection of a 780km-long, en echelon dike swarm parallel to the southeastern margin of Greenland. The swarm extends NE-SW along or east of the present continental slope, but in places runs onto the continental shelf where it is marked by a positive magnetic anomaly that "cuts directly across two [Precambrian] orogenic boundaries, and the major structures within these orogenic belts" (Larsen, op. cit.).

Ramberg and Larsen (1978), Ramberg and Spjeldnaes (1978), and Bockelie (1978) all note that the Permian Oslo graben mimics the tectonic trends of the late Precambrian-Cambrian Sparagmite Basin, though with deviations. The Basin is elongated NNW-SSW, and contains the early-Cambrian Fen alkaline igneous complex with typical rift geochemical traits, whereas the Oslo graben trends NNE. Whether a fundamental Proterozoic lineament underlies the Oslo graben has not been mooted.

In the eastern U.S.A., Triassic-Jurassic tilted blocks produced during the splitting of Europe from North America have their faults parallel to and, according to Lindholm (1978), controlled by metamorphic foliation trends in the early Paleozoic basement. The foliation dip is claimed to determine even the dip of the faults (cf. Freund and Merzer, 1976), and where foliation trends are erratic, no major block faulting is developed. Ratcliffe (1971) describes the contemporaneous, early Mesozoic Ramapo fault system of New York and New Jersey as a specific case of "tectonic heredity": from Precambrian to late Cainozoic, there have been various types of tectonic movement in east-central U.S.A., but all oriented along the same NNE-NE trends.

For the Rhine graben, Illies (1978) lists three main structural controls:

1. The Hercynian basement. The Neogene witnessed a reactivation of what Stille (1925) termed the Mjøsa-Mediterranean orogenic shear zone within this basement. Basement control must therefore influence deduction of Neogene stress-fields derived solely from rift fault patterns (cf. Illies, 1974).
2. A mantle diapir underlying the southern Rhine graben. But it is not known whether the graben has "passively followed" the crest of the uplift induced by the diapir, or whether "mantle upwelling was impelled by crustal rifting". (The old problem emerges of rifting or uplift as the prime tectonic phenomenon).
3. The regional stress field. Until the late Miocene, σ_1 was directed NNE-SSW, but from mid-Pliocene times on has been directed N-S.

Although a "fundamental" interdependence of old lineaments and young rift faults is observed in the Rhine region, Illies (1978) considers that we cannot yet answer whether the lineaments "caused" or merely had a passive control on the faults. It is Fuchs (1974) who goes to the heart of the matter: if the rift asthenolith and high heat-flow in the Rhine graben are caused by divergent convection in the underlying mantle, how did this viscous convection "know" the location of the overlying Hercynian crustal lineament? Could such lineaments possibly be common enough to make overprinting of any one of them by rift faults unavoidable?

A possible escape route here is to accept Osmaston's (1977) thesis of deep mantle roots coupled to the continental lithosphere. Longstanding linkages between lines of lithospheric weakness and occasional mantle perturbations then become feasible. Indeed, Neumann and Ramberg (1978) propose the Rio Grande rift to be a candidate for such a linkage, and if this relationship can be proven to exist then McConnell's 'resurgence hypothesis' remains viable. However, Cordell (1978) is able to speak of "the nebulous concept" of basement control on the Rio Grande rift. Nevertheless, Tweto (1979) shows that the northern segment of this rift has been tightly controlled by pre-existing structural orientations; but in the central segment of the N-S Rio Grande rift, obliquely trending Precambrian basement structures are related to rift offsets along possible proto-transform faults (Chapin, 1979; Baldridge, 1979).

Meanwhile, Sykes' 'passive hypothesis' is endorsed by Bailey (1978) from the viewpoint of igneous petrogenesis. Bailey denies a role for active asthenospheric processes under rifts, and contends that "horizontal movement" of anisotropic continental lithosphere reactivates earlier zones of weakness, typically the most recent orogenic belt girdling the old cratonic nuclei. One aspect where Bailey's thesis is weakened concerns the dichotomy of the Eastern and Western Rift Systems of Africa: the former manifests sublithospheric thermal perturbations not present beneath the latter, whereas the lower geothermal gradient for the latter is related to greater seismic activity. If lithospheric stresses alone are responsible for both these rifts, similar mantle patterns might have developed beneath them, given that the lithospheric thicknesses are similar (Fairhead, 1976).

The inconclusive and on-going debate on the role of antecedent structures in determining rift plans, stems largely from our present inability to formulate arguments and thus seek evidence that can decide between the contending theories. Sub-lithospheric geophysical soundings may offer a decisive avenue here, especially coupled to new insights into volcanic petrogenesis.

Petrogenetic aspects of rift volcanism

One of the hopes for distinguishing between passive and active rifts, that is, between rifts lacking or possessing an underlying mantle perturbation contributing to the regional lithospheric stress field, is through recent petrogenetic ideas. Schilling (1971, 1973 a,b, 1975) has consolidated a proposal, still subject to controversy, that passive and active rifts are characterised by quite distinct basalt geochemistries: basalts of passive rifts are depleted in large or highly charged ions, contrary to the enriched or undepleted basalts of active rifts. But we must note the wide range of magmatic activity that can be associated with rifts:

Wet and dry rifts

Continental rifts may be associated with profuse volcanism ('wet rifts'), or very little or none ('dry rifts'). Examples of the total volume of volcanic products associated with some continental rifts are:

Baikal	5000km^3	(Logatchev & Florensov, 1978)
Rio Grande	12,000km^3	(R.W. Decker, in Baldridge, 1978)
Eastern Rift (Africa)	500,000km^3	(Baker et al., 1972)

As the Baikal rift is certainly a trans-lithospheric rupture (Puzyrev et al., 1978), the paucity of volcanics there must reflect differing thermal conditions in the asthenosphere compared with eastern Africa. Yet heat-flow values in both rift systems are comparable (Lysak, 1978; Morgan et al., 1978). The meagre Baikalian volcanics mostly occur outside the rift floors, on the shoulders and flanks of the regional uplift; they are predominantly mildly alkaline basalts (Kisyelev et al., 1978). The Western Rift of Africa, a 'dry' continental rift that additionally lacks associated regional uplift contains, at strong structural nodes, localised young volcanics that are typically highly alkaline and in some instances unusually potassic (Holmes and Harwood, 1932; Harkin, 1960; Sahama, 1973; Ferguson and Cundari, 1975).

The virtual absence of volcanics and, according to gravity surveys, an absence also of upper crustal sheet intrusions in the Baikal and Tanganyika rifts may explain the very deep troughs there. By comparison, the Kenya and Ethiopian rifts have been filled with volcanic products to elevations of 2000m and 1700m respectively in their central sectors. The total volumes of magmatic material that have reached the surface are 144,000km^3 in Kenya and 345,000km^3 in Ethiopia. The proportions of mafic to silicic erupta in these two provinces are 1.3:1 and 6:1, respectively. In both provinces, volcanism has been most profuse where faulting has been most profound, in particular along faulted monoclines of the proto-rift margins. The rate of volcanic accumulation associated with the Eastern Rift has been quite rapid: for example, the Oligocene Simien shield of N. Ethiopia built up 2500m in 5 m.y., compared with 1-2 m.y. for a similar thickness in Hawaii (C.A. Wood, written communication, 1978).

Chemistry

Continental rift-valley magmatism is typically alkaline, over a basaltic spectrum from transitional to strongly alkaline. However, voluminous pre-graben basalts erupted around the present-day Afar region included continental-type tholeiites. The typical rift valley shows predominant alkali olivine basalts associated, sometimes intimately, with either basanites or olivine tholeiites, as for example in the Ethiopian, Rio Grande and Baikal (Tunka) rifts.

The alkaline basalts are notably enriched, relative to continental tholeiites and especially to oceanic floor tholeiites, in:

(i) Na + K, usually with Na $>$ K
(ii) incompatible trace elements (ITEs)
(iii) rare-earth elements (REE), with greater enrichment of light REE (but rift tholeiites can show a 'flat' REE pattern).
(iv) CO_2 and possible H_2O

Neumann and Ramberg (1978) reckon that the major chemical differences amongst continental rift basalts can stem from any of: "equilibrium and disequilibrium partial melting, fractional crystallisation, zone refining, magma mixing [and immiscibility], multiple-stage melting or enrichment, and crustal contamination". This is to be unduly pessimistic. We can acknowledge a compositional spectrum of basalts, and yet consider this in terms of depth of melting (Baker et al., 1978), possibly supplemented by high-level fractionation for the weakly alkaline basalts (O'Hara, 1977). But there exists a more fundamental division.

Two main types of mantle source for basalts can be distinguished, according to their ITE contents:
undepleted (or enriched) mantle sources, expressed in most continental and ocean-island basalts, and
depleted mantle sources, expressed in most mid-ocean ridge basalts (MORB).

Enriched sources beneath continents more likely stem from below the asthenosphere, rather than in typically depleted, granulitic lower crust.

In fact, the chemical and isotopic differences among terrestrial basalts require a heterogeneous mantle that has persisted in this state for several hundreds or even thousands of m.y. For example, a 2.5 G.a. age has been obtained for the source of the Snake River olivine tholeiites, erupted during the Tertiary (Leeman, 1977); A 1.9 G.a. age applies to the San Juan calc-alkalic volcanics, west of the northern Rio Grande rift (Lipman et al., 1978); A 0.5 G.a. age is yielded by xenoliths in lavas of Lashaine volcano, N. Tanzania (Hutchison and Dawson, 1970).

Further evidence for dual mantle sources, and for a spectrum of enrichment in non-depleted sources, comes from $^3He/^4He$ studies in active rifts (Craig and Lupton, 1978). Helium $R/R_{atmos.}$ values in Ethiopia (both rift valley and Afar), Iceland, Hawaii and Yellowstone are between 3 and 16, compared with typical crustal values of 0.1-1. This supports the concept of individual deep mantle plumes, whose central melting products are free from mixing with higher-level, depleted magmas.

Isotope studies of rift volcanics, in particular neodymium isotopes, are beginning to show a promise in petrogenetic applications that it is premature to evaluate here. I can merely list available data and make brief comment. Some alkali basalts from the Kenya rift indicate derivation from an originally ITE - depleted source, contaminated by "recent" enrichment in many trace elements including ITEs which has not yet affected isotope ratios (Norry et al., 1979). For these basalts; $^{87}Sr/^{86}Sr$ = .7035, and $^{143}Nd/^{144}Nd$ = .5126. The strontium ratio is close to the .7030 typical of the Ne-normative ITE-depleted basalts of Mull, derived from partially melted garnet lherzolite at more than 30kb. By contrast, Mull tholeiites have strontium ratios close to .7055, and show undepleted ITE patterns. These observations reverse the scheme of Schilling (1971, 1975), and imply a depleted deeper mantle and an enriched upper asthenosphere.

Afar is a small, quasi-mid ocean ridge sector at the northern termination of the Eastern Rift of Africa. The axial-range basalts of Erta-ali prove to have MORB-id petrology and geochemistry (CNR-CNRS Team, 1973; Barberi et al., 1974). $^{87}Sr/^{86}Sr$ averages .7022 but is rather variable. The mildly alkaline basalts of the Boina centre, erupted closer to the continental margin of Afar than Erta-ali, have distinctly higher $^{87}Sr/^{86}Sr$ = .7048. For the Boina fractionates, Barberi and coworkers show that the ratio increases to .7052 (dark trachytes), .7059 (comendites) and .7063 (pantellerites). These data strongly suggest contamination during fractionation, either from continental sialic rocks or circulating brines.

For a specified mantle source, the primary control on the composition of initial partial melts is depth of melting, and only secondarily degree of melting (Green and Ringwood, 1967; Smewing and Potts, 1976). REE patterns and absolute abundances confirm that parental alkalinity is a function of depth rather than amount of melting (Yoder, 1976; Baker et al., 1978). There is a greater degree of light REE enrichment for deeper sources, and plateau basalts show this enrichment relative to rift basalts in eastern Africa. Data concerning magmatic source and ascent processes are not yet available for African rift volcanics, but for the analogous early Tertiary Hebridean province, specifically the Skye Main Lava Series, Thompson et al. (1979) deduce a spinel lherzolite source subjected to 10% partial melting at 20kb, with equilibration occurring at 10kb prior to high-level fractionation.

Parental melts can also be affected by the presence of volatiles and their CO_2/H_2O ratio. Additional local influences are revealed for example in variable ITE versus $^{87}Sr/^{86}Sr$ relationships in some Afar Quaternary basalts (Barberi et al., 1975; Treuil and Joron, 1976). Local magmatic 'modifications' result where new melts rise into actively fractionating magma chambers; the ITE and isotope chemistries of subsequent basaltic eruptions thus become hybridised, with a concomitantly more silicic magmatic signature (O'Hara, 1977).

Bowen-type fractionation is generally accepted to be the essential, indeed the only important link between mafic and silicic igneous rocks in continental rifts (Baker et al., 1977, 1978), although Bailey (1978) remains a prominent dissenter in reiterating a role for lithospheric degassing and 'flux' melting of sialic crust. In the well-studied Kenya rift, two distinct fractionation lineages occur (Saggerson and Williams, 1964; Baker et al., 1978):

1. nephelinite/strongly alkaline basalt (> 30 kb) —— F=0.4 ——→ Ca-rich phonolite (>25%Ne) carbonatite

2. transitional/mildly alkaline olivine basalt (10-25 kb) ——→ Ca-poor phonolite (eg. plateau phonolites of central Kenya)
   —— F=0.3 ——→ trachyte —— F=0.07 ——→ pantellerite

where F expresses the proportion of remaining liquid compared with the initial volume of parental magma. The parent basalts of the above two lineages are considered to be derived from independent garnet peridotite sources at the indicated differing depths.

The evidence for closed-system, Rayleigh law fractionation as the single important process in deriving more-silicic magmas from parental basalts, without external influence from vaporphase infiltration or crustal contamination, stems from ITE covariances in mafic-intermediate-silicic suites from particular rift centres. In the Eastern Rift of Africa, ITE covariances with Zr (Weaver et al., 1972) and Ce (Treuil and Joron, 1976) yield remarkably linear plots, though different slopes may apply for different volcanoes. Among the ITEs, the best linearity occurs for ions with large charges, eg. Hf, Nb, Ta, Th, whilst good linearity is the case for lower charged ions with large radii, eg. Cs, Rb, La and Sm. The implication is that we are indeed dealing with closed cogenetic suites.

Further evidence that fractionation, and not partial melting, is the likely origin of rift silicic lavas comes from the voluminous and chemically homogenous rhyolites associated with the Eastern Rift. C_e/C_o v. F plots for ITEs in these rocks are linear for bulk distribution coefficients ($\bar{D}$) between 0.3 and 0.5 (see Treuil and Joron, 1976; Baker, 1978), consistent with fractional crystallisation.

An outline of this fractionation process is provided by the mineralogy and geochemistry of the axial range (quasi-ocean ridge) volcanoes in northern Afar, where a temporal sequence of progressively more-silicic lavas has been erupted (Barberi et al., 1975; Treuil and Joron, 1976). Commencing with a transitional basalt parent magma, there is successive precipitation of:

1. olivine,
2. plagioclase with olivine,
3. clinopyroxene + Fe-Ti oxides + sulphides,
4. alkali feldspar:

leading respectively to strong depletion in liquid of

1. Cr, Ni, Mg;
2. Al, Ca;
3. Fe, Ti, Cu, Zn, with accompanying SiO_2 enrichment;
4. Eu, with enrichment in alkalis; increased oxygen fugacity leads to increased Fe^3/Fe^2.

Barberi et al. (1975) consider that these four stages of fractionation are separated by discontinuities at F values of 0.65, 0.40 and 0.30. However, Baker et al. (1977) find that in the early stages of Kenya rift magmatic fractionation, there is "a smooth progression in cumulative amounts of phases crystallised and fractionated", and that "the proportions of plagioclase, olivine, clinopyroxene, and Fe-Ti oxides are essentially constant and imply prolonged crystallisation along a four-phase curve such as that in the system Di-Ab-Fo-An of Yoder and Tilley (1962)". The quantitative predominance of plagioclase amongst the fractionating minerals, once olivine has been precipitated, leads ultimately to alumina-depletion and accumulation of alkalis ("plagioclase effect" of Bowen, 1938). The results of Baker et al. and Barberi et al. concur in suggesting that Fe-Ti oxides begin to precipitate markedly at F = 0.4, with dramatic increase in SiO_2 content of the remaining liquid. A radical transition next occurs near F = 0.1 according to Baker et al., with precipitation of alkali feldspar leading to peralkalinity and depletion in Sr and Ba in the remnant melt. The fractionation from trachytic to pantelleritic liquid occurs in such a miniscule residual magma fraction (F = 0.03 leading to 0.007), that high volatile content (eg. halogens in peralkaline silicic magmas) is required to reduce viscosity and lower the freezing point. Indeed, Tazieff and Varet (1969) have described thin trachytes on Erta-ali volcano, Afar, which flowed with an exceptionally low viscosity akin to that of basalts.

Daly and other Gaps

The bimodal distribution of rift lavas in terms of their SiO_2 content is a well-known phenomenon that has attracted more speculation than careful experimentation. Recent studies (eg. De Fino et al., 1973), and especially those based on ITEs (Baker et al., 1978) show that the problem of a 'Daly Gap' derives largely from the assumption that SiO_2 content of a magma is proportional to the degree of differentiation. But SiO_2 percentage does not increase smoothly with decreasing F-factor. In the mildly alkaline olivine-basalt suite, extensive fractionation through hawaiite and mugearite can occur along cotectic lines without increase of SiO_2 (De Fino et al., 1973). At this point, the percentage of SiO_2 will suddenly increase to trachytic levels when titanomagnetite fractionation reaches and passes a maximum. Then, depending on the SiO_2 level, feldspar fractionation gives rise to either phonolite (with decreasing SiO_2), or rhyolite (with increasing SiO_2). However, if oxygen fugacity is low there is no titanomagnetite maximum, and a narrower Daly Gap should result. In fact, such a narrowing is not obvious in, for example, the Afar lavas (Barberi et al., 1975). The Daly Gap is only partly explained, but it does indeed seem that the Gap is an artifice resulting from the use of SiO_2 as an index of fractionation.

A further factor contributing to the Daly Gap could stem from crustal density filtering (Gill, 1973), with silicic rising more readily than intermediate magmas. However, it is likely that the density factor is more than countered by the much higher viscosity of silicic magmas compared with more mafic magmas, unless the former are 'superheated' in contact with the latter. Thus the volumes of erupted rock-types need not reflect the proportions of Bowen-series magmas evolving in deep magma chambers. Nevertheless, where such deep levels of ancient rift valleys now lie exposed, as in the Oslo region and off-axis in Iceland, the Daly Gap does not appear greatly narrowed; intermediate-term benmorites and icelandites remain relatively rare. The observed predominance of silicic over mafic lavas during the later evolution of the Eastern Rift of Africa cannot be ascribed to crustal anatexis (see below), and appears to reflect a new high-level generation of fractionates in zones of very steep geothermal gradient.

It is instructive to make rough calculations of the volumes of parental basalt magma required to yield the observed silicic volcanics in the Eastern Rift of Africa:

	vol. silicics	minimum required vol. parental basalt
Kenya	$9 \times 10^2 km^3$	$3 \times 10^4 km^3$
Ethiopia	$7 \times 10^4 km^3$	$3 \times 10^6 km^3$

leading to feasible dimensions of the parental basalt magma chamber as being:

	width	length	depth (km)
Kenya	15	100	20
Ethiopia	100	500	60

The Kenya data fit gravity and magnetic constraints quite well (Searle, 1970; Wohlenberg, 1975), though the depth through the chamber is perhaps excessive to be able to yield a homogenous magmatic character. The required Ethiopian magma chamber, however, is unacceptably large. Even if reduced to Kenyan dimensions, it would demand one hundred complete evacuations and re-charges during a period of only 2 m.y. Unless the volume of Ethiopian rift silicics has been grossly overestimated, there seems no accommodating the figures to fit with direct fractionation. An oft-proposed alternative hypothesis is for steady-state, intra-crustal asthenospheric fractionation acting over a long period and resulting in lithospheric underplating (Baker et al., 1972; Shackleton, 1973). Ensuing regional thermal pulses, allied to crustal dilatation episodes, could then result in massive remelting of this accreted material, leading to the rise and voluminous eruption of low-density silicic magmas accompanied by collapse of the rift valley floor.

Further objection to wholesale fractionation in a single, huge basaltic magma chamber under the rift valley stems from the likelihood that this will only occur once whilst the parental material gradually freezes. But silicic volcanics belonging to numerous, successive late Tertiary-Quaternary magmatic episodes are observed in eastern Africa. It has been claimed, on theoretical geophysical grounds, that magma chambers under very slowly spreading rifts are evanescent and never more than a few kilometres in width (Sleep 1975, Tapponnier and Fracheteau, 1978). On the other hand, the observed duration of the Quaternary eruptive cycle in southern Kenya, 0.75 m.y., matches closely the calculated freezing time for the large magma chamber ennumerated above (Baker et al. 1977). All in all, the successful comprehensiveness and elegant simplicity of the fractionation thesis, applied to particular suites of rift volcanics, should not blind us to observations that require other factors in order to fit the complete frame of rift magmatic history.

Could continental crustal anatexis be a significant source for the rift silicics? The relatively low $^{87}Sr/^{86}Sr$ ratios for Afar silicics (those for Boina have already been quoted) show that any anatexis must be restricted to the lower, granulitic crust. Only at Ma'alalta, near the western margin of Afar, is an upper crustal component indicated, with $^{87}Sr/^{86}Sr$ for a pantellerite being .7158 (Tazieff and Varet, 1969). Further south, in the northern sector of the Ethiopian rift, a trachyte-pantellerite suite at Fantali shows a progressive increase of $^{87}Sr/^{86}Sr$ with time (and F-factor) from .7041 to .7079 (Dickinson and Gibson, 1972). Peculiarly Ca and Sr-rich pantellerites from Hertali volcano have high ratios in the range .7082 to .7096 (Spies, 1975); if anatexis is required to explain these data, then it is correspondingly ruled out for the isotopically normal silicics of Afar. In fact, most petrologists consider anorthoclase fractionation to be the major factor responsible for raising $^{87}Sr/^{86}Sr$ ratio along the trachyte-to-pantellerite route (Dickinson and Gibson, 1972; Barberi et al., 1975). What is patently clear from the isotope data is that melting of Precambrian upper sialic crust has played no major role in the origin of the African Rift System silicic volcanics.

Finally, we can note examples of regional geochemical trends (mainly basaltic) associated with various continental rift systems:

1. decreasing alkalinity with time, in the Kenyan and Oslo rifts (but the converse is true for the E. Greenland dikes according to Nielsen, 1978). Baker et al. (1978) explain the decrease as due to rise of isotherms leading to progressively shallower levels of mantle melting.
2. increasing alkalinity, southwards along the African Rift System from Yemen through Ethiopia to Kenya (Harris, 1969). This may be expression of a progressively decreasing rate of crustal dilatation (McKenzie et al., 1970). Increasing alkalinity and decreasing magma production rate is claimed by Barberi and Varet (1977) to be a progressive feature south from neo-oceanic Afar into the continental Ethiopian rift. This may be true for erupted basalts, but not the silicic volcanics and their inferred parental basaltic magma chamber(s).
3. increasing alkalinity (and in some instances K/Na) away from the rift axis, in Ethiopia, Kenya and Iceland. In Ethiopia, mean % K_2O at 60% SiO_2 is 4.5 for plateau lavas, 2.7 for Afar margin lavas, and 2.3% for Afar floor lavas, in accordance with a riftward shoaling of magmatic sources.

These trends have a single implication concerning the mechanism of continental rifting, namely lithospheric dilatation.

Mechanism of Continental Rifting

We can first appropriately summarise some general observations and remaining questions on the stages of rift system evolution:

1. There is an early development of a broad, shallow proto-rift depression prior to any regional doming or volcanism.
2. The thermal pulse affecting rifted lithosphere reflects forceful diapiric asthenospheric injection. This rising injection undergoes decompression leading to melting and generation of enriched-type magmas passing up into continental crust.
   Is there a different thermal response of oceanic lithosphere-asthenosphere to passive rifting, such that it engenders depleted-type MORB magmatism? Or, do continents have deep roots of relatively enriched mantle?
3. Buoyant isostatic uplift of heated lithosphere leads to doming. This occurs episodically.
4. Crustal attenuation and lithospheric necking result both from the doming and more importantly from regional dilatation.
5. Episodic dike injection and volcanism alternate with crustal faulting in a manner that quickens with time. Volcanic volumes tend to decrease with time.
6. Rift valley formation is a late-stage phenomenon (Pliocene and Quaternary, for most active continental rifts), in some instanc-

es associated with profuse silicic volcanism. There is a possible global contemporaneity in Cainozoic continental rift development.

Despite intensive geophysical surveys (seismic, gravity, magnetic, thermal), the mechanism of continental rift formation remains sufficiently elusive that the following two statements could be made in conjunction at a recent symposium:

1. "The Cainozoic rift system of Africa is now in its dying stages" (King, 1978), and
2. "The Kenya rift is in a developing, waxing stage" (Lippard and Truckle, 1978)!

Earlier, Tazieff (1970) spoke of the Eastern Rift as a potential ocean, McCall (1972) as an aborted ocean. Such is the slow extension-rate for the African rifts, an integrated 3mm/yr or less according to Le Pichon and Francheteau (1978), that precise geodetic studies are of little assistance in studying the mechanics of rifting. Can such a slow rate lead into more rapid sea-floor spreading? Blundell (1978) observes, regarding the Tugtutoq rift of S. Greenland active some 1330 m.y. ago: "there need be no failed arm [at triple rift junctions]......... a continental rift can continue to develop without becoming a new ocean". So how do continental rifts develop into oceanic rifts? To give brief space to this question is, considering the volumes written on the geology of Atlantic-type continental margins, a near-impertinence. I will take the recent work of de Charpal et al. (1978) as a basis for discussion, because it concentrates specifically on the nature of the continental-oceanic crustal transition.

The north continental margin of the Bay of Biscay is one side of a rift that initially developed in a basement of Hercynian granites, overlain by Permo-Triassic sandstones and Up. Jurassic limestones. This rift opened as the start of a neo-oceanic basin during the Portlandian-Aptian interval. Seismic reflexion profiles show that rift margin faulting took a lystric form, with individual upper crustal blocks 10-30km wide being downdropped as much as 4km to the ocean side, and tilted 25-35º on curving fault planes. A prominent reflector at c. 9.5km depth coincides with the level of flattening of the lystric faulting. The Moho lies three kilometres lower at c. 12.5km depth, though an identity with the reflector cannot everywhere be ruled out.

The Biscay margin faulting represents a 10% extension of the brittle upper crust, according to the tilt-extension relations worked out by Morton and Black (1975) from Afar margin exposures. However, shoaling of the Moho indicates crustal thinning from an original 30km (presumed) to the present 6km thickness. De Charpal et al. therefore conclude that ductile thinning (Bucher, 1933) has occurred in the lower crust, between the lystric-fault bottoming level and the Moho. This requires not only a decoupling of upper from lower crust, but also mass transfer with concomitant crustal thickening elsewhere. Thickening of continental crust near the margin of the Ethiopian rift valley is mooted from independent geophysical studies (Searle & Gouin, 1971; Berckhemer et al., 1975). Within the rift, crustal attenuation has permitted rise of mantle material with resulting isostatic development of the rift depression. Francheteau and Tapponnier (1978) have recently formulated a theoretical basis for this concept of two-layer crustal rifting.

Seismic profiling around the Rhine graben adds further if somewhat contentious details to the crustal necking thesis, and suggests an alternative to lateral transfer of continental lower crust. Mueller (1978) recognises a typical continental crustal section in central Europe as comprising:

P velocity	thickness
4.0 km/s	3-4km
5.9-6.1	5-7
c. 5.5	3-8
6.3-6.4	3-12
(Conrad Discontinuity)	
c. 6.7	3-10
(Moho Discontinuity)	
8.0-8.2	

These layers are identified, from top downwards, as the near-surface sedimentary layer, upper crystalline basement, sialic low-velocity zone, intermediate crystalline basement, lower crystalline basement, and upper mantle.

Below the Upper Rhine graben, the Moho lies at 20 km depth with anomalous P_n = 7.0 km/sec, increasing gradually to 8.0 km/sec over 5km further depth (Bamford and Prodehl, 1977).

The initial thermal pulse from the mantle that precedes crustal rifting will, according to Mueller, lead to metamorphism in the lower crust and at the same time lower the P velocity there until a velocity-inversion develops. At the same time, a sialic low-velocity zone at 8-11km depth takes up initial rift extension through necking. This upper crustal zone "is identified with a semicontinuous (low viscosity) zone of (mushroom-shaped) granitic intrusions". Extensional fracturing of the overlying, uppermost brittle crust leads to graben formation, and a doming of 150-200 km width which reduces lithostatic pressure and facilitates rise of mantle basaltic magma. Sheet intrusion leads initially to thickening of the lower crust, but soon passes into a process of "oceanisation", with the seismic Moho being displaced from bottom to top of the lower crust as true sea-floor spreading commences. This questions a fundamental assumption made by de Charpal et al. (1978) and Fracheteau and Tapponnier (1978): that the Moho retains the same identity, separating the same rock layers, throughout the process of continental crustal attenuation and rifting.

Taking Afar as a present-day candidate for 'oceanisation' of continental crust, Mueller draws on profuse recent researches (Pilger and Rösler, 1975) to compare that region with the fully developed yet young ocean basins of the Red Sea and Gulf of Aden. Seismic refraction data (P velocities) have revealed the following crustal layerings;

Red Sea & Gulf of Aden:

P velocity	thickness
1.5 - 2.6 km/sec	1-2km
3.7 - 4.6	c. 2
6.6 - 7.2	7+

Afar:

P velocity	thickness
3.3 - 4.1 km/sec	2-3km
6.0 - 6.3	3-4
6.7 - 6.8	10-18
7.3 - 7.6	8+

Whereas the three Red Sea/Gulf of Aden layers can be assigned to oceanic crustal layers 1, 2 and 3 respectively, assignment of the Afar layers is more difficult. The 7.3-7.6 km/sec layer can confidently be equated with anomalous (thermally perturbed) upper mantle. Mueller (1978) takes the problematic 6.0-6.3 km/sec layer (see also several relevant papers in Pilger and Rösler, 1975) as continental crust that has been subject to attenuation, anatexis (yielding alkaline granites), and a final mafic phase of voluminous dike injection. Further attenuation will supposedly neck this layer to zero thickness.

However, geometric considerations of plate separations of Arabia from Africa, and Arabia from Somalia, suggest that the greater part of Afar must be underlain by new crust. Despite the convincing demonstration of Morton and Black (1975) that continental crustal attenuation has occurred at the margin of SE Afar, the 6.0-6.3 km/sec layer is not only too thick to be accommodated in a Red Sea - Gulf of Aden context, but thickens north in the direction of increasing attenuation! A likely alternative is that it is equivalent to oceanic layer 2B, a sheeted dike complex. In fact, Afar is not a good present-day example of processes marking the transitional stage from continental to oceanic rifting, owing to its highly 'magmatic' character (Schilling 1973b; cf. Barberi and Varet, 1977), akin to Iceland with respect to the mid-Atlantic ridge. Thus it is more appropriate to turn our attention to, say, the Eastern Rift of Africa.

The three-dimensional form of anomalous mantle, that extends to high levels beneath all active continental rifts, has been investigated by Long and co-workers (1976) in the central Kenya rift. Anomalous mantle with P_n = 7.3-7.5 km/sec reaches up to 20km beneath the floor of the rift valley. Its upper boundary initially plunges steeply (85º dip) westward from the western margin of the rift, but at fifty to a few hundred kilometres west of the rift maintains a shallow, regular dip (c. 35º) above 'normal' mantle velocities (P_n = 7.9-8.2). Long emphasises that any significant crustal dilatation in the Kenya rift is confined to the axial sector where the anomalous mantle is most shallow, and where a high-level axial intrusion (P = 6.4 km/sec) reaches to 3km below the rift floor (Griffiths et al., 1971). No azimuthal variation of P_n velocities occurs within this mantle (see also Bram and Schmeling, 1976; cf. Mueller, 1978).

An important conclusion from the seismic data of Long and Backhouse (1976) and Maguire and Long (1976) is that there is no crustal attenuation under the plateaux bordering the Kenyan rift, even up to the very margins of the rift itself, but only under the rift floor. The crustal thickness at the western margin of the central sector of this rift is c. 42km, concurring with an upper limit of 48km for the western margin of the Ethiopian rift at Addis Ababa (Searle and Gouin, 1971). Seismic refraction data for a latitudinal profile across the Ethiopian plateau through Addis Ababa, reveal a riftward deepening of the Moho (Berckhemer et al., 1975). No equivalent seismic data are available for the Western Rift but the rift block itself, at the northern end of the Tanganyika rift sector, is of continental character and is 32km thick upon 'normal' mantle with P_n = 8.0 km/sec and S_n = 4.6 km/sec (Bram and Schmeling, 1976). The southern end of the Kenya rift block is similarly c. 35km thick (Rykounov et al., 1972).

Teleseismic travel-time delays for stations on the margins of the Kenya rift are only partly accounted for by asthenospheric ray paths. The remainder can be accounted for either by higher asthenospheric values (relating velocity to density) than would be expected from the Nafe-Drake relationship (Bott, 1965; Fairhead, 1976), or by a thickened crust close to the rift. This second alternative certainly merits further attention, not only because of the seismic evidence already touched on, but because of geological evidence of crustal downwarping and thickening at some margin sectors of the Kenyan and Ethiopian rifts (Jepsen and Athearn, 1962; Baker et al., 1972), conflicting with the 'necking' hypothesis.

Gravity studies of the African rift valleys reveal a considerable variety of rock density/depth distributions. A common feature, however, is a broad c. 1000km-wide Bouguer anomaly 'negative' of 150-200 m. gal. amplitude coinciding with the area of regional uplift in both Ethiopia and East Africa (Gouin, 1970; Makris et al., 1975; Fairhead, 1976). Consensus exists that this association expresses the presence of hot, low-density asthenosphere beneath the East African lithosphere.

The Kenya rift shows a secondary gravity 'negative' of 150-350km width and 50-70 m.gal.amplitude superimposed on the broader gravity 'negative' of East Africa. It may express lithospheric attenuation and a shoaling of low-density asthenosphere. Normal lithospheric thickness in East Africa is 70-120 km (Knopoff and Schlue, 1972; Fairhead, 1976), with smaller values applying close to rift valley margins. However, the Ethiopian rift possesses a secondary gravity 'positive' of 50-60 m. gal. amplitude and 100 km wavelength (Gouin, 1970). Lithospheric attenuation is almost certainly greater in Ethiopia than Kenya, but why these two rifts have differing medium-term gravity signatures remains unexplained.

Third, there is a prominent 'positive' gravity anomaly running essentially unbroken but with varying amplitude (maximum 35 m.gal.) along the axis of the Kenya rift floor (Searle, 1970; Fairhead, 1976; Swain, 1976). For the Ethiopian rift floor, gravity data indicate that the crust is thinner than in the Kenya rift, and has been intruded by two or even three subparallel asthenospheric dikes (Searle and Gouin, 1972). This is consonant with plate tectonic dictates for greater dilatation in Ethiopia than in Kenya. At latitude 1.1°S in the Kenya rift, the axial positive anomaly is 55km wide and of 21 m.gal.amplitude, and density modelling leads to two alternative interpretations (or a compromise between the two):

1. a sill (of density 2.9 compared with 2.67 for the overlying volcaniclastic sediments), 2-5km thick with its surface 1-2km below the rift floor, or
2. a dike (utilising the same density contrasts as for 1.) 15km wide with its apex 3.5km below the rift floor.

The dike hypothesis for the initial splitting of continental lithosphere has been previously developed by Baker and Wohlenberg (1971), and endorsed from preliminary seismic refraction studies (Griffiths et al., 1971). Baker et al. (1978) briefly discuss the manner of emplacement of such a dike, and argue strongly for lithospheric dilatation as opposed to stoping because of two tight constraints imposed on the latter by required density contrasts:

1. the dike magma must be less dense than the country rock, to permit blocks of the latter to sink, and
2. the solidified dike must be appreciably more dense than the country rock, to produce the observed gravity anomaly.

However, giant dikes (up to 25km long and 500m wide) of middle Proterozoic age in Greenland show clear evidence of stoping (Bridgwater and Coe, 1970). Disregarding the twin problems of density and destiny of stoped country-rock, the dilatational theory is in any case more attractive in providing space for a giant magma chamber.

In summary, one's views on the mechanism of continental rifting must inevitably be colored by one's conclusions on other fundamental aspects of rifting, touched on in this essay: the passive or active role of antecedent crustal structures, the interaction of regional crustal uplift with rift subsidence, the apparently opportunistic association of volcanism with rifting, the nature of the magmatic source(s), and the thermal evolution of continental lithosphere and crustal low velocity layers being subjected to dilatation. Each of these aspects is a present bone of contention and focus of research. We can confidently expect some answers and additional fascinating new questions to inspire, tease and frustrate a new generation of 'rifters'. But if Hamilton (1969) is correct about the big-picture cause for the Rio Grande rift, if Tapponnier and Molnar (1976) are likewise correct about the Baikal rift, and McKenzie et al. (1970) for the African rift valleys, then we must seek more than one genesis for continental rift valleys.

Acknowledgments. Whatever merits this article might retain are largely inherited from conversations, collaborations with and criticisms from other rifters. In particular I owe a debt of gratitude to all my IGC Working Group 4 colleagues, with whom it has been such a pleasure to work. Dr. C. Chapin suggested numerous revisions to the penultimate typescript. Miss Lucy McCarthy typed the final manuscript. The author apologises for the absence of illustrations, consequence of an accident to a non-ambidextrous geologist.

References

Artemjev, M.E., and E.V. Artyushkov, J. Geophys. Res., 76, 1197-1211, 1971.

Bailey, D.K., In: Petrology and Geochemistry of Continental Rifts, Reidel, 1-13, 1978.

Baker, B.H., P.A. Mohr, and L.A.J. Williams, Geol. Soc. Am., Spec. Paper 136, 1972.

Baker, B.H., G.G. Goles, W. Leeman, and M.M. Lindstrom, Contrib. Min. Pet., 64, 303-332, 1977.

Baker, B.H., R. Crossley, and G.G. Goles, In: Petrology and Geochemistry of Continental Rifts, Reidel, 29-50, 1978.

Baker, B.H., and J. Wohlenberg, Nature, 229, 538-542, 1971.

Baldridge, W.S., In: Petrology and Geochemistry of Continental Rifts, Reidel, 71-78, 1978.

Baldridge, W.S., In: Rio Grande Rift: tectonics and magmatism (R.E. Riecker, ed.), Am. Geophys. Un., Washington, 323-353, 1979.

Bamford, D., and C. Prodehl, J. Geol. Soc. London, 134, 139-151, 1977.

Barberi, F., E. Bonatti, G. Marinelli, and J. Varet, Tectonophysics, 23, 17-29, 1974.

Barberi, F., G. Ferrara, R. Santacroce, M. Treuil, and J. Varet, J. Petrol., 16, 22-56, 1975.

Barberi, F., and J. Varet, Bull. Geol. Soc. Am., 88, 1251-1266, 1977.

Berckhemer, H., and others, In: Afar Depression of Ethiopia, Schweizerbart, 89-107, 1975.

Black, R., W.H. Morton, and D.C. Rex, In: Afar Depression of Ethiopia, Schweizerbart, 296-300, 1975.

Blundell, D.J., J. Geol. Soc. London, 135, 545-554, 1978.

Bockelie, J.F., In: Tectonics and Geophysics of Continental Rifts, Reidel, 195-202, 1978.

Bott, M.H.P., Geophys. J., 9, 275-277, 1965.

Bott, M.H.P., Tectonophysics, 36, 77-86, 1976.

Bowen, N.L., Am. J. Sci., 35A, 19-33, 1938.

Bram, K., and B.D. Schmeling, In: Afar between continental and oceanic rifting, Schweizerbart, 138-143, 1976.

Briden, J.C., and I.G. Gass, Nature, 248, 650-653, 1974.

Bridgwater, D., and K. Coe, In: Mechanism of

Igneous Intrusion, Seel House, 67-78, 1970.
Brown, P.E., Quart. J. Geol. Soc. London, 118, 295-317, 1962.
Bucher, W.H., The Deformation of the Earth's crust, Princeton, 518pp, 1933.
Burke, K., and A.J. Whiteman, In: Implications of Continental Drift to the Earth Sciences, Academic Press, 735-755, 1973.
Burke, K., and J.T. Wilson, Nature, 239, 387-390, 1972.
Carmichael, I.S.E., F.J. Turner, and J. Verhoogen, Igneous Petrology, McGraw-Hill, 739pp, 1974.
Chapin, C.E., In: Rio Grande Rift: tectonics and magmatism (R.E. Riecker, ed.), Am. Geophys. Un., Washington, 1-5, 1979.
Chapin, C.E., and W.R. Seager, New Mexico Geol. Soc. Guidebook, 26th Field Conf., Las Cruces Country, 297-322, 1975.
de Charpal, O., P. Guennoc, L. Montadert, and D.G. Roberts, Nature, 275, 706-711, 1978.
Cloos, E., Bull. Am. Assoc. Petrol. Geol., 52, 420-444, 1968.
Cloos, H., Geol. Rundschau, 30, 401-527, 1939.
CNR-CNRS Team, Rev. Geogr. Phys. Geol. Dynam., 15, 443-490, 1973.
Coney, P.J., and S.J. Reynolds, Nature, 270, 403-406, 1977.
Cordell, L., Bull. Geol. Soc. Am., 89, 1073-1090, 1978.
Craig, H., and J.E. Lupton, Trans. Am. Geophys. Un., 59, 1194 [Abstract], 1978.
Dainelli, G., Geologia dell'Africa Orientale, R. Accad. Ital., 3 vols. text, 1 vol. maps, 1943.
De Fino, M., L. La Volpe, L. Lirer, and J. Varet, Rev. Geogr. Phys. Geol. Dynam., 15, 373-386, 1973.
Dickinson, D.R., and I.L. Gibson, Bull. Geol. Soc. Amer., 83, 231-240, 1972.
Dixey, F., Bull. Colonial Geol. Mining Res., suppl. no. 1, 71pp, 1956.
Eaton, G.P., In: Rio Grande Rift: tectonics and magmatism (R.E. Riecker, ed.), 7-32, Am. Geophys. Un., Washington, 1979.
Evans, T.M., and H.Y. Tammemagi, Earth Planet. Sci. Lett., 23, 349-356, 1974.
Fairhead, J.D., Tectonophysics, 30, 269-298, 1976.
Faller, A.M., and N.J. Soper, J. Geol. Soc. London, 136, 737-744, 1979.
Ferguson, A.K., and A. Cundari, Contrib. Min. Pet., 50, 25-46, 1975.
Fischer, A.G., In: Petroleum and Global Tectonics, Princeton, 47-79, 1975.
Freund, R., and A.M. Merzer, Geol. Mag., 113, 561-568, 1976.
Fuchs, K., In: Approaches to Taphrogenesis, Schweizerbart, 420-432, 1974.
Fuchs, V.E., Phil. Trans. R. Soc. London, B 229, 219-274, 1939.
Gass, I.G., Phil. Trans. R. Soc. London, A 267, 369-381, 1970.
Gass, I.G., In: Afar Depression of Ethiopia, Schweizerbart, 10-18, 1975.
Gill, R.C.O., Nature Phys. Sci., 242, 41-42, 1973.
Gouin, P., Phil. Trans. R. Soc. London, A 267, 339-358, 1970.
Green, D.H., and A.E. Ringwood, Contrib. Min. Pet., 15, 103-190, 1967.
Griffiths, D.H., R.F. King, M.A. Khan, and D.J. Blundell, Nature, 229, 69-71, 1971.
Hamilton, W., Bull. Geol. Soc. Amer., 80, 2409-2429, 1969.
Harkin, D.A., Geol. Surv. Tanganyika, Mem. 2, 172pp, 1960.
Harris, P.G., Tectonophysics, 8, 427-436, 1969.
Hepworth, J.V., J. Geol. Soc. London, 135, 469, 1978.
Holmes, A., and H.F. Harwood, Q. J. Geol. Soc. London, 88, 370-440, 1932.
Holmes, A., Principles of Physical Geology, Nelson, 532pp, 1944.
Hutchinson, R.W., and G.G. Engels, Phil. Trans. R. Soc. London, A 267, 313-329, 1970.
Hutchinson, R.W., and G.G. Engels, Bull. Geol. Soc. Amer., 83, 2989-3002, 1972.
Hutchison, R., and J.B. Dawson, Earth Planet. Sci. Lett., 9, 87-92, 1970.
Illies, J.H., In: Graben Problems, Schweizerbart, 4-27, 1970.
Illies, J.H., In: Approaches to Taphrogenesis, Schweizerbart, 433-460, 1974.
Illies, J.H., In: Tectonics and Geophysics of Continental Rifts, Reidel, 63-71, 1978.
Jepsen, D.H., and M.J. Athearn, Geological Map of the Blue Nile Basin, Dept. of Water Resources, Addis Ababa (1:1 million), 1962.
Jones, W.B., and S.J. Lippard, J. Geol. Soc. London, 136, 693-704, 1979.
Jordan, T.H., Nature, 274, 544-548, 1978.
King, B.C., In: African Magmatism and Tectonics, Oliver & Boyd, 263-283, 1970.
King, B.C., In: Tectonics and Geophysics of Continental Rifts, Reidel, 347-350, 1978.
King, B.C., In: Geological Background to Fossil Man (W.W. Bishop, ed.), Geol. Soc. London, 29-45, 1978.
Kisyelev, A.I., Tectonophysics, 45, 49-59, 1978.
Knopoff, L., and J.W. Schlue, Tectonophysics, 15, 157-163, 1972.
Larsen, H.C., Nature, 274, 220-223, 1978.
Le Bas, M.J., Nature Phys. Sci., 230, 85-87, 1971.
Le Pichon, X., and J. Francheteau, Tectonophysics, 46, 369-406, 1978.
Leeman, W.P., In: Magma Genesis, Oregon Dept. Geol. Min. Indust., Bull. 96, 149-168, 1977.
Levitte, D., Geol. Surv. Ethiopia Rep., sheet ND37-11, 66pp, 1970.
Lindholm, R.C., Geology, 6, 365-368, 1978.
Lipman, P.W., B.R. Doe, C.E. Hedge, and T.A. Steven, Bull. Geol. Soc. Am., 89, 59-82, 1978.
Lippard, S.J., and P.H. Truckle, In: Petrology and Geochemistry of Continental Rifts, Reidel, 123-131, 1978.
Logatchev, N.A., In: Tectonics and Geophysics of Continental Rifts, Reidel, 351-366, 1978.
Logatchev, N.A., V.A. Rogozhina, V.P. Solonenko, and Yu.A. Zorin, In: Tectonics and Geophysics of Continental Rifts, Reidel, 49-61, 1978.

Logatchev, N.A., and N.A. Florensov, Tectonophysics, 45, 1-13, 1978.
Long, R.E., and R.W. Backhouse, Geophys. J., 44, 677-688, 1976.
Lowell, J.D., G.J. Genik, T.H. Nelson, and P.M. Tucker, In: Petroleum and Global Tectonics, Princeton, 129-153, 1975.
Lysek, S.V., Tectonophysics, 45, 87-93, 1978.
Maguire, P.K.H., and R.E. Long, Geophys. J., 44, 661-675, 1976.
Makris, J., H. Menzel, J. Zimmerman, and P. Gouin, In: Afar Depression of Ethiopia, Schweizerbart, 135-144, 1975.
Marsh, J.S., Earth Planet. Sci. Lett., 18, 317-323, 1973.
McCall, G.J.H., and G. Hornung, Tectonophysics, 15, 97-113, 1972.
McConnell, R.B., Bull. Geol. Soc. Am., 83, 2549-2572, 1972.
McConnell, R.B., Geol. Rundschau, 63, 389-430, 1974.
McConnell, R.B., J. Geol. Soc. London, 135, 470-473, 1978.
McKenzie, D.P., D. Davies, and P. Molnar, Nature, 226, 243-248, 1970.
Morgan, P., and others, Los Alamos Sci. Lab., Conf. Proc. LA-7487-C, 62-63 [Abstract], 1978.
Morris, W.A., and E.I. Tanczyk, Nature, 275, 120-121, 1978.
Morton, W.H., and R. Black, In: Afar Depression of Ethiopia, Schweizerbart, 55-65, 1975.
Mueller, St., In: Tectonics and Geophysics of Continental Rifts, Reidel, 11-28, 1978.
Neumann, E.-R., and I.B. Ramberg, In: Tectonics and Geophysics of Continental Rifts, Reidel, 409-424, 1978.
Nielsen, T.F.D., Contrib. Min. Pet., 67, 63-78, 1978.
O'Hara, M.J., Nature, 266, 503-507, 1977.
Osmaston, M.F., In: Developments in Petroleum Geology (G.D. Hobson, ed.), Applied Science Publishers, 1-52, 1977.
Pilger, A., and A. Rösler (editors), Afar Depression of Ethiopia, Schweizerbart, 416 pp, 1975.
Puzyrev, N.N., and others, Tectonophysics, 45, 15-22, 1978.
Quennell, A.M., Proc. Geol. Soc. London, 1579, 78-86, 1960.
Quennell, A.M., J. Geol. Soc. London, 135, 469-470, 1978.
Ratcliffe, N.M., Bull. Geol. Soc. Amer., 82, 125-141, 1971.
Ramberg, I.B., and B.T. Larsen, Norg. Geol. Unders. 337, 55-73, 1978.
Ramberg, I.B., and N. Spjeldnaes, In: Tectonics and Geophysics of Continental Rifts, Reidel, 167-194, 1978.
Rykounov, L.N., V.V. Sedov, L.A. Savrina, and V.Ju. Bourmin, Tectonophysics, 15, 123-130, 1972.
Saggerson, E.P., and L.A.J. Williams, J. Petrol., 5, 40-81, 1964.
Sahama, Th.G., J. Petrol., 14, 33-48, 1973.
Schilling, J.-G., Phil. Trans. R. Soc. London, 268A, 663-706, 1971.
Schilling, J.-G., Nature, 242, 565-571; and 246, 141-143, 1973a.
Schilling, J.-G., Nature Phys. Sci., 242, 2-5, 1973b.
Schilling, J.-G., J. Geophys. Res., 80, 1459-1473, 1975.
Searle, R.C., Nature, 227, 267-268, 1970.
Searle, R.C., and P. Gouin, Bull. Seism. Soc. Am., 61, 1061-1071, 1971.
Searle, R.C., and P. Gouin, Tectonophysics, 15, 41-52, 1972.
Sengör, A.M.C., and K. Burke, Geophys. Res. Lett., 1978.
Shackleton, R.M., In: Implications of Continental Drift to the Earth Sciences, Academic Press, vol. 2, 1091-1098, 1973.
Sleep, N.H., J. Geophys. Res., 80, 4037-4042, 1975.
Smewing, J.D., and P.J. Potts, Contrib. Min. Pet., 75, 243-258, 1976.
Sowerbutts, W.T.C., Nature, 235, 435-437, 1972.
Spies, O., In: Afar Depression of Ethiopia, Schweizerbart, 334-350, 1975.
Stille, H., Abh. Preuss, Geol. Landesanst., 95, 110-132, 1925.
Swain, C.J., In: Afar between Continental and Oceanic Rifting, Schweizerbart, 120-125, 1976.
Sykes, L.R., Rev. Geophys. Space Phys., 16, 621-688, 1978.
Tapponnier, P., and J. Francheteau, J. Geophys. Res., 83, 3955-3970, 1978.
Tapponnier, P., and P. Molnar, Nature, 264, 319-324, 1976.
Tazieff, H., Sci. Am., 222, 32-40, 1970.
Tazieff, H., and J. Varet, Rev. Geogr. Phys. Geol. Dynam., 11, 429-450, 1969.
Thompson, R.N., and others, J. Geol. Soc. London, 136, 257 [Abstract], 1979.
Treuil, M., and J.L. Joron, In: Afar between Continental and Oceanic Rifting, Schweizerbart, 26-79, 1976.
Tweto, O., In: Rio Grande Rift: tectonics and magmatism (R.E. Riecker, ed.), Am. Geophys. Un., Washington, 33-56, 1979.
Vail, J.R., Trans. Geol. Soc. S. Afr., 80, 87-92, 1977.
Weaver, S.D., J.S.C. Sceal, and I.L. Gibson, Contrib. Min. Pet., 36, 181-194, 1972.
Whiteman, A.J., Geol. Mag., 105, 231-246, 1968.
Williams, L.A.J., Tectonophysics, 15, 83-96, 1972.
Williams, L.A.J., In: Petrology and Geochemistry of Continental Rifts, Reidel, 101-121, 1978.
Wohlenberg, J., Geol. Jb. (Hannover), E4, 82 pp, 1975.
Yoder, H., Generation of Basaltic Magmas, Nat. Acad. Sci., Washington, 264 pp, 1976.
Yoder, H., and C.E. Tilley, J. Petrol., 3, 342-532, 1962.
Zanettin, B., and E. Justin-Visentin, Boll. Soc. Geol. Ital., 92, 313-327, 1973.
Zanettin, B., E. Justin-Visentin, M. Nicoletti, and C. Petrucciani, N. Jb. Geol. Paläont. Mh., 8, 473-490, 1978.